Lecture Notes in Networks and Systems

Volume 545

The series "Lecture Notes in Networks and Systems" publishes the latest developments in Networks and Systems—quickly, informally and with high quality. Original research reported in proceedings and post-proceedings represents the core of LNNS.

Volumes published in LNNS embrace all aspects and subfields of, as well as new challenges in, Networks and Systems.

The series contains proceedings and edited volumes in systems and networks, spanning the areas of Cyber-Physical Systems, Autonomous Systems, Sensor Networks, Control Systems, Energy Systems, Automotive Systems, Biological Systems, Vehicular Networking and Connected Vehicles, Aerospace Systems, Automation, Manufacturing, Smart Grids, Nonlinear Systems, Power Systems, Robotics, Social Systems, Economic Systems and other. Of particular value to both the contributors and the readership are the short publication timeframe and the world-wide distribution and exposure which enable both a wide and rapid dissemination of research output.

The series covers the theory, applications, and perspectives on the state of the art and future developments relevant to systems and networks, decision making, control, complex processes and related areas, as embedded in the fields of interdisciplinary and applied sciences, engineering, computer science, physics, economics, social, and life sciences, as well as the paradigms and methodologies behind them.

Indexed by SCOPUS, INSPEC, WTI Frankfurt eG, zbMATH, SCImago.

All books published in the series are submitted for consideration in Web of Science.

For proposals from Asia please contact Aninda Bose (aninda.bose@springer.com).

More information about this series at https://link.springer.com/bookseries/15179

Zdzislaw Kowalczuk
Editor

Intelligent and Safe Computer Systems in Control and Diagnostics

Springer

Editor
Zdzislaw Kowalczuk
Faculty of Electronics, Telecommunications
and Informatics
Gdansk University of Technology
Gdańsk, Poland

ISSN 2367-3370 ISSN 2367-3389 (electronic)
Lecture Notes in Networks and Systems
ISBN 978-3-031-16158-2 ISBN 978-3-031-16159-9 (eBook)
https://doi.org/10.1007/978-3-031-16159-9

This Springer imprint is published by the registered company Springer Nature Switzerland AG
The registered company address is: Gewerbestrasse 11, 6330 Cham, Switzerland

Preface

Since the end of the 1980s, technical diagnostics has been an area of great scientific interest and intensive research. It covers many already well-established topics as well as new developments in the fields of control and systems engineering, robotics, transport and mobile systems, the automotive and aerospace industries, as well as applied mathematics and statistics, decision-making sciences, signal processing, modeling, and artificial intelligence. Over the last three decades, the number of applications of various methods of fault diagnosis in computer science, medicine, and in industrial fields (electronic, electrical, mechanical, chemical) has increased significantly. In addition, the rapidly increasing complexity of industrial automation and the need to ensure reliability and safety at the highest possible level require continuous research and development of innovative approaches to diagnostics of failures and faults.

This book contains selected papers presented at the 15th International Conference on Diagnostics of Processes and Systems (DPS), held in Chmielno (near Gdańsk, Pomerania) on September 5–7, 2022. The conference (http://dps2022.konsulting.gda.pl/) was organized by the Gdańsk University of Technology, the Faculty of Electronics, Telecommunications and Informatics, the Department of Robotics and Decision Systems, with the support of the Warsaw University of Technology, and the University of Zielona Góra.

The series of DPS conferences has been gathering people interested in the subject of industrial process diagnostics, computer control systems, and expert systems as well as process monitoring for many years (since 1996). It is a meeting place for experts in the field of fault-tolerant diagnostics and control, researchers proposing new methods and technologies, users of control and IT systems, as well as representatives of industry and enterprises dealing with safety, security, environmental monitoring, signal processing, medical diagnostics, and the development and maintenance of safe and secure systems and software.

The previous conferences took place in Podkowa Leśna (1996), Łagów Lubuski (1997), Jurata (1998), Kazimierz Dolny (1999), Łagów Lubuski (2001), Władysławowo (2003), Rajgród (2005), Słubice (2007), Gdańsk (2009), Zamość (2011), and Łagów Lubuski (2013), Ustka (2015), Sandomierz (2017), and Zielona

Góra (2020) and attracted a large number of participants and internationally recognized speakers. In fact, the conference series is a two-year continuation of annual conferences on *Diagnostics of Industrial Processes* organized in the years 1996–1999 by the three Polish universities.

The series of conferences is an excellent forum for the exchange of knowledge and experience and sharing solutions in the academic and industrial environment. An important task of this forum is the integration of scientists and engineers from various industries, as well as producers of hardware and software for computer control and diagnostic systems.

The thematic scope of the DPS conference corresponds with the IFAC symposia on Fault Detection, Supervision and Safety of Technical Processes (SAFEPROCESS), as well as the international conference on Control and Fault-Tolerant Systems (SysTol).

From the outset, there are several main themes of DPS interest: (*i*) fault detection, isolation, and identification; (*ii*) fault-tolerant control systems; (*iii*) process safety, quality, and reliability; (*iv*) medical diagnostics; (*v*) general methodologies based on mathematical modeling, parameter identification and state estimation, qualitative models, statistical and signal processing, artificial intelligence, fuzzy logic and rough sets, expert systems, neural networks; and (*vi*) industrial applications of diagnostics in fault-tolerant problems, safety, monitoring and alarming, quality control, computer systems and networks, diagnostic software, software reliability, medicine and therapy, environment protection, production control, and other industries such as chemistry, electronics, and power systems.

This book is divided into six parts:

I. Artificial intelligence in medicine
II. Cybersecurity
III. Artificial neural networks
IV. Fault detection
V. Systems modeling
VI. Adaptive, robust, and FTC systems.

I sincerely thank all the participants and the reviewers of the articles from the International Program Committee for their personal scientific contributions to the conference. I extend special appreciation to the authors of the accepted articles that are published in this collective book by Springer, as well as to the speakers of the plenary and semi-plenary lectures:

- *Vicenç Puig*, Automatic Control Department and Institut de Robòtica i Informàtica Industrial, Universitat Politècnica de Catalunya (ES): *Security and Safety of Cyberphysical Systems*
- *Péter Bauer*, Institute for Computer Science and Control, ELKH (HU): *The "sense and avoid" aircraft system based on a monocular camera as the last chance to prevent accidents*

- *Eduardo F. Camacho*, Department of System and Automation Engineering, University of Seville (ES): *Distributed model predictive control of solar power plants*
- *Lizeth Torres*, Instituto de Ingeniería, Universidad Nacional Autónoma de México (MX): *Hydroinformatics tools for the diagnosis and monitoring of water networks*
- *Janusz Zalewski*, Professor Emeritus of Software Engineering, Florida Gulf Coast University, Ft. Myers, Florida; Professor of Informatics, Ignacy Mościcki State Professional College (US): *Fundamental concepts of modeling computer security in cyberphysical systems*
- *Michał Bartyś*, Institute of Automatic Control and Robotics, Warsaw University of Technology (PL): *Are logically correct fault diagnoses always consistent?*
- *Witold Byrski*, Department of Automatic Control and Robotics, Faculty of Electrical Engineering, Automatics, Computer Science and Biomedical Engineering, AGH University (PL): *Virtual sensor algorithms using advanced signal processing in moving windows for a model-based approach in FDI systems*
- *Maciej Bobowicz*, Department of Radiology, Medical University of Gdansk (PL): *The aspects of credibility and responsibility of medical AI solutions from the clinician's perspective.*

July 2022

Zdzisław Kowalczuk
International Program Committee,
Chairman of DPS 2022

Acknowledgment

I am grateful to the members of the DPS 2022 National Organizing Committee, especially to Chairman, *Marek Tatara*, for supervising many technical matters, to *Anna Witkowska* for organizational and administrative support, and to *Katarzyna Dorosz* for the secretariat.

As always, only the great synergistic effort of the organizers makes the conference a successful scientific event—especially in such difficult times of pandemic and war, when many scientific meetings are not taking place or are heavily disrupted.

Contents

Cybersecurity

Artificial Neural Networks

Fault Detection

Systems Modeling

Adaptive, Robust and FTC Systems

AI in Medicine

Trustworthy Applications of ML Algorithms in Medicine - Discussion and Preliminary Results for a Problem of Small Vessels Disease Diagnosis

Ferlin Maria[1], Klawikowska Zuzanna[1], Niemierko Julia[2], Grzywińska Małgorzata[2], Kwasigroch Arkadiusz[1], Szurowska Edyta[2], and Grochowski Michał[1](✉)

[1] Gdańsk University of Technology, Gdańsk, Poland
michal.grochowski@pg.edu.pl
[2] Medical University of Gdansk, Gdansk, Poland

Abstract. ML algorithms are very effective tools for medical data analyzing, especially at image recognition. Although they cannot be considered as a stand-alone diagnostic tool, because it is a black-box, it can certainly be a medical support that minimize negative effect of human-factors. In high-risk domains, not only the correct diagnosis is important, but also the reasoning behind it. Therefore, it is important to focus on trustworthiness which is a concept that includes fairness, data security, ethics, privacy, and the ability to explain model decisions, either post-hoc or during the development. One of the interesting examples of a medical applications is automatic SVD diagnostics. A complete diagnosis of this disease requires a fusion of results for different lesions. This paper presents preliminary results related to the automatic recognition of SVD, more specifically the detection of CMB and WMH. The results achieved are presented in the context of trustworthy AI-based systems.

Keywords: Machine learning · Artificial intelligence · Deep learning · Small vessels disease · Explainable AI · Trustworthiness

1 Introduction

Machine learning algorithms have achieved a tremendous success in various image processing tasks. In particular, they obtain state-of-the-art performance in exploring and analyzing huge and complex datasets, especially images. Ones of the most important for society image recognition applications are the medical ones as an imaging is an integral part of medical diagnostics [1]. They include analysis of various type of image data, including 2D and 3D, from ultasonography (USG), computed tomography (CT), magnetic resonance imagining (MRI), endoscopy and others. In addition to careful analysis, a description of the examination and conclusions are needed. Medical data analysis is tedious and difficult.

Z. Kowalczuk (Ed.): DPS 2022, LNNS 545, pp. 3–16, 2023.
https://doi.org/10.1007/978-3-031-16159-9_1

Moreover, the importance and sensitivity of this field makes it extremely important to describe it properly and clearly in order to make them trustworthy.

The clinician's description of the examination is biased in some ways. First of all, the analysis is subjective and based on the one's experience. Hence, sometimes it happens that two independent specialists assess the examination in different way, resulting in low observer reliability. Next, in some cases, there is a lack of standardized guidelines for evaluating the examination, while in others, despite many guidelines and rules, there are still some inconsistencies between descriptions provided by two specialists (e.g. for image data annotations). Another factor is the wide range of diagnostic equipment used, which varies depending on the manufacturer consequently influencing user's habits. Moreover, there are plenty of human-factors that strongly affect the evaluation process such as resting level, personal issues, and mood. Considering the above, the Clinical Decision Support System (CDSS) based on machine-learning [2]. Seems to be a great opportunity to support clinicians in their daily work and minimize negative effect of human-factors while following guidelines. Such solutions can also support diagnostics in areas where access to specialists is limited. What is more, recent studies proved that the level of agreement of the AI tools with the experts was at least as good as agreement between two experts [3], and therefore, although the ML algorithms cannot be considered as a stand-alone tool, it can certainly be a valuable medical support.

In general, humans are reticent to adopt techniques that are not directly interpretable, tractable and trustworthy even if they reduce the bias of human participation. Deep learning (DL) models are such techniques. In high-risk applications like medical ones ceding medical decision-making to DL models without understanding of diagnosis rationale may violate the principle of non-maleficence and expose patients to harm. Therefore, in such applications, it is important to pay attention to system trustworthiness. From the clinician's point of view, in order to be acceptable, ML-based systems should have a level of transparency that allows their decisions to be verified by medical specialists in their level of domain knowledge. Explainable AI (XAI) techniques can provide effective tools to support this [4].

Small Vessel Disease (SVD) is a term which encompasses a variety of changes in human brain which are attributed to pathological changes in the small vessels. Anatomically, small vessels are arterioles, capillaries and venules. These structures are too small to be visible in CT or MRI. For this reason we focus on evaluating the lesions which appear as a result of pathologic changes to the small vessels. Among the factors which can lead to SVD we can distinguish: arteriolosclerosis, cerebral amyloid angiopathy, inherited/genetic small vessel diseases, CNS vasculitis, venous collagenosis or radiotherapy [5].

Basic tool for diagnosing SVD is neuroimaging, which makes it a task that can be automated by machine learning algorithms analyzing the image. Neurological changes can be visualized in both CT and MRI, with the latter being the current gold standard for diagnosis. For many years, there were no structured guidelines for reporting the imaging findings which was affecting the communication between specialists. Comparing imaging studies without structured reporting caused difficulties in evaluating the progress and severity of the disease. In 2013, an international working group from the Centres of Excellence in Neurodegeneration published Standards for Reporting Vascular changes on Euroimaging (STRIVE). STRIVE provided a common advisory about terms and definitions for the features visible on MRI as well as with structured reporting of changes related to SVD on neuroimaging [6]. According to STRIVE, the MRI scans, can detect a spectrum of white matter lesions which include: recent small subcortical infarcts, lacunes, white matter hyperintensities, perivascular spaces and microbleeds. Brain atrophy is another pathology observed in the course of SVD [6,7]. Reviewing diverse imaging findings can sometimes be difficult for human eye to decide where the edges of the disease are and therefore to accurately monitor the disease and assess its severity. To help in this case, automatic solutions come to the rescue, which can provide expert's support and therefore speed up establishing patients diagnosis and providing appropriate treatment.

In this paper we highlight the challenge of creating an automatic tool for diagnosis of SVD and present some preliminary results regarding this topic. Moreover, together with medical practitioners, we have tried to cast this problem into the trustworthy AI framework so that the achieved analysis results can be accepted by the medical practitioners and applied to their practice.

2 Diagnosis of Small Vessel Disease - Fundamentals

SVD is a blanket term for lesions which appear in imaging studies secondarily to damage of the small vessels endothelium. Their descriptions along with a graphical representation (Fig. 1) are given below.

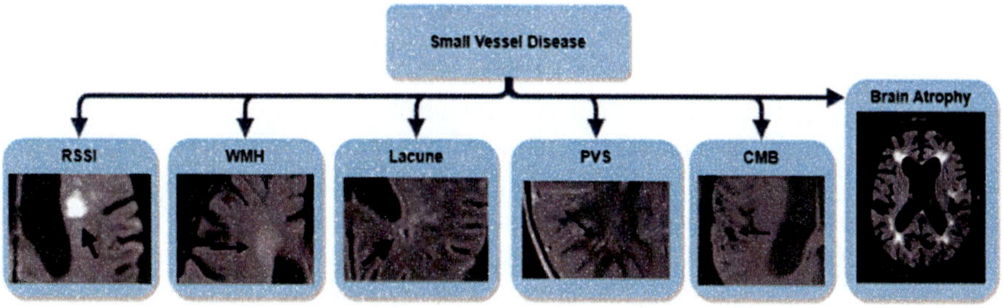

Fig. 1. Samples of SVD lesions

Recent small subcortical infarcts (RSSI) account for almost 25% of all ischemic strokes [5,8]. RSSIs occur in the areas supplied by a single perforating artery, which are devoid of collateral circulation [6,9]. According to STRIVE they are best identified on DWI and their diameter is usually smaller than 20 mm. On DWI RSSIs are hyperintense focal lesions which strongly restrict water diffusion. They are also hyperintense on T2 and FLAIR. RSSIs are often not visible on CT. Over time, they can evolve into white matter hyperintensities or lacunes, but complete regression is also a possibility [6,10].

White matter hyperintensities (WMH) are sometimes referred to as leukoaraiosis. They are usually symmetrical, variable in size lesions, which are hiperintense on T2 and FLAIR and isointense or slightly hipointense on T1. Aetiology of WMHs differs depending on location, which can be either periventricular white matter (PVWM) or the deep white matter (DWM). Lesions in PVWM were found to be the result of one of the following: ependymal loss, differing degrees of myelination in adjacent fiber tracts or cerebral ischemia with associated demyelination, whereas DWM lesions are ischemic in nature and their size corresponds with the increasing severity of tissue damage. To quantify the severity of WMHs Fazekas scale is used. It divides WMH lesions based on their location (either PVWML or DWML); lesions located in each of the areas receive a grade from 0 to 3 based on the size of lesions [11].

Lacunes are oval or round lesions which appear at the location of previous small subcortical infarcts or less frequently, microbleeds. They are hypointense on FLAIR and T1 and their diameter is 3–15 mm. Usually, lacunes have a hyperintense rim on FLAIR, which allows us to distinguish them from the perivascular spaces [6,10].

Perivascular spaces (PVS) are also known as Virchow-Robin spaces. They are fluid-filled spaces which surround arterioles, capillaries and venules in the brain. Once enlarged, PVSs become visible in imaging studies as a linear or round hypointense lesions on FLAIR and T1 with basal ganglia being the most common location. Rarely, they can become significantly enlarged and form tumefactive perivascular spaces which can cause mass-effect on surrounding brain tissue [12]. Although, PVSs are normal anatomical structures, it was observed that their number and size increases with the patients age and appearance of other lesions associated with SVD. It was also reported, that there is an association between PVS's and subsequent onset of dementia [13].

Cerebral microbleeds (CMB) are lesions which are hipointense on T2*/ SWI sequences and isointense on T1, T2 and FLAIR. They are usually round or ovoid and smaller than 10 mm. The size of microbleeds may vary due to "blooming artifact" which may cause cause micro bleeds to appear larger than they actually are. Differential diagnosis includes intracranial calcifications, metastases susceptible to bleeding and diffuse axonal injury [6,10].

Brain atrophy is a common outcome of the disease process which affects brain parenchyma and it can be either focal or generalised. Evaluating the change in the size of the brain might be a valuable tool in monitoring the progress of the disease [6,10]. It is important to note the variety of these lesions in terms of their nature, appearance, number of occurrences, and how they are diagnosed and described by medical specialists. Different sequences such as SWI or FLAIR are used to recognize them and different numbers of examinations are performed to visualize the changes over time as in the case of brain atrophy. In addition, with some lesions such as CMB the valuable information is the number of lesions, while with brain atrophy it is its volume. Due to the factors described above, in order to make a reliable diagnosis, many principles and standards must be considered.

Table 1. Simple and amended SVD score based on [14].

MRI feature	Quantity	Simple SVD score	Amended SVD score
Microbleeds	≥1	1	1
Lacunes	0	0	0
	1–2	0	1
	3–5	1	2
	>5	1	3
White matter hyperintensities (Fazekas score)	0	0	0
	1	0	1
	2	1	2
	3	1	3
Total SVD score (range)		0–3	0–7

Two scales, simple and amended SVD score, were developed to diagnose SVD, they are visible in Table 1. In the case of both scales, any score higher than 0 indicates SVD. The higher the score, the more severe SVD is. It is obvious that with such a complex and complicated issue, it is not easy to provide a clear-cut diagnosis. Such diagnosis should be evaluated through the use of various diagnostic tools and consultation with other specialists.

One of these tools may be AI algorithms, however, because of the diversity in how each SVD component is evaluated, it is not possible to use a single universal classifier for this purpose. A decision system that comprehensively diagnoses SVD and assesses its level of severity must consist of an assembly of multiple ML algorithms and a system that draws final conclusions based on it, such as a fuzzy inference system, or a neural classifier.

3 The Need for Trustworthiness of AI-Based Systems

ML-based CDSS systems, despite their many advantages, still have many significant flaws and weaknesses. The fully automated and complex data-driven nature of AI models especially deep learning seems to be a double-edged sword. First of all, the model performance strongly depends on provided training data. Unfortunately, the data are usually affected difficult to avoid bias [15]. Examples of such bias include the type and settings of the imaging machine, reasons for the examination, systematic errors by clinicians, and non-statistical appropriateness of the examination group for a given disease due to age, associated diseases, sex, race, etc. of the patients. Additionally, medical image data are usually high-resolution and multidimensional complex structure, while ML algorithms work with down-scaled input, resulting in blurring or even loss of important details [16]. Another data problem is that the publicly available benchmark data that would allow comparisons of the proposed approaches differ significantly from the

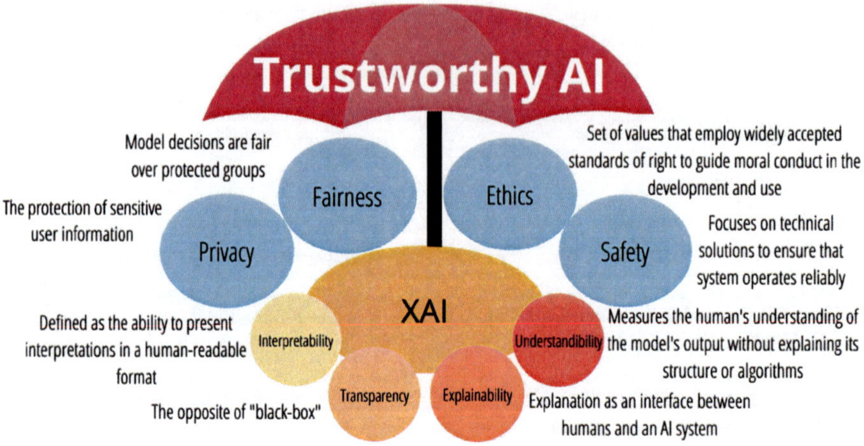

Fig. 2. Terms that are included in trustworthiness

raw data e.g. from the hospitals. They are often already prepossessed/balanced in some way or there is no information about the survey cohort and imaging parameters. Furthermore, DNN architectures consist of hundreds of layers and millions of parameters, resulting in a complex black-box model. Additionally, these architectures also, similarly to data, suffer from bias such as evaluation bias, deployment bias or illusion of control bias. Those pose problems related to their robustness, transparency of operation, and ability to generalize, which in turn implies problems related to their trustworthiness [17,18].

In the literature, the term trustworthiness includes many different terms. Their definitions are shown in Fig. 2. From the computer science point of view those terms means often something else then for the end users – the clinicians in this case. The developers for example need tools for sanity check the information system and in particularly AI models, mainly in terms of reliability, speed of operation and performance. Medical specialists, on the other hand, need to be shown and explained the links between the features extracted by AI algorithms from the data and their decisions, often giving less attention to the accuracy. It is important to pay attention to reliability of such a system because it might gain trust from the end-users, so it is a key driver for its deployment in clinical practice [18]. When thinking about the CDSS, an important question to consider is what can be done better by AI and what can be done better by the (human) clinician? Simple but time-consuming and tedious work should be left to the algorithms to not waste clinicians time, whereas complex, uncertain problems still require human expertise. In this regard, a reasonable approach is a fusion of both – interactive machine learning with a "human in the loop" that would combine the conceptual understanding and experience that clinicians have and automation of simple tasks. However, the simplicity and intuitiveness of CDSS should be taken into account so that the collaboration between the system and the clinician does not consume the clinician's mental resources [16].

The principle of non-maleficence in the context of medical, which states that clinicians have a fundamental duty not to harm their patients either intentionally or not, creates the need for explainability. This is necessary in the context of using these black-box models. In such vulnerable and fragile fields as medicine any mistakes, especially false negative results, are a significant problem. The ability of AI to explain its results enables disagreements between the system and human experts to be resolved. It allows the latter to make an informed decision whether or not to rely on the system's recommendations and to provide proper treatment, consequently increasing their trust in the system [17].

What is more, as a result of legal and forensic constraints, such as the European Union General Data Protection Regulations (GDPR), the Health Insurance Portability and Accountability Act (HIPAA), and the U.S. Food and Drug Administration (FDA), it is not allowed to use AI systems as black boxes [16,18]. The clinician must be able to understand why a certain decision has been reached. Due this future human AI interfaces should focus on explainability and interpretability.

A relatively new but rapidly growing field of AI addressing these problems is Explainable Artificial Intelligence (XAI). XAI covers a wide range of features, that can be analysed and explained in different ways. One of them is post-hoc explainability techniques, that suites bests for making AI-based medical tools more transparent and understandable for medicans not familiar with software and AI nuances. Post-hoc explainability include textual explanations, visual explanation, local explanations, explanations by example, explanations by simplification, and feature relevance explanation. These techniques can be applied to existing systems without changing their structure, algorithms, or other features. In our opinion, such techniques should be applied on a wide scale, both to systems already being in use and those being developed, especially in the context of medical applications to meet the described expectations of users and to solve the problems highlighted above - user-centric approach.

4 ML-Based System Development

Nowadays, systems employing machine learning algorithms, and in particular deep learning are black-box models. That means that we provide an input and get the output, but we are not sure, what happens inside - on what ground the decision is made.

Designing an algorithm for medical purposes is not particularly different from ML algorithms in other fields. It consist of standard steps like: problem definition, data preparation, model design, evaluation and trustworthy assessment. However, it requires special precautions, as it strongly affects a human health and life. In order to have a reliable and robust system, we must make every effort to eliminate as many biases as possible at each stage.

Problem Definition: This step requires a comprehensive analysis regarding the nature of the problem. Things that have to be considered are the solved task from the technical point of view e.g. classification, detection, segmentation or other; what data must be provided and in what form, what is the expected output etc.

In order to provide a responsible system it is crucial to consult with the specialists in the field. When the specific problem is concerned it is probable that ML specialist do not have enough knowledge to take into account the whole nature of the problem.

For instance, considering small vessel disease, we assume that we want a 0–100% score describing the probability of SVD presence. However, there are several markers that indicate SVD and each of them must be addressed. In case of cerebral microbleeds the required information is their amount in the brain. Therefore, a detection task is sufficient as each of CMBs must be found and counted. When it comes to recent small subcortial infarcts, lacunes or white matter hyperintensities there is a need for segmentation, because these abnormalities might be much bigger than CMB and have irregular shapes. Moreover, the size is crucial regarding the assessment of the patient's condition and it can not be calculated only based on the detection information. In contrast, a completely different approach should be taken to assess the presence of brain atrophy. This symptom cannot be quantified directly, but relatively to previous examinations. It is usually done by comparison of the white and grey matter volume from two scans.

An important issue at this point is also to make a choice between a 3D and 2D space. Although, MRI images are in 3D form, they actually consist of many 2D images merged together. Current 3D ML algorithms have high computation costs, so using a 2D space seems more suitable. However, it is important to provide information from adjacent slices as they carry valuable information, especially for distinguishing specific lesions from its mimics.

Another challenge is visibility of the lesions on different sequences of one scan. Depending on the lesion characteristic and stage, it is visible on specific sequence. For instance, to distinguish WMH from WML, a DWI sequence is crucial - only WMH will be visible. On the other hand, RSSI will be probably visible only for around 10 days at ADC map. Taking it into consideration, it occurs, that proper SVD diagnosis system design is actually a merge of several independent blocks.

Data Preparation: Normally, physicians diagnose SVD based on the MRI image analysis, particularly T2-weighted, T1-weighted and gradient echo/ T2*/ susceptibility-weighted sequences. In such case, the same images should be passed as an input to the system. While human can adapt and change the image properties during the analysis, like for example Gamma value. In the automatic process all the data have to be prepared before the process. It forces a careful data pre-processing, including normalization, brightness and contrast adjustment, resize etc. In case of 2D algorithm, this is also a step of introducing the knowledge from adjacent slices.

Moreover, ML algorithms require properly labeled data for pattern recognition. It is often a problem as data labeling is a very laborious and time-consuming process, however, it is a crucial step as it affects the whole training. Unfortunately, it should be performed by radiologists, as ML specialists simply do not have enough knowledge to create trustworthy annotations. Although it may seem a simple task, the annotating rules should be agreed: whether the lesion is annotated in every image that it is seen or only in one it is best seen; the level of preciseness; the agreement between several raters etc. Any mistakes in labels, new bias is introduced into the system and it leads to some incorrect features generation. Lack of databases is a major obstacle regarding ML system synthesis. Although medical facilities are in possession of huge amount of data, they are not annotated. To the best of our knowledge, there are no publicly available databases of SVD disease. Even for microbleeds - the simplest type of lesion in this disease - there are only few, small ones. Therefore, even after designing a system, it is hard to compare results with the state-of-the-art. Undeniably, medical databases creation is essential for development of automatic diagnosis.

It is also worth mention, that there are some benchmarks regarding image processing, that enable ML algorithms performance check. They mainly serve for algorithms development and comparison. In case of such specific tasks, they do not apply. Not only because of the problem nature, but also shortage in data and its weaker preparation. Therefore, achieving results similar to those achieved on benchmarks is extremely hard.

Model Preparation and Regularization: At this stage the actual machine-learning model - or models must be prepared. Firstly, the decision regarding the model has to be made. There is a number of already pre-trained models that can be taken advantage of or alternatively a custom model can be designed. The decision is usually made based on the problem specification. Some models deal better with small objects - like Faster RCNN, other do a precise segmentation and some are faster or have lower computational costs - like YOLO or MobileNet. There is a wide range of architectures for detection [19] and segmentation [20]. At the time of decision, all the features of the model have to be taken into account.

Next, the hyperparameters have to be carefully adjusted. Additionally, some regularization techniques should be applied to improve the system performance. There are plenty of solutions that can be added at this point. The base one is data augmentation - although usefull, should be performed carefully, to not produce images that has no connection with real examples, as it may introduce additional bias.

In case of lack of labeled data, self-supervised learning seems to be a promising approach [21]. There is also a branch of providing a domain knowledge into the system [22], which may be extremely useful, especially in medicine. It is obvious, that patient's medical history and his health state are crucial for diagnosis.

Model Evaluation: A proper model evaluation is integral with the system design process. There are a lot of metrics describing the system performance. They have to thoughtfully selected not only to show the results considering various aspect, but also to enable comparison between proposed solutions for a given problem. Different metrics are used for classification - accuracy, sensitivity and specificity; detection - sensitivity, precision, F1 score, mAP; segmentation - pixel accuracy, F1score or mAP.

Depending on the problem additional matrics may be informative and should be considered, like in case of microbleeds detection number of false positive predictions per scan and per one CMB is often reported.

Further important issue is the system balance. For instance, sensitivity and precision should be at the similar level. If sensitivity is relatively high and precision low, there is a high generation of false positive predictions. In other words, system is looking for the exact lesion and does not distinguish it from its mimics. On the other hand, when the precision is high and the sensitivity is low at thesame time, probably not many objects of interest are found. Such events may easily lead to radiologist's lost of trust in the system's performance.

It is worth remembering, that not only providing all the metrics is important, proper interpretation is even more valuable as it can point out some features of the system and be useful for improvement.

Regarding the limitations of training datasets, the good practice is to provide a nameplate with the characteristic of the system and data that it is intended for. It is obvious, that system trained on male Europeans in their 60s will not be satisfying enough in case of Asian women in their 40s.

Model Trustworthiness: In case of ML-based systems an inherent step in system's creation is providing its trustworthiness. Although it is an area that still requires a lot of research, consideration of aspects mentioned in Sect. 3 should be taken. Unfortunately, without this step, even accuracy close to 100% does not make a system able to be clinically used.

5 Small Vessels Disease Diagnosis - Preliminary Results

Our goal is to develop an automatic, trustworthy system for small vessel disease diagnosis. As described in previous sections it is very complex and challenging problem. One of the main elements of the system being developed is a system for cerebral microbleeds detection [23]. Our solution utilizes the Faster RCNN deep neural network, enhanced by the extra post-processing algorithm for false positive and false negative predictions reduction. The algorithm is based on comparison of predictions from adjacent slices and strongly improves system performance. Our solution achieved 92,62% sensitivity, 89,74% precision and 90,84% F1 score. In the Fig. 3a we present an example of microbleeds detection by the system.

Regarding issues related to the reliability and robustness of the system performance, we conducted a comprehensive study of the factors affecting the model development and learning process. The most important aspect is information from adjacent slices inclusion by merging three one-channel images into one three-channels. It enables using a two dimensional solution for three dimensional problem, whereas 3D neural networks are much more computational demanding. We also found out that, in case of CMBs, applying only one label in the slice, where the lesion is the most visible, is more effective than multiple labels in every slice, where the lesion is visible. Further, a larger image size improves

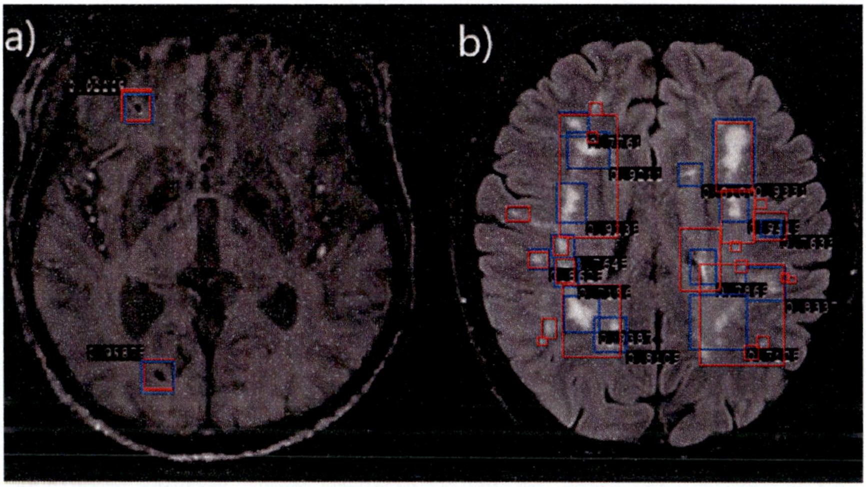

Fig. 3. Samples of detected: a) CMB b) WMH. Ground truth – red, prediction – blue. (Color figure online)

the sensitivity of detection, as the lesion occupies a larger absolute area. Unfortunately, such training has higher computational cost, so the balance between these two aspects must be maintained. Next, by the number of experiments and domain knowledge a proper threshold of prediction confidence score to maintain a balance between sensitivity and precision.

Our current research focuses on a system for diagnosing white matter hyperintensities (WHM). We approached the problem using a similar method to that used for the detection of CMBs. The preliminary results are presented in Fig. 3b. However, it is clearly seen, that detection is not sufficient for this task. Although areas of interest are found with quite good confidence score, any volume and metrics count is inadequate in the current state. One reason is the lack of sufficient amount of labeled data. However, it seems that a much more appropriate approach is to use segmentation instead of detection.

Probably, lacunes, PVS or RSSI also will have to be segmented as WMH, as they have similar geometrical properties, however, different sequences will be considered. While, brain atrophy is a entirely different case - here the volume of brain should be calculated and compared with previous examination.

6 Concluding Remarks

The CDSS is first of all designed for clinicians, therefore it should be simple and intuitive in usage, but also reliable and trustworthy.

In recent years, many research report improving metrics of examined algorithms. Obviously, this is an crucial factor regarding AI development. However, discussions with medical specialists suggest that they care more about reliability, transparency, and intuitiveness of that systems than about their, sometimes only seemingly, high performance.

Biases are inevitable in ML systems, but all measures should be taken in order to limit their influence. Interpretation of the system and all the rules standing behind enables understanding of decision process making. Moreover, explanation of a specific decision may ease the radiologist work as it not only suggest the diagnosis, but shows the critical areas in the image.

Next essential issue states for results presentation. As long as there are medical regulations regarding disease classification and progression assessment, the system output should be presented in the same way - for example in case of SVD in STRIVE scale. Hence, when designing an applicable diagnosis system we need to pay a special attention to the matter of responsibility if we want it to be used in medical practice.

There are a set of recommendations for ensuring designing of responsible and trustworthy AI systems [24]: use a human-centered design approach; identify multiple metrics to assess training and monitoring; directly examine your raw data, understand the limitations of the dataset and the model; conduct rigorous unit tests to test each component of the system in isolation and as a whole; together with a field specialists design the model using concrete goals for fairness and inclusion; use representative datasets to train and test the model, check the system for biases; analyze performance by using different metrics; treat interpretability as a core part of the user experience, understand the trained model; provide explanations that are understandable and appropriate for the user.

We strongly believe, that matter of trustworthiness still requires a lot of research and methods development, nevertheless it will finally enable wide usage of ML algorithms in medicine.

Acknowledgement. This work was supported by the Ministry of Science and Higher Education in the years 2017–2022, under Diamond Grant DI2016020746.

References

1. Bali, J., Bali, O.: Artificial intelligence applications in medicine: a rapid overview of current paradigms. EMJ Innov. 73–81 (2020). https://doi.org/10.33590/emjinnov/19-00167. ISSN 2513-8634
2. Sutton, R.T., Pincock, D., Baumgart, D.C., Sadowski, D.C., Fedorak, R.N., Kroeker, K.I.: An overview of clinical decision support systems: benefits, risks, and strategies for success. NPJ Digit. Med. **3**(1), 1–10 (2020). https://doi.org/10.1038/s41746-020-0221-y. ISSN 2398-6352
3. Chen, L., et al.: Rapid automated quantification of cerebral leukoaraiosis on CT images: a multicenter validation study. Radiology **288**(2), 573–581 (2018). https://doi.org/10.1148/radiol.2018171567. ISSN 0033-8419, 1527-1315
4. Angelov, P.P., Soares, E.A., Jiang, R., Arnold, N.I., Atkinson, P.M.: Explainable artificial intelligence: an analytical review. WIREs Data Mining Knowl. Discov. **11**(5) (2021). https://doi.org/10.1002/widm.1424. ISSN 1942-4787, 1942-4795

5. Pantoni, L.: Cerebral small vessel disease: from pathogenesis and clinical characteristics to therapeutic challenges. Lancet Neurol. **9**(7), 689–701 (2010). https://doi.org/10.1016/S1474-4422(10)70104-6. ISSN 1474-4422
6. Wardlaw, J.M., et al.: Neuroimaging standards for research into small vessel disease and its contribution to ageing and neurodegeneration. Lancet Neurol. **12**(8), 822–838 (2013). https://doi.org/10.1016/S1474-4422(13)70124-8. ISSN 1474-4422
7. Rosenberg, G.A., et al.: Consensus statement for diagnosis of subcortical small vessel disease. J. Cereb. Blood Flow Metab. **36**(1), 6–25 (2016). https://doi.org/10.1038/jcbfm.2015.172. ISSN 0271-678X, 1559-7016
8. Moran, C., Phan, T.G., Srikanth, V.K.: Cerebral small vessel disease: a review of clinical, radiological, and histopathological phenotypes. Int. J. Stroke **7**(1), 36–46 (2012). https://doi.org/10.1111/j.1747-4949.2011.00725.x. ISSN 1747-4930, 1747-4949
9. Wardlaw, J.M., Smith, C., Dichgans, M.: Mechanisms of sporadic cerebral small vessel disease: insights from neuroimaging. Lancet Neurol. **12**(5), 483–497 (2013). https://doi.org/10.1016/S1474-4422(13)70060-7. ISSN 1474-4422
10. Shi, Y., Wardlaw, J.M.: Upyear on cerebral small vessel disease: a dynamic whole-brain disease. BMJ **1**(3), 83–92 (2016). https://doi.org/10.1136/svn-2016-000035. ISSN 2059-8688, 2059-8696
11. Fazekas, F., Chawluk, J., Alavi, A., Hurtig, H., Zimmerman, R.: MR signal abnormalities at 1.5 T in Alzheimer's dementia and normal aging. Am. J. Roentgenol. **149**(2), 351–356 (1987). https://doi.org/10.2214/ajr.149.2.351. ISSN 0361-803X, 1546-3141
12. Potter, G.M., Marlborough, F.J., Wardlaw, J.M.: Wide variation in definition, detection, and description of lacunar lesions on imaging. Stroke **42**(2), 359–366 (2011). https://doi.org/10.1161/STROKEAHA.110.594754. ISSN 0039-2499, 1524-4628
13. Ding, J., et al.: Large perivascular spaces visible on magnetic resonance imaging, cerebral small vessel disease progression, and risk of dementia: the age, gene/environment susceptibility-reykjavik study. JAMA Neurol. **74**(9), 1105 (2017). https://doi.org/10.1001/jamaneurol.2017.1397. ISSN 2168-6149
14. Al Olama, A.A., et al.: Simple MRI score aids prediction of dementia in cerebral small vessel disease. Neurology **94**(12), e1294–e1302 (2020). https://doi.org/10.1212/WNL.0000000000009141. ISSN 0028-3878, 1526-632X
15. Mikołajczyk, A., Grochowski, M., Kwasigroch, A.: Towards explainable classifiers using the counterfactual approach - global explanations for discovering bias in data, no. arXiv:2005.02269 (2020). https://doi.org/10.48550/arXiv.2005.02269
16. Sorantin, E., et al.: The augmented radiologist: artificial intelligence in the practice of radiology. Pediatr. Radiol. (2021). https://doi.org/10.1007/s00247-021-05177-7. ISSN 0301-0449, 1432-1998
17. The Precise4Q consortium, Amann, J., Blasimme, A., Vayena, E., Frey, D., Madai, V.I.: Explainability for artificial intelligence in healthcare: a multidisciplinary perspective. BMC Medical Inform. Decis. Mak. **20**(1), 310 (2020). https://doi.org/10.1186/s12911-020-01332-6. ISSN 1472-6947
18. Liu, Q., Hu, P.: Extendable and explainable deep learning for pan-cancer radiogenomics research. Curr. Opin. Chem. Biol. **66**, 102111 (2022). https://doi.org/10.1016/j.cbpa.2021.102111. ISSN 1367-5931
19. Zaidi, S.S.A., Ansari, M.S., Aslam, A., Kanwal, N., Asghar, M., Lee, B.: A survey of modern deep learning based object detection models. Digit. Signal Process. **126**, 103514 (2022). https://doi.org/10.1016/j.dsp.2022.103514. ISSN 1051-2004

20. Hao, S., Zhou, Y., Guo, Y.: A brief survey on semantic segmentation with deep learning. Neurocomputing **406**, 302–321 (2020). https://doi.org/10.1016/j.neucom.2019.11.118. ISSN 0925-2312
21. Jaiswal, A., Babu, A.R., Zadeh, M.Z., Banerjee, D., Makedon, F.: A survey on contrastive self-supervised learning. Technologies **9**(1), 2 (2020). https://doi.org/10.3390/technologies9010002. ISSN 2227-7080
22. Schuster, D., van Zelst, S.J., van der Aalst, W.M.: Utilizing domain knowledge in data-driven process discovery: a literature review. Comput. Ind. **137**, 103612 (2022). https://doi.org/10.1016/j.compind.2022.103612. ISSN 0166-3615
23. Ferlin, M.A., et al.: A comprehensive analysis of deep neural-based cerebral microbleeds detection system. Electronics **10**(18), 2208 (2021). https://doi.org/10.3390/electronics10182208. ISSN 2079-9292
24. GoogleAI: Responsible AI practices. https://ai.google/responsibilities/responsible-ai-practices?category=general. Accessed 14 May 2022

Machine-Aided Detection of SARS-CoV-2 from Complete Blood Count

Barbara Klaudel[1(✉)], Aleksander Obuchowski[1], Małgorzata Dąbrowska[2,3], Kornelia Sałaga-Zaleska[2,3], and Zdzisław Kowalczuk[1]

[1] Gdańsk University of Technology, Gdańsk, Poland
barbara.klaudel@student.pg.edu.pl
[2] Medical University of Gdańsk, Gdańsk, Poland
[3] University Clinical Centre in Gdańsk, Gdańsk, Poland

Abstract. The current gold standard for SARS-CoV-2 detection methods lacks the functionality to perform population screening. Complete blood count (CBC) tests are a cost-effective way to reach a wide range of people – e.g. according to the data of the Central Statistical Office of Poland from 2016, there are 3,000 blood diagnostic laboratories in Poland, and 46% of Polish people have at least one CBC test per year. In our work, we show the possibility of machine detection of SARS-CoV-2 virus on the basis of routine blood tests. The role of the model is to facilitate the screening of SARS-CoV-2 in asymptomatic patients or in the incubation phase. Early research suggests that asymptomatic patients with COVID-19 may develop complications of COVID-19 (e.g., a type of lung injury). The solution we propose has an F1 score of 87.37%. We show the difference in the results obtained on Polish and Italian data sets, challenges in cross-country knowledge transfer and the selection of machine learning algorithms. We also show that CBC-based models can be a convenient, cost-effective and accurate method for the detection of SARS-CoV-2, however, such a model requires validation on an external cohort before being put into clinical practice.

Keywords: Deep learning · Computer-aided diagnosis · COVID-19

1 Introduction

As of this writing, 472 million people have been infected with the SARS-CoV-2 virus and 6 million have died from the disease caused by the COVID-19 virus [24]. Effective pandemic mitigation requires early virus detection tools. We currently lack cost-effective screening methods. Reverse transcription polymerase chain reaction (RT-PCR) is a laboratory technique considered as the gold standard for the detection of SARS-CoV-2. However, it is an expensive solution, requiring specialized equipment and trained personnel, which limits its effectiveness.

In this work, we examined the feasibility of predicting COVID-19 disease with machine learning models based on complete blood count (CBC) data.

Z. Kowalczuk (Ed.): DPS 2022, LNNS 545, pp. 17–28, 2023.
https://doi.org/10.1007/978-3-031-16159-9_2

The CBC test shows the general health of patients and is often ordered to check the extent of the disorder. It is also used to monitor your health and response to treatment. Due to widespread use, CBC data from a wide range of populations are available and provide insight into assessing the actual spread of the virus.

Therefore, in this article, we introduce the SARS-CoV-2 detection model based on blood count data. In addition, we compare the performance of tree-based models and artificial neural network approaches, and analyze the effectiveness of knowledge transfer between different countries.

2 Related Work

Due to the limited availability and long response time of RT-PCR tests, chest imaging can be used to detect symptoms related to COVID-19 [6]. Other imaging methods such as computed tomography and lung ultrasound have also been studied for this purpose. Detection of COVID-19 without imaging (such as CBC) is a much less explored research area than imaging-based detection. The following sections describe the different modalities used to diagnose COVID-19 and present the most important research findings in these areas.

Chest X-Ray (CXR): X-ray imaging is the most common radiography method due to its cost-effectiveness. Unlike computed tomography and magnetic resonance imaging, X-ray imaging equipment is not stationary and can be used to diagnose critical patients without moving them from their beds and its disinfection is less troublesome. However, its sensitivity is limited and may fail to provide useful information for early stage of SARS-CoV-2 infection or mild cases [25]. Thanks to the availability of publicly available datasets, machine-aided SARS-CoV-2 detection based on X-rays has been a popular research topic. The datasets present the cases of solely COVID-19 patients [3,4], patients with COVID-19 and other viral diseases (SARS, MERS, ARDS) [8] or they collect the patients with COVID-19, bacterial and viral pneumonia or without a disease [2]. In Das et al. [9], they created an ensemble neural network to detect COVID-positive patients. The ensemble model achieved the classification accuracy of 91.6% and was tested on a dataset of 57 images of patients infected with COVID-19 and 60 images of healthy patients.

Chest Computed Tomography (CT): Computed Tomography has better sensitivity than X-ray, but is more expensive and involves higher doses of radiation. High-resolution CT is considered the primary means of determining COVID-19 from radiography. The level of suspicion of COVID-19 infection is determined by the 6-point CO-RADS scale [19]. Values 4–6 indicate a high level of suspicion of COVID-19 infection. MosMedData [16] additionally provides masks showing consolidations and opacities of ground-glass (symptoms typical of COVID-19). In Zhao et al. [27], the authors used a deep neural network to classify COVID-19, pneumonia and healthy patients. The architecture of their network was based on ResNet-v2 [11] and reached an accuracy of 99.2% on the test dataset.

Lung Ultrasonography (LUS): Lung Ultrasonography is the least common research area to predict SARS-COV-2 infection. It turned out to be useful during the influenza epidemic in 2009 [17], but its use in the treatment of COVID-19 is still unrecognized [23]. To the best of our knowledge, there are no publicly available datasets for this task. Roy et al. [20] developed a deep learning model for the classification and location of COVID-19 markers. Their architecture was based on the Spatial Transformer [12] and consistency loss [10]. The model obtained the F1 score at 61%.

Complete Blood Count (CBC) Clinical Data: Cabitza et al. [7] analyzed hematochemical data from 1,684 patients to create a model for the detection of COVID-19. Their dataset was made available for public use, and its subset of complete blood count (CBC) was used as a supplementary dataset in this paper. Several machine learning models were tested, including: logistic regression, naive Bayes, k-nearest neighbours, random forest and support vector machines (SVM). Analysis of the CBC data set by the K-nearest neighbors method gave the highest accuracy (at 86%), with sensitivity 82%, a specificity of 89%, and AUC (Area Under receiver operating characteristic Curve) of 86%.

In the study [14] of Kukar et al., based on data from 5333 patients (3% positive), an XGBoost model was developed. This model reached the sensitivity of 81.9%, specificity 97.9% and AUC indicator at 97%. However, the imbalance between classes and the low availability of COVID-19 positive samples mean that perhaps the results may differ for a wider validation cohort.

Soltan et al. [22] created 2 models: model for the emergency department (ED) and the admissions model for the hospital. The ED model achieved 77.4% sensitivity and 95.7% specificity, while the admissions model achieved 77.4% sensitivity and 95.8% specificity. Both models were implemented using XGBoost.

In Soares et al. [21], the ER-CoV model was created with 16 input features from the most common examinations. It was implemented as a Support Vector Machine. SMOTEBoost was used to minimize data imbalance. The purpose of this model was to determine whether a suspicious patient coming to the ED might have a negative for COVID-19 test result. In such a situation, high specificity is crucial, and the sensitivity should be of a reasonable value. The indicated model achieved a specificity of 86% and a sensitivity of 70.3%.

In Wu et al. [26], 11 CBC features served as inputs to a random forest model. The dataset consisted of 259 samples from 169 patients, of which 105 (62%) tested positive for COVID-19. The testing was done using a 10-fold cross-validation. The model achieved a sensitivity of 97% and a specificity of 96%.

In all the studies mentioned above, the COVID-19-positive group (cohort) was selected on the basis of the RT-PCR test result, which is considered the gold standard for the detection of SARS-CoV-2. Current work in a very similar direction is presented in Klaudel et al. [13]. The dataset that was used there to analyze COVID-19-positive patients was also used in this study.

3 Our Solution

Our goal is to detect patients infected with COVID-19. As input, the model takes complete blood counts and basic patient information. Below, we describe the model building process, from collecting and preprocessing data to training.

3.1 Data Collection

The data analyzed in this article comes from two sources. The first dataset was collected at the University Clinical Center (UCC) in Gdańsk, Poland. The second dataset (Zenodo) was made public in an article by Cabitza et al. [7].

UCC Dataset: The UCC dataset consists of the medical records of 22,463 patients admitted between March 25, 2019 and December 16, 2020. The data used in this study comes from routine blood tests, RT-PCR tests and patient care cards. The positivity of COVID-19 was determined on the basis of the RT-PCR test result performed within 1 day before or after the CBC test. Only patients with unambiguous positive/negative results were enrolled in the study, and patients with low positive results were excluded.

Zenodo Dataset: The dataset of [7] comprises of 1,624 patients admitted to San Raffaele Hospital in Italy between February 19, 2020 and May 31, 2021. There are 786 (48%) positive and 838 (52%) negative cases. The dataset has two subsets: COVID-specific and CBC features. Only the CBC features subset was selected for the experiments. SARS-CoV-2 infection was established based on the molecular test performed by RT-PCR. In the case of ambiguous cases, the RT-PCR result was compared with the chest radiograph and X-rays. The inconclusive category corresponded to patients with positive 72 h post-negative or patients with a hematological profile similar to patients with COVID-19. Using the criteria given in Table 1, the final patient cohort is 1624 patients.

Table 1. Feature pairs from the Zenodo dataset with a correlation greater than 50%. A detailed description of the features can be found in Table 2.

Feature 1	Feature 2	Correlation
HGB	HCT	0.97
HGB	RBC	0.82
MCHC	MCH	0.54
LY	NE	0.95
LY	NET	0.57
EO	EOT	0.92
LY	LYT	0.64
BA	BAT	0.64
NE	NET	0.59

3.2 Data Preprocessing

Robust data handling required the pre-processing of two separate datasets: UCC and Zenodo. The Zenodo dataset provided fewer features than the UCC dataset. Therefore, in order to have a larger number of patients included in the analysis, the Zenodo dataset was only used as a source when selecting the input features for the SARS-CoV-2 detection model. In the Zenodo dataset, all features had a missing rate of less than 20%, therefore no features were rejected due to the high scrap/missing rate.

Table 2. CBC features selected as model input.

Feature	Description
Hemoglobin (HGB)	A dye that transports oxygen. HGB combines with oxygen in pulmonary alveoli and releases oxygen in tissues
Mean corpuscular volume (MCV)	A parameter describing the average volume of a red blood cells
Mean corpuscolar hemoglobin concentration (MCHC)	A parameter describing the average hemoglobin concentration per unit volume of red blood cells
Platelets (PLT)	Platelets participate in the blood clotting, store and transport selected substances
White Blood Cells (WBC)	Cells which role is to protect the organism against pathogenes
Basophils count % (BA)	White blood cells that support the immune system and produce a substance that reduces blood clotting
Lymphocytes count (LY)	White blood cells that detect and neutralize pathogenes
Monocytes count (MO)	White blood cells that transform into macrophages and absorb microorganisms and damaged cells
Eosinophils count (EO)	White blood cells that help fight parasites, participate in decomposition of histamine that controls allergic symptoms

In order to select the input features of the models, a correlation analysis was performed. Table 1 lists parameter pairs with absolute correlation coefficients greater than 50%. For each pair, the feature from the second column was excluded from further analysis. The order of the traits in the pair was determined on the basis of the following rules: (1) the feature having correlation with the target variable was preferred, and (2) if the traits describe the same variable (with different units: metric and percentage), the feature described as a percentage was selected. In addition, the suspicion trait that indicates that a patient may be suffering from specific symptoms of COVID-19 during segregation has been removed from the dataset as such information is not available for patients included in the UCC dataset.

Table 2 presents selected input features with their description, standards and units. Additionally, sex and age were included as input features. The UCC dataset has been narrowed down to patients with the selected input features. The missing values of the continuous features were imputed on the basis of the k-nearest neighbours algorithm. Sex was the only categorical feature. There were no missing values here. Finally, the features were standarized by removing the mean value and scaling to unit variance.

3.3 Architectures

Models based on traditional machine learning and deep neural networks were tested. The models tested are listed below.

XGBoost: Extreme Gradient Boosting (XGBoost) [1] is the most popular model used in Kaggle competitions with tabular data. However, it is a tree-based algorithm recommended for small tabular datasets. In our model, we used empirically selected hyperparameters: colsample by tree 0.5, learning rate 0.001, max depth 5, alpha 5, and 2 estimators.

CatBoost: Categorical Boosting (CatBoost) [18] is a gradient-boosting model for numeric, categorical and textual data. It introduces a separate categorical data processing algorithm, which also prevents leakage of targets [18]. The categorical data algorithm transforms the categorical features so that they can be mathematically compared, which is based on the concept of introducing artificial "time". As the loss function, we used categorical cross-entropy with a learning rate of 0.001.

FC ANN: In our work, we used a Fully-Connected Artificial Neural Network (FC ANN) with 5 (fully-connected) layers and separating dropout layers (in-between) applied to reduce overfitting. We also used sparse categorical cross-entropy as a function of loss.

TabNet. TabNet [5] is a state-of-the-art deep neural network that surpasses boosting algorithms previously considered the leading solution for tabular data. Unlike tree-based algorithms, models based on deep neural networks do not require pre-processing of features and can learn from structurally different types of data (like images with captions) [5]. The loss function assigned class weights to the samples. The class weights were calculated from the inverse ratio of the samples in all classes. We chose the following values of the hyper-parameters: Optimizer: Adam; Learning rate: 0.01; Maximum epoch: 1000; Patience: 60; Batch size: 256/64; Loss: Categorical Cross Entropy; Scheduler by Step Learning Rate; Scheduler step size: 10; Scheduler Gamma: 0.9.

We used the Synthetic Minority Oversampling Technique (SMOTE) to account for class imbalance in the XGBoost and CatBoost models. In contrast, we did not use SMOTE for the ANN and TabNet models because it did not improve the results.

4 Experiments and Results

In this section, we describe the experimental setup common to all experimental studies and the details of the individual tests.

4.1 Experimental Setup

The experiments allowed us to test different approaches to the detection of SARS-CoV-2. The training dataset consisted of 80% of the samples and the validation dataset had 20% of the samples. We compared the performance of four models: XGBoost, CatBoost, fully-connected Artificial Neural Network (ANN) and TabNet. We tested the impact of data imbalance and knowledge transfer.

The following sections describe the results in terms of Accuracy, Precision, Recall, AUC and F1 score.

4.2 Baseline Training on UCC and Zenodo Datasets

To establish a baseline for further experiments, each detection model was trained on both available datasets: UCC and Zenodo. The results obtained with the models trained with each of the datasets are collected in Table 3. In the UCC dataset, the TabNet model scored the best of all models for accuracy, sensitivity, F1 score and AUC, but its specificity came third. Based on Zenodo, the results turned out to be lower than those reported in the paper providing this dataset [7]. The models presented in that paper were, however, constructed on the basis of a wider range of input features than the one used in this work. Because here we narrowed down the selection of features to the most common parameters checked during the complete blood count, which clearly made model training difficult, but facilitated the implementation and population study.

4.3 Unbalanced vs Balanced Training

The COVID-19 positive samples account for 20% of the UCC dataset, while the remaining 80% are COVID-19 negative samples. Such a relationship between classes constitutes a class imbalance, which can challenge the quality of the models and even result in poor performance in predicting minority class. The goal of the experiment described in this section is to see if training from a balanced dataset improves network performance. Therefore, a balanced dataset was created according to the UCC dataset, in which the number of COVID-19 negative samples was limited to the number of positive samples. Table 4 shows the results of the models trained on this dataset. An upward pointing arrow indicates

Table 3. Results of the models for SARS-CoV-2 detection trained on the UCC dataset (unbalanced dataset) and Zenodo.

UCC dataset					
Model	Accuracy	Sensitivity	Specificity	F1 score	AUC
XGBoost	76.82%	77.26%	59.77%	11.39%	68.52%
CatBoost	75.56%	75.50%	70.11%	12.51%	72.91%
ANN	72.73%	72.77%	71.26%	11.52%	71.94%
TabNet	81.81%	82.21%	66.67%	15.94%	74.44%
Zenodo dataset					
Model	Accuracy	Sensitivity	Specificity	F1 score	AUC
XGBoost	75%	78.57%	71.01%	73.07%	74.83%
CatBoost	75.86%	76.92%	74.7%	74.7%	75.81%
ANN	73.56%	71.98%	75.3%	73.1%	73.64%
TabNet	74.14%	75.54%	72.56%	72.56%	74.05%

that the value of the metric has increased from the base value (Table 3), while an arrow pointing down means that the value has decreased. The absence of an arrow indicates that the value remained unchanged. Accuracy and sensitivity for the balanced dataset have either slightly increased or decreased compared to the unbalanced baseline dataset. The major difference between the imbalanced and balanced methods can be seen in the sensitivity and F1 score. Training with the balanced dataset significantly improved specificity and maintained a relatively high sensitivity, which substantially increased the value of the F1 metric.

Since the F1 score is the most important metric for this kind of task, we can conclude that limiting the number of negative samples in the dataset is a better approach than taking all of the available samples. The TabNet model continued to be the best of all. For the balanced dataset, all metrics were the highest for the TabNet model, and the spread between TabNet and other models widened.

Table 4. Results of the models for SARS-CoV-2 detection trained on UCC balanced dataset.

Model	Accuracy	Sensitivity	Specificity	F1 score	AUC
XGBoost	70.43% ↓	77.78% ↑	62.07% ↑	66.26% ↑	69.92% ↑
CatBoost	72.58% ↓	74.75% ↓	70.11% ↑	70.52% ↑	72.43% ↓
ANN	74.19% ↑	76.77% ↑	71.26%	72.01% ↑	74.02% ↑
TabNet	87.1% ↑	84.95% ↑	89.25% ↑	87.37% ↑	87.1% ↑

4.4 Impact of Joined Learning with an Additional Dataset

The balanced dataset was chosen as the new benchmark because it gave better training results than the unbalanced dataset.

In this experiment, we train the model jointly on Zenodo and UCC dataset. We keep the test samples separately in order to later test the effect of such joint learning on the analysis of both datasets. We create 2 scenarios: (a) training on both the UCC and Zenodo dataset and testing on the UCC dataset and (b) training on the UCC and Zenodo dataset and testing on the Zenodo dataset.

Typically, a larger set of training data provides better performance. However, in this case, the datasets come from hospitals from two different countries, so the population characteristics of the datasets may be different. For example, the lower value of a variable may be due to the overall characteristic of population 1, not due to disease. Moreover, in population 2, the value of a given parameter may be higher for healthy patients, and only decreases after the infection with the virus. Regardless of this, the values of the features are usually measured with a variety of devices. Such differences can distort the phenomenon of generalization.

Table 5 shows the results of the models trained on this dataset. Arrows indicate the change in value compared to the UCC balanced (Table 4) and Zenodo (Table 3) datasets, respectively. In the case of testing of UCC datasets, the metrics values of all models decreased. In the Zenodo validation dataset, the sensitivity of the models decreased after combination, however, the specificity value increased for all models. The F1 score increased slightly for all. The overall impact of knowledge transfer was negative for UCC and positive for Zenodo.

Table 5. Results of SARS-CoV-2 detection models trained on UCC and Zenodo datasets, and validated separately on UCC (a) and Zenodo (b) datasets.

(a) Training on UCC+Zenodo, testing on UCC					
Model	Accuracy	Sensitivity	Specificity	F1 score	AUC
XGBoost	60.75% ↓	63.64% ↓	57.47% ↓	57.8% ↓	60.55% ↓
CatBoost	63.98% ↓	63.64% ↓	64.37% ↓	62.57% ↓	64% ↓
ANN	68.82% ↓	69.7% ↓	67.82% ↓	67.05% ↓	68.76% ↓
TabNet	70.43% ↓	71.72% ↓	68.97% ↓	68.57% ↓	70.34% ↓
(b) Training on UCC+Zenodo, testing on Zenodo					
Model	Accuracy	Sensitivity	Specificity	F1 score	AUC
XGBoost	72.41% ↓	61.54% ↓	84.34% ↑	74.47% ↑	72.94% ↓
CatBoost	73.28% ↓	63.74% ↓	83.74% ↑	74.93% ↑	73.74% ↑
ANN	72.41% ↓	68.68% ↓	76.5% ↑	72.57% ↑	72.59% ↓
TabNet	75% ↑	71.43% ↓	78.92% ↑	75.07% ↑	75.17% ↑

5 Discussion

Experiments show that a complete blood count supported by basic demographic data provides sufficient information about SARS-CoV-2 infection. It is clear, however, that due to the unsatisfactory quantity and variety of samples available for our research, the models still require careful evaluation before being put into clinical practice.

We have shown that models trained on Zenodo or UCC datasets have a relatively high F1 score of 74.7% and 87.37%, respectively. However, training a model on both datasets does not necessarily contribute positively to better predictive ability when validating on one of them. This effect can be attributed to the fact that there are differences in population characteristics, diagnostic equipment or internal policies (thus the datasets are not completely equivalent). The lack of a positive impact of knowledge transfer in this case suggests that the network modeled for clinical use should be trained on data from many institutions and parts of the country. Otherwise, the model may fail for people from different regions or institutions using equipment from a different manufacturer.

Moreover, we have shown that all models perform significantly better when trained on a balanced dataset. ANN approaches proved to be more robust to class-unbalanced datasets than the tree-based solutions that relied heavily on SMOTE (without SMOTE, they classified all samples to the majority class).

The TabNet model turned out to be the most reliable model in both tasks and in all variations of the (training) datasets. This network outperformed both the tree-based methods and the plain neural network in both unbalanced and balanced classification and all variants of the UCC-based dataset. The only experiment in which TabNet did not achieve the best F1 score is the case of training exclusively on the Zenodo dataset. However, the difference between the F1 score of TabNet and other models did not exceed 1.5%.

The target users of the SARS-CoV-2 detection model are patients with no clear suspicion of virus infection. The role of the model is to facilitate the screening of SARS-CoV-2 in asymptomatic patients or in the incubation phase. Early research suggests that asymptomatic patients with COVID-19 may also develop complications related to COVID-19 (e.g. lung damage) [15].

Hence, examining asymptomatic patients is also important as they may have complications and spread the virus without even knowing about the infection. The model presented in this work can be used to predict the risk of infection in patients undergoing routine blood tests. This screening method has limited additional cost for the laboratory. It uses data that needed to be collected for a different purpose anyway, requires no additional pretreatment, and produces results within seconds. The model can also be used as an additional preselection step prior to RT-PCR testing, e.g. for testing with RT-PCR only those patients for whom the ML model has returned uncertain prediction scores (e.g. 60% - COVID-19, 40% - no COVID-19).

6 Conclusion

The results of a complete blood count can be used to detect of potential SARS-CoV-2 infection and to extend the diagnosis to RT-PCR in a smaller group of patients. However, before implementing such models in clinical practice, they should be trained on a broader dataset coming from different populations and validated on multiple external cohorts. If validated, these models can serve as a cost-effective screening method, analyzing up to half of the population annually.

In the case of tabular data, models based on neural networks can outperform state-of-the-art tree-based approaches. They are more resistant to imbalance among data and give better results in all experiments but one (joint training on Zenodo + UCC and testing on the UCC dataset).

Experiments have shown that including additional data from a different population does not necessarily result in a gain in knowledge transfer. Therefore, all machine learning models should be carefully validated against multiple external datasets before being released into clinical practice. Overall, models based on complete blood counts can be a convenient, cost-effective and accurate method of screening patients potentially infected with SARS-CoV-2.

References

1. Python package index, xgboost. https://pypi.org/project/xgboost/. Accessed 29 Oct 2021
2. Coronahack - Chest X-Ray-Dataset (2020). https://www.kaggle.com/praveengovi/coronahack-chest-xraydataset?select=Chest_xray_Corona_Metadata.csv. Accessed 29 Oct 2021
3. COVID-19 dataset collected by Societa Italiana di Radiologia Medica e Interventistica (2020). https://sirm.org/category/covid-19/. Accessed 29 Oct 2021
4. COVID-19 dataset made available by a twitter user (2020). https://twitter.com/ChestImaging/status/1243928581983670272. Accessed 29 Oct 2021
5. Arık, S.O., Pfister, T.: Tabnet: attentive interpretable tabular learning. In: AAAI, vol. 35, pp. 6679–6687 (2021)
6. Bernheim, A., et al.: Chest CT findings in coronavirus disease-19 (COVID-19): relationship to duration of infection. Radiology 200463 (2020)
7. Cabitza, F., et al.: Development, evaluation, and validation of machine learning models for COVID-19 detection based on routine blood tests. Clin. Chem. Lab. Med. (CCLM) **59**(2), 421–431 (2021)
8. Cohen, J.P., Morrison, P., Dao, L., Roth, K., Duong, T.Q., Ghassemi, M.: COVID-19 image data collection: prospective predictions are the future. arXiv preprint arXiv:2006.11988 (2020)
9. Das, A.K., Ghosh, S., Thunder, S., Dutta, R., Agarwal, S., Chakrabarti, A.: Automatic COVID-19 detection from X-ray images using ensemble learning with convolutional neural network. Pattern Anal. Appl. **24**(3), 1111–1124 (2021). https://doi.org/10.1007/s10044-021-00970-4
10. French, G., Mackiewicz, M., Fisher, M.: Self-ensembling for visual domain adaptation. arXiv preprint arXiv:1706.05208 (2017)

11. He, K., Zhang, X., Ren, S., Sun, J.: Deep residual learning for image recognition. In: Proceedings of the IEEE Conference on Computer Vision and Pattern Recognition, pp. 770–778 (2016)
12. Jaderberg, M., Simonyan, K., Zisserman, A., Kavukcuoglu, K.: Spatial transformer networks. In: Advances in Neural Information Processing Systems, vol. 28, pp. 2017–2025 (2015)
13. Klaudel, B., Obuchowski, A., Karski, R., Rydziński, B., Jasik, P., Kowalczuk, Z.: COVID-19 severity forecast based on machine learning and complete blood count data. In: 2022 15th International Conference on Diagnostics of Processes and Systems (DPS), Chmielno/Gdańsk (Poland) (2022, submitted)
14. Kukar, M., et al.: COVID-19 diagnosis by routine blood tests using machine learning. Sci. Rep. **11**(1), 1–9 (2021)
15. Long, Q.X., et al.: Clinical and immunological assessment of asymptomatic SARS-CoV-2 infections. Nat. Med. **26**(8), 1200–1204 (2020)
16. Morozov, S., et al.: MosMedData: chest CT scans with COVID-19 related findings dataset. arXiv preprint arXiv:2005.06465 (2020)
17. Poggiali, E., et al.: Can lung US help critical care clinicians in the early diagnosis of novel coronavirus (COVID-19) pneumonia? Radiology **295**(3), E6–E6 (2020)
18. Prokhorenkova, L., Gusev, G., Vorobev, A., Dorogush, A.V., Gulin, A.: CatBoost: unbiased boosting with categorical features. arXiv preprint arXiv:1706.09516 (2017)
19. Prokop, M., et al.: CO-RADS: a categorical CT assessment scheme for patients suspected of having COVID-19 - definition and evaluation. Radiology **296**(2), E97–E104 (2020)
20. Roy, S., et al.: Deep learning for classification and localization of COVID-19 markers in point-of-care lung ultrasound. IEEE Trans. Med. Imaging **39**(8), 2676–2687 (2020)
21. Soares, F.: A novel specific artificial intelligence-based method to identify COVID-19 cases using simple blood exams. MedRxiv (2020)
22. Soltan, A.A., et al.: Artificial intelligence driven assessment of routinely collected healthcare data is an effective screening test for COVID-19 in patients presenting to hospital. MedRxiv (2020)
23. Syeda, H.B., et al.: Role of machine learning techniques to tackle the COVID-19 crisis: Systematic review. JMIR Med. Inform. **9**(1), e23811 (2021)
24. WHO: WHO coronavirus (COVID-19) dashboard. https://covid19.who.int/
25. Wong, M.D., Thai, T., Li, Y., Liu, H.: The role of chest computed tomography in the management of COVID-19: a review of results and recommendations. Exp. Biol. Med. **245**(13), 1096–1103 (2020)
26. Wu, J., et al.: Rapid and accurate identification of COVID-19 infection through machine learning based on clinical available blood test results. MedRxiv (2020)
27. Zhao, W., Jiang, W., Qiu, X.: Deep learning for COVID-19 detection based on CT images. Sci. Rep. **11**(1), 1–12 (2021)

Automatic Breath Analysis System Using Convolutional Neural Networks

Zdzisław Kowalczuk(✉), Michał Czubenko, and Michał Bosak

Faculty of Electronics, Telecommunications and Informatics, Department of Robotics and Decision Systems, Gdansk University of Technology, 80-233 Gdańsk, Poland
{kova,micczube,michal.bosak}@pg.edu.pl

Abstract. Diseases related to the human respiratory system have always been a burden for the entire society. The situation has become particularly difficult now after the outbreak of the COVID-19 pandemic. Even now, however, it is common for people to consult their doctor too late, after the disease has developed. To protect patients from severe disease, it is recommended that any symptoms disturbing the respiratory system be detected as early as possible. This article presents an early prototype of a device that can be compared to a digital stethoscope that performs auto-breath analysis. So apart from recording the respiratory cycles, the device also analyzes them. In addition, it also has the functionality of notifying the user (e.g. via a smartphone) about the need to go to the doctor for a more detailed examination. The audio recording of breath cycles is transformed to a two-dimensional matrix using mel-frequency cepstrum coefficients (MFCC). Such a matrix is analyzed by an artificial neural network. As a result of the research, it was found that the best of the obtained solutions of the presented neural network achieved the desired accuracy and precision at the level of 84%.

1 Introduction

Currently, the standards of medical practice are systematically changing, and computers are increasingly used in medical diagnostics. In recent years, artificial intelligence (AI) has started to play a special role in medicine, mainly due to its usefulness and the observed dynamic development [17]. There are different types of autonomous diagnosis projects, e.g. those based on decision trees, automated X-ray annotations, fMRI image segmentation, and many other contemporarily workable ideas that can be mentioned here [7,14]. It is estimated that in the near future, artificial intelligence will have an even greater impact on medicine [3,20]. Major progress will also be driven by electronic collection of all kinds of medical data in all countries of the world. Such shared medical data and their digital imaging will likely contribute to the creation of more effective artificial intelligence tools.

A doctor must possess great knowledge to become a specialist in the field of medicine. Medical studies sometimes last up to 12 years, if you take into account the internship and specialization. Moreover, even this is not enough, and more

Z. Kowalczuk (Ed.): DPS 2022, LNNS 545, pp. 29–41, 2023.
https://doi.org/10.1007/978-3-031-16159-9_3

years of medical practice are needed. Artificial intelligence is also developing in a similar way, the implementation of which is based on knowledge and exercises that require a lot of time and high computing power. For example, if you want to train an artificial bot to run on a classic real-time strategy game like Starcraft II, you will need to run a large number of full games [10, 21], which can take up to several years. Currently, AI solutions must be adapted to cooperation with doctors. In some tasks, such solutions can compete with and even surpass human expertise [17], as is the case, for example, in image analysis in radiology [5].

This article presents a prototype device for the detection of certain abnormal breaths using a Breathing Analysis System (BAS) based on artificial neural networks. This device is intended to assist the patient in deciding whether a visit to the doctor is necessary to deepen the diagnosis. The prototype BAS system is based on a Raspberry Pi and a microphone attached to the stethoscope head. The recorded audio is pre-processed and converted into a Mel-Frequency Cepstral Coefficient Matrix (MFCC). Data in this format is fed to the input of an artificial neural network.

Below we consider some possible solutions. Then we present the architecture of our prototype system and the principles of its operation. Conclusions resulting from the system operation tests performed can be found at the end of the article.

2 A Brief Overview of Similar Systems

Respiratory disease is the third largest cause of death for people and a huge burden on the national healthcare system. That is why the world pays a lot of attention to the engineering of early diagnosis solutions.

It is known that breathing sounds can be classified as normal or abnormal. Auscultation diagnosis looks for characteristic symptoms: discontinuous (crackling) or continuous (wheezing). Crackling is commonly associated with cardiovascular disease. A distinction can be made between smooth or rough crackles, depending on their length, pitch and respiratory cycle. Wheezes are typical of patients with asthma or chronic obstructive pulmonary disease [2]. One wheezing usually lasts longer than 250 ms of breath recording.

Pramono *et al.* [16] identified the features of the most appropriate types of respiratory sound analysis, which can be divided into two stages: feature extraction and analysis. Typical useful features of the analyzed respiratory sounds (RS) are: MFCC, spectral density, entropy spectrum and detected ripples [22]. Most of the results available, unfortunately, come from a small sample of patients. Usually, such a training sample is not sufficient to compete with an experienced physician.

Automated breath analysis has long been at the heart of medical research. It has great potential for detecting disease in the early stages. The *International Conference on Biomedical and Health Informatics* has even created a scientific challenge for developing algorithms for breath analysis. The dataset provided contains 920 recordings from 126 patients [18]. Each patient's audio recordings come from 7 different areas: trachea, left and right anterior sides, and lateral and posterior lungs. Several models of stethoscopes were used. In addition, all samples were annotated by three specialists.

Pinho *et al.* [15] proposed a method for automatically detecting crackles using this database. The solution developed by them covers three stages. The first is to extract the window with a potential crackle. This window is then verified by computerized RS analysis using the established criteria. The third step concerns the characterization and extraction of the crackle parameters. Their crack detection method achieved 89% for sensitivity and 95% for precision.

The sputum level assessment [13] is another method of monitoring different aspects of the respiratory system. The hardware part of such a system consists of the following elements: sound sensor, audio card, waterproof tube, sound amplifier and power supply. The sound sensor is embedded in a tube to reduce the effect of parasitic noise, and the amplifier increases the signal-to-noise ratio (SNR). In the first stage of data analysis, audio recordings are divided into chunks related to individual respiratory cycles, which are analyzed using the autocorrelation method.

After autocorrelation and short-time Fourier transform (STFT), interesting signal features can be computed in the form of the gray-level co-occurrence matrix (GLCM). Then such features are filtered with the Pearson correlation coefficient. These features emphasize the difference between signals with and without sputum, which allows for a simple dichotomous classification derived in the last step of the analysis. As used herein, the dataset includes 272 audio recordings from 12 patients. In practice, the best classifier achieved 85% efficiency in terms of accuracy.

Asthma can be described as a chronic disease of the respiratory system. Scientists estimate that more than 10% of the North American population suffers from it. During an asthma attack, our airways become swollen and the muscles around them weaken. Pharmaceutical treatment depends on the stage of the disease. A number of different types of research studies have already been initiated in this field.

Asthma severity can be divided into four levels/categories: mild intermittent (MI), mild persistent (MP), moderately persistent (MOP) and severe persistent (SP). One of the solutions to determine the severity of asthma is the use of fuzzy systems [23]. Instead of four categories, the authors of this method propose using a scale from 0 to 10 (the higher the number, the stronger the symptoms), where the range 0–2 is assigned to MI, 2–5 to MP, 5–7 to MOP and 7–10 to SP. The operation of their fuzzy expert system [23] was tested on 28 patients with asthma. The obtained results in terms of the severity of the disease did not differ much from the actual state of the disease in individual patients. Note that such more precise diagnostic information can greatly support a physician's decision to issue a more appropriate prescription.

3 Datasets

The data used to train the presented neural network was collected from two resources. The first dataset, called the *Respiratory Sound Database*[1], was produced by two research teams from Portugal and Greece. This dataset contains 920 annotated records that were retrieved from 126 patients. In total, this data set offers 5.5 h of recording of 6,898 breathing cycles. These cycles are divided into four classes: healthy, crackles, wheezes, and both crackles and wheezes. To simulate real conditions, some distortions such as recording noise are introduced here. Many very different patients were diagnosed - from children to the elderly. On the basis of this dataset, we were able to build a relatively good model.

Due to the fact that our prototype stethoscope is different from those used by professionals, we also collected our own data set. The data collection procedure was carried out during the COVID-19 pandemic. Therefore, public contact was very limited here. However, we managed to collect breath sound samples from 13 people. No specialist physician was involved in the collection of this dataset, so we decided to divide the sound samples into two classes: normal and abnormal, the latter class containing crackling and wheezing. Of course, without medical experience, it is very difficult to classify the respiratory cycles, and therefore the synthesis of good machine learning models requires the correct labeling of the data samples used.

4 Breath Analysis System

The main purpose of the BAS prototype is to continuously monitor and detect if the patient has any breathing problems. This system is not intended to replace the doctor, but only to support him. It helps to detect significant objective breathing disorders in the early stages, and thus protect patients from the severe states of this disease. In the difficult time of the pandemic and the blockade, people are afraid to go to hospital for treatment. The BAS prototype could reduce the health debt that accumulates during such a pandemic.

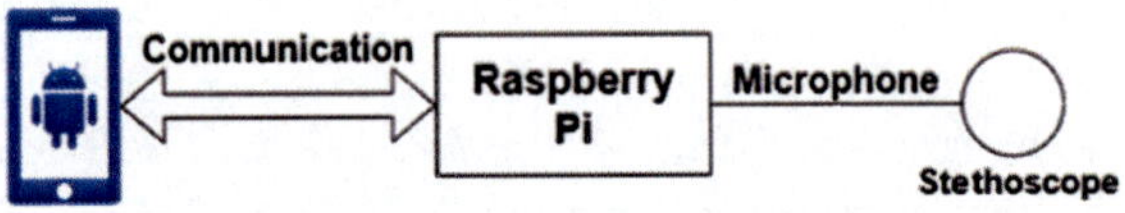

Fig. 1. BAS prototype scheme.

The BAS prototype can be connected to a simple mobile application (via TCP/IP) which is only a front-end to a core Python system run on the Raspberry Pi. The hardware part of the BAS consists of a Raspberry Pi, a USB microphone, a rubber tube, a stethoscope head, and a power supply. The stethoscope part

[1] https://www.kaggle.com/datasets/vbookshelf/respiratory-sound-database.

consists of a signal receiver (head, resonator), a rubber tube and a microphone that is directly connected to the stethoscope head by this tube. Since the rubber tube contributes significantly to signal attenuation, its length should be as short as possible. The BAS prototype is shown in Fig. 1 where the user application is implemented on an Android smartphone.

The algorithm implemented on the microcomputer is characterized by three stages of operation: collecting (using Pyaudio and wave modules), pre-processing, and analyzing data (using artificial neural networks).

The first line of research concerned the use of a recursive neural network to analyze the respiratory cycle on the basis of normalized data (matched to the network input format). However, this approach did not bring satisfactory results. The second approach was to use STFT (Short-Time Fourier Transform), which is widely used to search for time-frequency representations of local intervals/segments/chunks of time-varying signals with the time window function [9]. Also, this direction of searching for an effective solution did not meet our expectations.

On the other hand, the application of the MFCC method referred to in Sect. 2 provided the best performance in the considered data analysis. This method can be broken down into seven [12] computational steps. The first step (1) is to pass the signal through a filter, emphasizing its useful characteristics and increasing the signal energy over a certain (fixed) bandwidth. Then (2), this signal is split into small frames of N samples, where adjacent frames are superimposed on M samples ($M < N$). Then, on the basis of this window function, the signal features are extracted (3). Step 4 is to use FFT to transform the time domain into the frequency domain. Then (5), triangular filters are applied to make the processing result close to the Mel scale. Step 6 is to transform the spectrogram obtained by means of a discrete cosine transformation. It is at this stage that the Mel frequency cepstrum coefficients (MFCC) are obtained. Finally (7), the features related to the change of the cepstral coefficients due to the signal and frame changes are derived. The entire 7-step transformation process is visualized in Fig. 2.

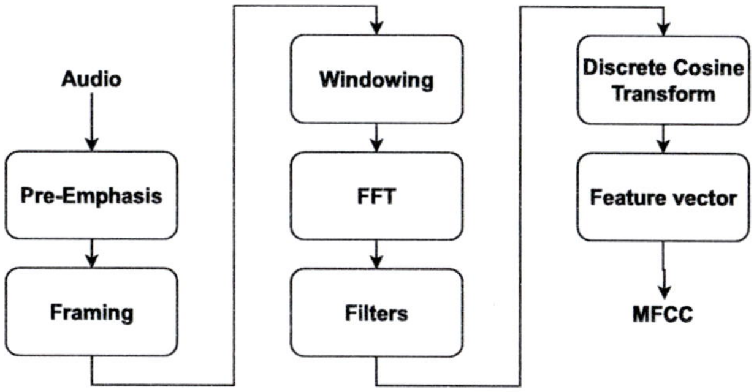

Fig. 2. MFCC block diagram.

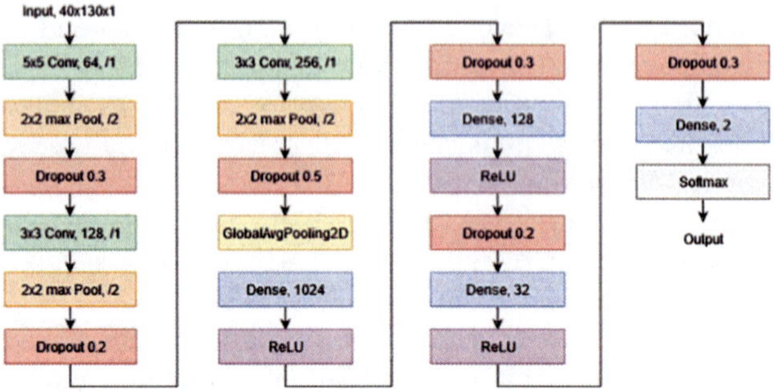

Fig. 3. Neural network diagram.

As a result of the above signal processing, a two-dimensional matrix appears as the output product. It is also possible to extract a spectral image as an output, which can then be analyzed with any known neural network architecture such as Resnet [1]. Unfortunately, as we will show later, it is not possible to meet our requirements in this way.

On the other hand, it turns out that the use of a convolutional neural network offers greater possibilities in this respect. At the network input we use a 40×150 matrix consisting of MFCC features. This data is passed through the three convolutional layers and the pooling is done in the next layer. A dropout layer is applied after each pooling to reduce the risk of overfitting. The basic part of the neural network is characterized by dense connections (of three layers). One hot vector representing the encoded four classes was designed as the output of the entire developed structure: healthy, crackling, wheezing and both (crackles and wheezes).

In some practical cases, and mainly due to the lack of adequate amount of data from hospitals, the neural network can be used to process data divided into only two classes: normal and incorrect.

Figure 3 shows the structure of the applied neural network, which we call BA-CNN (Breath Analysis via CNN). Each convolutional layer has the stride parameter of one. It is worth noting that thanks to the dropout method, the obtained accuracy increased by several percent. As you can see, the structure uses seven different dropout layers. The dropout layers block certain weights, thus the network acquires greater generalizability, and thus the spread between the effectiveness measured on the training and validation data increases.

The data used to train our BA-CNN system (part of BAS) was taken from the Kaggle website [18]. This dataset originally contained 920 audio (wav) files with annotations, demographics and diagnosis for each patient. To enlarge the dataset, each audio file has been divided into several subfiles consisting of exactly one breath cycle. This resulted in a dataset of over 6,800 samples.

The data recorded by the BAS prototype we constructed differs from the data provided by Kaggle (based on different microphones). Consequently, a new dataset was created and collected using the BAS prototype hardware. It is clear that with the learning transfer approach any previously pre-trained neural network can be used. Due to the COVID situation in Poland, collecting large amounts of data has, however, been significantly hindered. Therefore, the BAS system was actually trained on only 700 samples taken at home conditions.

5 Tests

The tests were divided into two parts related to hardware and software. In the first stage, the part of the equipment responsible for sound recording was checked. Three microphones with different parameters were tested. The most satisfactory result was obtained with instruments with higher sensitivity to low frequencies. The use of such a microphone allows you to clearly record the breathing or heartbeat of the patient. One should also remember about the most important parameters of microphones, which in this case are the signal-to-noise ratio and frequency bandwidth.

The great challenge was to develop an appropriate machine learning algorithm to analyze audio recordings. As mentioned earlier, one of the solutions was based on ResNet, which has a 224×224 matrix as input. To get the right fit, all training data had to be scaled (typically, we increase the amplitude and reduce the frequency resolution; in our case we make 224×224 from the 300×1000 spectrogram).

Figure 4c is an image of the respiratory cycle spectrogram of a patient with chronic obstructive pulmonary disease before rescaling. An example of healthy human breathing is shown in Fig. 4d, where it can be seen that the peaks are lower and smoother. Of course, each data collecting device has a specific effect on the received signal. Before conversion to an image, each signal was extended to the same length of six seconds.

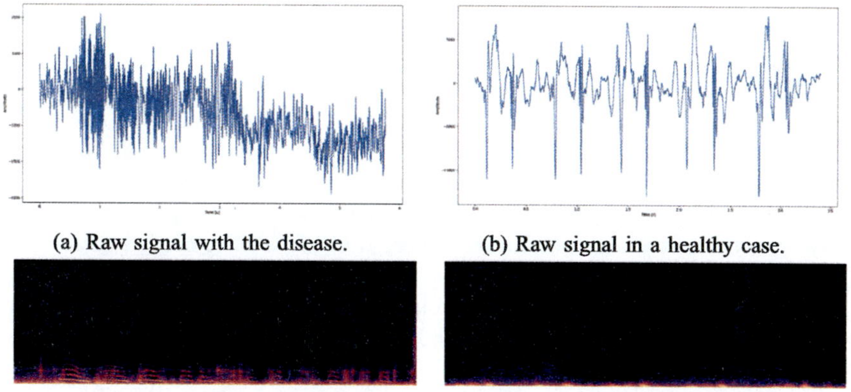

(a) Raw signal with the disease. (b) Raw signal in a healthy case.

(c) Spectrogram with the disease. (d) Spectrogram in a healthy case.

Fig. 4. Respiratory signals and power spectrograms with linear frequency [8] in chronic obstructive pulmonary disease in healthy cases.

Table 1. Comparison of the considered architectures: stock (VGG16, ResNet50, MobileNet) and own designs (ConvNets).

	VGG16	ResNet50	MobileNet	ConvNet (v1)	ConvNet (v2)
Accuracy	76%	75%	67%	80%	82%
Precision	27%	34%	38%	80%	85%
Recall	22%	27%	35%	81%	82%
F1-score	24%	30%	37%	78%	83%

After the previously described processing, the resized image is transferred through the BA-CNN network. Three different stock architectures were considered and tested: VGG16, ResNet50, MobileNet [4,6,19]. These architectures were considered because they require less processing power to calculate the weights. We used batch normalization for each of these networks. It is a technique used in training deep neural networks in which each mini-batch of data delivered to the input layer is normalized. Data augmentation was not included here due to the high specificity of the data samples and the lack of involvement of physicians in the collection of the dataset, i.e. we simply did not modify the data to be as reliable as possible.

Note that in the case of the stock architectures, transfer learning was based on the stock weights of the backbones trained on the ImageNet problem. In some structures, only a small subset of the layers have been fine-tuned. The best performance was achieved with ResNet50 with all layers retrained. However, the obtained results were not satisfactory. Accuracy was only around 75%. It is likely that changing the size of the spectrum image could have had a negative effect. Typically, the width of a spectrogram image is much greater than its height. Squeezing it into a square shape (due to the scaling described above) could result in the loss of valuable spectral information. Therefore, after several attempts to use known available stock architectures, a self-constructed BA-CNN neural network was used.

Table 1 lists some standard performance metrics for the tested neural network architectures. Two subclasses were taken into consideration in this table. It seems that accuracy should not be considered the most appropriate criterion. Although the three known stock architectures were developed by experienced professionals, they may perform worse in some tasks than simple convolutional neural networks. It is also worth noting here that MobileNet needed many more epochs to learn than other architectures. Moreover, the described base architectures are mainly used for non-spectral images. For this reason, their performance may be worse than that of simple convolutional networks.

In the analyzed case, the use of a two-dimensional matrix (MFCC result) as input data to a hand-designed convolutional neural network (BA-CNN) gives positive results. Both architectures of our own design consist of three convolutional layers followed by max-puling and dropout. The first layer has 64 filters, the second – 128, and the third – 256 filters. The convolution kernel dimension is

3×3. On the head of BA-CNN we have a flattened layer and three dense layers, the last of which has four neurons (representing healthy, wheezing, crackling, and both) or two neurons (healthy or unhealthy) that represent the patient's breathing state. This simpler (binary) variant serves to detect if there is any abnormal breathing condition. This variant provides better performance and may be more useful for the patient, as for him/her it is not necessary to distinguish between crackles and whistles. The BA-CNN training results for two binary classes are presented in Fig. 5. In this case, we selected 5% of the training set to be used in the validation process.

The accuracy plot observed during BA-CNN training for the four classes (healthy, wheezing or crackling, and both) was very similar to that shown in Fig. 5a for the two-class case. The same is true for the loss function. Interestingly, after a certain epoch, the BA-CNN loss starts to increase, as shown in Fig. 5b. It is noteworthy that the discrepancy between training and validation results for the four classes was even greater.

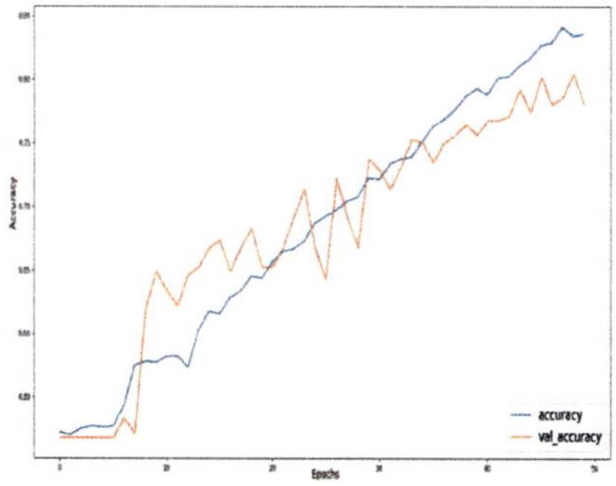

(a) Accuracy of training (blue line) and validation (orange line).

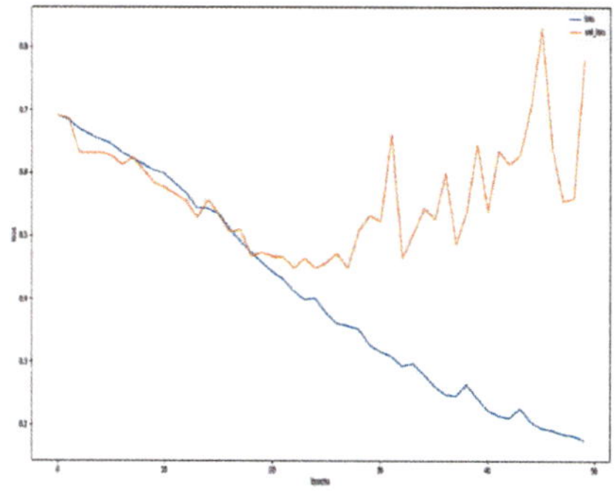

(b) Losses on training (blue line) and validation (orange line).

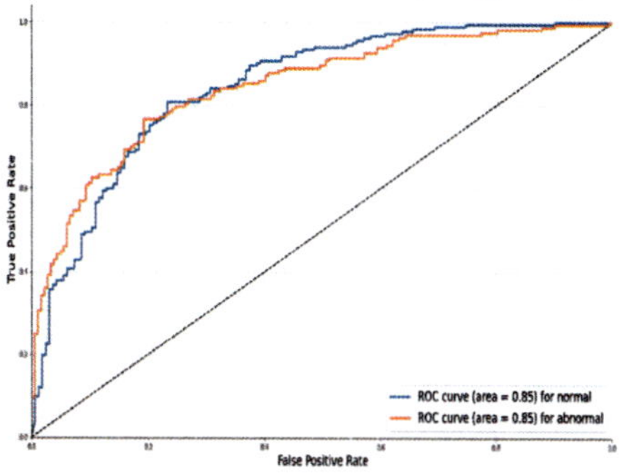

(c) Receiver operating characteristic curve (ROC) for normal breathing (note that the area under the curve (AUC) is 0.85).

Fig. 5. BA-CNN training rates for two cases (normal and abnormal breathing) obtained on the Kaggle dataset.

In the case of four classes, the neural network was only 70% accurate. The above-introduced ROC curve, which expresses the ratio of the true positive indicator to the false positive indicator, very often plays an important role in medical analysis.

As is clear from the analyzes of both cases shown in Fig. 5c and 6, the results in ROC terms are relatively good.

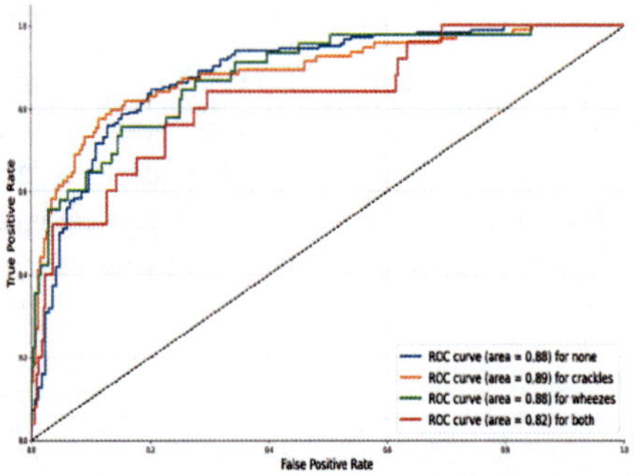

Fig. 6. ROC curves for the case of detection of four classes. The blue line shows healthy breathing at AUC 0.88. The orange line represents crackling (AUC = 0.89) and the green line represents wheezing (AUC = 0.88). The red line shows the mixed case (both crackling and wheezing) with AUC = 0.82. (Color figure online)

Note that the case shown above represents the metrics of the BA-CNN network trained on the Kaggle dataset. As mentioned in the Sect. 4, transfer learning seems necessary due to the differences between the microphones used. The first, rough learning, is about the Kaggle dataset, and the second, fine-tuning is based on our dataset.

As for the metrics obtained using our own dataset, the results seem promising. We are aware, however, that the amount of data is insufficient for the final verdict on the neural network structures used and their performance. For example, in Fig. 7a, the level of precision does not change over several epochs. This may be because the training and validation datasets are too small. In this case, 10% of the training set was extracted and used as a validation set. It is worth mentioning that when evaluated on this test set, this network achieved an accuracy of 88.5%. Figure 7b proves, in turn, that the BA-CNN architecture is created correctly, because the learning losses converge to approximately 0. Figure 7c shows that this neural network (even with our limited datasets) can distinguish between healthy and unhealthy breathing almost perfectly.

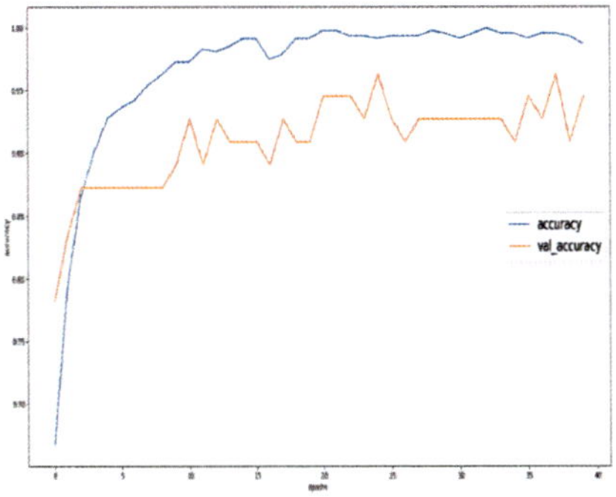

(a) Accuracy of training (blue) and validation (orange).

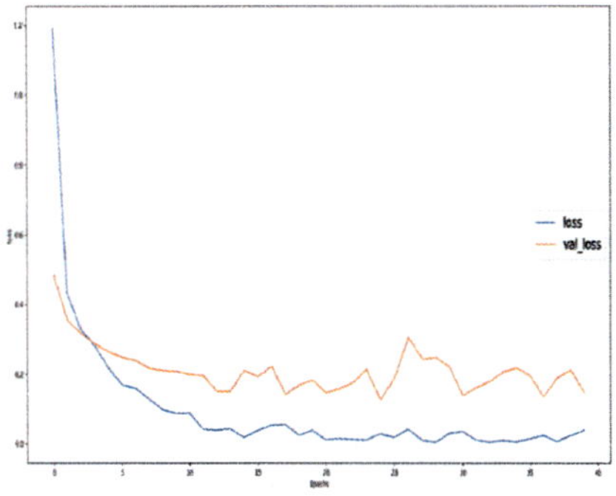

(b) Losses in training (blue) and validation (orange).

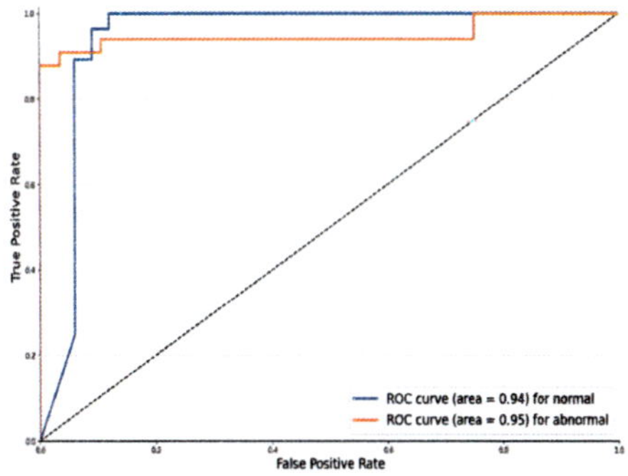

(c) Final ROC curve score for BA-CNN in healthy (blue) and abnormal (orange) cases. The area under the curve (AUC) for healthy breathing is 0.93.

Fig. 7. Final results obtained after BA-CNN fine-tuning with the binary classifier (normal and abnormal breathing).

6 Conclusions

Human respiratory diseases place a huge burden on national health systems. According to the World Health Organization (WHO) records, several million people die each year from respiratory diseases. The COVID-19 pandemic is a very contemporary representative example of such a scourge. People usually inform their doctors too late about their own health problems. In this way, there is an excessive number of severe hospitalizations and the world-known overload of hospitals. For the above reasons, any damage to health should be detected as early as possible for a positive and efficient recovery process.

The BAS prototype presented in this paper can at least partially contribute to this by keeping people informed about their health. It can also tell doctors what problems the patient is struggling with. Of course, the system is not a substitute for a physician in delivering the final professional verdict.

In order to further improve the performance of the developed prototype, first of all, much more data needs to be collected. Since the accuracy of the training so far has already reached a high level (above 95%), it can be assumed that the adopted structure of the neural network is appropriate. To increase the validation precision, you should use data normalization (as in this solution) or a method of extending the datasets [11].

The human body is a very complex system, and all subsystems interact with each other. Therefore, the proposed BAS prototype should be further developed by adding new functions and equipment, e.g. a heart rate sensor, enriching the scope of collected medical data. Collecting data such as a patient's breathing, pulse and pressure can go a long way in helping a doctor diagnose and identify a specific disease.

In addition, this type of instrument allows you to think about a wider and more effective online consultation. It is not difficult to predict that the whole world will be constantly moving to virtual reality, and this is what our BAS device, among others, serves.

References

1. Duan, J., Shi, T., Zhou, H., Xuan, J., Wang, S.: A novel ResNet-based model structure and its applications in machine health monitoring. J. Vib. Control **27**(9–10), 1036–1050 (2020). https://doi.org/10.1177/1077546320936506
2. Grotberg, J.B.: Crackles and wheezes: agents of injury? Ann. Am. Thorac. Soc. **16**(8), 967–969 (2019)
3. Hamet, P., Tremblay, J.: Artificial intelligence in medicine. Metabolism **69**, 36–40 (2017). https://doi.org/10.1016/j.metabol.2017.01.011
4. He, K., Zhang, X., Ren, S., Sun, J.: Deep residual learning for image recognition. In: Proceedings of the IEEE Conference on Computer Vision and Pattern Recognition, pp. 770–778 (2016)
5. Hosny, A., Parmar, C., Quackenbush, J., Schwartz, L.H., Aerts, H.J.: Artificial intelligence in radiology. Nat. Rev. Cancer **18**(8), 500–510 (2018)
6. Howard, A.G., et al.: MobileNets: efficient convolutional neural networks for mobile vision applications. arXiv preprint arXiv:1704.04861 (2017)
7. Khan, R.S., Zardar, A.A., Bhatti, Z.: Artificial intelligence based smart doctor using decision tree algorithm. arXiv preprint arXiv:1808.01884 (2018)
8. Khunarsa, P., Lursinsap, C., Raicharoen, T.: Impulsive environment sound detection by neural classification of spectrogram and mel-frequency coefficient images. In: Zeng, Z., Wang, J. (eds.) Advances in Neural Network Research and Applications, pp. 337–346. Springer, Heidelberg (2010). https://doi.org/10.1007/978-3-642-12990-2_38
9. Kim, B., Kong, S.H., Kim, S.: Low computational enhancement of STFT-based parameter estimation. IEEE J. Sel. Top. Signal Process. **9**(8), 1610–1619 (2015)
10. Kowalczuk, Z., Cybulski, J., Czubenko, M.: JamesBot-an intelligent agent playing StarCraft II. In: 2019 24th International Conference on Methods and Models in Automation and Robotics (MMAR), pp. 105–110. IEEE (2019)
11. Kowalczuk, Z., Glinko, J.: Training of deep learning models using synthetic datasets. In: Kowalczuk, Z. (ed.) DPS 2022. LNNS, vol. 545, pp. 141–152. Springer, Cham (2022)

12. Muda, L., Begam, M., Elamvazuthi, I.: Voice recognition algorithms using mel frequency cepstral coefficient (MFCC) and dynamic time warping (DTW) techniques. arXiv preprint arXiv:1003.4083 (2010)
13. Niu, J., et al.: Detection of sputum by interpreting the time-frequency distribution of respiratory sound signal using image processing techniques. Bioinformatics **34**(5), 820–827 (2018)
14. Pakdemirli, E.: Artificial intelligence in radiology: friend or foe? Where are we now and where are we heading? Acta Radiologica Open **8**(2), 2058460119830222 (2019)
15. Pinho, C., Oliveira, A., Jácome, C., Rodrigues, J., Marques, A.: Automatic crackle detection algorithm based on fractal dimension and box filtering. Procedia Comput. Sci. **64**, 705–712 (2015)
16. Pramono, R.X.A., Bowyer, S., Rodriguez-Villegas, E.: Automatic adventitious respiratory sound analysis: a systematic review. PLoS **12**(5), e0177926 (2017)
17. Richens, J.G., Lee, C.M., Johri, S.: Improving the accuracy of medical diagnosis with causal machine learning. Nat. Commun. **11**(1), 1–9 (2020)
18. Rocha, B., et al.: A respiratory sound database for the development of automated classification. In: Maglaveras, N., Chouvarda, I., de Carvalho, P. (eds.) Precision Medicine Powered by pHealth and Connected Health, pp. 33–37. Springer, Singapore (2017). https://doi.org/10.1007/978-981-10-7419-6_6
19. Simonyan, K., Zisserman, A.: Very deep convolutional networks for large-scale image recognition. arXiv preprint arXiv:1409.1556 (2014)
20. Sparrow, R., Hatherley, J.: High hopes for "deep medicine"? ai, economics, and the future of care. Hastings Cent. Rep. **50**(1), 14–17 (2020)
21. Vinyals, O., et al.: Grandmaster level in StarCraft II using multi-agent reinforcement learning. Nature **575**(7782), 350–354 (2019)
22. Zheng, F., Zhang, G., Song, Z.: Comparison of different implementations of MFCC. J. Comput. Sci. Technol. **16**(6), 582–589 (2001)
23. Zolnoori, M., Zarandi, M.H.F., Moin, M., Teimorian, S.: Fuzzy rule-based expert system for assessment severity of asthma. J. Med. Syst. **36**(3), 1707–1717 (2012)

Bridging Functional Model of Arterial Oxygen with Information of Venous Blood Gas: Validating Bioprocess Soft Sensor on Human Respiration

Benas Kemesis[1](✉), Renaldas Urniezius[1], Tomas Kondratas[2], Lina Jankauskaite[2], Deividas Masaitis[1], and Povilas Babilius[1]

[1] Kaunas University of Technology, Studentu 50, 51368 Kaunas, Lithuania
benkem@ktu.lt

[2] Lithuanian University of Health Sciences Hospital Kauno Klinikos (LSMU KK), Eiveniu Street 2, 50161 Kaunas, Lithuania

Abstract. Oxygen and carbon dioxide gas exchange are one of the principal indicators of a microorganism state. In bioprocesses, cultivated cell oxygen consumption and carbon dioxide production are descriptors of process quality. This paper presents how a soft-sensor for gas analysis from biotechnology also applies to macroorganisms. The study combines information from venous blood gas analysis and expiratory gasses to estimate partial pressures of oxygen and carbon dioxide in the venous blood of children in the pediatric intensive care unit. Observed data were from three patients with monitoring intervals ranging from 6 to 13 days. Presented models had the lowest mean average error of 3.17 mmHg for carbon dioxide $PvCO_2$ and 1.64 mmHg for oxygen PvO_2. Additionally, the carbon dioxide model proposes a critical flow of inspiratory gas at which no carbon dioxide should accumulate in the respiratory system. The paper lays a basis for further research on the noninvasive monitoring of breath data and its applicability in the medical field.

Keywords: Oxygen consumption · Oxygen uptake rate · Human respiratory system · Sensor fusion

1 Introduction

1.1 Historic Context

In March 2020, with the start of the global pandemic, an SME, *Cumulatis* (Kaunas, Lithuania), under the curation of the *Science, Innovation and Technology Agency* (Vilnius, Lithuania), analyzed human exhaled gases for checking *in-vivo* a therapeutic effect when estimating and correcting dissolved oxygen in the human body. The *Cumulatis* team designed and tested the system, containing sensor fusion and soft sensors to assess the overall aerobic capacity state of the human body. The project results led to further

Z. Kowalczuk (Ed.): DPS 2022, LNNS 545, pp. 42–51, 2023.
https://doi.org/10.1007/978-3-031-16159-9_4

official clinical study under permit number *BE-2-92* of the *Kaunas Regional Biomedical Research Ethics Committee* [1] for intubated children in the pediatric intensive care unit (PICU). The study's goal is early pathogen identification. The candidate descriptors are changes in oxygen consumption and volatile substances released during ventilatory lower respiratory tract infections (VA-LRTI).

As the clinical study is in its early stage, the noninvasive viral or bacterial infection diagnosis needs more data. However, current gas analysis technology and sensor-fusion routines assess gas metabolism in the respiratory tract, especially oxygen and carbon dioxide gasses. This paper presents how noninvasive gas analysis and fundamental knowledge of microorganisms' behavior apply to the human body. The desirable purpose was the overall health condition monitoring and noninvasive tracking of gas exchange within the respiratory system.

1.2 Related Work

Research has proved that the measurement and analysis of human exhaled gases and their components have the potential to be successfully used in medical practice [2]. Information on respiratory physiology and pathology is critical, especially when dealing with severely ill patients whose condition might change in minutes [3]. Over the years, the development and research of breath analyzers have increased [4–6]. These devices provide information about human breath temperature, humidity, O_2, and CO_2 concentrations.

Despite the potential of expiratory gas analysis, breath analyzing devices in the medical field are still in their infancy. It faces obstacles of standardization of measurements and transition from pre-clinical laboratory-scale to actual clinical use [7]. Therefore, blood gas analysis remains the most frequent diagnostic tool when treating mechanically ventilated patients with VA-LRTI [8]. However, blood gasses are not sensitive or specific enough to represent microbiota changes in the respiratory tract, particularly in the early onset of VA-LRTI. Moreover, children's respiratory tract microbiota is highly versatile and not consistent. Thus, specific breath analysis or so-called breath-biopsy is still not well defined for the pediatric population [9]. The partial pressures of oxygen (PO_2) and carbon dioxide (PCO_2) obtained from the blood gas analysis allow assessing the patient's oxygenation, ventilation status, and acid-base balance [10]. Arterial blood gas analysis (ABG) has been the standard measurement. However, arterial blood sampling may cause pain or discomfort for the patient and is riskier to perform [11]. Therefore, there has been a shift to using venous blood gas analysis, where blood sampling is more convenient and accessible [12]. The correlation between these two measurements is relatively good for clinical acceptance of acid-base parameters of pH and PCO_2 (correlation of around 0.9). However, a complete oxygenation assessment of the patient still requires ABG analysis of PO_2 [13, 14].

There have been few attempts to non-invasively and continuously estimate the PO_2 or PCO_2 in arterial or venous blood for long periods. Researchers have used single breath gas samples, or the trial conditions were based on exercise and rest states [15–17]. Prior work showed how noninvasive off-gas analysis technology and fundamental cell laws successfully tracked PO_2 in arterial blood [18]. This paper continues the research on the correlation between expired gasses and respiratory system gas exchange and presents an alternative fundamental model for estimating PCO_2 in venous blood and PO_2 in arterial blood using information from VBG.

2 Methods

2.1 Clinical Study Conditions and Hardware

The analyzed up to the date data belonged to three infants. All three patients were treated in PICU. The *Maquet Servo-U* ventilated the intubated children. Ventilation parameters were monitored and corrected based on *pH* analysis from venous blood samples. Radiometer ABL90Flex served for VBG analysis. The time interval between VBG analysis ranged from 4 to 48 h for all three patients (Table 1).

Table 1. Main patient characteristics (* is the volunteer from previous work)

Patient	Age	Gender	Weight, kg	Primary condition	Duration of data monitoring, days
1	Four months	Male	5.7	COVID-19, pneumonia	6.75
2	Five months	Male	4.6	Bronchopulmonary dysplasia	13.54
3	Nine months	Female	5	Bronchiolitis, pneumonia	7.37
0*	45 years	Male	90	–	0.08

The *Cumulatis*'s device *OUR120421A1* monitored and tracked data from exhaled breath, including exhaled oxygen and carbon dioxide concentrations, temperature, pressure, and humidity. The central communication unit of the *OUR120421A1* device also collected data about inspiratory information (flow, oxygen concentration) from Servo-U mechanical ventilator. A flex tube linked the gas outlet and inlet of both devices.

2.2 Model for Partial Pressures of Oxygen and Carbon Dioxide

Partial Pressure of Oxygen in Arterial Blood. The model of partial pressure of oxygen in arterial blood PaO_2 originates from the equation of alveolar partial pressure of oxygen PAO_2 [19]. As the arterial-alveolar oxygen gradient is relatively minor for new-borns, the equality assumption ($PAO_2 \cong PaO_2$) is valid, therefore:

$$PaO_2^* \equiv PaO_2 \cong (P_{atm} - P_{H2O}) \cdot FiO_2 - PACO_2/RQ, \tag{1}$$

where P_{atm} is atmospheric pressure (760 mmHg in standard atmosphere conditions), P_{H2O} is the water vapor pressure, usually around 46 mmHg, FiO_2 is the inspired oxygen fraction, $PACO_2$ is the partial pressure of carbon dioxide in alveoli, and RQ is the respiratory quotient, equal to the ratio of the amount of carbon produced by oxygen consumed. The maximal solubility (PaO_2^*) restricts the arterial partial pressure of oxygen in the human body, including the effects of pressure and inspiratory oxygen concentration.

Respiratory quotient expressed as partial pressures of CO_2 and O2 gases in the alveoli:

$$RQ \cong \frac{PACO_2}{PAO_2} = \frac{PACO_2}{\left(PO_{2,insp} - PO_{2,exp}\right)} \tag{2}$$

where $PO_{2,insp}$ and $PO_{2,exp}$ are inspired and expired partial pressures of oxygen. If alveolar carbon dioxide is approximately equal to pulmonary end-capillary ($PACO_2 \cong PvCO_2$) and the patients had not experienced any shock or extreme acid-base abnormalities, Eq. (2) becomes [20, 21]:

$$RQ = \frac{PvCO_2}{\left(PO_{2,insp} - PO_{2,exp}\right)}. \tag{3}$$

Previous work expressed PaO_2^* through atmospheric oxygen partial pressure, therefore Eq. (1) and Eq. (3) become [18]:

$$PaO_2^* = (P_{atm} - P_{H2O}) \cdot FiO_2 - PvCO_2 \cong PO_{2,atm} \cdot \frac{P_{atm} \cdot 0.94}{760} \cdot \frac{FiO_2}{0.2097} - PvCO_2, \tag{4}$$

where the partial pressure of oxygen in standard atmosphere conditions ($PO_{2,atm}$) is equal to 160 mmHg, factors of 0.94 and 0.2097 stand for water vapor pressure loss and inspired oxygen fraction at standard atmospheric conditions.

As the blood gas analysis during the study yields information about oxygen in venous blood, the arterial dissolved oxygen (PaO_2) depends on PvO_2 and P_{offs}. The latter is assumed to be the perfusion gradient of partial pressure of oxygen in tissues. Therefore, PaO_2 takes the following form:

$$PaO_2 = PvO_2 + P_{offs}. \tag{5}$$

With the addition of fundamental knowledge of microorganisms in bioprocesses, PaO_2 becomes [18]:

$$\frac{dPaO_2(t)}{dt} = K_L a \times \left(PaO_2^*(t) - PaO_2(t)\right) - k_{OUR} \cdot OUR(t) \tag{6}$$

where *OUR* stands for oxygen uptake rate and coefficient $K_L a$ is oxygen mass transfer capacity, vital for any aerobic organism [22, 23]. The parameter (k_{OUR}) is related to the overall calibration of the *OUR120421A1* device, and its value is the same for all three patients.

The multidisciplinary approach from biotechnology and medical fields (Eqs. 4, 5, and 6) yields the final PaO_2 expression:

$$\frac{d\left(PvO_2(t) + P_{offs}(t)\right)}{dt} + k_{OUR} \cdot OUR(t) =$$

$$K_La(t) \cdot \left(\left(PO_{2,atm} \cdot \frac{p(t) \cdot 0.94}{P_{atm}} \cdot \frac{O_{2,inh(t)}}{0.2097} - PvCO_2(t)\right) - (PvO_2(\mathrm{t}) + P_{offs}(t))\right) \tag{7}$$

The main difference from the previous work is that the oxygen transfer coefficient K_La varies in time due to the extended monitoring period during which the patient's heart and lung capabilities and health conditions may change. Similarly, variable ($P_{offs}(t)$) can also alter throughout time because of the change of oxygen demand in tissues.

$PvCO_2$ estimation. As exhaled carbon dioxide mixes with air, its concentration is lower in the exhaled breath analyzer. Therefore, for partial pressure of carbon dioxide in the venous blood, the effect of dilution is included, commonly used in bioprocesses modeling [24]:

$$\frac{\mathrm{d}C}{\mathrm{d}t} = k \cdot \mathrm{C} - \frac{F}{V} \tag{8}$$

where F is expiratory flow, V is the volume of the system or object in question, C is the wanted concentration, and k is the specific reaction rate. Applying the dilution equation to the partial pressure of carbon dioxide in venous blood results in:

$$\frac{\mathrm{d}PvCO_{2,out}(t)}{\mathrm{d}t} = K(t) \cdot \left(PvCO_{2,est}(t) - PvCO_{2,out}(t)\right) - PvCO_{2,out}(t) \cdot \frac{\mathrm{d}F(t)}{\mathrm{d}t \cdot V} \tag{9}$$

where $PvCO_{2,est}(t)$ is the estimated partial pressure of carbon dioxide in venous blood, the partial pressure of carbon dioxide ($PvCO_{2,out}(t)$) is the partial pressure measured by the breath analysis device. V is the total volume of the patient lungs, the connected exhaust tube, and the measurement chamber of *OUR120421A1*. *K(t)* is a time-varying factor. Multiplying it with the total volume mimics a critical flow of inspiratory gas during which no carbon dioxide accumulates in the human respiratory system. Such a proposition is one of the novelties of this paper and, if proved correct, has the potential to treat hypercapnia conditions.

3 Results

The optimization criterion for partial pressure of CO_2 was the residual sum of squares *RSS*. The mean absolute error (*MAE*) and the average error (*AE*) of estimated partial pressure of CO_2 in venous blood for all three patients are presented in Table 2, together with the identified total volumes V and critical inspiratory gas flows F:

Table 2. Estimation results of $PvCO_2$

Patient	*MAE*, mmHg	*AE*, mmHg	*RSS*	*V, liters*	*F, liters per hr*
1	5.95	3.63	1554	0.558	75.6
2	3.17	0.89	1680	0.63	89.6
3	11.94	10.85	2121	0.60	72.87

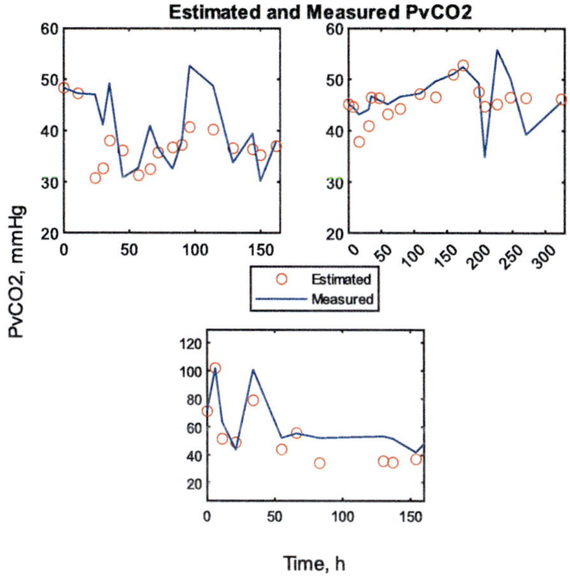

Fig. 1. Estimated $PvCO_2$ (from left to right: Patient 1, 2, and 3)

Below are the measured and estimated $PvCO_2$ curves for all patients:
The model (PaO_2) hypothesized three case scenarios:

1. K_La is constant and using offline measurements of $PvCO_2$.
2. K_La varies in time and using offline measurements of $PvCO_2$.
3. K_La varies in time and uses estimated $PvCO_2$.

As no measurements of PaO_2 were taken during the study, a bicriterion (BC) for model fitting was used. The first criterion was the sum of absolute model fitting errors between measured and estimated PvO_2 (SAE) throughout the whole time (T). The second criterion was a soft constraint on the variable ($P_{offs}(t)$), which helped emulate the inertia of partial oxygen pressure in tissues. The proposed expression of BC is

$$BC = SAE + \frac{\left|\int_0^T \left(\overline{P_{offs}} - P_{offs}(t)\right)dt\right|}{T}. \quad (10)$$

Numeric results of BC, MAE, and AE of PvO_2 estimation for each patient:

Table 3. Estimation results of PvO_2 in distinct model case scenarios. * Are the results of PaO_2 from previous work

Patient	Cases								
	MAE, mmHg			*BC*			*AE*		
	1	2	3	1	2	3	1	2	3
1	12.42	11.33	10.85	189.6	175.54	168.16	3.63	−1.56	−1.12
2	1.65	1.64	1.64	27.23	27.18	27.13	0.82	0.67	0.68
3	4.89	4.35	4.36	62.87	56.72	52.83	3.27	1.59	1.84
*0**	*0.51**	–	–	–	–	–	*–0.31**	–	–

Table 3 indicates that making the oxygen transfer coefficient time-dependent reduced the *MAE* for all patients, indicating that patients' oxygen transfer potentially has a role in treatment. Addition of estimated $PvCO_2$ also minimizes the MAE and the result of model fitting bicriterion (Fig. 2).

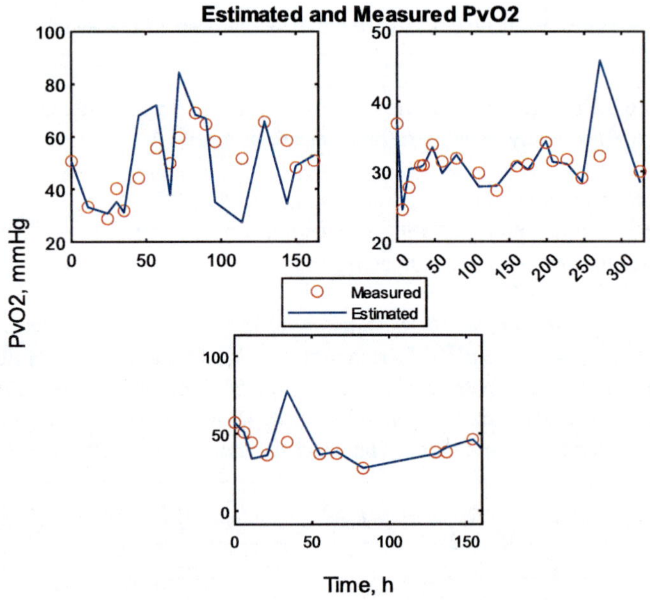

Fig. 2. Results of estimated PvO_2 (from left to right: Patient 1, 2, and 3)

4 Conclusions

This study presents a soft-sensor technology for the noninvasive estimation of partial pressures of oxygen and carbon dioxide in the venous blood of patients in the pediatric intensive care unit. Data consisted of information from three patients with monitoring periods ranging from 6 to 14 days. Presented models originate from the laws describing bioprocesses and formulae used in medicine. Such technique lead to an average error as low as 3.17 mmHg of $PvCO_2$ and 1.64 mmHg of PvO_2 for 13 plus days of monitoring. In the worst case, the *MAE* of estimated venous oxygen (PvO_2) was 12.42 mmHg and *MAE* of estimated $PvCO_2$ reached 11.94 mmHg. Although the estimation errors are higher than in the previous work, one should consider the factors of higher discretion of the dissolved gas measurements and the longer duration of the experiments.

Additionally, the model of $PvCO_2$ identified a critical flow of inspiratory gas that prevents the accumulation of CO2 in the respiratory system of patients. If proven accurate, the model is a promising tool for the medical application, capable of helping the medical staff treat patients with hypercapnia or hypercapnia-like conditions. Moreover, medical data analysis helps validate oxygen transfer coefficient assumptions without actual cultivation with microorganisms. The physiological parameter of the oxygen transfer coefficient is significantly more stable in macro-organisms than its variation in cultivations of microorganisms. In bioreactors, it strongly depends on the agitator speed, pressure, and oxygen concentration in the injection gas and the flow rate of the injection gas mass.

Collecting more data will improve the models' applicability in medicine, especially when dealing with boundary condition cases. However, the results already show that noninvasive exhaled gas analysis technology has the potential to monitor patients' health and gas exchange in the respiratory system. The developing technology and its noninvasive origin with constant monitoring of respiratory dynamics seek to assist in adapting treatment and diagnosing health conditions for ill patients. From a biotechnology perspective, the development of noninvasive gas exchange monitoring allows for future elimination of dissolved oxygen and dissolved carbon electrodes or other point-invasive measuring devices in fully autonomous sensor-less bioreactors.

Funding. This project received funding from the European Regional Development Fund (project no. 01.2.2-LMT-K-718-03-0039) under a grant agreement with the Research Council of Lithuania (LMTLT).

Legal Issues. The medical study is carried out under the permit BE-2-92 granted by the *Kaunas Regional Biomedical Research Ethics Committee.*

References

1. Lithuanian National Register of Biomedical and Drug Clinical Trials. http://bioetika.sam.lt/index.php?1102490711
2. Kharitonov, S.A., Barnes, P.J.: Exhaled markers of pulmonary disease. Am. J. Respir. Crit. Care Med. **163**, 1693–1722 (2001). https://doi.org/10.1164/ajrccm.163.7.2009041
3. Folke, M., Cernerud, L., Ekström, M., Hök, B.: Critical review of noninvasive respiratory monitoring in medical care. Med. Biol. Eng. Comput. **41**, 377–383 (2003). https://doi.org/10.1007/BF02348078
4. Zegdi, R., et al.: Exhaled carbon monoxide in mechanically ventilated critically ill patients: influence of inspired oxygen fraction. Intensive Care Med. **26**(9), 1228–1231 (2000). https://doi.org/10.1007/s001340000590

5. Smallwood, C.D., Kheir, J.N., Walsh, B.K., Mehta, N.M.: Accuracy of oxygen consumption and carbon dioxide elimination measurements in 2 Breath-by-Breath devices. Respir Care **62**, 475–480 (2017). https://doi.org/10.4187/respcare.05115
6. Chen, H.-Y., Chen, C.: Development of a breath analyzer for O_2 and CO_2 measurement. TOBEJ **13**, 21–32 (2019). https://doi.org/10.2174/1874120701913010021
7. Lawal, O., Ahmed, W.M., Nijsen, T.M.E., Goodacre, R., Fowler, S.J.: Exhaled breath analysis: a review of 'breath-taking' methods for off-line analysis. Metabolomics **13**(10), 1–16 (2017). https://doi.org/10.1007/s11306-017-1241-8
8. Castro, D., Patil, S.M., Keenaghan, M.: Arterial blood gas. In: StatPearls. StatPearls Publishing, Treasure Island (FL) (2022)
9. Pulvirenti, G., et al.: Lower airway microbiota. Front. Pediatr. **7**, 393 (2019). https://doi.org/10.3389/fped.2019.00393
10. Rieser, T.M.: Arterial and venous blood gas analyses. Top. Companion Anim. Med. **28**, 86–90 (2013). https://doi.org/10.1053/j.tcam.2013.04.002
11. Thangaraj, R.K., Chidambaram, H.H.S., Dominic, M., Chandrasekaran, V.P., Padmanabhan, K.N., Chanjal, K.S.: A comparison of arterial and venous blood gas analysis and its interpretation in emergency department: a cross-sectional study. Eurasian J. Emerg. Med. **20**, 178–182 (2021). https://doi.org/10.4274/eajem.galenos.2021.85520
12. Razi, E., Nasiri, O., Akbari, H., Razi, A.: Correlation of arterial blood gas measurements with venous blood gas values in mechanically ventilated patients. Tanaffos **11**, 30–35 (2012)
13. Schütz, N., Roth, D., Schwameis, M., Röggla, M., Domanovits, H.: Can venous blood gas be used as an alternative to arterial blood gas in intubated patients at admission to the emergency department? A retrospective study. OAEM **11**, 305–312 (2019). https://doi.org/10.2147/OAEM.S228420
14. Shirani, F., Salehi, R., Naini, A.E., Azizkhani, R., Gholamrezaei, A.: The effects of hypotension on differences between the results of simultaneous venous and arterial blood gas analysis. J. Res. Med. Sci. **16**, 188–194 (2011)
15. Epstein, M.F., Cohen, A.R., Feldman, H.A., Raemer, D.B.: Estimation of $Paco_2$ by two noninvasive methods in the critically ill newborn infant. J. Pediatr. **106**, 282–286 (1985). https://doi.org/10.1016/S0022-3476(85)80306-1
16. Prisk, G.K., West, J.B.: Deriving the arterial PO_2 and oxygen deficit from expired gas and pulse oximetry. J. Appl. Physiol. **127**, 1067–1074 (2019). https://doi.org/10.1152/japplphysiol.01100.2018
17. Bissonnette, B., Lerman, J.: Single breath end-tidal CO_2 estimates of arterial PCO_2 in infants and children. Can. J. Anaesth **36**, 110–112 (1989). https://doi.org/10.1007/BF03011429
18. Survyla, A., et al.: Noninvasive continuous tracking of partial pressure of oxygen in arterial blood: adapting microorganisms bioprocess soft sensor technology for holistic analysis of human respiratory system. In: 2021 IEEE International Conference on Multisensor Fusion and Integration for Intelligent Systems (MFI), pp. 1–5. IEEE, Karlsruhe, Germany (2021)
19. Messina, Z., Patrick, H.: Partial pressure of carbon dioxide. In: StatPearls. StatPearls Publishing, Treasure Island (FL) (2022)
20. Forster, R.E.: Can alveolar pCO_2 exceed pulmonary end-capillary CO_2? No. J. Appl. Physiol. **42**, 326–328 (1977). https://doi.org/10.1152/jappl.1977.42.3.326
21. Sharma, S., Hashmi, M.F., Burns, B.: Alveolar gas equation. In: StatPearls. StatPearls Publishing, Treasure Island (FL) (2022)
22. Urniezius, R., Survyla, A., Paulauskas, D., Bumelis, V.A., Galvanauskas, V.: Generic estimator of biomass concentration for Escherichia coli and Saccharomyces cerevisiae fed-batch cultures based on cumulative oxygen consumption rate. Microb. Cell Fact. **18**, 190 (2019). https://doi.org/10.1186/s12934-019-1241-7

23. Donatas, L., Rimvydas, S., Vytautas, G., Renaldas, U.: Simple control systems for set-point control of dissolved oxygen concentration in batch fermentation processes. Chem. Eng. Trans. **74**, 127–132 (2019). https://doi.org/10.3303/CET1974022
24. Survyla, A., Levisauskas, D., Urniezius, R., Simutis, R.: An oxygen-uptake-rate-based estimator of the specific growth rate in Escherichia coli BL21 strains cultivation processes. Comput. Struct. Biotechnol. J. **19**, 5856–5863 (2021). https://doi.org/10.1016/j.csbj.2021.10.015

COVID-19 Severity Forecast Based on Machine Learning and Complete Blood Count Data

Barbara Klaudel, Aleksander Obuchowski(✉), Roman Karski, Bartosz Rydziński, Patryk Jasik, and Zdzisław Kowalczuk

Gdańsk University of Technology, Gdańsk, Poland
aleksander.obuchowski@student.pg.edu.pl

Abstract. Proper triage of COVID-19 patients is a key factor in effective case management, especially with limited and insufficient resources. In this paper, we propose a machine-aided diagnostic system to predict how badly a patient with COVID-19 will develop disease. The prognosis of this type is based on the parameters of commonly used complete blood count tests, which makes it possible to obtain data from a wide range of patients. We chose the four-tier nursing care category as the outcome variable. In this paper, we compare traditional tree-based machine learning models with approaches based on neural networks. The developed tool achieves a weighted average F1 score of 73% for a three-class COVID-19 severity forecast. We show that the complete blood count test can form the basis of a convenient and easily accessible method of predicting COVID-19 severity. Of course, such a model requires meticulous validation before it is proposed for inclusion in real medical procedures.

Keywords: Deep learning · Computer-aided diagnosis · COVID-19

1 Introduction

The uncontrolled spread of the 2019 coronavirus (COVID-19) pandemic, which is rapidly depleting the available hospital resources of the national health service, has created the need to search for new solutions for effective segregation of patients. At present, we still lack the tools to effectively predict the severity of COVID-19.

The application of machine learning to healthcare is a widely studied research topic and the number of new articles is growing exponentially [14]. While the number of algorithms certified for use in clinical practice is currently quite limited, it is expected that they will play a significant role in the near future. The advantage of artificial intelligence models is the ease of implementation, fast response time and the possibility of continuous operation without time constraints. If thoroughly tested and positively verified, they can become a valuable tool for clinicians.

Z. Kowalczuk (Ed.): DPS 2022, LNNS 545, pp. 52–62, 2023.
https://doi.org/10.1007/978-3-031-16159-9_5

In this article, we propose a computer-aided diagnostic system for predicting the severity of COVID-19 from popular complete blood count parameters. In practice, the role of the system is to predict the severity of a disease that a patient may develop, as determined by the four-grade scale of nursing care.

2 Related Work

As mentioned above, the uncontrollable spread of the COVID-19 pandemic and the depletion of available resources have resulted in the need for new solutions for predicting the severity of the COVID-19 disease course. Various machine learning models have been proposed as a way to solve this problem.

The severity of COVID-19 can be described by a number of parameters, such as the need of a ventilator, length of hospital stay, stable or critical hospitalization, etc. The input features of such models also vary and are usually a set of multimodal values (e.g. demographic, clinical and imaging in one model). To the best of our knowledge, there are no publicly available datasets for this task.

Fang et al. [6] used a deep learning model to predict the rapid progression of COVID-19, which applies to patients who are admitted to the Intensive Care Unit (ICU). The forecast is based on statistical clinical data and the dynamic sequence of chest computed tomography (CT) and laboratory characteristics. Chest CT scans were processed using a sequential modeling method with a Long Short-Term Memory network [9]. Prior to uploading the features to such a network, they are extracted from the images using the 3D ResNet [8] encoder, which creates the appropriate data vector.

Zhang et al. [18] created their COVID-19 progression model. Clinical data and features of lung lesions were used as input. Features of lung lesions were extracted from segmentation masks. The segmentation masks have captured consolidation shadows and ground-glass opacities. Their model can distinguish critical and non-critical disease. Critical illness indicates death, clinical need for ventilation, or transfer to the ICU. Light Gradient Boosting machines with a Cox proportional-hazards regression model were used for this task. The model achieved a sensitivity of 86.7% and a specificity of 80% on a test data set of 158 observations with 133 high-risk patients and 274 observations with 37 low-risk patients.

Gao et al. [7] created a machine learning system to predict the risk of death 20 days in advance for COVID patients. This model integrates four machine learning methods: logistic regression, support vector machine, gradient-boosted decision tree, and neural network. Their system was tested on two external validation cohorts and achieved an average sensitivity of 43.5% and a specificity of 98.3%. The input data consisted of 14 features extracted from electronic health records to detect mortality risks on admission. Eight of these traits correlated positively with mortality, and 6 - negatively. The system output is a normalized risk of death ranging from 0 to 1, with a threshold of 0.6 selected for the assessment of high risk of death.

Liang et al. [13] created a logistic regression model to predict the critical illness risk of hospitalized COVID-19 patients. Critical illness was associated with admission to the ICU, the need for ventilation, or death. The risk score was calculated on the basis of 10 predictors that were selected from 72 available variables with the Least Absolute Shrinkage and Selection Operator (LASSO) regression. The model was tested on 4 external cohorts with a total of 724 patients (87 with critical illness) and achieved an average area under receiver operating characteristic (AUC) of 88%.

Ji et al. [11] created a regression model to predict the risk progression in COVID-19 patients. High-risk patients were defined as having at least one of the following symptoms: respiratory rate of 30 or more breaths per minute, resting oxygen saturation of 93% or less, arterial oxygen partial pressure, or oxygen concentration of 300 mmHg or less, or requiring mechanical ventilation. The model was based on the Multivariate Cox regression, a method of analyzing the effects of several variables over the time an event might take place. Only 4 features were used as input. The model was validated on the training set, therefore the results of this model may be unreliable. The model achieved 95% sensitivity and 78% specificity.

Of all the models mentioned above, only Gao et al. [7], determined how many days in advance they could generate predictions. Most of the models were validated on external cohorts. External validation assesses the performance of models on data from a different hospital, which gives good insight into the generalization of the model.

The comparative results of the work related to this study are presented in Table 1. When training and validating, each model used a different dataset, not available for public use, and used different target criteria and different metrics, therefore it is difficult to fully and consistently compare their performance.

Table 1. Comparison of the performance of selected models for COVID-19 forecasting (the results were averaged for multi-cohort validation).

Model	AUC	Accuracy	Sensitivity	Specificity
Fang	87.4%	80.9%	75%	84.9%
Zhang	90.93%	Unknown	86.7%	80%
Gao	95.03%	91.7%	43.5%	98.3%
Liang	88%	Unknown	Unknown	Unknown
Ji	91%	Unknown	95%	78%

3 Our Solution

The aim of this work was to create a machine learning system that predicts the severity of the course of COVID-19 in patients infected with the virus. The prognosis is made on the bases of the complete blood count and basic demographic information.

3.1 Data Collection

Table 2. The 3-step scale of care applied in Polish hospitals (according to [1]).

Care determinant	Category		
	I Minimal	II Moderate	III Intensive
Motion	Completely unassisted	Patient moves with a cane or a walker, requires some help with getting up from bed, armchair, spends most of the time in bed	Lying patient, does not leave bed, able to change position by himself or with the help of a nurse. Transport only on a stretcher or a wheelchair
Hygiene	Completely unassisted	Patient requires a little help with cleaning, getting in a bathtub, requires help with washing hair and cleaning his back	Patient requires help with all cleaning activities, underwear change. Cleaning in bed, help with oral cavity cleaning, anti-bedsore treatment
Nutrition	Completely unassisted	Patient requires a little help: passing a tray, cutting up the meal and control over food consumption	Patient requires being fed, cutting up the meal or is nourished through a feeding tube
Defecation	Completely unassisted	Patient requires a little help (walking to the lavatory)	Patient usually signals his needs, uses a urine bottle or does not control urinating and defecation
Range of observation	Pulse and temperature measured twice a day	Pulse, pressure and temperature measured more than twice a day, observation after a diagnostic test	Complete observation at intensive care unit. Monitoring, pulse and respiratory rate measured every hour. Blood tests, urine tests more than twice a day

The dataset was collected at the University Clinical Center (UCC) hospital in Gdańsk, Poland. This dataset was also used in our other work [12], but was used there for another task – the detection of SARS-CoV-2. The dataset includes the medical records of 22,463 patients admitted between March 25, 2019 and December 16, 2020. The data used in this study comes from routine blood tests, RT-PCR tests, and patient care cards. The presence of COVID-19 was determined on the basis of the RT-PCR test result. Only patients with conclusive results were enrolled in the study. Records of patients with low positive results were discarded from further analysis. We chose patients who had CBC and RT-PCR tests in no more than one day. Patients not infected with SARS-CoV-2 were excluded from the analysis. The category of medical care was adopted as the target variable. A detailed comparison of the categories in the 3-point scale of care used in Polish hospitals is provided in Table 2. Additionally the dataset also contains patients with positive results who had no need for hospitalization at UCC (therefore had no care determinant). To account for such cases we created an additional category "0" as no need for hospitalization constitutes even less severe COVID-19 course than category 1.

Table 3. Number of entries for each feature in the COVID-19 severity prediction task according to the UCC dataset.

Feature	0	I	II	III
Hemoglobin (HGB)	544	7	47	376
Platelets (PLT)	544	7	47	376
White Blood Cells (WBC)	544	7	47	376
Lymphocytes count (LY)	541	7	46	376
Age	541	7	47	376
Sex	541	7	47	376

3.2 Data Preprocessing

Due to the limited availability of training data, only 4 CBC features were selected as input data for the synthesized model. We combine the data into 3 classes: A, B, and C. Class A applies to patients who were not admitted to the hospital. Due to the small number of samples in groups I and II, they were merged into one class B. An additional motivation here was the fact that both categories I and II describe patients who hardly need significant assistance and that they consume much less hospital resources than category III. This new class (B) can therefore be described as "non-intensive care". In contrast, class C was made up of the most critical patients (category 3). Since only 6 input features were available for this task, none of them were rejected even though they could have had a relatively high correlation with another feature. Missing values of lymphocyte count, age and sex were supplemented with the k-nearest neighbors algorithm. The dataset has been scaled using a Standard Scaler. The number of entries for each input model feature is given in Table 3.

3.3 Architectures

In this work we compared tree-based machine learning methods with neural network approaches. Machine learning models for tabular data are dominated by tree-based models, which typically outperform deep learning solutions [16].

However, recently it has been reported that new transformer-based deep neural network models [17] outperform state-of-the-art tree-based models [4,10]. We used 2 tree-based models: XGBoost [5] and CatBoost [15], and 2 models based on artificial neural networks: a fully-connected neural network and TabNet [4].

XGBoost: Extreme Gradient Boosting (XGBoost) is the most popular model for tabular-data Kaggle competitions [2]. It is a tree-based algorithm recommended for smaller tabular datasets. XGBoost was implemented with the xgboost Python Library module [3], XGBClassifier version 1.5.0. In our model, we used empirically selected hyperparameters: colsample by tree 0.5, learning rate 0.001, max depth 5, alpha 5 and 2 estimators.

CatBoost: Categorical Boosting (CatBoost) [15] is a gradient-boosting model for categorical, numerical and text data. It introduces a separate categorical data processing algorithm that also prevents leakage of targets [15] (when a variable unavailable for the target dataset is used to train the model). The algorithm for handling categorical data transforms the categorical features to be mathematically comparable. It is based on the concept of introducing artificial "time" into the model. We used categorical cross-entropy as a loss function with a learning rate coefficient of 0.001.

Fully-Connected Artificial Neural Network. A fully-connected artificial neural network used 5 fully-connected layers with separating dropout layers (in between) applied to decrease overfitting. We used categorical cross-entropy as the loss function.

TabNet. TabNet [4] (shown in Fig. 1) is a state-of-the-art deep neural network that surpasses boosting algorithms previously considered the leading solution for tabular data. Unlike tree-based algorithms, models based on deep neural networks do not require pre-processing of features and can learn from structurally different types of data (e.g. images with captions) [4]. The loss function assigned class weights to the samples. The class weights were calculated on the basis of the proportion of samples in all classes. We chose the values of the hyperparameters as follows: Optimizer: Adam; Learning rate: 0.01; Maximum epoch: 1000; Patience: 60; Batch size: 256/64; Loss: Categorical Cross Entropy; Scheduler by Step Learning Rate; Scheduler step size: 10; Scheduler Gamma: 0.9.

We used the Synthetic Minority Oversampling Technique (SMOTE) to account for class imbalance in the XGBoost and CatBoost models. We did not use SMOTE for the ANN and TabNet models as this did not improve the results.

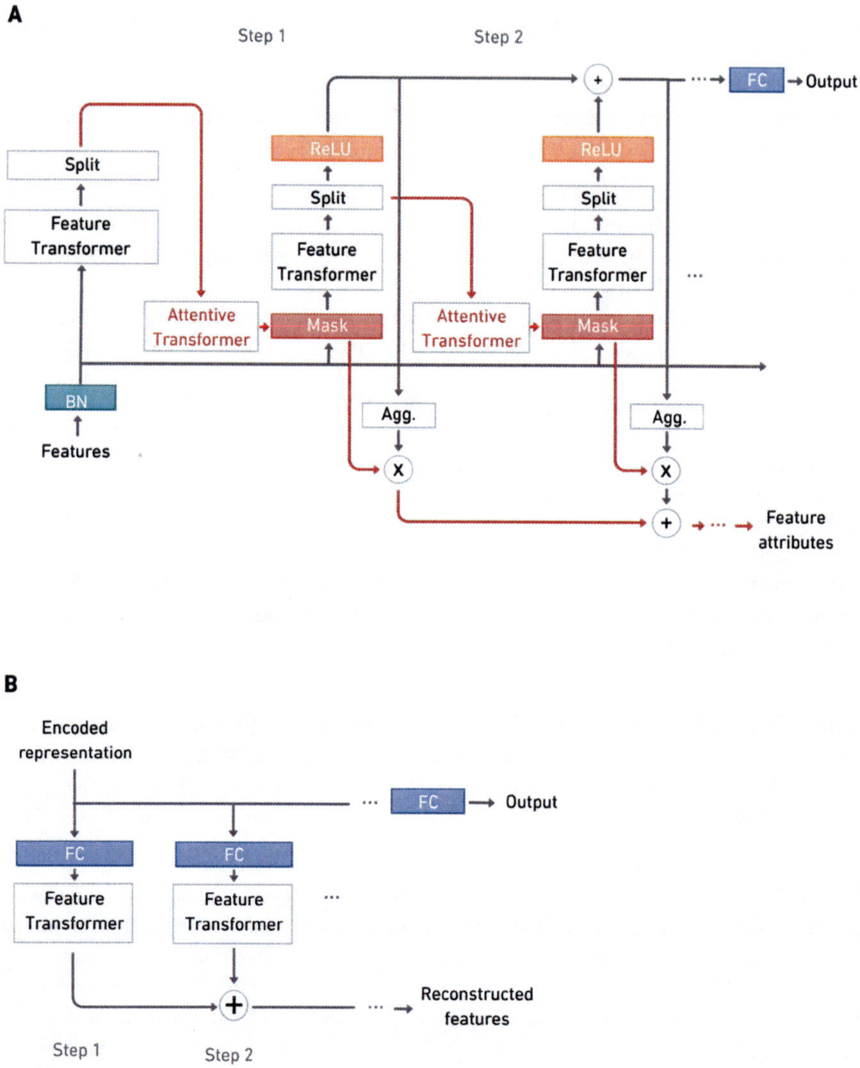

Fig. 1. TabNet architecture divided into (A) encoder and (B) decoder.

4 Experiments and Results

The training dataset consisted of 80% of the samples and the validation dataset had 20% of the samples. We employed 5-fold cross-validation due to limited data availability. We compared the performance of four models: XGBoost, CatBoost, fully-connected Artificial Neural Network (ANN), and TabNet. Models were assessed with a class-appropriate accuracy, recall and F1 score, and overall accuracy. Severity prediction distinguishes between 3 classes. For each class, precision, recall and F1 should be calculated individually (i.e. sample belongs to that class, sample does not belong to that class). The reason for computing a class-specific metric is to select a model that correctly distinguishes all classes.

Table 4 shows the results for the XGBoost, CatBoost, ANN, and TabNet models, respectively. The models distinguish between 3 classes: class A (no hospitalization), class B ("non-intensive" care), and class C (intensive care). The table shows class-specific precision, recall and F1 and overall accuracy. It also presents these metrics (precision, recall and F1) with macro average and weighted average.

The ANN model achieved the highest accuracy, but at the same time it was characterized by a very low recall (sensitivity) for class B, which was the most difficult to predict for all models because we had the lowest samples here. The TabNet model came second for accuracy and also had the highest recall in class B of all models. Only TabNet had a recall for class B at a similar level to other classes; while in the other cases of models we had a significant gap between recalls from classes A and C and class B.

5 Discussion

The results of the experiments described in the previous section show that we can predict the potential development of a critical COVID-19 illness from complete blood count data and basic demographic information. Nonetheless, the models presented in this paper should be considered only as a proof of concept and more training and validation data is needed before these models can be implemented in clinical practice.

All models showed high efficiency in the classification of two extreme cases – critical patients and patients in good health. However, the classification of the intermediate states (class B) turned out to be much more difficult for all models. Class B has significantly fewer samples than classes 0 and 3. This situation suggests that collecting more samples for class B should increase the predictive ability of the analyzed methods. This solution (collecting as many samples as possible) may also allow the training of models to recognize both category of care 1 and 2 individually.

The TabNet model turned out to be the most reliable, as only it was able to recognize class B, and at the same time was able to classify classes A and C with decent performance.

The novelty of the approach presented in this paper comes from the use of only basic data from Complete Blood Count with easy-to-obtain demographic data. To the best of our knowledge, previous studies have reported the results of models that also used less frequently studied blood characteristics. Limiting the input features to the most frequently measured and easiest features should make them accessible to a wider range of patients.

Moreover, earlier models for COVID-19 patients working with tabular data relied on classical machine learning algorithms rather than artificial neural networks. Additionally, this study proposes a clear 3-step categorical variable (nursing care category) as the target for predicting the severity of COVID-19. Previous

Table 4. Comparison of the results (means of 5-fold cross-validation) of different models of predicting the severity of COVID-19 disease.

	Precision	Recall	F1 score	Samples
XGBoost				
Class A	71.75%	57.91%	63.78%	109
Class B	16.67%	51.82%	25.13%	11
Class C	57.16%	55.06%	55.93%	75
Accuracy	56%			
Macro average	48.53%	54.93%	48.28%	195
Weighted average	63.07%	56.47%	58.6%	195
CatBoost				
Class A	72.22%	70.04%	71.02%	109
Class B and 2	22.72%	64.73%	33.26%	11
Class C	69%	52.12%	59.27%	75
Accuracy	63%			
Macro average	54.65%	62.3%	54.52%	195
Weighted average	68.24%	62.83%	64.4%	195
ANN				
Class A	78.75%	86.04%	82.17%	109
Class B and 2	53.96%	37.27%	43.69%	11
Class C	79.34%	72.33%	75.5%	75
Accuracy	78%			
Macro average	70.68%	65.22%	67.12%	195
Weighted average	77.61%	78.03%	77.46%	195
TabNet				
Class A	80.41%	72.06%	75.96%	109
Class B and 2	43.96%	68.36%	52.45%	11
Class C	68.87%	72.35%	70.45%	75
Accuracy	72%			
Macro average	64.41%	70.93%	66.29%	195
Weighted average	73.95%	71.97%	72.54%	195

work on the COVID-19 forecast predicted a less informative binary result. This paper proposes to predict more nuanced outcomes.

The COVID-19 severity prediction model can be used to triage patients even before their condition worsens. Detecting patients at highest risk of developing severe disease at an early stage can reduce mortality and ensure faster access to diagnostic methods for endangered patients. Early diagnosis may contribute to reducing the number of patients admitted to the ICU, reducing hospitalization cost and optimizing the usage of available resources.

The shortcomings of our models are mainly related to the limited availability of data. Another potential challenge relates to the inclusion criteria of patients with COVID-19 severity class 0. We assume that these patients did not develop severe disease. However, we only know that they were not admitted to the UCC hospital with COVID-19. In fact, these patients may have been admitted to another hospital or died at home. Moreover, we do not anticipate any time frame for the development of severe disease. A model estimating the severity level along with the expected time frame would certainly be more informative.

6 Conclusion

Complete blood count results can be a valuable resource in predicting COVID-19 severity. Before such a model is implemented in clinical practise, it should be validated on many external cohorts from different populations. Prognostic models can help select patients and provide faster access to diagnosis for those patients most likely to develop critical illness.

In this work, models based on neural networks performed better than the state-of-the-art approaches to tabular data based on trees. The ANN model in particular has proven successful in generalizing even under difficult conditions with limited data.

Overall, COVID-19 severity predictions based on complete blood count data offer a fast and accurate way of triaging patients.

References

1. Dziennik Ustaw Rzeczypospolitej Polskiej, rozporządzenie Ministra Zdrowia z dnia 28 grudnia 2012 r. w sprawie sposobu ustalania minimalnych norm zatrudnienia pielęgniarek i położnych w podmiotach leczniczych niebędących przedsiębiorcami. https://oipip.opole.pl/wp-content/uploads/2014/04/nz_rozporzadzenie.pdf. Accessed 19 Nov 2021
2. Kaggle, data science trends on Kaggle. https://www.kaggle.com/shivamb/data-science-trends-on-kaggle#1.-Linear-Vs-Logistic-Regression. Accessed 19 Nov 2021
3. Python package index, xgboost. https://pypi.org/project/xgboost/. Accessed 29 Oct 2021
4. Arık, S.O., Pfister, T.: Tabnet: attentive interpretable tabular learning. arXiv (2020)
5. Chen, T., Guestrin, C.: XGBoost: a scalable tree boosting system. In: Proceedings of the 22nd ACM SIGKD International Conference on Knowledge Discovery and Data Mining, pp. 785–794 (2016)
6. Fang, C., et al.: Deep learning for predicting COVID-19 malignant progression. Med. Image Anal. **72**, 102096 (2021)
7. Gao, Y., et al.: Machine learning based early warning system enables accurate mortality risk prediction for COVID-19. Nat. Commun. **11**(1), 1–10 (2020)
8. He, K., Zhang, X., Ren, S., Sun, J.: Deep residual learning for image recognition. arXiv 2015. arXiv preprint arXiv:1512.03385 (2015)

9. Hochreiter, S., Schmidhuber, J.: Long short-term memory. Neural Comput. **9**(8), 1735–1780 (1997)
10. Huang, X., Khetan, A., Cvitkovic, M., Karnin, Z.: Tabtransformer: tabular data modeling using contextual embeddings. arXiv preprint arXiv:2012.06678 (2020)
11. Ji, D., et al.: Prediction for progression risk in patients with COVID-19 pneumonia: the CALL score. Clin. Infect. Dis. **71**(6), 1393–1399 (2020)
12. Klaudel, B., Obuchowski, A., Dąbrowska, M., Sałaga-Zaleska, K., Kowalczuk, Z.: Machine-aided detection of SARS-CoV-2 from complete blood count. In: Kowalczuk, Z. (ed.) DPS 2022. LNNS, vol. 545, pp. 17–28. Springer, Cham (2022)
13. Liang, W., et al.: Development and validation of a clinical risk score to predict the occurrence of critical illness in hospitalized patients with COVID-19. JAMA Internal Med. **180**(8), 1081–1089 (2020)
14. Meskó, B., Görög, M.: A short guide for medical professionals in the era of artificial intelligence. NPJ Digit. Med. **3**(1), 1–8 (2020)
15. Prokhorenkova, L., Gusev, G., Vorobev, A., Dorogush, A.V., Gulin, A.: CatBoost: unbiased boosting with categorical features. arXiv preprint arXiv:1706.09516 (2017)
16. Shwartz-Ziv, R., Armon, A.: Tabular data: deep learning is not all you need. arXiv preprint arXiv:2106.03253 (2021)
17. Vaswani, A., et al.: Attention is all you need. In: Advances in Neural Information Processing Systems, pp. 5998–6008 (2017)
18. Zhang, K., et al.: Clinically applicable AI system for accurate diagnosis, quantitative measurements, and prognosis of COVID-19 pneumonia using computed tomography. Cell **181**(6), 1423–1433 (2020)

Computer Diagnosis of Color Vision Deficiencies Using a Mobile Device

Natalia Wcisło(✉), Michał Szczepanik, and Ireneusz Jóźwiak

Department of Applied Informatics, Faculty of Information and Communication Technology, Wroclaw University of Science and Technology, Wybrzeże Wyspiańskiego 27, 50-370 Wrocław, Poland
240255@student.pwr.edu.pl,
{michal.szczepanik,ireneusz.jozwiak}@pwr.edu.pl
https://kis.pwr.edu.pl/

Abstract. This paper presents an application for iOS devices that will allow users to diagnose the degree of their color vision defect The main functionalities of the application are the diagnosis of color vision disorders using the Ishihara and 100 Hue test. Ishihara test is used to detect a green-red defect and includes a full set consisting of 38 boards. 100 hue test check special spectrum patient's defect. This test includes 44 hues, which user must arrange the pieces from the lightest to the darkest color. The result shows the defect in a specific color area. Due to the similar parameters of the models of mobile devices, the application was written for iOS.

Keywords: Medical diagnostics · Mobile solution · Color vision · Color blindness · Daltonism

1 Classification of Color Vision Deficiencies

Color vision deficiency (CVD) is also called color vision defect [2]. CVD is a very popular disease in the sociality of different ages. There are two different types of color vision deficiencies into acquired and congenital [5,13]. The cause of CVDs is missing or damaged photoreceptor cones [8]. CVDs are classified into three levels of vision anomalies: tetrachrome, dichromization, and monochrome. In the case of trichrome, one of the photoreceptor cones is damaged. In the case of dichrome, one of the cones is completely missing. In the case of monochrome, at least two photoreceptor cones are absent. Monochrome is the rarest type of CVD. The results of the clinical trials (ClinicalTrials.gov) and the National Institutes of Health (nih.gov) show that 1 in 12 men and 1 in 200 women suffer from the red-green recognition defect. Furthermore, 1 in 10,000 people worldwide cannot differentiate between blue and yellow, and 1 in 100,000 worldwide has blue-cone deficiency Fig. 1.

Prevalence of these CVD types in the population is presented in Table 1.

Supported by organization Wroclaw University of Science and Technology.

Z. Kowalczuk (Ed.): DPS 2022, LNNS 545, pp. 63–70, 2023.
https://doi.org/10.1007/978-3-031-16159-9_6

Table 1. Prevalence of CVD types. (*source:* [11])

CVD form	Type	Prevalence [%]
2*Monochromacy	Achromatopsia	0.003
	Blue cone monochromacy	0.001
3*Dichromacy	Protanopia	1.01
	Deuteranopia	1.27
	Tritanopia	0.2
3*Anomalous trichromacy	Protanomaly	1.08
	Deuteranomaly	4.63
	Tritanomaly	0.2

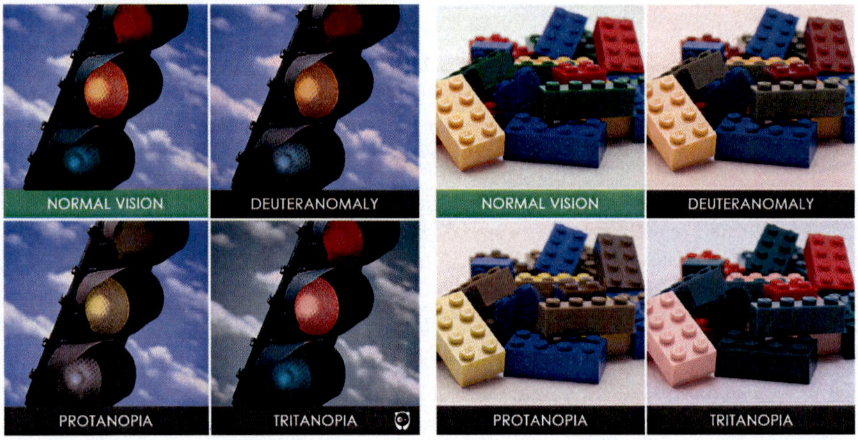

(a) Stoplight. (b) Lego.

Fig. 1. Examples of color perception by people with color blindness in images. (*source:* [15])

2 Computer Test for CVD

Diagnostic tests are generally available on the Internet, on the websites of ophthalmology salons, in shops with mobile applications, and other sources. There are also desktop applications that are available for purchase. Computer testing is a technology that has only yet to be implemented in many units in the world.

A common problem with computerized diagnostic tests is their reliability and precision due to displaying or other adversities that may affect the test result. CVD studies were conducted on 267 volunteers using Ishihara plates and a computerized test in Egypt. The results were summarized and published in an article [12]. The sensitivity of the test was calculated at 100 of the conducted tests without negative errors when examining cases with CVDs. New technology using the computer version of the test is able to replace traditional clinical tests.

Ishihara tests are not the only ones developed as a computer version. Using a Macintosh computer that was calibrated to a standard white display (D65), the effectiveness and specificity of conventional and computerized CVD tests were compared. Computer emulations of the City University Color Vision Test (CUT) [7], Ishihara plates and the American Optical Hardy-Rand-Rittler (AO-HRR) have been created. The results of the test subjects prove that computer emulations of the CUT, Ishihara, AO-HRR tests, and the results of a person with normal color vision are at a high level of specificity, and the determination of inborn color defects is very sensitive and comparable to their conventional counterparts.

3 Solution

The purpose of the work is to design an application for iOS devices, which will allow patient and employers in clinic to diagnose CVD. Two of the most common and relevant tests for CVD diagnosis are included in the program.

A common problem with computerized diagnostic tests is their reliability and precision due to displaying or other adversities that may affect the test result. The selection of screen parameters has a significant impact on the course of the test and its result. The application was designed for Apple devices such as iPhone and iPad, as their displays do not differ significantly from each other when it comes to displaying colors. The new SwiftUI technology allows it to adapt to different iPhone models, regardless of the differences in their display sizes. The devices's parameters are presented in Table 2.

Table 2. Comparison of different models of telephone displays. (*source: own*)

Parameters	iPhone 12	iPhone X	iPhone 8
Year of the premiere:	2021	2018	2017
Display:	Super Retina XDR	Super Retina HD	Retina HD
Maximum brightness of:	625 nits	625 nits	625 nits
Wide color display (P3):	Yes	Yes	Yes
True Tone display:	Yes	Yes	Yes
Contrast ratio:	2,000,000:1	1,000,000:1	400:1

The selection of screen parameters has a significant impact on the course of the test and its result, that is why the difference in PPI and max screen brightness are very important features during the choice of device to test. According to the Table 3, the difference in PPI and max screen brightness in different iPhone models is less widespread than in other producers mobile devices. That is why the differences during the 100 Hue test on Apple devices would be less visible and the test more accurate [9,10].

Table 3. List of parameters of various phone companies. (*source: own*)

Model	PPI	Max brightness
Xiaomi Civi Pro	673	950 nits
Xiaomi Redmi Note 8 2021	409	559 nits
Google Pixel 6 Pro	512	497 nits
Huawei Mate 40 Pro 4G	456	476 nits
Samsung Galaxy Z Flip3 5G	426	935 nits
Motorola Moto G31	411	700 nits
Vivo S10	409	646 nits
iPhone 12	460	625 nits
iPhone 11	326	625 nits
iPhone X	458	634 nits
iPhone 8	326	625 nits

The interface was built using SwiftUI, which is a relatively new technology and declarative framework for building Apple applications. Moreover, SwiftUI comes with a state-based declarative approach. The user no longer uses the storyboard but the declarative UI structure, consequently, it simplifies the interface implementation. Exits and actions are effectively checked at compile-time, reducing the risk of UI failure at runtime [6]. The visual layer, which is what the user sees on the screen, is new-fashioned, intuitive. In addition the interface is simple in use, and follow Apple's Human Interface Guidelines guidelines.

The diagnosing functionality consists of 2 screening tests for the patient's CVD. The first is the Ishihara test, which is the most popular topic among the public, as it is widely used for the diagnosis of CVD when visiting an ophthalmologist. Test diagnosed regarding the recognition of green and red color (deuteranopia and protanopia). Each of them contains a number, one or more curved lines or nothing particular at all. These objects are drawn on the plate using many multi-colored circles. The choice of colors on each board is different. This makes it possible to diagnose various visual disturbances. The test detects two of the most common types of color vision defects in humans i.e. deuteranopia and protanopia. In order for the test to be carried out reliably, the application uses a timer—the patient sees the test only for a few seconds so that the ophthalmologist can give an appropriate result.

The second test is the 100 hue test, also called an arrangement test. Its purpose is to diagnose and detect a specific wavelength that the user is unable to see. The test contains a set of 44 colors, based on which the application can determine the appropriate range of colors. This is a practical test when the disease is acquired or is an uncommon form of CVD [5,13]. The test results are shown in the form of a radial bar, which is used to visualize the color palette as well as the user's mistake. Its goal is to diagnose and detect a wavelength that the user is unable to notice. In clinics, the 100-color test is used in the

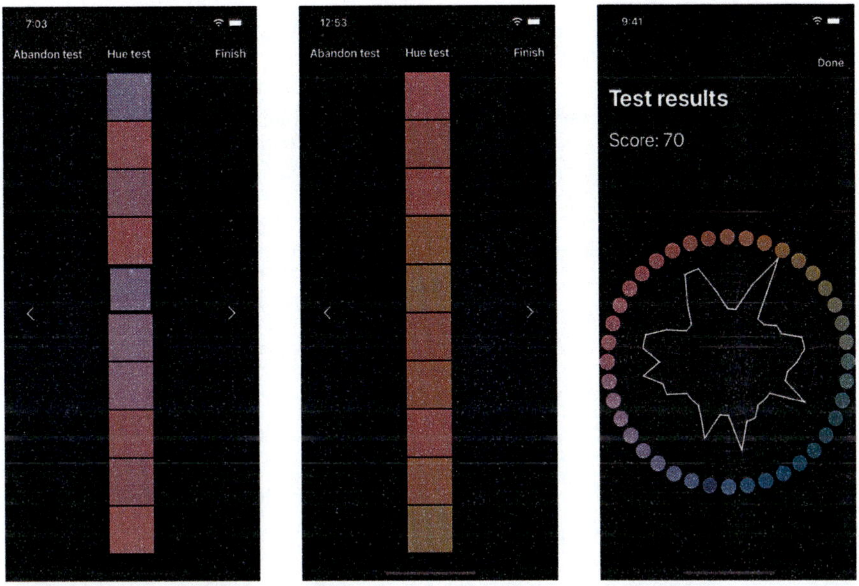

(a) The home view with the 100 hue test results.

(b) The details of the recognized color.

(c) The settings screen for deuteranopia.

Fig. 2. Screenshots of views related to the tests and settings. (*source: own work*)

form of puzzles, which the patient arranges in order from lightest to darkest. After the test is completed, the employee turns the puzzle over and checks the order in which the instructions are followed. Because the process of checking the results is laborious and time-consuming, the computer calculation of the results will improve the diagnostic process. Additionally, the radial strip is a simpler form for presenting the result to the patient. The application is intended for people of all ages, which is why, before each test, a tutorial is presented. Ishihara plates displayed in the Ishihara tests are not regular raster images. Instead, each plate is defined as a set of circles, each with a position, radius, and a color. To draw the Ishihara plate, the application first has to load the CSV file of an appropriate plate, parse it, and draw each circle on a canvas. Images of Ishihara plates retrieved from materials available on the Internet [1] proved to have too low resolution for the purpose of this project. Additionally, because of the JPG format, they contain a lot of noise and have a white background, which would be unacceptable, given that the application provides a non-white background to users. For these reasons, the plates have been created by hand with the use of a specialized editor created just for this purpose. The editor was implemented as a simple web application with the use of the p5.js library and JavaScript. The application allows the user to set the background image of the plate to serve as a reference. The user then traces out the circles on the plate with the mouse by clicking on origins of circles and dragging the mouse to their circumferences. The colors are automatically picked from the original plate. After tracing all circles on the image, the user has the option to export all circles in JSON and CSV data formats. The test screen is shown in in Fig. 2. The self-generated table can adapt to any size of the display such as iPad mini, iPhone X or iPhone 12.

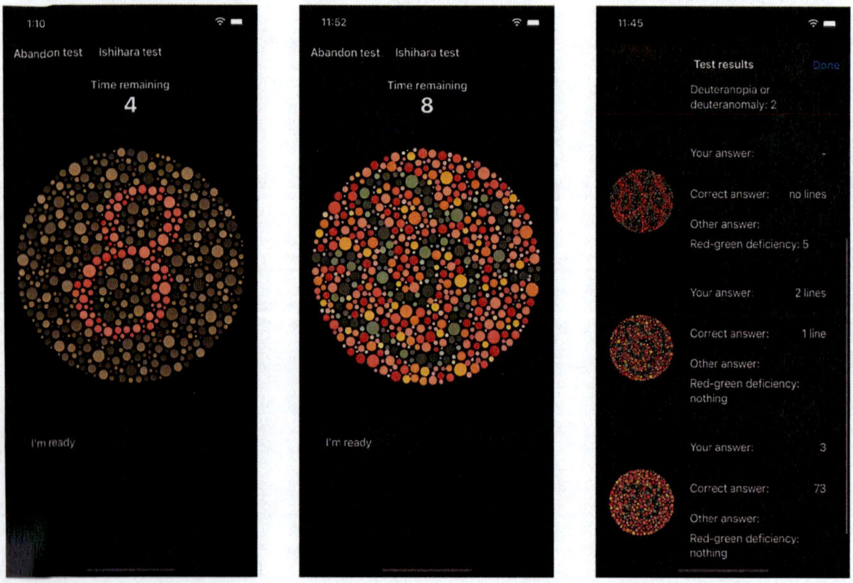

(a) The home view with the 100 hue test results.

(b) The details of the recognized color.

(c) The settings screen for deuteranopia.

Fig. 3. Screenshots of views related to the tests and settings. (*source: own work*)

The hue test is divided into four color groups. These groups are defined as lists of colors, where each color c gets associated with an index i—the "correct" position in the color gradient. At the start of the test, the list of (c, i) pairs gets shuffled randomly. The user then has the ability to swap the pairs on the list by clicking on the colored blocks. After finishing the test, each (c, i) pair is compared to its final position on the list. This way, the application computes a score for each pair, which is essentially the distance from the final position of the pair to its "correct" i value [3]. These scores are then presented to the user in the form of a radar chart. The test screen is shown in Fig. 3.

4 Conclusion and Future Work

The results of the clinical trials (ClinicalTrials.gov) and National Institutes of Health (nih.gov) [9], present that 1 in 12 men and 1 in 200 women suffer from the red-green recognition defect. Furthermore, 1 in 10,000 people worldwide cannot differentiate between blue and yellow, and 1 in 100,000 people worldwide has blue cone deficiency.A common problem with computerized diagnostic tests is their reliability and precision due to variety of type displays or other adversities that may impact the course of the test and its result. The selection of screen parameters has a significant impact on the patient's test results. Computer testing is needed because it speeds up the work of ophthalmology clinics when performing diagnostic tests and calculating test results, and reduces the workload of clinic staff. The process of checking the results of tests for CVD is laborious and time-consuming, the computer calculation of the results will improve the diagnostic process. Additionally, the present results on the screen device are more comfortable for the patient. Moreover Computer testing are a convenient form for patients of all ages.

The next stage of diagnostic support is the development of the application. As a consequence, the application will be enriched with 2 new diagnostic tests, which are aimed at improving the diagnostic process by ophthalmologists and optometrists, by providing a more accurate examination of the patient's color vision defect, acceleration of the examination performed by specialists, and streamlining the process of obtaining results.

The first test to be implemented is the D-15 test, which is a smaller version of the 100 hue test. This is an arrangement test. The purpose of the test is to quickly check for a color vision defect. Due to the fact that the 100 hue test is time-consuming, the d-15 test will enable the initial diagnosis of the patient. In the end, the patient can save time and start with a test containing a smaller set of colors [14].

The HRR (Hardy Rand and Rittler) Standard Pseudoisochromatic Test, 4th Edition is to detect three different type of defect Protan, Deutan, Tritan. The figures used by the plates are independent of language and suitable for both adults and children.The plates are language-independent and suitable for people of different ages, for both children and adults. The HRR is able to detect several characteristic features for advanced color vision defects: congenital and acquired testing, identification of the type of defect, and diagnosis of the extent of the defect, as well as quick positive classification of normals. The test HRR consists of 18 plants. The first four plates are used to show the patient how the test works. The subsequent 14 plates are the diagnostic series and chech diagnostic as to the extent (mild, medium, or strong) and type of defect (Protan, Deutan, Tritan). Test HRR eliminates the potential for memorization and malingering results. That is why it is the more effective version for detecting color vision defects than the traditional Ishihara test [4].

References

1. Colblindor: Ishihara's Test for Colour Deficiency: 38 Plates Edition. https://www.color-blindness.com/ishiharas-test-for-colour-deficiency-38-plates-edition/. Accessed 9 Apr 2022

2. Colour Blind Awareness: Inherited colour vision deficiency. https://www.colourblindawareness.org/colour-blindness/inherited-colour-vision-deficiency/. Accessed 18 Apr 2022
3. Esposito, T.: An adjusted error score calculation for the farnsworth-munsell 100 hue test. LEUKOS **15**(2–3), 195–202 (2019)
4. Foote, K.G., Neitz, M., Neitz, J.: Comparison of the Richmond HRR 4th edition and Farnsworth-Munsell 100 hue test for quantitative assessment of tritan color deficiencies. J. Opt. Soc. Am. **31**(4), A186 (2014)
5. Hasrod, N., Rubin, A.: Defects of colour vision: a review of congenital and acquired colour vision deficiencies. Afr. Vis. Eye Health **75**, 1 (2016)
6. Hudson, P.: SwiftUI lets us build declarative user interfaces in Swift. https://www.hackingwithswift.com/articles/191/swiftui-lets-us-build-declarative-user-interfaces-in-swift. Accessed 11 Mar 2022
7. Ing, E.B., Parker, J.A., Emerton, L.A.: Computerized colour vision testing. Can J. Ophthalmol. **29**(3), 125–128 (1994)
8. Michael, K., Charles, L.: Color perception (2011). https://webvision.med.utah.edu/. Accessed 29 Nov 2020
9. PhonesData: phonesdata. https://phonesdata.com/en/best/screenppi/2021/. Accessed 12 Apr 2022
10. Pixensity. pixensity. https://pixensity.com/list/phone/. Accessed 10 Apr 2022
11. Salih, A.E., Elsherif, M., Ali, M., Vahdati, N., Yetisen, A.K., Butt, H.: Ophthalmic wearable devices for color blindness management. Adv. Mater. Technol. **5**(8), 1901134 (2020)
12. Semary, N.A., Marey, H.M.: An evaluation of computer based color vision deficiency test: Egypt as a study case. In: 2014 International Conference on Engineering and Technology (ICET). IEEE, April 2014
13. Turgut, B.: Discriminators for optic neuropathy and maculopathy. Adv. Ophthalmol. Vis. Syst. **7**, 7 (2017)
14. University, H: Color Sorting Exercise. https://sites.harding.edu/gclayton/Color/Assignments/C06_ClrSort.html. Accessed 16 Apr 2022
15. Vaičiulaitytė, G.: You'll be amazed how people with color blindness see the world. Bored Panda (2017)

Cybersecurity

Simulation Model and Scenarios for Testing Detectability of Cyberattacks in Industrial Control Systems

Michał Syfert, Jan Maciej Kościelny, Jakub Możaryn, Andrzej Ordys(✉), and Paweł Wnuk

Institute of Automatic Control and Robotics, Warsaw University of Technology, 02-525 Warszawa, Poland
andrzej.ordys@pw.ed.pl

Abstract. This article concerns the detection of cyberattacks in industrial installations. A comprehensive experimental stand, and any associated simulator, have been developed in the Institute of Automatic Control and Robotics, Warsaw University of Technology. This article focuses on the simulator, its properties and its functionalities. The simulator can work in standalone mode or hardware-in-the-loop mode. For testing purposes, a wide range of cyberattacks can be designed and injected in the simulator. The approach is illustrated by a numerical example of a replay attack.

Keywords: Modelling and simulation · Cybersecurity · Cyberattacks · Fault diagnosis

1 Introduction

There are various anomalies in the functioning of Industrial Control Systems (ICS). They are manifested by various changes in the operation of the control system and in the course of the process, deviating from its normal state. The causes of the anomalies are:

1. Interferences, i.e. independent factors affecting the controlled process. Properly designed control systems should eliminate interference.
2. Damage/defects - destructive events resulting in the loss or deterioration of the system devices' ability to perform their functions. Both components of the technological installation (e.g. pipeline cracks) and elements of the control system (measuring and actuating devices, communication networks, control units and human-system interface devices) can be damaged.
3. Human errors, including operator's errors.
4. Sabotage actions - deliberate human actions that have a detrimental effect on the course of the process.

Z. Kowalczuk (Ed.): DPS 2022, LNNS 545, pp. 73–84, 2023.
https://doi.org/10.1007/978-3-031-16159-9_7

5. Cyberattacks - deliberate disturbance of the proper functioning of the control system carried out in virtual space (usually via the Internet), the purpose of which is to take control of the control system.

The prevention and counteraction of the first three types of threats are referred to as safety problems, while hacker attacks on control systems and sabotage activities are the area referred to as security issues—the main difference between these threats concerns their source of origin. Cyber attacks and sabotage actions are carried out intentionally by humans, in a deliberate manner, to cause specific physical, economic or public relations losses [7]. In contrast, damage, human errors and disruptions arise due to destructive physical processes occurring in devices or incorrect (unintentional) handling, or the occurrence of changes in external factors influencing the process.

Despite the different causes of damage (safety) and cyber attacks (security), they can produce identical or similar symptoms. The effects of these anomalies, such as fire, explosion, environmental contamination, destruction of the installation, and process shutdown, can also be equally dangerous. It applies especially to critical processes occurring in chemical, energy and other processes.

To be able to prevent the effects of anomalies, it is necessary to detect and recognize them early. In the current practice, the issues of detecting attacks and damages are carried out independently as Intrusion Detection Systems (IDS) or Advanced Diagnostic Systems (ADS). Many control system applications do not have these solutions, and the only method of anomaly detection is to control the limits of the process variables. In IDS systems, attack detection is carried out based on network traffic exploration or network protocol analysis. Whereas, in the case of ADS, the reference model of the object being diagnosed is used for fault detection. This type of approach has recently also been used to detect cyberattacks under the name of Intrusion detection based on process data analysis [3]. The attack detection schemes presented in [3, 4] are identical to the damage detection scheme used for a long time in FDI methods.

Therefore, in our opinion, it is not justified to separate the issues of detecting attacks and detecting damages or other anomalies. It is necessary to break with these „bad habits" which have been shaped by the separation of IT specialists and automation specialists. The overall security strategy should include and integrate security issues in the sense of safety and security, including the problems of detection and isolation of these anomalies.

In this postulated approach, the two systems (IDS and ADS) would work in parallel and exchange information as needed [6]. The features of ADS would enhance the methods of IDS to improve the detectability of cyber-attacks. Hence, the characteristics of the industrial process, the process model, and the controller's properties would be considered, enriching the information contained in the networking data flow. Such an integrated approach can open-up vast new opportunities to elaborate algorithms for detection (and potentially also isolation) of anomalies. In the first instance, the algorithms traditionally used to detect faults could be assessed from the perspective of the ability to detect cyberattacks. Hence, there is a need for a proper assessment and evaluation of such algorithms and suitable test beds [1]. There has been research concerning benchmarks and evaluation of fault diagnosis systems (e.g. [5]). For an integrated approach, such benchmarks could perhaps be adopted. However, an important aspect is that, unlike for purely damage detection tasks, the information technology aspect plays equally important role as the operation/automation technology aspect. Hence, it is postulated that the test environment should be relatively simple in terms of the process model but equipped with suitable data transmission facilities. In general, such a testing environment should consist of: the process itself, an industrial controller, sensors, actuators, an operator's station, and communication links between all the components – communication via a digital network with an internet/cloud access. The IDS would be mainly concerned with the communication link with the external world and, to some extent, with the data flow between the system's components. The ADS will work in parallel with the operator's station, collecting and analyzing the signals.

Furthermore, it is highly desirable that the tests can be performed on a physical process but also on a process simulator. It is a standard development path to test algorithms on a simulator first before applying them to a physical system. It shortens the development time and substantially reduces costs. Hence, building a "digital twin" of the testing environment would facilitate any testing procedures. Moreover, testing abilities, possibilities of injecting faults/anomalies and simulated cyber-attacks would be greatly enhanced if selected parts of the test-bed could be interchanged with their simulation models (so-called "hardware-in-the-loop simulation"). Hence, for the proposed configuration, the components interchanged with their software models would be the process, the controller, the sensors, and the actuators. The connection between the hardware part and the software part would be realized by means of networking data transmission.

This paper presents a proposal of a test-bed for evaluation of algorithms for detection of cyber-attacks and an associated simulator. First, the experimental stand is briefly described. This is followed, in Sect. 3, by a more detailed description of the simulator with hardware-in-the-loop facilities. Some examples of simulated cyber-attacks and observed system's responses are presented in Sect. 4.

2 Description of the Experimental Stand

The process part of the experimental laboratory stand is presented in Fig. 1, whereas the schematic diagram is given in Fig. 2.

Components of the stand are:

- Connected tanks T1,T2;
- Reservoir tank T3;
- The pump P1 (the pump capacity 0 - 6.5 [l/min]),
- Pressure transducers LT_1, LT_2 (range 0–500 mm $H_2 0$) for measuring a liquid level in each of the tanks (H_1, H_2);
- The elastic tube W1 introduces the delay in the system. It is operated by two electro-mechanical shut-off valves V_{d1}, V_{d2};
- The tanks are connected T_1, T_2 with the control valve V_2;
- The outlet of each tank is connected with the main tank with control valves: V_1 (tank T_1), and V_3 (tank T_2);
- Two shut-off electro-mechanical valves (VE1, VE2) are also used to introduce disturbances into the process. D_1 denotes leakage from the tank Z1 (opening the valve VE_1), and D_2 denotes leakage at the pump outlet (opening the valve VE_2).

Depending on the configuration of valve states V1, V2, V3, Vd1, Vd2, various properties of the control object can be realized (e.g., First Order Lag/Second Order Lag/Delay) [1]. The set of available process variables in the experimental stand is given in Table 1.

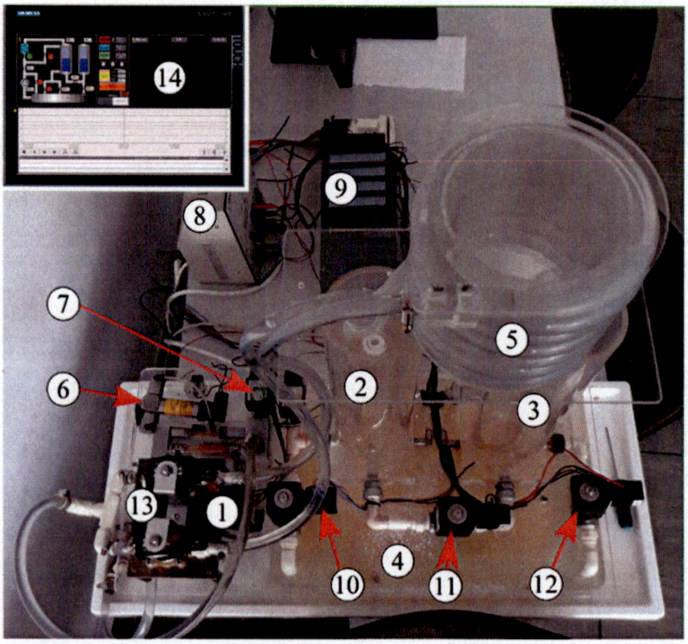

Fig. 1. The experimental laboratory stand: 1) pump: P1, 2) tank: T1, 3) tank: T2, 4) tank: T3, 5) elastic tube: W1, 6) Venturi flowmeter: O1, 7) shut-off electro-mechanical valves VE_1, VE_2, 8) control interface cabinet, 9) SIMATIC PLC S7-1500, 10)-12) control valves V_1, V_2, V_3, 13) shut-off valves V_{d1}, V_{d2}, 14) HMI Panel simulation in the TIA Portal programming environment.

Table 1. Set of available process variables.

Name	Description	Units	Range	Signal
SP_L	Tank level setpoint	%	$\langle 0, 100\rangle$	–
CV_V	Control signal	%	$\langle 0, 100\rangle$	$\langle 4, 20\rangle$ $[mA]$
F_1	Water flow at the inlet to the tank I	$\frac{l}{min}$	$\langle 0, 6.5\rangle$	$\langle 4, 20\rangle$ $[mA]$
L_1	Tank level I	m	$\langle 0, 0.5\rangle$	$\langle 4, 20\rangle$ $[mA]$
L_2	Tank level II	m	$\langle 0, 0.5\rangle$	$\langle 4, 20\rangle$ $[mA]$
PV_L	Controlled level: 100 [%] = 0.374 [m] for L_1 control 100 [%] = 0.374 [m] for L_2 control	%	$\langle 0, 100\rangle$	–

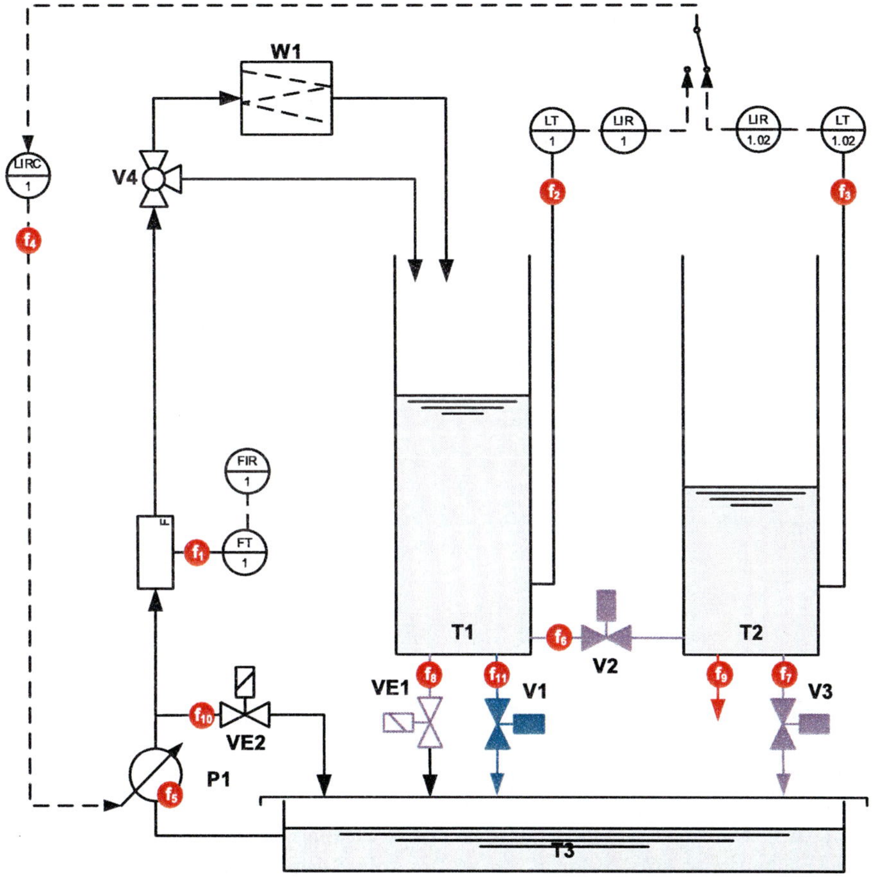

Fig. 2. A schematic diagram of the laboratory test stand. Additionally, available measurements and symbolic places of introduction of process faults (description in Sect. 3.2) are depicted.

The control system is based on the Siemens SIMATIC S7-1500 controller [2]. This controller has a modular design and is equipped with Profinet/Industrial Ethernet interfaces integrated with support for TCP/IP, ISO-on-TCP and S7 protocols. The controller can diagnose and monitor software through the Ethernet port and communicate via RS-232, RS-485 and Modbus RTU protocols. Configuration, programming, and monitoring of the SIMATIC S7-1500 controller is performed using the TIA-Portal environment (version 15.1). It integrates the STEP 7 Basic software (application for PLC programming) and WinCC Basic (application for HMI programming), allowing data exchange between them. The TIA Portal environment was also used for the research project to simulate the HMI panel (Fig. 2:14) for the data monitoring and create the data logs for further analysis.

3 Description of the Simulator

3.1 Overall Structure

The simulator was developed in MATLAB/Simulink. The general block diagram of the simulator is presented in Fig. 3.

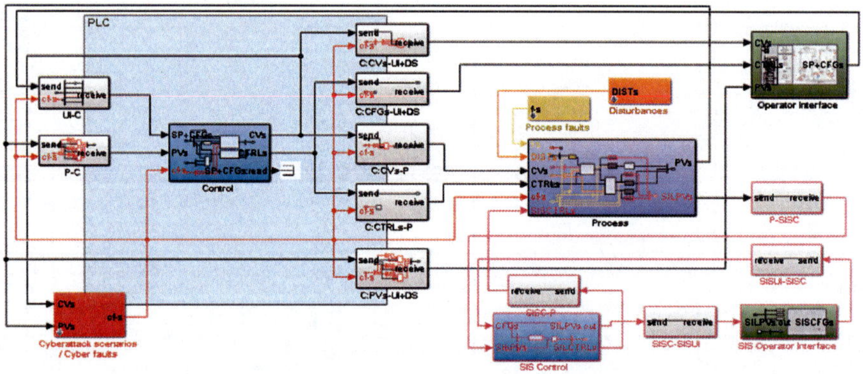

Fig. 3. General block diagram of the simulator.

In the simulator, one can distinguish the following main subsystems:

- **Process**. A subsystem represents the physical process components together with measuring devices.
- **Control**. A subsystem representing the PLC controller, on which signals controlling the configuration are developed, and the control system is implemented.
- **Operator interface**. The subsystem corresponds to the HMI/SCADA system on which the operator interface is implemented.

Vectors of setpoint values and configuration options (SP + CFGs), the control and configuration signals (CVs, CTRLs) and process variables (PVs) are exchanged between subsystems. It is realized by blocks that symbolize specific communication standards.

Additional simulator subsystems are representing the **safety system** (SIS Control + SIS Operator Interface).

A safety system has been added to be able to consider configurations where an additional measurement and control system that is physically separate from the basic automation equipment is available. Thus, it is possible to consider such attack scenarios where the attacker gains access only to selected resources and communication channels.

The parameters of individual process components, including actuating and measuring devices, were selected experimentally to reflect the behaviour of real components as faithfully as possible, including such elements as non-linearities, dynamics, and the amplitude of measurement noise.

3.2 Disturbances, Process and Cyber Faults Simulation

Disturbances. The simulator also takes into account the possibility of simulating process disturbances in the form of fluctuations in the pump efficiency and outflow from tank II. The list of disturbances and their parameters are presented in Table 2.

Table 2. Set of available process disturbances

Name	Description	Range
d_1	Fluctuations in the pump efficiency	$\langle -0.5, 0.5\rangle$ V
d_2	Obstruction in the pipe out of tank T2	$\langle -20, 20\rangle$%

Process Faults. Another important aspect is simulating process faults of measuring devices, actuators, and process components. The place where the faults are introduced is shown in Fig. 2. The list of faults and their parameters is presented in Table 3.

Table 3. The set of process faults and their parameters.

f_k	Description	Range
f_1	The fault in the measurement channel F_1	$\langle -7.5, 7.5\rangle \left[\frac{l}{min}\right]$
f_2	The fault in the measurement channel L_1	$\langle -100, 100\rangle$[%]
f_3	The fault in the measurement channel L_2	$\langle -0.5, 0.5\rangle$ [m]
f_4	The fault in the transmission of CV_V	$\langle -100, 100\rangle$[%]
f_5	The fault of the pump P (change in pump flow)	$\langle -100, 0\rangle$%
f_6	The obstruction in the pipe between tanks T1 and T2	$\langle 0, 100\rangle$%
f_7	The obstruction in the pipe out of tank T2	$\langle 0, 100\rangle$%
f_8	The leak from tank T1	$\langle 0, 100\rangle$%
f_9	The leak from tank T2	$\langle 0, 100\rangle$%
f_{10}	The leak from the inlet to tank T1	$\langle 0, 100\rangle$%
f_{11}	The obstruction in the pipe out of tank T1	$\langle 0, 100\rangle$%

Faults of clogging and leakages are simulated by changing the linear valve positions or changing the PWM signal that controls the opening of ON/OFF valves between tanks and on the outlets. In the case of fault f_9, the size of the fault corresponds to the degree of opening of the cross-section simulating the leak.

The option to enable/disable simulation of disturbances and process faults allows for testing cyberattack detection algorithms in more complex situations particularly, for the assessment of the possibility of distinguishing the effects of cyberattacks from the impact of process faults.

Process fault and disturbances are introduced and controlled by separate subsystems (Fig. 3):

- **Disturbances.** Process disturbance generation subsystem.
- **Faults.** Process fault generation subsystem.

Cyber Faults. Each cyberattack is conducted according to the designed scenario - a specific attack method. Such a scenario defines independent sub-activities influencing individual system components and communication channels. Each activity is determined by the place of influence, its nature and variability over time. In this article, these individual activities are called *cyber faults,* similarly to process faults.

Attack Vector. For the purpose of this simulator, the vector of places which a cyber attack can influence process is called the *attack vector,* whereas the particular activities are *cyber faults.* Hence, an *attack vector* may contain several *cyber faults.*

The cyber faults predefined in the simulator are presented in Table 4.

Table 4. The set of cyber-faults.

Name	Description
cf_F^C	Modification of the measurement signal F at the input to the control system
cf_F^{UI}	Modification of the F measuring signal at the input to the operator station
cf_{L1}^C	Modification of the L_1 measuring signal at the input to the control system
cf_{L1}^{UI}	Modification of the L_1 measuring signal at the input to the operator station
cf_{L2}^C	Modification of the L_2 measuring signal at the input to the control system
cf_{L2}^{UI}	Modification of the L_2 measuring signal at the input to the operator station
cf_{CV}^C	Modification of the CV control signal at the input to the process (actuator)
cf_{CV}^{UI}	Modification of the CV control signal at the input to the operator station
cf_{SP}^C	Modification of the SP set point at the input to the control system
cf_{PID}^{MODE}	Changing the controller's operating mode
cf_{PID}^{SET}	Modification of the controller settings

The lower index of the cyber fault *cf* denotes the affected signal or component, and the upper one denotes the group of signals or performed action.

The following designation of signal groups was used:

- $\ldots^P$ – input/output signals to/from the process,
- $\ldots^{UI}$ – input/output signals from/to the operator interface,
- $\ldots^C$ – input/output signals from/to the controller.

The cyber faults can be divided into two groups affecting:

- transmitted signals (F, L1, L2, CV, SP),
- controller (PID).

The symbolic place of transferring setpoints (SP), control values (CV) and process values (PV) from the groups mentioned above is shown in Fig. 4.

3.3 Possible Hardware in the Loop Configurations

Individual subsystems are connected with blocks that symbolize specific communication standards (Fig. 3). These blocks are possible places of "disconnection" of the simulator and enable the replacement of a given subsystem by its hardware version. In this way, different hardware-in-the-loop systems can be realized. The three basic types are: (a) virtual controller, (b) virtual process, and (c) real process. Physical and virtual (simulated) components in those configurations are shown in Fig. 5.

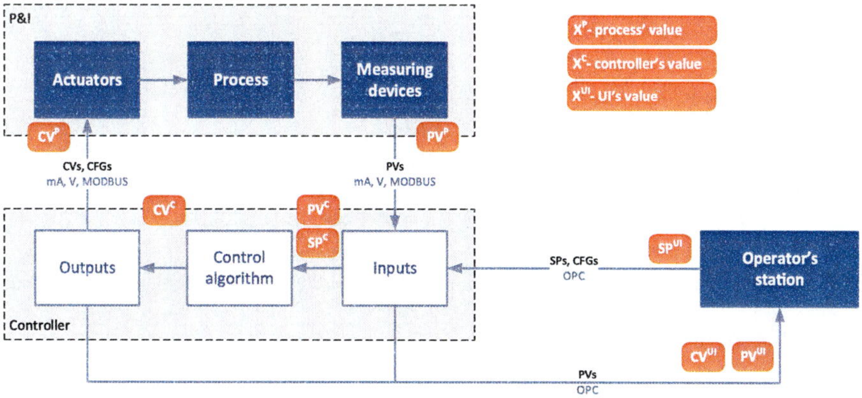

Fig. 4. Group of signals (type and communication channels) used in cyber faults definitions regarding process, control and operator's interface subsystems.

4 Example of Cyber-Attack and Simulation of the System Performance

A cyberattack involving replaying a looped record of the historical measurement values to reach an unacceptable process state, e.g. emptying or overfilling the tank 1 or 2, is presented as an example. This is a "replay attacks/in open-loop" type of an attack [8]. The version of the attack in which only the value of the controlled process quantity (L2), which is sent to the controller and the operator, is falsified is considered.

Thus, the attack vector is: $\left[cf_{L2}^{C}, cf_{L2}^{UI}\right]$. The places of the introduction of cyber faults are depicted in Fig. 6.

The time-line of the scenario is as follows:

- from a given moment $t_{from} = 600\,[s]$ process variable L_2 sent to the controller (cyber fault cf_{L2}^{C}) and operator interface (cyber fault cf_{L2}^{UI}) is replaced with the historical values saved in the time window of the indicated width $L = 200$ samples,
- playback is looped from the beginning after reaching the end of the buffer.

An operation of the process for a predefined control scenario (changes of SP value) without a cyberattack is shown in Fig. 7.

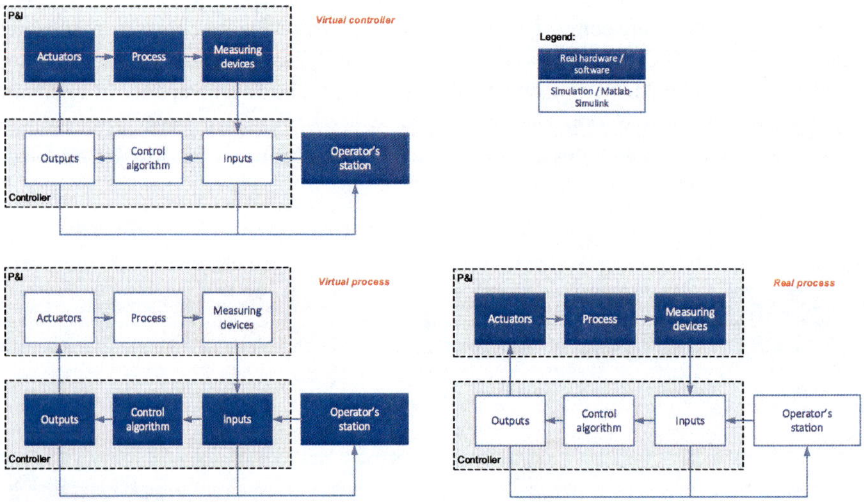

Fig. 5. Types of hardware-in-the-loop configuration with real (blue) and simulated (transparent) components.

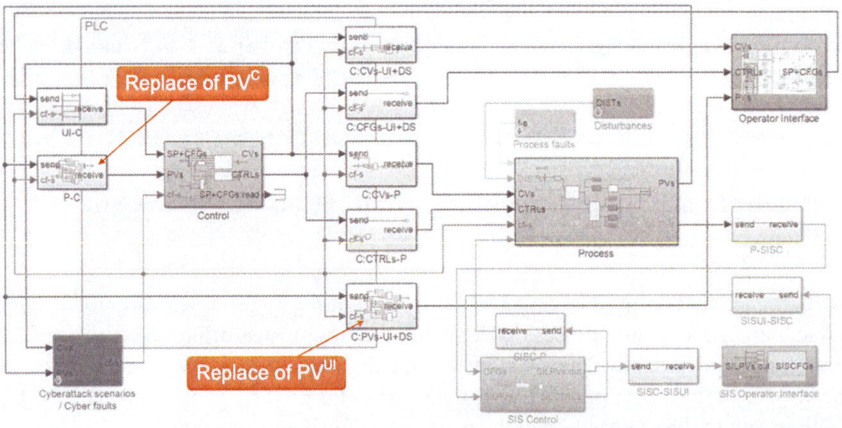

Fig. 6. Places of introducing cyber-faults in Scenario 1.

An operation of the process under cyberattack for the same control scenario is shown in Fig. 8.

It can be seen that tank 2 becomes empty due to the attack. The attack may go unnoticed as the value of the liquid level in tank 2 displayed on the operator's interface comes from the recording and does not follow the actual value.

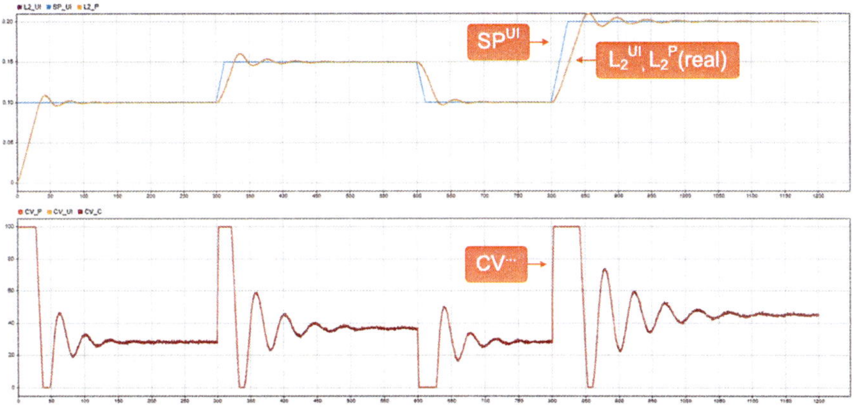

Fig. 7. Courses of the signals during normal process operation: (a) selected process variables, (b) control signals.

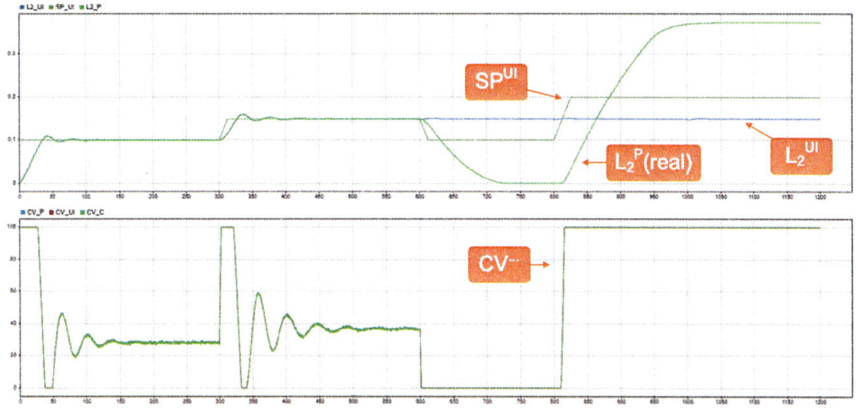

Fig. 8. Courses of the signals during cyber-attack: selected process variables, (b) control signals.

5 Conclusions

In this article, we have proposed an environment for testing algorithms to detect cyber-attacks in industrial installations. The test bed consists of an experimental stand and a validated stand simulator. The communication channels are also modelled, and these channels are the places where cyberattacks can be injected. The system can also work in hardware-in-the-loop mode. In this article, we focus on the description of the simulator's functionalities. We introduce the term of cyber fault, analogous to process fault. Several cyber faults can be agglomerated together, forming an attack vector of a cyber attack. The layout of the simulation environment, enhanced by the ability to work in hardware-in-the-loop mode, provides significant opportunities for reproducing various cyberattacks and testing detection algorithms.

Additionally, system disturbances and process faults can be simulated. In the first instance, themethods used in FaultDetection algorithms can be used. Moreover, because of the simulator's ability to inject process faults, the issue of distinguishing between cyber faults and process faults can be researched.

Acknowledgments. The authors acknowledge support from the POB Research Centre Cybersecurity and Data Science of Warsaw University of Technology within the Excellence Initiative Program-Research University (ID-UB).

References

1. Możaryn, J., Ordys, A., Stec, A., Bogusz, K., Al-Jarrah, O.Y., Maple, C.: Design and development of industrial cyber-physical system testbed. In: Bartoszewicz, A., Kabziński, J., Kacprzyk, J. (eds.) Advanced, Contemporary Control. AISC, vol. 1196, pp. 725–735. Springer, Cham (2020). https://doi.org/10.1007/978-3-030-50936-1_61
2. Stenerson, J., Deeg, D.: Siemens Step 7 (TIA Portal) Programming, a Practical Approach, CreateSpace Independent Publishing Platform (2015)
3. Hu, Y., Li, H., Yang, H., Sun, Y., Sun, L., Wang, Z.: Detecting stealthy attacks against industrial control systems based on residual skewness analysis. EURASIP J. Wirel. Commun. Netw. **2019**(1), 1–14 (2019). https://doi.org/10.1186/s13638-019-1389-1
4. Urbina, D.I., et al.: Limiting the impact of stealthy attacks on industrial control systems. In: CCS 2016: Proceedings of the 2016 ACM SIGSAC Conference on Computer and Communications Security October 2016, pp. 1092–105 (2016) https://doi.org/10.1145/2976749.2978388
5. Bartyś, M., Patton, R., Syfert, M.: Salvador de las Heras, Joseba Quevedo, Introduction to the DAMADICS actuator FDI benchmark study. Control. Eng. Pract. **14**(6), 577–596 (2006). https://doi.org/10.1016/j.conengprac.2005.06.015
6. Koscielny, J., et al.: Towards a unified approach to detection of faults and cyber-attacks in industrial installations. In: A: European Control Conference. "ECC 2021: 19th European Control Conference, June 29–July 2 2021, pp. 1839–1844. Institute of Electrical and Electronics Engineers (IEEE), Rotterdam, The Netherlands (2021). https://doi.org/10.23919/ECC54610.2021.9655212
7. Pan, X., Wang, Z., Sun, Y.: Review of PLC security issues in industrial control system. J. Cybersecur. **2**(2), 69–83 (2020)
8. Teixeira, A., Pérez, D., Sandberg, H., Johansson, K.H.: Attack models and scenarios for networked control systems. In: Proceedings of the 1st international conference on High Confidence Networked Systems (HiCoNS 2012), pp. 55–64. Association for Computing Machinery, New York, NY, USA (2012). https://doi.org/10.1145/2185505.2185515

Functional Safety Management in Hazardous Process Installations Regarding the Role of Human Operators Interacting with the Control and Alarm Systems

Kazimierz T. Kosmowski(✉)

Gdansk University of Technology, G. Narutowicza 11/12, 80-233 Gdańsk, Poland
kazkosmo@pg.edu.pl

Abstract. This article addresses selected issues of the functional safety management of a hazardous process installation. An important role in reducing risks plays nowadays a safety-related control system (SRCS) as a part of the industrial automation and control system (IACS). Responsible tasks in abnormal and accident situations execute the human operators that make use of an alarm system (AS) and its interface within the human system interface (HSI). In this article an approach is outlined for evaluating the human error probability (HEP) interacting with AS. It includes determining the required risk reduction expressed by relevant safety integrity level (SIL). Determined SIL of given safety function to be implemented in the basic process control system (BPCS) and/or the safety instrumented system (SIS) must be then verified for their architectures considered. The HEP for relevant human operator behaviour type is evaluated using the human cognitive reliability (HCR) model.

Keywords: Industrial installations · Risk reduction · Functional safety · Industrial control system · Alarm system · Human factors · Human cognitive reliability

1 Introduction

Human operators must continuously supervise changing states of technological processes and the integrity of industrial installations. They interact when required, through relevant interfaces, with the industrial automation and control system (IACS) and alarm system (AS) according to operation goals and predefined procedures [5, 7, 17]. Their performance in some conditions can be challenging, especially when an abnormal state or accident occurs, and the time window for necessary reaction is relatively short [4, 10]. It can contribute significantly to human errors [11, 23, 25] to be committed during the diagnosis of such situations and/or actions undertaken.

Human factors and cognitive engineering [2, 8, 26] are treated nowadays as important multidisciplinary domains that focus on improving interactions between human and system using relevant interfaces, e.g., a human machine interface (HMI) and/or a human

Z. Kowalczuk (Ed.): DPS 2022, LNNS 545, pp. 85–99, 2023.
https://doi.org/10.1007/978-3-031-16159-9_8

system interface (HSI) at relevant levels of the distributed control system (DCS) and supervisory control and data acquisition (SCADA) system [12, 17].

The industrial control system (ICS), especially its safety-related part, perform nowadays various important safety functions. The safety-related control systems (SRCSs) are often designed and operated according to general requirements given in functional safety standards [13, 14]. The main objective is to maintain high performance and productivity of the plant and reduce risks related to potential hazards and threats. Therefore, the issue of human factors should be carefully considered in the reliability and functional safety evaluation to develop modern solutions [4, 5, 17].

In second editions of the functional safety standards [13, 14] the importance of human factors (HF) and human reliability analysis (HRA) is emphasized. However, there are no clear indication how these issues should be treated in the functional safety management in entire life cycle of industrial installations.

This article is structured as follows. In Sect. 2 defining safety functions for reducing risk of accidents in hazardous plant is outlined. Sect. 3 describes typical layered protection system. Sect. 4 outlines incorporating cognitive aspects in human reliability analysis in the context of functional safety solutions. In Sect. 5 a case study based on author's publication [19] is presented to enhance previous approach proposed for applying of a cognitive HRA in the context of functional safety analysis. In final part, some alarm system design issues to meet functional safety requirements in context of the human error probability (HEP) evaluated are discussed.

2 Defining Safety Functions for Reducing Risks

The functional safety a part of general safety, which depends on the proper functioning in time of the programmable control and protection systems. General concept of the functional safety was formulated in an international standard IEC 61508 [13]. It includes defining, for given hazardous installation, a set of safety functions (SF) that are implemented in properly designed the electric, electronic, and programmable electronic (E/E/PE) systems, or so-called safety instrumented systems (SIS) [14] in the process industry sector.

Two different requirements have to be specified to ensure required safety level:

- the requirements imposed on performance of the safety function considered,
- the safety integrity requirements, understood as the probability that given safety function will be performed in a satisfactory manner within a specified time.

These requirements are specified regarding hazards identified and potential accident scenarios defined. The safety integrity level (SIL) requirements stem from the results of the risk assessment considering the risk criteria to be specified [13, 14].

Two categories of operation modes are to be considered in the functional safety analysis, namely: (1) low, and (2) high or continuous [13]. A low demand mode is typical for the process industry protection systems, e.g., within protection layers. A high or continuous demand mode is encountered in many systems for monitoring and control, for instance in the production or transportation sectors.

The E/E/PE systems or SIS have to be appropriately designed to perform specified functions to ensure that relevant risks are reduced to fulfil specified criteria at the plant design stage and then verified periodically during operation, especially when the operating conditions change. The risk criteria are not specified generally in standards [13, 14]. Only some examples of risk graphs are presented, but with remarks, that specific criteria should be defined for particular process installation. It is suggested to consider three types of losses, namely health, environmental, and material [13, 14].

Allocation of requirements for the safety functions and safety-related systems is illustrated in Fig. 1. It starts with the hazard identification and the risk evaluation to determine required safety-integrity level (SIL) of the safety function (SF) to be implemented for the risk reduction. The risk acceptance criteria are to be defined for the individual risk or societal risk [13].

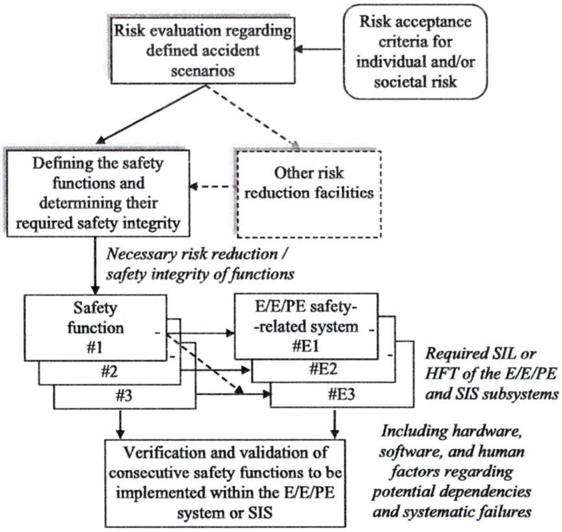

Fig. 1. Allocation of requirements for safety-related systems

As it was mentioned, generally the societal risk evaluation can be oriented on three categories of losses: health, environmental, and material/economic [18]. Assigning of the SIL required is illustrated on an example of the risk graph shown in Table 1 for six categories of severity (C) and six categories of likelihood categories expressed by six ranges of integer numbers from 1–2 to 11–12. An appropriate range is indicated for the sum of risk related parameters F+P+W [14].

Categories of the demand rate parameter W is determined as explained in Table 2. Categories of parameters for the occupancy (F) and avoidance (P) take values of 0, 1 or 2 depending on characteristics of the industrial plant considered. The rationale of such approach based on categories was explained in standard [14] and publication [18]. For F+P+W > 10 and catastrophic severity C_f a layered protection (LP) should be designed.

Consecutive safety functions are implemented in the SRCS, e.g., a safety instrumented system (SIS) [14]. The SIL to be achieved by designed system of the architecture proposed is verified using relevant probabilistic models [13, 18]. It is important to include potential dependencies between failure events in the logical and probabilistic models, in particular potential common cause failures (CCF) in the SRCS comprising redundant subsystems. The architecture of SRCS (hardware and software) must be then verified as regards potential systematic failures and deteriorating contribution of human factors resulting in potential human errors of relevant types [18].

Table 1. An example of risk graph matrix for determining required SIL (based on [14])

Consequence		Determining likelihood category for evaluated **(F+P+W)**					
Severity categories	**C**	**1-2**	**3-4**	**5-6**	**7-8**	**9-10**	**11-12**
Catastrophic	C_f	NR	SIL 1	SIL 2	SIL 3	SIL 4	LP
Extensive	C_e	NR	NR	SIL 1	SIL 2	SIL 3	SIL 4
Serious	C_d	OK	NR	NR	SIL 1	SIL 2	SIL 3
Considerable	C_c	OK	OK	NR	NR	SIL 1	SIL 2
Marginal	C_b	OK	OK	OK	NR	NR	SIL 1
Negligible	C_a	OK	OK	OK	OK	NR	NR

Table 2. Categories of demand rate frequency (based on [14])

Categories of demand rate parameter W			
Estimated SIF frequency $[a^{-1}]$			**W**
W_9	Often	≥ 1	9
W_8	Frequent	$[3 \cdot 10^{-1}, 1)$	8
W_7	Likely	$[10^{-1}, 3 \cdot 10^{-1})$	7
W_6	Probable	$[3 \cdot 10^{-2}, 10^{-1})$	6
W_5	Occasional	$[10^{-2}, 3 \cdot 10^{-2})$	5
W_4	Remote	$[3 \cdot 10^{-3}, 10^{-2})$	4
W_3	Improbable	$[10^{-3}, 3 \cdot 10^{-3})$	3
W_2	Incredible	$[10^{-4}, 10^{-3})$	2
W_1	Inconceivable	$<10^{-4}$	1

3 Layered Protection System in Hazardous Industrial Plants

In some hazardous plants a single SRCS is not sufficient to reduce the risk as required and the safety-related system has to be designed according to a concept of defence in depths (D-in-D) using several protection layers. In Fig. 2 typical protection layers in a process plant [14] are presented. Appreciated approach for the preliminary risk analysis and safety-related decision making in industrial process installations is the layer of protection analysis (LOPA) methodology [20].

The protection layers include: the basic process control system (BPCS), the alarm system (AS) and human-operator interactions, and the safety instrumented system (SIS) numbered respectively as layers: 2, 3 and 4. These protection layers should be physically and functionally independent, however, it is not always achievable in industrial practice. They are denoted in Fig. 3 respectively as:

- PL_2 – basic process control system (BPCS),
- PL_3 – team of human operators supervising the process, installation and control systems and intervening, according to prescribed principles and rules, in cases of abnormal situations or accidents, using current information from relevant interfaces, including the alarm system (AS) interface, in transients or changing operation conditions, e.g., during plant reconfiguration,
- PL_4 – safety instrumented system (SIS) performing an emergency shutdown (ESD) function when justified.

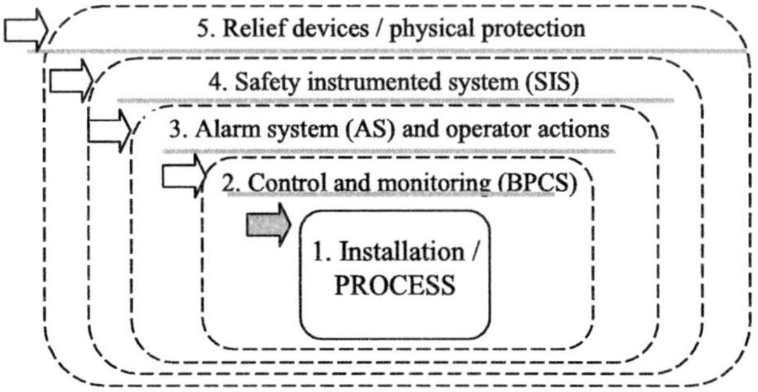

Fig. 2. Typical protection layers in hazardous industrial installation (based on [14])

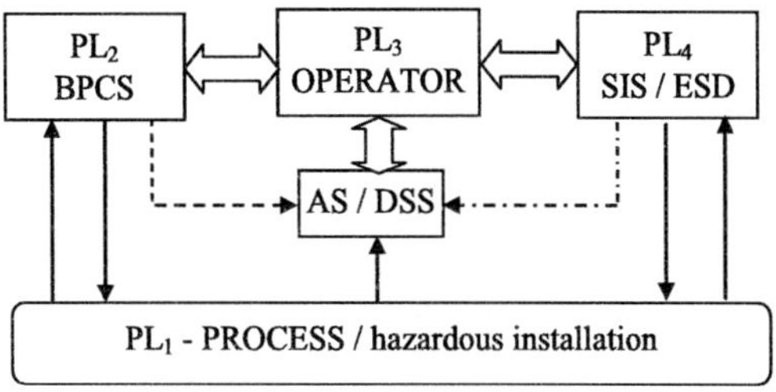

Fig. 3. Human operator and alarm system (AS) as elements of the protection layers

Achieving independency of protection layers requires appropriate technical and organizational solutions. In case of PL_2 and PL_4 it can be achieved using separate sensors and input elements, input modules for information processing in PLCs and actuators (final elements). Required SIL of BPCS and SIS for given safety function is to be achieved using appropriate architectures of their subsystems considering the probabilistic criteria for verifying the SIL to be achieved by SIS [18]. The safety function can be also implemented in BPCS, but due to its complexity the SIL achieved is usually not higher than SIL 1.

It is worth to mention that the functional safety evaluation framework including assumptions and technical specification concerning the hardware and software solutions of SRCS, AS and safety related DSS with relevant interfaces within overall HSI, provide useful insights for contextual human reliability analysis (HRA) [19].

4 Incorporating Cognitive Aspects in Human Reliability Analysis

4.1 Human Factors and Systems Cognitive Engineering

The domain of systems engineering (SE) is traditionally focused on the technological aspects of the system design, such as hardware, software, and automation, while ignoring sometimes the fact that these systems will ultimately be used, operated, and maintained by humans to meet the mission and production goals [12]. It is widely known that human factors significantly influence the operation, reliability, safety, and security of technical systems. Therefore, they should be properly incorporated in industrial practice applying, for instance, a systems engineering (SE) general concept for effective management of installations including the control systems and communication networks in entire lifetime of industrial plant [2, 22, 26].

The goal of cognitive engineering (CE), or more generally cognitive systems engineering (CSE), is to develop advanced systems, training programs, and other products that support cognitive functions in decision-making, abnormal situation management, course-of-action selection, resource allocation and other information processing tasks [2].

The CE approaches are classified regarding their focus and application purpose. Following categories of the CE research areas may be distinguished that concern [19]: (1) Analysis of system oriented human interactions, (2) Task analysis and identification of critical tasks, (3) Behavioural processes, (4) Cognitive processes, and (5) Identification of erroneous diagnosis and following actions. Each of these areas can be divided into several subareas.

An approach outlined below is focused on area (5). For the purpose of a case study an event tree (ET) method is used for defining accident scenarios including human related events for identified actions and potential errors, similarly as in an operator action event tree (OAET) [6, 9, 15]. The failure events and success events in the ET are also defined for the safety-related control system and human operator actions that are evaluated using selected methods of the human reliability analysis (HRA). Human actions and potential errors are considered in the context of safety functions implemented using relevant systems: BPCS, AS, and/or SIS. Cognitive aspects of human operator behaviour are incorporated into probabilistic model using the Rasmussen's skill-rule-knowledge (SRK) conceptual framework [21].

4.2 Human Behaviour Types

Some HRA methods are based on conceptual frameworks that include the human behaviour types and distinguished categories of human unsafe acts [23]. Rasmussen

[21] proposes the distinction of three types of human behaviour. His known framework assumes following cognitive levels of human behaviour:

- skill-based (S), highly practiced tasks that can be performed as more or less subconscious routines governed by stored patterns of behaviour,
- rule-based (R), performance of less familiar tasks in which a person follows some common-sense rules and previously developed procedures, and
- knowledge-based (K), rough analysing a system state in unknown situation when familiar patterns and rules cannot be applied directly, and actions that follow include information processing with the inclusion of diagnosis, planning and decision making directed towards reducing losses.

Figure 4 illustrates this concept that is useful in the analysis of human behaviour during abnormal situations and accidents to identify potential *human errors*.

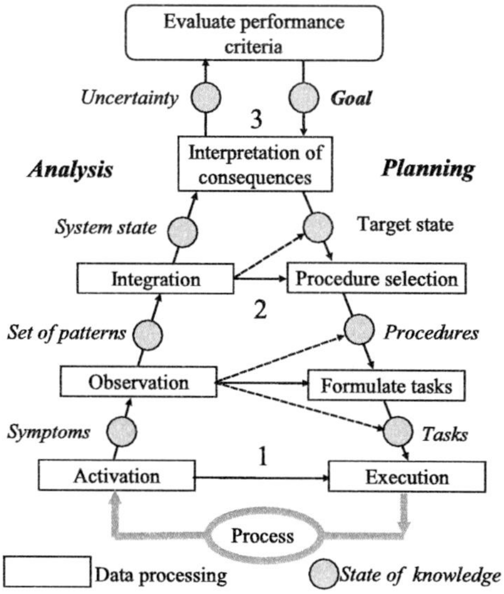

Fig. 4. Schematic representation of information processing by operators in context of human behaviour types (1 - skill, 2 - rules, 3 - knowledge)

4.3 Including Cognitive Aspects in Human Reliability Analysis

Highly appreciated HRA method developed for dealing with cognitive aspects in evaluating HEP for activity considered is HCR (human cognitive reliability), based on a model developed by Hannaman et al. [10]. The HEP is treated as the probability of an event to be assigned in the event tree developed for defining accident scenarios.

Time-dependent error probability, treated as an event of non-response in the situation considered, is calculated using the Weibull distribution from the formula [10]

$$Q^{X}(t) = \exp\left\{-\left[\frac{t/t_{0.5} - a}{c}\right]^{b}\right\} \tag{1}$$

where: *a, b, c* - are behaviour type coefficients specified below for behaviour type X (S, R, K) in given situation as explained below, and $t_{0.5}$ is the median value of time required to perform required task by a crew of human operators to be evaluated as follows

$$t_{0.5} = t_{0,5/nom} \prod_{i=1}^{3} (1 + k_i) \tag{2}$$

where:

k_1 – coefficient of operator experience: $k_1 = 0.22$ (expert, well-trained), $k_1 = 0$ (average knowledge training), or $k_1 = 0.44$ (novice, minimum training);
k_2 – coefficient of stress level: $k_2 = 0.44$ (situation of grave emergency), $k_2 = 0.28$ (situation of potential emergency), $k_2 = 0$ (normal activity, no emergency), or $k_2 = 0.28$ (low activity, low vigilance);
k_3 – coefficient for quality of operator/plant interface: $k_3 = -0.22$ (excellent), $k_3 = 0$ (good), $k_3 = 0.44$ (fair), $k_3 = 0.78$ (poor), or $k_3 = 0.92$ (extremely poor).

The behaviour-type coefficients (*a, b, c*) in formula (1) are to be evaluated as follows:

(0.7, 1.2, 0.407) for skill-based behaviour (S),
(0.6, 0.9, 0.601) for rule-based behaviour (R),
(0.5, 0.8, 0.791) for knowledge-based behaviour (K).

The results of HEP evaluations based on HCR model are presented and discussed below in context of accident scenarios of a case study. For the time window allowable $t = t_x$ and behaviour type X (S, R, K) the human error probability is to be evaluated according to (1) as follows

$$Q^{X}(t_x) = \exp\left\{-\left[\frac{t_x/t_{0.5} - a}{c}\right]^{b}\right\} = \mathrm{HEP}^{X} \tag{3}$$

If time window t_x allowable is relatively short for a dynamic process in an industrial installation during accident, then HEP^{X} is high and can take in some cases a value of close to 1. For illustrating evaluation issues of HEP using this model, three cases are considered below for behaviour types S, R, and K to show changes of HEP^{X} for some assumptions made.

Behaviour type S (skill):

(1) $\mathrm{HEP}^{S} = 0.0178$, for: $t_x = 2$ min, $t_{0.5/nom} = 1$ min, $k_1 = k_2 = k_3 = 0.0$ (normal or good cases), $t_{0.5} = 1$ min, $t_x / t_{0.5} = 2.0$,

(2) $HEP^S = 0.922$, for: $t_x = 2$ min, $t_{0.5/nom} = 1$ min, $k_1 = 0.44$, $k_2 = 0.28$, $k_3 = 0.44$, $t_{0.5} = 2.65$ min, $t_x / t_{0.5} = 0.75$.

Behaviour type R (rules):

(1) $HEP^R = 0.118$, for: $t_x = 6$ min, $t_{0.5/nom} = 3$ min, $k_1 = k_2 = k_3 = 0.0$ (normal or good cases), $t_{0.5} = 3$ min, $t_x / t_{0.5} = 2.0$,
(2) $HEP^R = 0.751$, for: $t_x = 6$ min, $t_{0.5/nom} = 3$ min, $k_1 = 0.44$, $k_2 = 0.28$, $k_3 = 0.44$, $t_{0.5} = 7.96$ min, $t_x / t_{0.5} = 0.75$.

Behaviour type K (knowledge):

(1) $HEP^K = 0.189$, for: $t_x = 10$ min, $t_{0.5/nom} = 5$ min, $k_1 = k_2 = k_3 = 0.0$ (normal or good cases), $t_{0.5} = 5$ min, $t_x / t_{0.5} = 2.0$,
(2) $HEP^K = 0.672$, for: $t_x = 10$ min, $t_{0.5/nom} = 5$ min, $k_1 = 0.44$, $k_2 = 0.28$, $k_3 = 0.44$, $t_{0.5} = 13.3$ min, $t_x / t_{0.5} = 0.75$.

The results of calculations as above illustrate the significant influence of performance shaping factors k_1, k_2, *and* k_3 and parameters t_x and $t_{0.5/nom}$ to be assigned by the HRA expert on HEP for specific situation considered. In some cases, the values of HEP are high, e.g., 0.672 or 0.922 for parameters assigned. It contrasts with the value of HEP = 0.1 suggested for assuming in the functional safety standard IEC 61511.

4.4 Human Reliability Analysis in Context of Accident Scenarios

HRA methods are used in industrial practice for assessing the contribution of potential human errors treated as failure events in the logical and probabilistic models of accident scenarios identified. In conventional, first generation HRA methods, it was assumed that applying the concept of human error is justified to estimate human error probability (HEP) based on relevant models [1, 3, 24]. Lately, it is emphasised that there is a need to develop and use for solving engineering problems a next generation of HRA methods. The objective is to incorporate in probabilistic models some cognitive aspects of human behaviour, especially those contributing significantly to erroneous diagnosis of situation and following action [10].

Obviously, potential human errors should be considered in given context, i.e., the installation complexity, dynamic of technological process, time limitation for reaction, solutions of control and protection systems, quality of the HMI/HSI, availability and quality of procedures, etc. Some categories of potential human errors and their consequences are presented in Fig. 5.

Several traditional HRA methods are appreciated for using in PSA practice, e.g., Technique for Human Error Rate Prediction (THERP) method [25], developed for the nuclear industry, applied also in various industrial sectors. Other HRA methods more often used in practice are as follows: Accident Sequence Evaluation Procedure (ASEP), Human Error Assessment and Reduction Technique (HEART), and Success Likelihood Index Method (SLIM). These conventional HRA methods are characterized and evaluated in various papers, monographs, and reports [1, 3].

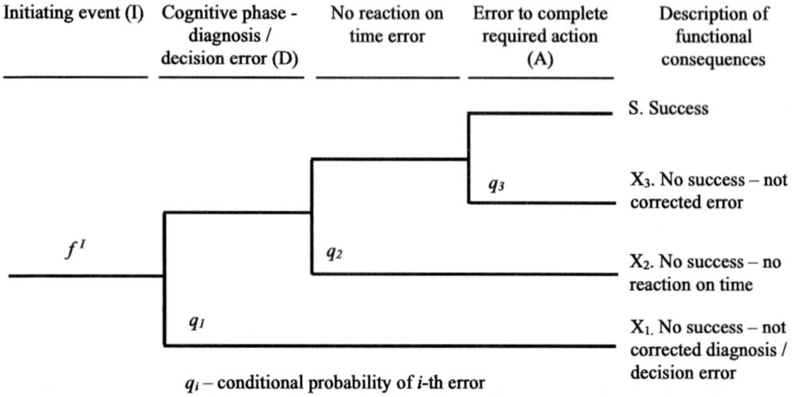

Fig. 5. Typical human-operator errors and their consequences

A considerable interest has been lately noticed to apply in industrial practice the Cognitive Reliability and Error Analysis Method (CREAM) developed by E. Hollnagel [11], and lately also simplified, but easy to apply the SPAR-H method [24]. However, the SPAR-H method is a second-generation method, not suitable for dealing with cognitive aspects of human operators' behaviour during diagnosis of abnormal or emergency situation. In a case study below the human cognitive reliability (HCR) model was applied that prove to be useful for the layered protection analysis.

5 Case Study

5.1 Defining Accident Scenarios in Layered Protection System

An industrial installation and process is considered according to the author's publication [19] denoted as variant B. It comprises of a pressurized vessel containing a mixture of gas and volatile flammable liquid with necessary instrumentation. Process control is handled using a basic process control system (BPCS) that monitors the signal from the flow transmitters and makes the flow control to operate a control valve. Additional layers of protection are also considered: an alarm system (AS) with relevant interface to human operators, a safety instrumented system (SIS) performing the emergency shutdown (ESD) of the installation, and a pressure relief valve certified for SIL 2.

Relevant event tree including the operator action (response is required when the flow control loop fails) as it is shown in Fig. 6. The results of the safety related analyses and calculations of the frequency for consecutive scenarios are presented in Table 3. The frequency of an initiating event was evaluated for a danger failure of the BPCS (SIL1), approximately at the level of $f^I = 10^{-5}\ \text{h}^{-1} = 0.0876\ \text{a}^{-1} \cong 0.1\ \text{a}^{-1}$ (per year).

In the case considered the human error probability was evaluated as equal $q_2 = \text{HEP}^\text{R} = 0.5$, i.e., for the rule-based behaviour (R) due to a greater complexity of that process installation and performing tasks according to procedures of moderate quality. Relevant probabilistic model of the HCR method for the rule-based behaviour was applied as described in the Sect. 4.3. It is worth to mention that if AS would be implemented within

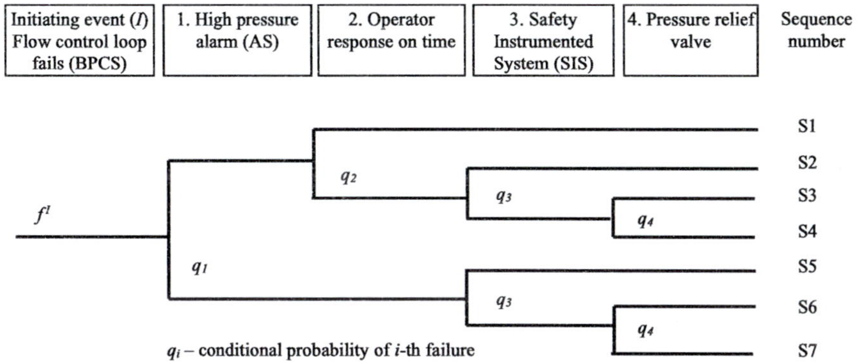

Fig. 6. An event tree for the pressure vessel with protection layers

BPCS, the danger failure of BPCS (DCS/SCADA) could cause a danger situation due to unavailability of alarms and therefore human error probability close to 1.

Table 3. Description of the event tree sequences

Sequence number	Frequency $[a^{-1}]$	Data: $q_1 = 0.1$; $q_2 = \text{HEP}^{\text{R}} = 0.5$; $q_3 = 0.01$; $q_4 = 0.01$ Consequences and remarks
S1	$0.45 \cdot 10^{-1}$	No release to the flare
S2	$0.45 \cdot 10^{-1}$	No release to the flare
S3	$0.45 \cdot 10^{-3}$	Release to the flare
S4	$0.45 \cdot 10^{-5}$	Release to the environment (a failure of the vessel)
S5	$0.99 \cdot 10^{-2}$	No release to the flare
S6	$0.99 \cdot 10^{-4}$	Release to the flare
S7	$1.00 . 10^{-6}$	Release to the environment (a failure of the vessel)

The cumulative frequency of release to the environment (sequence S4 and S7) is equal $0.55 \cdot 10^{-5}$ $[a^{-1}]$, i.e., 36 times lower than for the case A evaluated in publication [19]. If conditional probability $q_2 = \text{HEP}^{\text{X}} = 1$ (e.g., for knowledge-based behaviour X = K in complex situation of multiple failures), then the release frequency to the environment will be $1.0 \cdot 10^{-5}$ $[a^{-1}]$, i.e., approximately 2 times higher. It shows the importance of careful evaluation of HEP^{X} in probabilistic modelling of physically and functionally dependent systems. It should be emphasised that potential dependent failures in probabilistic modelling of complex architecture systems is challenging and require multidisciplinary research including current achievements of the systems engineering and human factors engineering [12, 19].

5.2 Alarm System Design Issues to Meet Functional Safety Criteria in Context of Human Reliability Analysis

Probabilistic models described above enable to effectively perform the functional safety analyses and evaluations during the safety-related control system design regarding protection layers to verify the risk criteria defined by the regulatory institutions in consultation with safety experts. It shows the importance of defence in depth (D-in-D) solutions using, e.g., BPCS, the human operators interacting with AS, and SIS for the risk reduction to meet criteria defined.

Expected reaction time of human operators cannot be too short, i.e., less than 2 - 5 min depending on the system complexity and dynamic of processes during transients, abnormalities occurring or major accidents. In such situations the human error probability HEP^X, for $X = R$ or K can be high (close to 1), especially for knowledge-based behaviour ($X = K$). In addition, if the risk reduction required is high, a separate alarm system (AS), physically and functionally independent from the BPCS, is suggested to be considered at the design stage of the layered protection system. Basic safety integrity requirements for the alarm systems are discussed in Table 4.

Table 4. Safety integrity requirements for the alarm system (based on [5])

Claimed *PFD*avg	AS safety integrity requirements	Human operators' reliability and organizational requirements
$\geq 10^{-1}$	Standard AS, may be integrated into BPCS	No special requirements, however, AS should be designed, operated, and maintained according to a good engineering practice
$[10^{-2}, 10^{-1})$	AS is designated as safety-related, SIL1, it should be independent from BPCS (unless BPCS is also designed as safety-related)	The alarm presentation arrangement should make the claimed alarm obvious to the operator. The operator must be trained for specific plant failures that the alarm system indicates. The operator should have written clear procedure to support responding correctly and on time to alarms. The required operator response should be simple, obvious, and invariant. The claimed operator performance should be audited and verified in time
$<10^{-2}$	AS designated as safety related for SIL2 or higher	It is not recommended to assume HEP below 10^{-2} for any operator action even if there is multiple alarming and task is relatively simple

6 Conclusions

The human factors engineering, and cognitive engineering are important interdisciplinary domains that offer methodological support and new possibilities to deal more systematically with issues of integrated analyses of interactions between humans and complex systems. The industrial control systems (ICS) in hazardous plants perform nowadays crucial safety functions. They are designed and operated according to the functional safety concept. The objective is to maintain high performance and productivity of industrial installations, and to reduce risks due to existing or emerging hazards and threats.

In the second edition of the functional safety standards IEC 61508 and IEC 61511 the importance of human factors and the human reliability analysis (HRA) is emphasized. These issues should be carefully considered at the design stage of industrial installations and during operation, regarding functions and architectures of the safety-related control systems (SRCSs). However, there are no clear indications in these standards how to support the functional safety analysis and management in this respect. Some characteristic HEP range values are given in IEC 61511 to be considered by HRA analysts, e.g., from 10^{-4} to 10^{-2} for trained operators in no stress conditions, and from 0.5 to 1.0 under stress. Typical HEP for the operator response to alarms was often evaluated in these standards as equal 10^{-1}.

In the article an approach is outlined how to apply the human cognitive reliability (HCR) method for evaluating HEP and verifying the design of layered protection system regarding the functional safety requirements. These issues are considered in the context of the ICS architecture and alarm system (AS) solution. As example of the initiating event the danger failure of BPCS was in particular considered. The HCR method for HEP evaluation for selected behaviour types of human operators ($X = S$, R or K) seems to be well suited for dealing with probabilistic modelling of the layered protection systems to be designed regarding the functional safety concept and relevant requirements.

An important issue that requires further research concerns the functional and probabilistic modelling of human operators interacting with the alarm system (AS) in dynamic situations when many messages and alarms can appear in relatively short time, for instance, several in the period of 10 min. The alarm system design should be suited for intelligent support of human operators during accidents. It concerns, for instance, selecting more important, possibly root messages and ranking them regarding importance for safety, but without confusion with predefined procedures (symptom or root failure oriented) to be followed by human operators during diagnosis and recovery actions. It should cover situations of multiple failures and/or errors.

These issues can not be satisfactorily solved without applying theoretical findings of system oriented cognitive engineering and human cognitive reliability. Applying the HRA cognitive method, such as HCR or CREAM, is undoubtedly a valuable step in this direction. Some new methods to be developed should be directed towards integrating theoretical findings of the systems engineering (SE), human factors engineering (HFE), cognitive engineering (CE), cognitive reliability (CR), and cyber physical systems (CPS). Thus, it requires further interdisciplinary research.

References

1. Adhikari, S., et al.: Human Reliability Analysis: A Review and Critique. Final report of the EPSRC funded project "Rethinking Human Reliability Analysis Methodologies", Manchester Business School Working Paper No 589 (2009)
2. Bonaceto, C., Burns, K.: Using Cognitive Engineering to Improve Systems Engineering. MITRE Corporation, Bedford (2005)
3. Bell, J., Holroyd, J.: Review of human reliability assessment methods, Prepared by the Health and Safety Laboratory for the Health and Safety Executive (2009)
4. Carey, M.: Proposed Framework for Addressing Human Factors in IEC 61508. A Study prepared by Amey VECTRA Ltd. for Health and Safety Executive (HSE), U.K., Research Report 373 (2001)
5. EEMUA Publication 191: Alarm Systems, A Guide to Design, Management and Procurement (Edition 2). London: The Engineering Equipment and Materials Users' Association (2007)
6. Embrey, D.: Task analysis techniques. Human Reliability Associates Ltd. (2000)
7. Froome, P., Jones, C.: Developing Advisory Software to comply with IEC 61508. Contract Research Report 419. HSE Books (2002)
8. Gersh, J.R., McKneely, J.A., Remington, R.W.: Cognitive Engineering: Understanding Human Interaction with Complex Systems, vol. 26, no. 4. John Hopkins Technical Digest (2005)
9. Gertman, I.D., Blackman, H.S.: Human Reliability and Safety Analysis Data Handbook. A Wiley-Interscience Publication, New York (1994)
10. Hannaman, G.W., Spurgin, A.J., Lukic, Y.D.: Human cognitive reliability model for PRA analysis. Report NUS-4531, EPRI Project RP2170-3 (1984)
11. Hollnagel, E.: Cognitive Reliability and Error Analysis Method. Elsevier (1998)
12. IAEA: Human Factors Engineering Aspects of Instrumentation and Control System Design. Nuclear Energy Series No. NR-T-2.12 (2021)
13. IEC 61508: Functional Safety of Electrical/Electronic/Programmable Electronic Safety-Related Systems, Parts 1–7. International Electrotechnical Commission, Geneva (2010)
14. IEC 61511: Functional safety: Safety Instrumented Systems for the Process Industry Sector. Parts 1–3. International Electrotechnical Commission, Geneva (2016)
15. Kirwan, B.: A Guide to Practical Human Reliability Assessment. CRC Press, London (1994)
16. Kosmowski, K.T.: Functional safety concept for hazardous system and new challenges. J. Loss Prev. Process Ind. **19**(1), 298–305 (2006)
17. Kosmowski, K.T.: Functional safety analysis including human factors. Int. J. Perform. Eng. **7**(1), 61–76 (2011)
18. Kosmowski, K.T.: Functional safety and reliability analysis methodology for hazardous industrial plants. Gdańsk University of Technology Publishers (2013)
19. Kosmowski, K.T.: Human factors and cognitive engineering in functional safety analysis. In: Kościelny, J.M., Syfert, M., Sztyber, A. (eds.) DPS 2017. AISC, vol. 635, pp. 434–448. Springer, Cham (2017). https://doi.org/10.1007/978-3-319-64474-5_37
20. LOPA: Layer of Protection Analysis, Simplified Process Risk Assessment. Center for Chemical Process Safety. American Institute of Chemical Engineers, New York (2001)
21. Rasmussen, J.: Skills, rules, knowledge; signals, signs and symbols and other distinctions on human performance models. IEEE Trans. Syst. Man Cybern. SMC-13/3 (1983)
22. Rasmussen, J., Svedung, I.: Proactive Risk Management in a Dynamic Society. Swedish Rescue Services Agency, Karlstad (2000)
23. Reason, J.: Human Error. Cambridge University Press (1990)
24. SPAR-H: Human Reliability Analysis Method, NUREG/CR-6883, INL/EXT-05-00509, US NRC (2005)

25. Swain, A.D., Guttmann, H.E.: Handbook of Human Reliability Analysis with Emphasis on Nuclear Power Plant Applications. NUREG/CR-1278. US Nuclear Regulatory Commission, Washington (1983)
26. Whaley, A.M., et al.: Cognitive Basis for Human Reliability Analysis. NUREG-2114, US NRC (2016)

Check for updates

Controller Modelling as a Tool for Cyber-Attacks Detection

Anna Sztyber(✉), Zuzanna Górecka, Jan Maciej Kościelny, and Michał Syfert

Warsaw University of Technology, Warsaw, Poland
anna.sztyber@pw.edu.pl

Abstract. This paper aims to develop a method for detecting intrusions into the control system, consisting of malicious modifications of the control algorithm or its parameters. Data from the control loops are routinely collected and archived. Therefore we propose a data-based diagnosis method consisting of modelling the controller using data from normal operation and calculating residual as a difference between model output and real control signal. Larger residual values indicate a change in the controller behaviour, possibly resulting from the malicious intervention. We use two modelling approaches: linear models and neural networks. The linear model additionally gives estimates of PID controller parameters. The methods are evaluated and compared on simulated and real control data. Finally, a superheaters system case study presents the method's effectiveness for detecting intrusions into the control system.

Keywords: Fault diagnosis · Cyber attack · Controller diagnosis · PID controller · Modelling · Neural networks

1 Introduction

Recently, apart from hazards related to equipment faults and human errors, there are also new threats [7,13,17], related to destructive targeted activities such as cyber-attacks and sabotage actions. Both types of risks are particularly dangerous for the critical infrastructure, such as the chemical industry, power plants, power grids, and water supply [7]. Despite various reasons, the effects of severe failures and attacks may be the same, e.g., fire, explosion, environmental contamination, destruction of the installation, and process stop.

In the FDI methods of fault detection and isolation [12], developed for over 40 years, the analysed scope of diagnosis included: process components, sensors, and actuators. Faults of these elements were a hazard to the proper functioning of the control systems. The correctness of the control algorithms in the FDI approach was not checked. On the other hand, faults of the control units implementing the control algorithms were detected independently by the diagnostic software of these units using computer systems diagnostics methods.

Z. Kowalczuk (Ed.): DPS 2022, LNNS 545, pp. 100–111, 2023.
https://doi.org/10.1007/978-3-031-16159-9_9

Regardless of the FDI methods, another direction of research was developed, which can be described as the assessment of control loop performance, initiated by the work [8]. State of the art has been described in [3,11]. Many indicators have been developed to diagnose the most common problems in the control circuits.

The following practical problem is essential: how to increase the security of industrial control systems (ICS) in the case of threats related to faults of technological equipment, sensors, and actuators, human errors, and destructive targeted actions such as cyber-attacks and sabotage actions. The current solutions are not satisfactory because they relate to specific threats. There are no holistic solutions that would comprehensively address security issues. One of the partial elements of this problem is finding effective methods of detecting intrusions into the control system, consisting of malicious modifications of the control algorithm or its parameters.

Cyber-attacks entering the control system through the protection rings may destructively distort the measurement and control signals and the control algorithm itself or its parameters. Therefore, real-time supervision of the control loops becomes necessary to detect intrusions into the control system early. The diagnostic system should detect and isolate both faults and attacks.

This work aimed to develop and test algorithms to detect cyber-attacks aimed at control algorithms. The paper shows that using attack detection methods based on controller models effectively detects malicious cyber-attacks on control algorithms. The attacks may, for example, consist in modifying the parameters of the algorithm. The change of the normal-reverse regulator operation mode leads to positive feedback. Changes in the settings may also result in the loss of system stability, and the modification of the setpoint value may result in the process entering the emergency area.

A case study was used as the research method. The possibilities of solving the formulated research problems were analysed on the example of the superheaters system.

The structure of the paper is as follows: Sect. 2 presents state of the art in cyber-attack detection in control systems, Sect. 3 introduces proposed modelling approaches and shows modelling quality for systems without intrusions. Section 4 presents case study and Sect. 5 concludes the paper.

2 Cyber-Attack Detection in Control Systems

The protection of the control system against cyber-attacks is achieved primarily by the use of demilitarised zones, firewalls, data encryption, VPN (IPsec) networks, network segmentation, identity verification, access authorisation, and password management. Cyber-attack detection is carried out by monitoring network traffic that does not allow detection of all anomalies [4]. The solutions used in IT systems do not guarantee the lack of possibility of an attack getting into the control system, thus its destructive impact on the controlled process [6]. Early detection of anomalies and their isolation is essential for an effective response to emerging hazards and threats.

The issues of cyber attack detection in industrial control systems (ICS) are developed very intensively [6,10,18]. Intrusion detection techniques against cyber-attacks are mainly divided into two categories: based on signatures and anomalies. Signature-based methods require a model of the system's functioning in the case of a given type of attack. Anomaly-based attack detection techniques detect deviations from normal system behaviour. Similar approaches are used in fault diagnosis.

In terms of the detection of cyber-attacks in ICS, the following research directions can be distinguished: detection based on network traffic exploration, the analysis of network protocols, and the analysis of process data. ICS attacks often cause unusual network traffic or violate network protocol specifications. The methods of detecting these anomalies are derived from the methods used in IT systems. The above approaches are most often used in Intrusion Detection Systems (IDS) [15].

If the attacks penetrate the control system through the security measures applied, they affect the functioning of the control systems. Therefore, they can be detected as a discrepancy between the observed and reference activity, represented by models characterising the normal state of the controlled process. The detection scheme presented in [9,20] is identical to the fault detection scheme used for a long time in FDI methods.

There are passive and active approaches to detecting cyber-attacks. Passive methods are based on operating signals, while active methods require introducing an appropriate input signal and the analysis of its influence on the output signals of control systems. The active approach was presented in [19]. Observers [16] and Kalman filters [5] were used in passive detection approaches. The use of neural models to detect anomalies was the subject of works [1,21]. For anomaly detection, quantitative and qualitative models can be used. The works [13,14] present models in the form of the rules that detect faults and attacks that cause the feedback sign to change from negative to positive or the actuator block.

The detection methods derived from the approaches known in IT networks have already reached a high level of advancement, measured by many publications and the offered IDS systems using these techniques. The solutions derived from the approaches developed in the automation community, including intrusion detection based on process data analysis and control loop supervision, are more recent and less documented in terms of publication. In the opinion of the authors of this paper, these methods that control the integrity of process data collected from measuring devices, actuators and controllers may be of great practical importance. Methods developed based on fault diagnostics and models linking process variables can be used for this task.

The introduction of diagnostic systems that recognise both faults and cyber-attacks will constitute an additional layer of functional security (Layer of Protection) and another Ring of Protection in ICS.

3 Controller Modelling

The main idea of this paper is to build a model of a controller (control signal (CV) - output, control error (e) - input) and calculate the residual as a difference between model output and real control signal. Larger residual values indicate a change in the controller behaviour, possibly resulting from the malicious intervention.

In practical applications, above 90% of controllers are PID. PID controller is linear. Thus, it can be modelled by a linear model, and its parameters can be easily identified. However, in practice, there are different structures of PID controllers, and there are possible non-linearities as an effect of saturation, anti wind-up, deadband, and progressive or regressive action, where controller parameters are modified depending on the value of control error. Some other types of controllers like MPC or fuzzy controllers are also used. Therefore there is also a need for non-linear modelling.

We propose two modelling approaches: linear and neural network. The linear model is computationally and conceptually simple, allows for identification of controller parameters, and is dedicated to linear PID controllers with standard structure. The neural network works with any controller and allows for modelling controller output without parameter identification.

3.1 Linear Model

Calculations were based on the model of the ideal PID controller in an incremental version:

$$\Delta CV(k) = k_p \left(e(k) - e(k-1) + \frac{T_s}{T_i} e(k-1) + \frac{T_d}{T_s} \left[e(k) - 2e(k-1) + e(k-2) \right] \right), \quad (1)$$

where: k - time step, $\Delta CV(k) = CV(k) - CV(k-1)$, $e(k)$ - control error, k_p - proportional gain, T_i - integral time, T_d - derivative time, T_s - sampling time.

For different controller settings (P, PI, PD, PID), different input and model structures can be used:

- P controller: input: $e(k) - e(k-1)$, model: $\Delta \hat{CV}(k) = a_1(e(k) - e(k-1))$;
- PI controller: inputs: $e(k)$, $e(k-1)$, model: $\Delta \hat{CV}(k) = a_1 e(k) + a_2 e(k-1)$;
- PD or PID controller: inputs: $e(k)$, $e(k-1)$, $e(k-2)$, model: $\Delta \hat{CV}(k) = a_1 e(k) + a_2 e(k-1) + a_3 e(k-2)$,

where $\Delta \hat{CV}(k)$ - model output, a_1, a_2, a_3 - model coefficients. The coefficients are estimated minimizing mean squared error.

In the case of lack of knowledge on controller type general PID scheme should be used.

Controller parameters can be identified based on model coefficients:

- For P controller: $k_p = a_1$,
- For PI controller: $k_p = a_1$, $T_i = \frac{T_s}{\frac{a_2}{k_p}+1}$
- For PD controller: $k_p = a_1 - a_3$, $T_d = \frac{a_3 \cdot T_s}{k_p}$
- For PID controller: $k_p = a_1 - a_3$, $T_d = \frac{a_3 \cdot T_s}{k_p}$, $T_i = \frac{T_s}{\frac{a_2}{k_p}+1+2\frac{T_d}{T_s}}$

The program processing data and creating models was written in Python using the NumPy, Pandas, Matplotlib, and scikit-learn libraries. Before modelling, data from the controller was split into train and test datasets. The linear model was created using scikit-learn library.

Numerical results from linear modelling are summarised in Table 1 and Table 2. The Tables show respectively quality of fit metrics (r^2 - coefficient of determination, $RMSE$ - root mean squared error, MAE - mean absolute error) and estimated controller parameters (k_p, T_i - real settings values, $\hat{k}_p$, $\hat{T}_i$ - values estimated based on model coefficients). Datasets 1–8 were simulated using Matlab Simulink, and in that case, controller type and parameters were known. Dataset 9 is from PID control loop performance monitoring repository[1] [2]. The data are from a hydrogen reformer unit belonging to a SE Asian refinery. The process variable is temperature. Dataset 10 is from a simulated control system with a fuzzy controller. Datasets with the suffix "sat." included controller saturation and thus non-linearity.

Table 1. Results for the linear model.

		Train			Test		
	Dataset	r^2	$RMSE$	MAE	r^2	$RMSE$	MAE
1	P	1.0000	$3 \cdot 10^{-6}$	$2 \cdot 10^{-6}$	1.0000	$3 \cdot 10^{-6}$	$2 \cdot 10^{-6}$
2	P sat.	0.5957	0.0317	0.0238	0.6428	0.0329	0.0235
3	PI	1.0000	10^{-5}	$8 \cdot 10^{-6}$	1.0000	10^{-5}	$9 \cdot 10^{-6}$
4	PI sat.	0.8792	0.0149	0.0102	0.8825	0.0162	0.0108
5	PD	1.0000	$3 \cdot 10^{-6}$	$2 \cdot 10^{-6}$	1.0000	$3 \cdot 10^{-6}$	$2 \cdot 10^{-6}$
6	PD sat.	0.3534	0.1591	0.1065	0.3246	0.1748	0.1078
7	PID	1.0000	$2 \cdot 10^{-5}$	$9 \cdot 10^{-6}$	1.0000	$2 \cdot 10^{-5}$	10^{-5}
8	PID sat.	0.8485	0.0400	0.0258	0.8259	0.0460	0.0269
9	Real PID sat.	−0.4169	0.0890	0.0699	−2.6837	0.1380	0.1109
10	Fuzzy	0.8155	0.0476	0.0315	0.8772	0.0379	0.0290

Table 2. Controller parameter estimation for the linear model.

	Dataset	k_p	$\hat{k}_p$	T_i	$\hat{T}_i$	T_d	$\hat{T}_d$
1	P	4.12	4.1200	–	–	–	–
2	P sat.	4.12	2.3113	–	–	–	–
3	PI	2.3	2.3575	10	10.2500	–	–
4	PI sat.	2.3	1.7904	10	58.7560	–	–
5	PD	2.8	2.8000	–	–	3	3.0000
6	PD sat.	2.8	1.6215	–	–	3	1.9838
7	PID	1.2	1.2600	5	5.2501	2	1.9047
8	PID sat.	1.2	−0.1917	5	−1.9678	2	−10.1332
9	Real PID sat.	–	−0.4808	–	−325.1327	–	−12.8453

[1] https://sacac.org.za/resources, file "saturation-T-oil-gas-thornhill-2002.csv".

3.2 Neural Network

The inputs for the neural network are $e(k)$, $e(k-1)$, $e(k-2)$ and $CV(k-1)$. $CV(k-1)$ values are needed for accurate saturation model. The output is $\Delta\hat{CV}$. Before modelling, data from the controller was split into train, validation and test datasets. Afterwards, the data were scaled so that all input signals were within <0, 1>. The neural network was created with library Keras and KerasTuner for hyperparameters selection. The following configurations were tested: number of hidden layers: <1, 5>; number of neurons in each layer <15, 60>; activation functions: {ReLU, leaky ReLU, tanh}, optimizers: {Adam, Nadam, RMSprop, Adadelta}; loss function: $\{MSE, MAE\}$. The selection criterion was minimal MSE on the validation set. The selected model was fitted to the training data using validation $RMSE$ as an early stopping criterion. All models had in the output layer one neuron with a linear activation function.

The selected model had the following structure: the number of hidden layers: 4, the number of neurons: 4, 50, 50, 33, 1, activation function for all layers except the output: leaky ReLU, optimizer: Adam, cost function: MSE.

Numerical results from neural model are summarised in Table 3. Datasets are the same as for Table 1.

Table 3. Results for the neural model.

		Train			Validation			Test		
	Dataset	r^2	$RMSE$	MAE	r^2	$RMSE$	MAE	r^2	$RMSE$	MAE
1	P	0.9996	0.0009	0.0003	0.9972	0.0024	0.0004	0.9992	0.0013	0.0004
2	P sat.	0.9850	0.0078	0.0048	0.9829	0.0084	0.0055	0.9800	0.0087	0.0053
3	PI	0.9989	0.0014	0.0008	0.9977	0.0023	0.0009	0.9981	0.0020	0.0009
4	PI sat.	0.9349	0.0118	0.0083	0.9281	0.0132	0.0095	0.9363	0.0122	0.0089
5	PD	0.9994	0.0015	0.0009	0.9983	0.0027	0.0010	0.9992	0.0019	0.0009
6	PD sat.	0.9985	0.0097	0.0066	0.9979	0.0112	0.0069	0.9984	0.0098	0.0064
7	PID	0.9990	0.0019	0.0012	0.9975	0.0032	0.0014	0.9986	0.0023	0.0014
8	PID sat.	0.9849	0.0134	0.0087	0.9790	0.0159	0.0105	0.9811	0.0148	0.0095
9	Real PID sat.	0.7970	0.0487	0.0325	0.8476	0.0708	0.0537	0.8615	0.0673	0.0542
10	Fuzzy	0.9467	0.0268	0.0160	0.9597	0.0174	0.0122	0.9687	0.0194	0.0130

3.3 Comparison of the Models

For data without saturation from the control system with PID controller, the linear model correctly identifies controller parameters with a maximum relative error of 5%. This model very accurately predicts the ΔCV signal value for simulated data without saturation - better than the neural model. On the other hand, for data with saturation, the prediction of the ΔCV signal value is significantly worse, and the settings are incorrectly identified - even the result may be a negative value. The model is also worse in the task of simulating the fuzzy controller. In comparison, the neural network predicts the ΔCV signal value satisfactorily for all data. However, this model does not identify the controller parameters.

4 Case Study

The developed modelling approach was tested for a superheaters system under malicious interventions.

The study used a model of superheaters of 3rd and 4th stages of steam line in the boiler of the power unit. The process schematic diagram is given in Fig. 1.

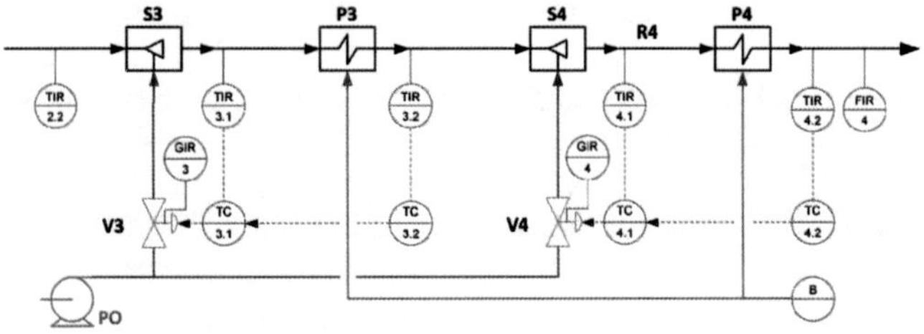

Fig. 1. A schematic process diagram with designated control loops and process variables.

The modelled part of the process consists of 3rd and 4th steam line sections. In each of them, there are: attemperator, superheater and cascade control system – the main controller controls the temperature of the steam behind the superheater, while the auxiliary controller controlling the injection water valve controls the temperature behind the cooler. The notation will be as follows: 3.1, 3.2 - respectively auxiliary and main control loops of 3rd stage, 4.1, 4.2 - respectively auxiliary and main control loops of 4th stage.

The process was modelled in a simulation environment using a simplified linearised (at an operating point) model, reflecting the fundamental relationships between physical quantities. A specific variability of input quantities was assumed in the conducted experiments (temperature and steam flow at the inlet to 3rd stage and the contractual value of the fuel flow fed to the boiler) and cycles of the variability of setpoint (SP) values of the main controllers of the stage 3rd and 4th: independent and related changes of $SP\_4.2$ and $SP\_3.2$.

The study considered three different types of cyber-attacks related to the different impacts on the control system. Each cyber-attack is conducted according to a designed scenario defining independent activities conducted by the attacker influencing particular system signals or component parameters. The particular cyber-attack scenarios are presented in Table 4.

The symbolic place of introduction of the changes done during cyber-attacks (changes of SP, CV and PID mode and parameters) is shown in the Fig. 2.

Table 4. The set of cyber-faults.

Cyber-attack	Description
$CAS_{1.B}^{CON}$	Change of the operation mode of the main controllers 3.2 and 4.2 to "manual", setting a constant values of CVs
$CAS_{2.B}^{CON}$	Change of the operation mode of the main controllers 3.1 to "reverse" (positive feedback)
$CAS_{3.A}^{CON}$	Change of the settings (k_p, T_i) of the main 3.2 and auxiliary 3.1 controllers to more aggressive

In this study, we only consider intrusions directed at the controller (marked in red in Fig. 2).

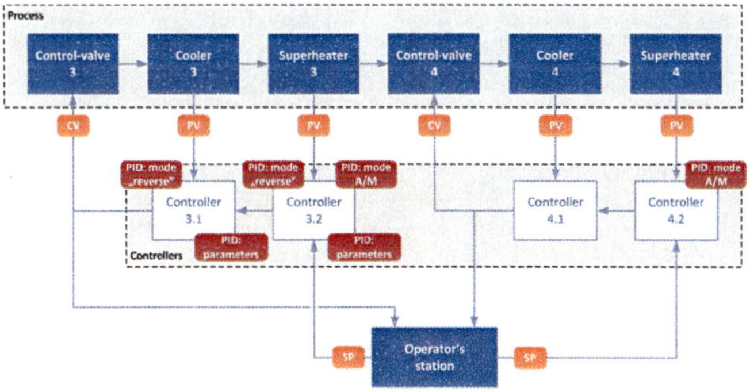

Fig. 2. Places of influence on the process during the implementation of cyberattacks.

For cyber-attack detection we calculate the residuals for both neural and linear model:

$$r_{linear} = \Delta\hat{CV}_{linear} - \Delta CV \qquad r_{nn} = \Delta\hat{CV}_{nn} - \Delta CV, \tag{2}$$

where r_{linear}, r_{nn} - residuals for linear and neural models, ΔCV - real value, $\Delta\hat{CV}_{linear}$, $\Delta\hat{CV}_{nn}$ predicted values for linear and neural model.

Additionally, a linear model is fitted in the sliding window for controller parameters estimation. Therefore for each control loop we obtain four diagnostic symptoms: $\Delta\hat{CV}_{linear}$, $\Delta\hat{CV}_{nn}$, $\hat{k}_p$, $\hat{T}_i$. Data from the normal system operation are used to train the models (both neural and linear) and to estimate the residuals values in the case without intrusion. For alarm generation the residuals are squared and averaged in sliding window of length 1000, giving indices $rmse_{linear}$ and $rmse_{nn}$. The mean value and the standard deviation of the indices are estimated using normal operation data. Index value above mean plus three standard deviations leads to alarm. The alarm signals activation and deactivation contains deadband (20% of the range of three standard deviations from the mean) to avoid flickering alarms.

Exemplary data from normal system operation were shown in Fig. 3a and Fig. 3b. These datasets were used for the training of the linear and neural models. The last 10% of data was used for testing. Comparison of real and predicted values is presented in Fig. 3c and Fig. 3d respectively for linear and neural model. According to conclusions from Sect. 3, the linear model utilizes only data without saturation (both training and prediction), whereas the neural model also works with controller saturation. Saturation is detected if $CV \geq 90\%$ or $CV \leq 10\%$. Additionally, for better prediction, the data at the time of step changes in the SP were not taken into account. Both models achieved satisfactory quality of fit.

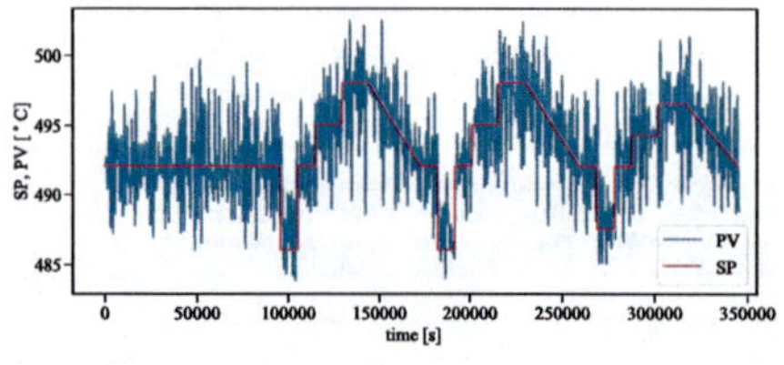

(a) PV and SP, loop 3.2

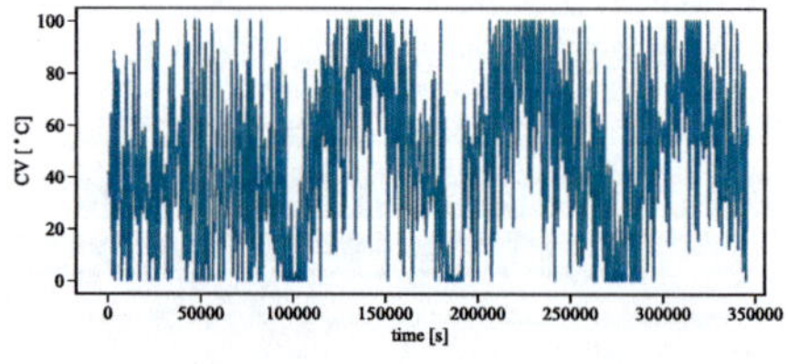

(b) Controller output CV, loop 3.2

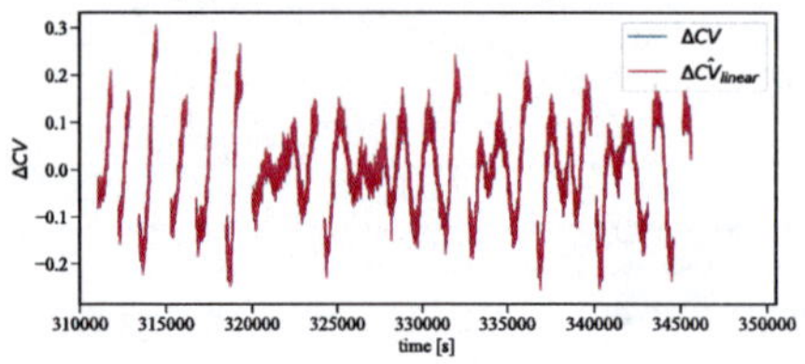

(c) Linear model fit, loop 3.2, test data.

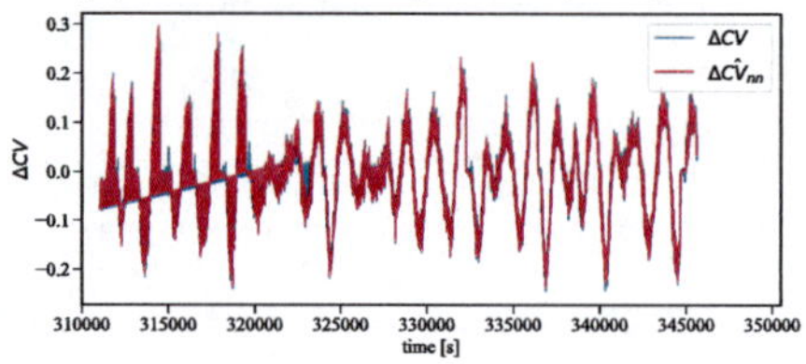

(d) Neural model fit, loop 3.2, test data.

Fig. 3. Normal operation data and modelling results.

Indices values and estimated controller parameters for considered scenarios are shown in Fig. 4. The plots contain: the values of indices $rmse_{linear}$ and $rmse_{nn}$, and estimated controller parameters $\hat{k}_p$, $\hat{T}_i$. The vertical red line marks the beginning of the attack. The red rectangles show alarm activation. Due to the lack of space only selected affected loops were presented.

Scenario $CAS_{1.B}^{CON}$ (Fig. 4a) contains the change of the operation mode of the controller to "manual" and setting $CV = 100\%$. In this scenario the linear model is switched of because of the saturation. The neural model correctly indicates the change in controller behaviour.

Scenario $CAS_{2.B}^{CON}$ (Fig. 4b) contains the change of controller operation mode to "reverse" in loop 3.1. It can be observed that the change of the sign of k_p is correctly identified. The linear model is switched off after the initial period because of the saturation, but the neural model indicates the change in controller action.

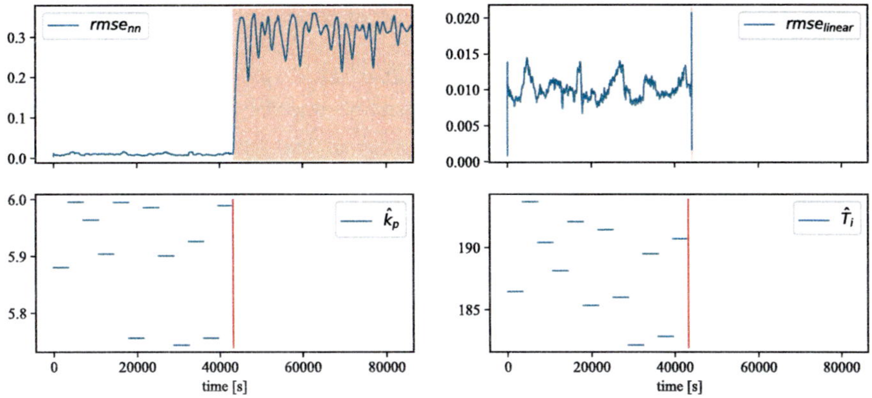

(a) Scenario $CAS_{1.B}^{CON}$, loop 3.2

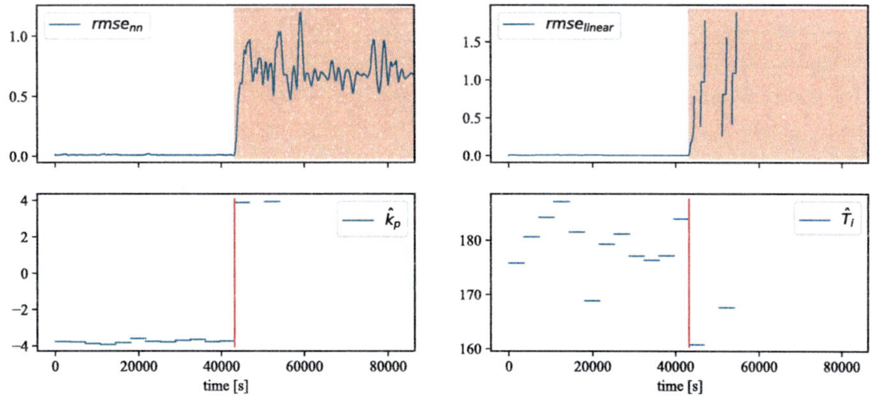

(b) Scenario $CAS_{2.B}^{CON}$, loop 3.1

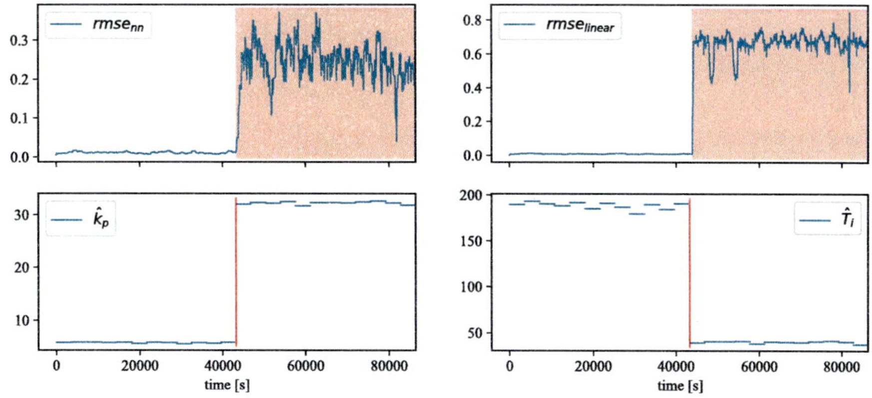

(c) Scenario $CAS_{3.A}^{CON}$, loop 3.2

Fig. 4. Example scenarios.

Scenario $CAS_{3.B}^{CON}$ (Fig. 4c) contains the change of controller settings in loop 3.2: k_p from 6 to 30 and T_i from 190 to 38. The estimated values of the parameters are close to the real ones and the change after attack is clearly visible. Square roots of residuals for both models increase their values in the moment of attack leading to alarms.

All scenarios were successfully detected and the controller parameter estimation helps identify the kind of the change in the controller.

5 Conclusions

In this paper, we proposed controller modelling as a tool for cyber-attack detection. We considered two modelling approaches: linear and neural network. The linear model is appropriate for PID controllers without saturation and allows for identifying controller settings. The neural model can be applied to a broader range of controllers. Both modelling approaches showed good estimation quality on various datasets representing normal conditions.

We tested residuals generated by the models on cyber-attack scenarios for superheaters system. All considered scenarios were effectively detected, and the linear model can identify the change in the controller settings. Therefore we conclude that controller modelling is a promising tool for cyber-attack detection and is a valuable step toward more secure process control.

In this paper, we considered only malicious intrusions directed at the controller. Further works will contain an analysis of loop performance indices and process models for widening the range of attack detection and isolation and more advanced methods of diagnostic reasoning.

Acknowledgements. The work was partially supported from the POB Research Centre for Artificial Intelligence and Robotics of the Warsaw University of Technology in Poland within the Excellence Initiative Program—Research University (ID-UB). And partially financed from the project "Advanced early warning system for damage, cyber attacks, sabotage and operator errors" granted by agreement POIR.01.01.01-00-0541/20 funded by European Funds.

References

1. Abbaspour, A., Sargolzaei, A., Yen, K.: Detection of false data injection attack on load frequency control in distributed power systems. In: 2017 North American Power Symposium (NAPS), pp. 1–6 (2017)
2. Bauer, M., Auret, L., le Roux, D., Aharonson, V.: An industrial PID data repository for Control Loop Performance Monitoring (CPM). IFAC-PapersOnLine **51**(4), 823–828 (2018)
3. Bauer, M., Horch, A., Xie, L., Jelali, M., Thornhill, N.: The current state of control loop performance monitoring - A survey of application in industry. J. Process Control **38**, 1–10 (2016)
4. Cardenas, A., Amin, S., Sinopoli, B., Giani, A., Perrig, A., Sastry, S.: Challenges for securing cyber physical systems. In: Workshop on Future Directions in Cyber-physical Systems Security (2009)

5. Cong, T., Tan, R., Ottewill, J.R., Thornhill, N.F., Baranowski, J.: Anomaly detection and mode identification in multimode processes using the field Kalman filter. IEEE Trans. Control Syst. Technol. **29**(5), 2192–2205 (2021)
6. Ding, D., Han, Q.L., Xiang, Y., Ge, X., Zhang, X.M.: A survey on security control and attack detection for industrial cyber-physical systems. Neurocomputing **275**(C), 1674-1683 (2018)
7. Genge, B., Kiss, I., Haller, P.: A system dynamics approach for assessing the impact of cyber attacks on critical infrastructures. Int. J. Crit. Infrastruct. Prot. **10**(C), 3-17 (2015)
8. Harris, T.J.: Assessment of control loop performance. Can. J. Chem. Eng. **67**, 856–861 (1989)
9. Hu, Y., Li, H., Yang, H., Sun, Y., Sun, L., Wang, Z.: Detecting stealthy attacks against industrial control systems based on residual skewness analysis. EURASIP J. Wirel. Commun. Netw. **2019**(1), 1–14 (2019). https://doi.org/10.1186/s13638-019-1389-1
10. Jang-Jaccard, J., Nepal, S.: A survey of emerging threats in cybersecurity. J. Comput. Syst. Sci. **80**(5), 973–993 (2014)
11. Jelali, M.: Control Performance Management in Industrial Automation: Assessment Diagnosis and Improvement of Control Loop. Springer, London (2013). https://doi.org/10.1007/978-1-4471-4546-2
12. Korbicz, J., Kowalczuk, Z., Kościelny, J.M., Cholewa, W. (eds.): Fault Diagnosis: Models, Artificial Intelligence Methods, Applications. Springer, Heidelberg (1998). https://doi.org/10.1007/978-3-642-18615-8
13. Kościelny, J.: Towards a unified approach to detection of faults and cyber-attacks in industrial installations. In: European Control Conference, pp. 1839–1844 (2021)
14. Kościelny, J.M., Syfert, M., Wnuk, P.: The idea of on-line diagnostics as a method of cyberattack recognition. In: Kościelny, J.M., Syfert, M., Sztyber, A. (eds.) DPS 2017. AISC, vol. 635, pp. 449–457. Springer, Cham (2018). https://doi.org/10.1007/978-3-319-64474-5_38
15. Mitchell, R., Chen, I.R.: A survey of intrusion detection techniques for cyber-physical systems. ACM Comput. Surv. **46**(4), 1–29 (2014)
16. Pasqualetti, F., Dörfler, F., Bullo, F.: Attack detection and identification in cyber-physical systems. IEEE Trans. Autom. Control **58**(11), 2715–2729 (2013)
17. Rubio, J.E., Roman, R., Lopez, J.: Analysis of cybersecurity threats in industry 4.0: the case of intrusion detection. In: D'Agostino, G., Scala, A. (eds.) CRITIS 2017. LNCS, vol. 10707, pp. 119–130. Springer, Cham (2018). https://doi.org/10.1007/978-3-319-99843-5_11
18. Sánchez, H.S., Rotondo, D., Escobet, T., Puig, V., Quevedo, J.: Bibliographical review on cyber attacks from a control oriented perspective. Annu Rev. Control **48**, 103–128 (2019)
19. Trapiello, C., Puig, V.: Replay attack detection using a zonotopic KF and LQ approach. In: 2020 IEEE International Conference on Systems, Man, and Cybernetics (SMC), pp. 3117–3122 (2020)
20. Urbina, D.I., et al.: Limiting the impact of stealthy attacks on industrial control systems. In: Proceedings of the 2016 ACM SIGSAC Conference on Computer and Communications Security, pp. 1092-1105 (2016)
21. Wu, Z., Albalawi, F., Zhang, J., Zhang, Z., Durand, H., Christofides, P.D.: Detecting and handling cyber-attacks in model predictive control of chemical processes. Mathematics **6**(10), 173 (2018)

Comparison of Traditional and Elliptic Curves Digital Signatures Providing the Same Security Level

Maria Baczyńska-Wilkowska(✉)

The Automatics and Robotics Institute, Warsaw University of Technology, Warsaw, Poland
maria.wilkowska@pw.edu.pl
https://iair.mchtr.pw.edu.pl/

Abstract. The paper presents the efficiency comparison of different digital signature schemes taking into account both digital signatures based on algebraic groups of numbers and digital signatures based on elliptic curves. This efficiency comparison of two types of digital signatures is possible under the condition that for each scheme the appropriate key length was taken ensuring the same level of security. The paper show that the key length for classic digital signature needs to be greater than the key length for elliptic curve digital signature and thus different processing speed is observed.

Keywords: Digital signature · Elliptic curves · Cryptography · Security

1 Introduction

The digital signatures based on Elliptic Curve Cryptography gain popularity and now has several applications in industrial cryptographic systems. The most attractive property of ECC is that so far all known attacks on this system have complexity much larger than in the case of classical signature algorithms such as RSA, DSA or ElGamal and thus enable to use shorter key lengths to achieve the same security level [2].

This paper presents selected ElGamal-like digital signature schemes based on elliptic curves and evaluates their efficiency with comparison to classical ElGamal digital signature on cyclic groups.

The general idea of a digital signature is quite simple. First a message is digested with a hash function. Then it is signed with the use of the private key.

Similarly, a verification procedure is performed. The message digest, public key and a signature are input for the verification. An output is a confirmation of the signature or rejection if the verification gives negative result [6].

2 Digital Signature Algorithms

One of the unsolved problems on computer science deal with integer factorization in polynomial time. The decomposition of a composed integer into some

Z. Kowalczuk (Ed.): DPS 2022, LNNS 545, pp. 112–123, 2023.
https://doi.org/10.1007/978-3-031-16159-9_10

smaller divisors seems to be quite simple. However when the composed number is very large it become hard to factor it into some non-trivial factors as no efficient factorization algorithm is known. Moreover, when this very large number is semiprime (in other words is the product of two primes) and composed of two large primes of about the same size, it is not efficient enough to decompose it. If the effective algorithm will be found, the RSA-based public-key cryptography will be not secure any more.

In addition to integer factorization problem, there are other problems difficult or impossible to solve in reasonable time. One of the problems used for cryptography purposes is the discrete logarithm problem. It is a basis of El-Gamal and Nyberg-Rueppel digital signatures security [7].

All these algorithms work on basic algebraic structures like groups. The elliptic curves digital signatures also use groups but those groups are composed of points coordinates on elliptic curves [5]. Thus in order to comply with the typical notation for the first two sections a notation where the group operation is denoted as multiplicity will be used and for the last section the group operation is denoted as addition.

2.1 Schemes Based on Discrete Logarithm Complexity

ElGamal. The ElGamal problem is stated for an arbitrary finite cyclic group G with generator α. It is based on Diffie-Hellman problem in G [3]. It is supposed that there is an efficiently computable bijection $g : \mathcal{M} = G \rightarrow \{0, \ldots, |G| - 1\}$, where $\mathcal{M} = G$ is a message space.
Key generation is performed according to the following procedure:

1. Choose a cyclic group G and a generator α of G,
2. Choose randomly a random integer k, k is a private key and α^k is a public key.

Signing a message m ($m \in \mathcal{M}$) is performed in the following steps [3]:

1. Calculate $r = \alpha^{k'}$, where k' is randomly chosen integer coprime to $|G|$.
2. Solve the congruence

$$g(m) \equiv kg(r) + k's \pmod{|G|}. \tag{1}$$

   Assuming that k' and $|G|$ are comprime, the congruence has a unique solution $s \in \{0, \ldots |G| - 1\}$.
3. Signature of the message m is the pair $(r, s) \in G \times \mathbb{Z}_{|G|}$.

Verification of the signature (r, s) for the message m is performed as follows:

1. Using available data: public key α^k as well as signature components (r, s) and the message m, calculate two values $\alpha^{g(m)}$ and $(\alpha^k)^{g(r)} r^s$.
2. If the two values agree accept signature, otherwise reject it.

As it can be easily seen for the correct signature we have:

$$(\alpha^k)^{g(r)} r^s = \alpha^{kg(r)+k's} = \alpha^{g(m)} \tag{2}$$

which shows the correctness of the scheme.

Important security factor is the selection of distinct k' for different messages [3]. In addition most often instead of working on original messages (which are long), the message is partitioned into blocks $m_1, m_2, \ldots, m_n$ and a hash function

$$h : \mathcal{M}^{(\mathbb{N})} \to \mathcal{H} \tag{3}$$

is used ($\mathcal{M}^{(\mathbb{N})}$ is the set of finite sequences with entries in $\mathcal{M}$, $\mathcal{H}$ is a finite set and a hash function can take form $h(m_1, m_2, \ldots, m_n)$). As it can be seen the message we drop here our initial assumption that m must be an element of the group used in signing.

Nyberg-Rueppel. The Nyberg-Rueppel signature scheme was designed to overcome shortcomings of the discreet logarithm based signatures (ElGamal and DSA), and provide a possibility not only to verify the signed message using the public key, but also to extract the original message from the very signature (which is actually possible using RSA) [7].

As in the ElGamal case it is supposed that there is an efficiently computable bijection $g : \mathcal{M} = G \to \{0, \ldots, |G| - 1\}$.
Key generation is performed according to the following procedure:

1. Choose a cyclic group G and a generator α of G,
2. Choose randomly a random integer k, so that $0 < k < |G|$, k is a private key and α^k is a public key.

Signing a message m ($m \in \mathcal{M}$ and m is an integer $0 \leq m < |G|$) is performed in the following steps [4]:

1. Choose a random integer k' with $0 < k' < |G|$.
2. Compute and integer $r \equiv g(\alpha^{k'}) + m \pmod{|G|}$, if $r = 0$ go back to step 1.
3. Compute another integer $s \equiv k' - kr \pmod{|G|}$.
4. The pair (r, s) is the generated signature.

Verification of the signature (r, s) for the message m is performed as follows [4]:

1. If either r or s is not in the interval $\{1, \ldots, |G| - 1\}$, reject signature.
2. Compute a value $P = \alpha^s (\alpha^k)^r$ (remember that α^k is the public key). If $P = \alpha^0$ then reject signature.
3. Compute $\tilde{m} \equiv r - g(P) \pmod{|G|}$. If $\tilde{m} = m$ accept signature.

It is easy to check that for a correctly generated signature the output is actually the signed message. See that

$$P = \alpha^s (\alpha^k)^r = \alpha^{k'-kr} \alpha^{kr} = \alpha^{k'} \tag{4}$$

and therefore

$$\tilde{m} \equiv r - g(P) \equiv g(\alpha^{k'}) + m - g(\alpha^{k'}) \equiv m \pmod{|G|}. \quad (5)$$

The original version of Nyberg-Rueppel as given [8] (actually specified for the finite field $\mathbb{F}_p$) differs from the IEEE specification in details of signature generation and verification schemes.

3 Elliptic Curve Specific Signature Schemes

3.1 ElGamal Digital Signature Based on Elliptic Curves

Elliptic Curve Domain Parameters over F_p. Every elliptic curve over F_p for further cryptographic considerations needs a specification of domain parameters. The set of domain parameters comprises from eight values $(p, s, a, b, x_G, y_G, n, h)$ which are described below:

- p - a field size which defines the underlying finite field F_p, it is prime and bigger than 3
- s - a bit string seed of length at least 160 bits, presence of this value is optional
- a, b - two integer elements of F_p which define the elliptic curve $E : y^2 = x^3 + ax + b$
- x_G, y_G - two integer elements of F_p which define point $G = (x_G, y_G)$ of prime order on E, $G \neq \infty$
- n - the order of point G, $n > 2^160$, $n > 4\sqrt{p}$ (ANSI X9.62)
- h - the cofactor $h = \#E(F_p)$, this cofactor is not used in ECDSA but is described in ANSI X9.62 to keep the compatibility with ANSI X9.63

All that domain parameters can be verified with the use of the following algorithm mentioned in ANSI X9.62:

1. Verify that p is odd prime number.
2. Verify that $a, b, x_G, y_G \in [0, p-1]$.
3. If the elliptic curve was randomly generated by appropriate elliptic generator, verify that seed value s is a bit string of at least 160 bits and a and b were appropriately derived from it.
4. Verify that $(4a^3 + 27b^2) \equiv 0(\mod p)$.
5. Verify that $y_G^2 = x_G^3 + ax_G + b(\mod p)$.
6. Verify that n is prime, $n > 2^160$, $n > 4\sqrt{p}$.
7. Verify that $nG = \infty$.
8. Verify that $h = \lfloor(\sqrt{p} + 1)^2/n\rfloor$.
9. Verify that (non-) MOV and Anomalous conditions hold.

If any of the above verifications is false, then all set of domain parameters is invalid. If all the verifications are true, then the set of domain parameters is valid.

Choice of Key Length. All cryptographic operations on elliptic curves starts from the choice of the elliptic curve E and some point G on that curve called the base point. The point G order r is a large prime. The number of points on the fixed curve E is given from the equation $n = f \cdot r$, where f is the integer cofactor not divisible by the order r. Usually f is takes as small as possible as it makes calculations more efficient. All recommended curves have cofactor f equal 1, 2 or 4. Thus the length of private and public key is similar.

Choice of Curves. There are two groups of elliptic curves:

- pseudo-random curves,
- special curves.

Pseudo-random curves are generated with the use of seeded cryptographic hash. Basically, their coefficients are produced from hash output. It is very easy to verify it a curve was generated by this method if its coefficients are provided together with seed value. In case of special curves, no cryptographic hash is needed. The coefficients of those curves are selected to maximize the efficiency of calculations.

Choice of Base Points. Base point is a fixed point on the chosen elliptic curve. Its coordinates are used for cryptographic operations. Any point on the curve whose order is equal r can be chosen as a base point. In case of recommended elliptic curves, a sample base point is usually specified.

NIST Recomended Elliptic Curves over Prime Fields. For any prime p it is possible to define a curve

$$E : y^2 \equiv x^3 + ax + b \pmod{p} \tag{6}$$

of prime order r. Let take $a = -3$ for efficiency reason as it was suggested by IEEE P1363 standard. Then we have w pseudo-random curve:

$$E : y^2 \equiv x^3 - 3x + b \pmod{p}. \tag{7}$$

NIST suggests five elliptic curves over prime field which are P-192, P-224, P-256, P-384 and P-512.

4 Security Level of Signature Schemes

Considering efficiency comparison of standard signature schemes and schemes based on elliptic curves it is crucial to define the basic conditions under which the comparison is performed. One of the most important aspects of the signature scheme is the level of security that this scheme can provide. Thus it is necessary to consider the algorithms for solving the discrete logarithm problem and their complexity. In case of elliptic curves signature schemes the complexity of discrete logarithm problem depends on the size of field element n, where

$$n = \lceil \log_2 q \rceil. \tag{8}$$

Complexity of discrete logarithm problem in F_p^* group is subexponential:

$$N = \lceil \log_2 p \rceil. \tag{9}$$

In other words the complexity of DLP algorithms for elliptic curves is proportional to

$$C_{EC}(n) = 2^{n/2} \tag{10}$$

while the complexity of DLP algorithms for standard signature schemes is proportional to

$$C_{ST}(N) = \exp(c_0 N^{1/3} (\log(N \log 2))^{2/3}) \tag{11}$$

where $c_0 = (64/9)^{1/3} \approx 1,92$ as described with details in [1]. Comparing C_{EC} and C_{ST} we obtain the relation between n and N when the same security level is assured:

$$n = \beta N^{1/3} (\log(N \log 2))^{2/3} \tag{12}$$

where

$$\beta = 2c_0/(\log 2)^{2/3} \approx 4,91. \tag{13}$$

The above relation was used to distinguish security levels. For further considerations and for experiments purposes five different elliptic curves suggested by NIST were taken into account: $P-192$, $P-224$, $P-256$, $P-384$, $P-521$. Then keys lengths for signatures based on elliptic curves are as follows: 192, 224, 256, 384 and 521. Then for each of them a key length for classical signature scheme was calculated according to the formula above. In this way it was possible to define five security levels as presented in the Table 1.

Table 1. Security levels and appropriate key lengths

Security level	Elliptic curve	Key length for elliptic curve signature	Key length for classical signature
I	$P-192$	192	1294
II	$P-224$	224	1853
III	$P-256$	256	2538
IV	$P-384$	384	6708
V	$P-521$	521	14144

This relation between n (key length for signature schemes based on elliptic curves) and N (key length for classical signature schemes) can be easily shown in Fig. 1.

For RSA purposes usually 1024 and 4096 bits keys are used. They are equivalent to 173 and 313 bits keys respectively in elliptic curves cryptosystems. This shows that difference in keys length for the same security level is significant and increases exponentially.

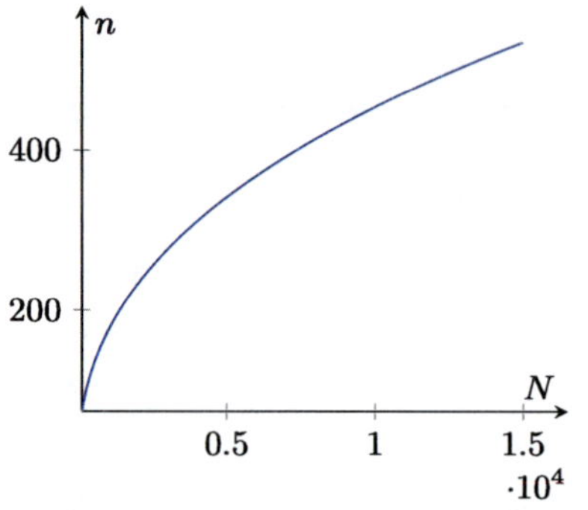

Fig. 1. Comparison of key lengths for elliptic curve signatures (n) and standard signatures (N) providing corresponding security level

5 Experimental Comparison of Signature Schemes

5.1 Experiment Setup

Experiments was done in such a way to experiment with three different signature schemes, two of them were based on elliptic curves:

- classical ElGamal,
- ElGamal over elliptic curves,
- Nyberg Rueppel over elliptic curves.

For both kinds of algorithms (classical and these over elliptic curves) tests were done in different way. At first experiments with ElGamal over elliptic curves and Nyberg Rueppel over elliptic curves were done. Tests on signatures based on elliptic curves were done on curves suggested by the standard: $P-192$, $P-224$, $P-256$, $P-384$, $P-521$. For each of these elliptic curves, a random key for standard ElGamal should be taken with the length that assures similar level of security. In this way for the project purposes five levels of security were defined and briefly described in the table below. Results are shown in the Table 1.

In case of experiments with classical ElGamal signature only first three levels of security were tested. It is due to the complexity of calculations and amount of time it was taking. For the purposes of these tests a small Java application was created that was generating keys of a given length. It was possible to generate keys of 1294, 1853 or 2538 bits of length.

In case of experiments with signature schemes based on elliptic curves all five discussed security levels were tested. It was done with the use of five elliptic curves with keys of 192, 224, 256, 384 and 521 bits of length. For each key length keys were generated five times with the generator created as a part of the developed digital signature tool. Then, having five different keys of a given length, for each of them both digital signature over elliptic curves (ElGamal and Nyberg Rueppel) were generated one thousand times and the number of cycles necessary for these operations were counted. On this basis the average number of cycles necessary to perform one signature generation was calculated. It might be important to notice that having one key of a given length generated over some elliptic curve, signature generation was performed on both ElGamal and Nyberg Rueppel schemes.

5.2 Results

First, the tests were performed on elliptic curves signatures. For each suggested curve random private and public key were generated. For this pair of keys the signature process was repeated 1000 times and the time of calculations was measured. The whole procedure was repeated 5 times for each elliptic curve. Obtained results for elliptic curve P-192 are presented in the Table 2. Then the procedure was repeated for other elliptic curves suggested by American National Institute of Standards and Technology (NIST): P-224, P-256, P-384 and P-521. Finally for each of the curves the average time in seconds necessary to perform 1000 signatures over that elliptic curve was calculated. These results are presented in Table 4.

Table 2. Time in seconds necessary to perform 1000 digital signatures over $P - 192$ curve

Test no	No of repetitions	Time in seconds to perform ElGamal on elliptic curves	Time in seconds to perform Nyberg Rueppel on elliptic curves
1	1000	9.590	9.521
2	1000	9.601	9.524
3	1000	9.620	9.521
4	1000	9.662	9.549
5	1000	9.671	9.510

In comparison some tests on standard ElGamal digital signature scheme were performed. To keep the same security level the appropriate key lenght had to be used. Thus to comply with elliptic curve tests, for each equivalent key lengths the keys were generated five times and then for every key signature was calculated 1000 times. The obtained results are given in the Table 3.

5.3 Results Analysis

Knowing the results presented in Sect. 5.2 it was possible to calculate the average time necessary for a single signature. First on the basis of data from Table 2 to Table 3 the average time necessary to perform 1000 signatures over elliptic curves was calculated. Results are presented in Table 4.

Comparing ElGamal signature over elliptic curves and Nyberg Rueppel signature over elliptic curves it is visible that time necessary to perform signing

Table 3. Time in seconds necessary to perform 1000 classical ElGamal digital signature

Test no	No of repetitions	Time in seconds to perform ElGamal signature with key lenght equal to 1292	Time in seconds to perform ElGamal signature with key lenght equal to 1853	Time in seconds to perform ElGamal signature with key lenght equal to 2538
1	1000	25.299	62.685	156.103
2	1000	25.103	61.644	156.228
3	1000	25.085	61.687	156.277
4	1000	25.414	61.766	156.477
5	1000	25.082	62.428	156.223

Table 4. Average time in seconds necessary to perform 1000 signatures over elliptic curves

Key length	ElGamal on elliptic curves	Nyberg Rueppel on elliptic curves
192	9.6288	9.5250
224	13.8560	13.7930
256	17.9214	17.8008
384	51.0686	50.8808
521	89.6622	89.3654

process is similar in both cases, however it usually takes a little longer to make Nyberg Rueppel signature than ElGamal signature.

In a similar way it is possible to calculate average time of a single signature for classical Elgamal scheme. At first on the basis of Table 3 the average time of 1000 signatures for different key length was calculated. Results are presented in Table 5 (Fig. 2).

Table 5. Time in seconds necessary to perform 1000 classical ElGamal signatures of a given key length

Key length	Time to perform 1000 classical ElGamal signatures
1292	25.1966
1853	62.0420
2538	156.2616

As it was mentioned in previous sections, the tests with longer key length were not performed due to long time necessary to generate key with appropriate length. For better results comparison the average times of all considered signature schemes are presented together in the Table 6. Security levels and appropriate key lengths were presented in the Table 1. The comparison of results is also presented in a form of a diagram in Fig. 3.

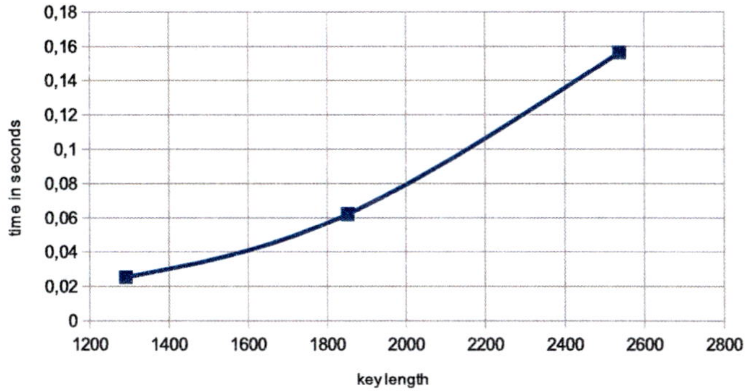

Fig. 2. Time in seconds necessary to perform classical ElGamal signature (with interpolation)

Table 6. Average time in seconds necessary to perform signature with given security level

Security level	Time for ElGamal signature	Time for ElGamal signature over elliptic curve	Time for Nyberg Rueppel signature over elliptic curve
I	0.0251966	0.0096288	0.0095250
II	0.0620420	0.0138560	0.0137930
III	0.1562616	0.0179214	0.0178008
IV	–	0.0510686	0.0508808
V	–	0.0896622	0.0893654

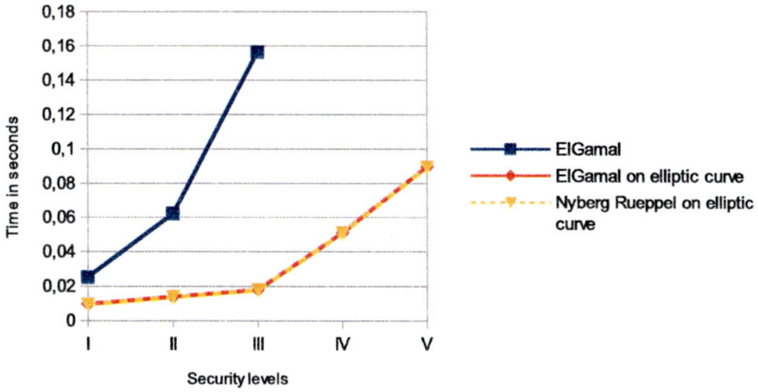

Fig. 3. Time in seconds necessary to perform different digital signatures (with interpolation)

The difference between ElGamal on elliptic curves and Nyberg Rueppel on elliptic curves was also investigated. The difference between the time needed to perform them in seconds is very small and do not present the difference in a relevant way. Thus the difference in percentages is much better as it shows the relative difference between them (see Table 7).

Table 7. Differences in ElGamal on elliptic curves and Nyberg Rueppel on elliptic curves in seconds and in percentages of the time necessary for ElGamal signature on elliptic curves

Key length	ElGamal signature time	Nyberg Rueppel signature time	Difference in seconds	Difference in %
192	0.0096288	0.0095250	0.0001038	1.0780
224	0.0138560	0.0137930	0.0000630	0.4547
256	0.0179214	0.0178008	0.0001206	0.6729
384	0.0510686	0.0508808	0.0001878	0.3677
521	0.0896622	0.0893654	0.0002968	0.3310

Similarly the investigation was done on the difference between ElGamal and ElGamal on elliptic curves. The difference in time necessary to perform both signatures was very big and thus significant. Thus instead of percentage difference the proportion of time necessary for these signatures was calculated (Table 8).

Table 8. ElGamal signatures

Security level	EC ElGamal [s]	ElGamal [s]	Proportion of time of ElGamal/EC ElGamal
I	0.0096288	0.0251966	2.6168
II	0.0138560	0.0620420	4.4776
III	0.0179214	0.1562616	8.7193

6 Conclusions

In this paper three signature schemes were investigated and implemented and their mathematical background was analyzed. The choice of schemes was not accidental, discussed signatures were chosen on purpose. All three schemes were first analyzed theoretically in terms of the security they provide and appropriate key length that should be taken into considerations. Then they were implemented in similar way to avoid optimization that might influence the comparison of time necessary to perform each signature. Finally the experiments were done with keys parameters suggested by NIST and an analysis of the gathered results was done.

As it was shown in the previous chapter in Table 7 there is no significant difference between time necessary to perform elliptic curves ElGamal digital signature and elliptic curves Nyberg Rueppel digital signature. It takes a little bit longer to sign a message with ElGamal signature but the difference is neglectable and even decreases with the key length.

The ElGamal signature on elliptic curves was also compared with classical ElGamal on groups. As mentioned above there is not significant difference in time necessary for ElGamal on elliptic curves and for Nyberg Rueppel on elliptic curves. Thus for further comparison only two ElGamal signature schemes were taken. The analysis was done on keys with different length but appropriate to provide the same security level.

It was observed that the time necessary to perform classical ElGamal signature was longer than time necessary to perform ElGamal signature on elliptic curves. As stated previously the time was measured and compared for the same security level. It has been shown that the time spend on classical ElGamal signature is significantly longer and increases nearly exponentially when the key length decreases according to assumed security levels. It turns out that more complex elliptic curve arithmetic does not deny efficiency gains resulting from application of shorter keys. It is worth to recall here that only the first three security levels were taken into account as it was impossible to generate long keys for the classic signatures in reasonable time on the machine used for the tests.

The use of signature schemes based on elliptic curves allow for keys shortening maintaining the same security level, which influence requirements for memory and processors. We may conclude that elliptic curve cryptography is a valuable tool especially for the systems that require very high performance or offer very limited computing environment (i.e. microprocessor cards, systems for industry sector).

References

1. Blake, I., Seroussi, G., Smart, N.: Krzywe eliptyczne w kryptografii (Elliptic Curves in Cryptography). Wydawnictwa Naukowo-Techniczne (2004)
2. Cohen, H., Frey, G., et al.: Handbook of Elliptic and Hyperelliptic Curve Cryptography. Chapman & Hall/CRC, London (2006)
3. Enge, A.: Elliptic Curves and Their Applications to Cryptography: An Introduction. Kluwer Academic Publishers, Norwell (1999)
4. IEEE: IEEE standard specifications for public-key cryptography. IEEE STD 1363-2000 p. i (2000). https://doi.org/10.1109/IEEESTD.2000.92292
5. Koblitz, N.: Algebraiczne aspekty kryptografii. Wydawnictwa Naukowo-Techniczne (2000)
6. Menezes, A.J., van Oorschot, P.C., Vanstone, S.A.: Handbook of Applied Cryptography (2018)
7. Nyberg, K., Rueppel, R.A.: A new signature scheme based on the DSA giving message recovery. In: Proceedings of the 1st ACM Conference on Computer and Communications Security, CCS 1993, pp. 58–61. ACM, New York (1993). https://doi.org/10.1145/168588.168595
8. Nyberg, K., Rueppel, R.A.: Message recovery for signature schemes based on the discrete logarithm problem. Des. Codes Cryptogr. **7**(1–2), 61–81 (1996). https://doi.org/10.1007/BF00125076

Fundamental Concepts of Modeling Computer Security in Cyberphysical Systems

Janusz Zalewski[1,2(✉)]

[1] Department of Software Engineering, Florida Gulf Coast University, Ft. Myers, FL 33965, USA
janusz.zalewski@puzim.edu.pl, zalewski@fgcu.edu

[2] Ignacy Mościcki State Professional College, 9 Narutowicza Street, 06-400 Ciechanów, Poland

Abstract. The ability to detect cyberattacks in industrial installations depends heavily on in-advance learning about potential threats and vulnerabilities, which is best done through extensive modeling. Three general types of modeling approaches exist, which are based on three pillars of science: theory, experiments, and simulation. The paper reviews the author's take to integrate all three views.

Using the theoretical approach, the author with coworkers previously applied the Non-Functional Requirements (NFR) method to security analysis of SCADA installations. The objective of the current work is to complement and enhance it with the use of simulation and practical experiments. With respect to simulation, building models with the Monterey Phoenix tool has been applied to an IEEE standard related to SCADA security. Experimental approaches to cybersecurity rely on applying penetration testing, with tools such as Nmap or Shodan that can be useful in studying security vulnerabilities. Here, we advocate a comprehensive approach, where software tools, such as those mentioned above, could complement theoretical analysis. Work is reported on building an NFR model for SCADA security for the laboratory example with three kinds of devices (valves, flowmeters and sensors), in terms of architectural properties of the SCADA system. A practical NFR model with the use of both Monterey Phoenix tool and respective penetration experiments has been developed.

Keywords: Computer security · Security modeling · Cyberphysical systems · SCADA security

1 Introduction

As anyone can tell, computer security is of overwhelming importance in a contemporary society. Although practically every computing profession is involved in making our computers secure, it is the particular job of software engineers to design defenses against security violations. To do this effectively and develop appropriate methods of computer security, we need to have a model of what to protect and against what kind of threats. These models can vary based on specifics of the application domain, but one sufficiently general domain of cyberphysical systems can be used as a template for many others.

Z. Kowalczuk (Ed.): DPS 2022, LNNS 545, pp. 124–138, 2023.
https://doi.org/10.1007/978-3-031-16159-9_11

A respective model has been in existence for decades, since at least 1991 [1], although it is rarely used or even referred to. It is derived from an international standard known as ITU-T Recommendation X.800 "Security Architecture for OSI". It relies on a realization that security vulnerabilities are exploited by attacks, making an attack a fundamental concept for security considerations.

Consequently, a security attack is a central notion of this model, which leads to an assumption that before any security related activities are initiated, a clear understanding of an *attack surface* is necessary. With this knowledge, the second concept is introduced: *security services.* A security service can be described in brief [1, 2] as "a service [...] that ensures adequate security of the systems or of data transfers". Examples of such services include: confidentiality, integrity, availability, non-repudiation, authentication, authorization, etc., and form in fact respective security requirements.

Furthermore, as stated in the Recommendation X.800 and follow-up publications, the third concept is essential in the OSI Security Architecture, that is, *protection mechanisms* that "implement those services." A definition of this concept in [2], although actually applied to the term "security mechanism", reads as follows: "A process (or a device incorporating such a process) that is designed to detect, prevent, or recover from a security attack." One illustrative example of such a protection mechanism is encryption, as it addresses providing confidentiality, integrity and authentication services.

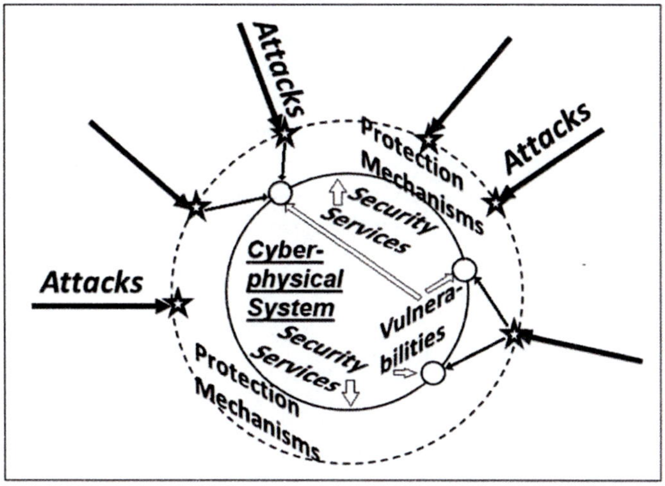

Fig. 1. The OSI security architecture model for a cyberphysical system.

To summarize, the adopted OSI security architecture model for a cyberphysical system (represented by an inner circle in Fig. 1) involves three essential elements:

1. Identifying the attacks, which has been reflected in the concept of an *attack surface*, that is, a collection of entry points to exploit potential vulnerabilities (shown as little circles on the edges of a cyberphysical system).

2. Defining *security services*, addressed here by focusing on counteracting the selected components of the attack surface.
3. Developing *protection mechanisms* to provide security services for detection and prevention of security attacks, illustrated in the diagram as a layer separating the attacks from the cyberphysical system.

From this perspective, the objective of this paper is to review the problems of modeling protection mechanisms viewed from three angles: theory, experimentation and simulation, as explained in the following sections. The rest of the paper is structured as follows. Section 2 discusses the attack surface of a cyberphysical system, Sect. 3 presents the modeling approach, Sect. 4 presents a laboratory case study, and Sect. 5 concludes the paper.

2 Identifying the Attack Surface

2.1 Basic Terminology

The job of an engineer in providing computer security is to design the security measures. These are the protection mechanisms as per the Recommendation X.800 discussed in the previous section. Knowing what to protect against and where the attacks will be coming from is essential to the effectiveness of the designs. This section takes a closer look at the identification of the attack surface of cyberphysical systems.

The model is based on a generic pattern of the architecture of a cyberphysical system as outlined, for example, in [3]. It includes interactions not only with the external Process (e.g. the Plant and its related devices), but also interfaces to a Human Operator (HMI), local or remote database storing information, as well as takes care of network connectivity. Internal hardware and software can be also included in security considerations, but in considered out of scope in the current research (see Fig. 2).

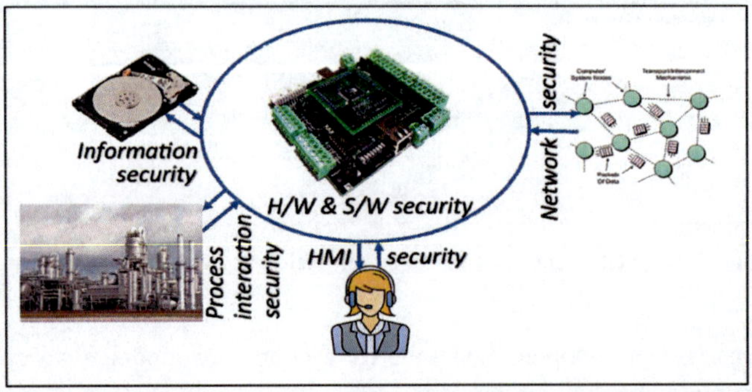

Fig. 2. Interactions of a cyberphysical system with external environment.

Consequently, a security attack is a central notion of this model, which leads to the necessity of identification of potential entry points that adversaries could use to violate security. It is through these entry points that the interactions shown in Fig. 2 can take the form of attacks. An abstraction of a pictorial diagram from Fig. 2 is shown in Fig. 3, where the four types of interactions between the controller and its environment constitute formally an attack surface, marked on the diagram as red dotted line.

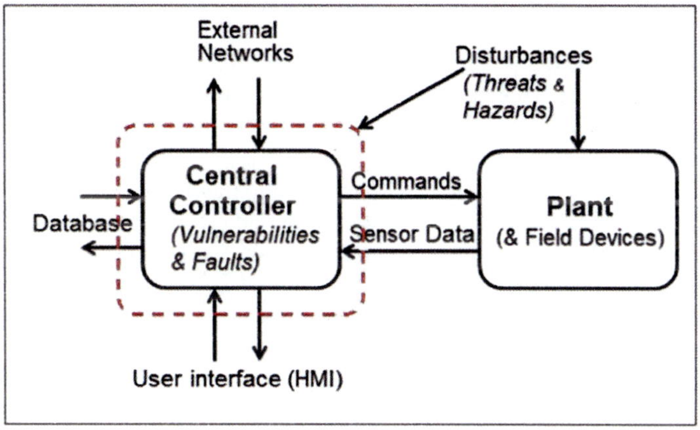

Fig. 3. Abstraction of an architecture of a cyberphysical system including all related interfaces.

An important consideration reflected on this diagram is the existence of disturbances, as another abstraction. Typically, in control engineering, this is the reason why we need a controller to apply feedback control to adjust plant parameters due to their unwanted and unpredictable changes caused by external disturbances.

Thus, in control terms, the attack surface is subjected to disturbances, which can take the form of intentional or unintentional violations. The former are called *threats* and are important to security, and the latter are called *hazards* and are important from the safety perspective. Internally, the controller can fail to perform its required functions from either the security or the safety perspective only if it has some deficiencies, which are traditionally called *vulnerabilities* in the security community and *faults* in the safety community. Acknowledging the duality of the terminology, in this paper we are interested only in computer security.

2.2 Defining Security Services

With the attack surface established and sources of threats being identified as leading to attacks, the question arises: what specific services are to be provided to meet the security requirements that would protect the cyberphysical system against these threats and attacks?

It is essential to realize that the disturbances are represented by threats from external entities (attackers) trying to exploit vulnerabilities in the system. This view is shown in a model illustrated in Fig. 3, where all the concepts are lumped together: attack surface, shown by a red line indicating the interfaces vulnerable to attacks, threats as external disturbances, and vulnerabilities that can be exploited by attackers.

This abstraction can be now used for security modeling, because:

- It uniquely identifies the attack surface.
- It shows a clear distinction between threats and vulnerabilities.
- Most importantly, it defines the assets to be considered in providing security services to be used in modeling.

Examples of the assets suitable for modeling are different for each interface from Fig. 3 and depend on the application domain. As an example, for SCADA security they may include:

- for the Plant interface: SCADA field communication protocols
- for the External Networks interface: a web server and its internal facilities
- for the HMI interface: procedures for authentication and authorization
- for the Database interface: stored information and its security properties, such as confidentiality, integrity and availability, and
- hardware and software of the Controller itself.

In the following, we apply these concepts in practical modeling of SCADA security.

3 Modeling Approach

3.1 An Overview

Taking into consideration the OSI security architecture and its consequences for cyber-physical systems, as outlined in the previous sectio4ns, the approach to modeling protection mechanisms is based on the three-pillar method in making discoveries, a view widely accepted in research, as stated – for example – by Glimm and Sharp [4]: "It is an old saw that science has three pillars: theory, experiment, and simulation." Thus, ideally, a theoretical model of security vulnerabilities could be used for security assessment by employing software tools for both practical experiments and simulation. With this assumption, two kinds of tools were taken into consideration to fit into the theory based on the NFR approach [5]:

- Monterey Phoenix [6] for simulation modeling, and
- Shodan [7] for conducting pentesting experiments.

In previous work [5, 8], we have successfully used the NFR approach to determine safety and security levels of cyberphysical system architectures, but with certain limitations. In particular, the methodology was used for simple, small scale, applications, and its usefulness was not compared with other modeling techniques. These deficiencies are addressed in the current section by using a realistic SCADA configuration as a case study and incorporating two other modeling techniques by blending them with the NFR approach.

In the sequel, we outline briefly the NFR method, then present the simulation of SCADA behavior with the use of a formal tool Monterey Phoenix (MP) [6] and finally show the results of using a Shodan penetration tool [7] for obtaining additional insight into security.

3.2 The NFR Approach

This section introduces briefly the Non-Functional Requirements (NFR) approach to behavioral modeling of security. The entire approach has been extensively explained in previous publications [5, 8], so here only a short mention of the method is given. It relies on considering system properties, such as security or safety, and objectives (that is, reaching certain levels of these properties) as softgoals that can be satisfied not necessarily in the absolute sense but only within a certain range (that is, approximately). The ability to satisfy within a range is referred to as *satisficing*, borrowing the term from economics, first used by Herbert Simon in 1956 [9]. It means "finding a choice mechanism that will lead it to pursue a 'satisficing' path, a path that will permit satisfaction at some specified level...".

Technically, the NFR relies on building a special graph named the Softgoal Interdependency Graph (SIG) for capturing the relationships between different components of the system. In a typical SIG, shown in Fig. 4, NFR softgoal hierarchy appears at the top, operationalizing softgoal hierarchy appears at the bottom, while contribution hierarchy appears in-between.

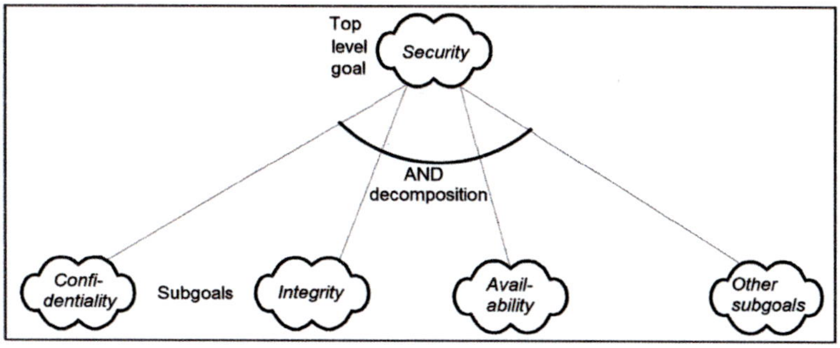

Fig. 4. The principle of building NFR SIG models.

As an illustrative example, Fig. 4 shows the template for decomposing security property into its constituents, chosen – for simplicity – to be the C-I-A triad, extendable both horizontally, by adding other security constituents, and vertically, by relating (mapping) the constituents to the actual components of the architecture. The AND clause is an example how the child constituents can be related to the parent across the entire SIG diagram. There is also an OR decomposition, which means any of the low-level constituents can contribute to the upper level. It would be marked by a double arc stretching across the connections.

3.3 Simulation Modeling with Monterey Phoenix

Monterey Phoenix (MP) [6] is a software tool developed to address modeling of system behavior based on a formal description of its architecture. It has been designed to reflect the interactions between the system and its environment. Its objective is to uncover the unintended behaviors to manifest defects in the design. As such, it seems suitable for modeling security vulnerabilities, as long as they can be reflected as behavioral features of the system under consideration.

Formally, a behavior in MP is represented textually by a concept of a schema, which can be viewed as an equivalent of a computer program. Within a schema, the behavior is modeled by the concept of events that are organized in a collection of rules reflecting actions of system components, external actors and their interactions. The execution of rules, which is done by MP tool based on the schema, produces event traces, which are the subject of analysis for desirable and/or undesirable behavior of the system being modeled.

There are three essential steps to do the complete MP modeling:

1) Develop a graphical model of system behavior. This step is optional and not included in the MP framework, but it is helpful to start with the representation of a system that is familiar to the modeler.
2) Convert the graphical model into to a schema. Knowledge of MP syntax is necessary at this stage. In case the modeler becomes fluent with MP, modeling can begin at this stage.
3) Executing the schema to produce event traces and do their manual analysis.

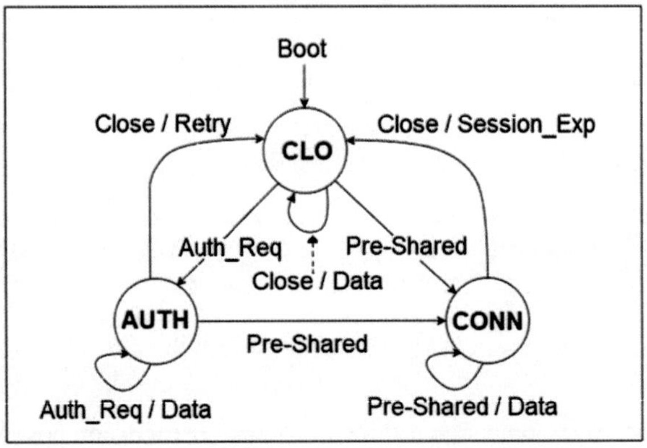

Fig. 5. Simplified IEEE Std 1711.2 Slave state transition diagram adopted for modeling.

For the purpose of this project, the functionality of the Plant interface has been considered with the SCADA example from the IEEE Std 1711.2 "Secure SCADA Communications Protocol (SSCP)" [10]. A simplified Slave state transition diagram has been adopted for modeling, with the three states represented in Fig. 5:

- CLO representing the Closed state in SSCP. In this state, the device does not have an active communication session with the remote device.
- AUTH representing the Authentication state in SSCP. A slave has received an AUTHENTICATION_CHALLANGE (see description below) but a key exchange has not been initiated so the slave should continue to try to authenticate itself.
- CONN representing the Connected state in SSCP. The device has successfully performed a key exchange with a communicating device and all session data have been established. The two communicating devices can freely exchange authenticated data.

A similar list of state transitions had to be produced for all the signals in Fig. 5.

The schema is shown in Fig. 6 and includes three major sections, each describing the behavior of one of the nodes (states) from Fig. 5, CLO, AUTH and CONN, with respective transitions, according to the restrictions outlined above.

```
/*  SSCP_Slave_V0.mp */

SCHEMA SSCP_Slave

/*=====================================================================*/
CLO_to_CONN: a;
CLO_to_AUTH: (a | b);
CLO_to_CLO: b;

ROOT CLO_behavior:  Start CLO
                    (* (CLO_to_CONN | CLO_to_AUTH) AUTH_to_CLO CLO *)
                    (* (CLO_to_AUTH | CLO_to_CONN) CONN_to_CLO CLO *)
                    (CLO_to_CONN | CLO_to_AUTH) CLO End ;

/*=====================================================================*/
AUTH_to_CLO: b;
AUTH_to_CONN: a;
AUTH_to_AUTH: b;

ROOT AUTH_behavior:  (* AUTH_to_CLO AUTH_to_CONN *) AUTH ;

CLO_behavior, AUTH_behavior SHARE ALL AUTH_to_CLO;

/*=====================================================================*/
CONN_to_CONN: a;
CONN_to_CLO: b;

ROOT CONN_behavior: (* (CLO_to_CONN | AUTH_to_CONN) CONN
                    (* (CONN_to_CLO | CONN_to_CONN ) CONN *) *) ;

CONN_behavior, CLO_behavior SHARE ALL CLO_to_CONN;
CONN_behavior, AUTH_behavior SHARE ALL AUTH_to_CONN;
CONN_behavior, CLO_behavior SHARE ALL CONN_to_CLO;

/*=====================================================================*/
SAY("The path: ");
/* Sort state visits and transition events according to
the BEFORE relation and then show them */
COORDINATE    <SORT> $a: (CLO | AUTH | CONN | a | b | Start | End)
       DO SAY($a); OD;
```

Fig. 6. Schema corresponding to the State Transition Diagram from Fig. 5.

This schema executes currently within the MP tool, producing 22 traces for the simplest scope. Executing it for more comprehensive scope causes a state explosion problem, as it would generate orders of magnitude more traces and may become unmanageable for manual analysis. A sample trace involving transitions between all three states is shown in Fig. 7.

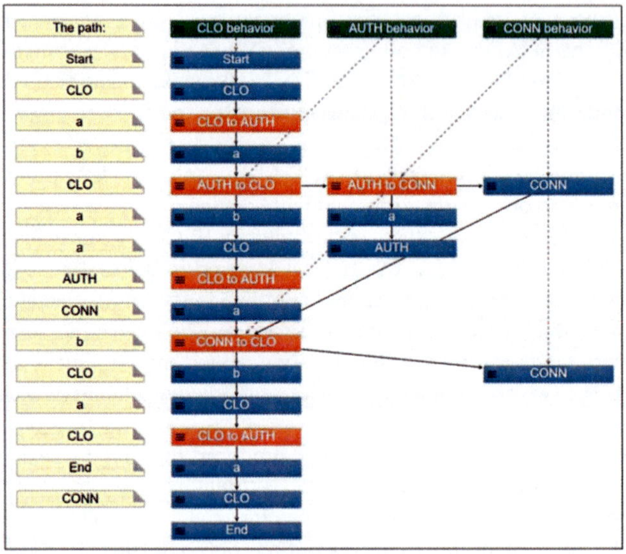

Fig. 7. Sample Trace for the Schema of Fig. 6 showing authentication in the Connected state.

3.4 Penetration Testing with Shodan Internet Search Engine

A brief analysis of the available pentesting tools that would be suitable for use in this project led to the conclusion that the adoption of a Shodan search engine is the most desirable for two essential reasons:

- First, the use of Shodan is non-invasive from the perspective of a user. It conducts searches and pentesting on its own and stores the results in a vast database, which then can be mined by a user. This is an important advantage, since the user is not required to do invasive testing by themselves.
- Second, Shodan produces data in different file formats, to the extent that they can be made directly accessible to other tools, which can conduct automatic data analysis. This feature is also very advantageous, because it facilitates writing scripts that would do the required data analytics offline.

A closer look at the capabilities of Shodan reveals its suitability to study vulnerabilities of various kinds of computer networks and their related network services. In particular, Industrial Control Systems have been studied, since Shodan first appeared and researchers and security engineers realized its capabilities. As early as in 2011, Leverett [11] investigated the use of Shodan to study the attack surface of industrial systems to prevent remote attacks on selected devices and identify networks for further reconnaissance and exploitation.

Querying Shodan for generic information on the hosts located in the author's own town, Estero, Florida, returned results for one of the hosts data shown partially in a report in Fig. 8. The report includes some basic administrative information on the host's name, owner, location, etc., but also lists all identified ports used by this host for communication. In this case, the ports listed are: 22 for SSH protocol, 5353 for multicast DNS, and 20000 for DNP3, which is a SCADA specific protocol.

What is extremely important, however, is that Shodan automatically searched the known databases for potential vulnerabilities in the context of specific ports discovered and current versions of the protocol software used. In this specific case, as shown in Fig. 8, seven vulnerabilities in the OpenSSH Version 7.1 and higher have been reported based on information from the Common Vulnerabilities and Exposures database: https://cve.mitre.org/. It is amazing that for the other host from a different provider located in Estero, which used jQuery and PHP technologies, nearly fifty CVE vulnerabilities have been reported. These experiments with Shodan confirmed its suitability for use in more comprehensive security modeling.

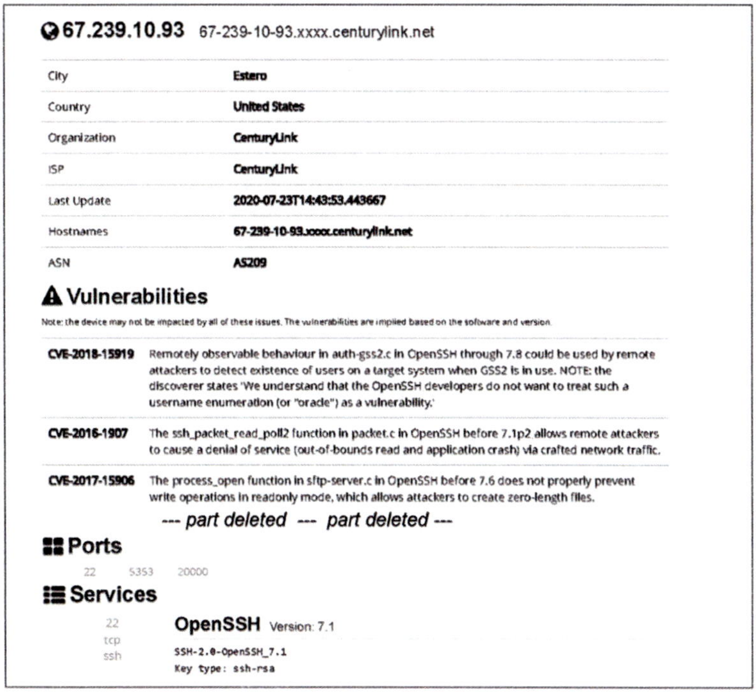

Fig. 8. Abbreviated report returned by Shodan for one of the investigated hosts.

4 Integrating the Simulation and Pentesting into the NFR

4.1 Laboratory SCADA Equipment

A small laboratory SCADA system, such as the one illustrated in Fig. 9 has been used for the NFR model building with incorporating data from MP tool and Shodan. It contains interfaces to all the typical devices listed in Fig. 3 for the architecture of a generic cyberphysical system. The mapping of respective components is as follows:

- Master plays the role of a Controller and connects to the Plant's sensors and actuators through the RTU
- Historian is a typical SCADA database
- Server connects via the Network to several potential Masters, and
- Human Operator with direct access to the Master (not shown).

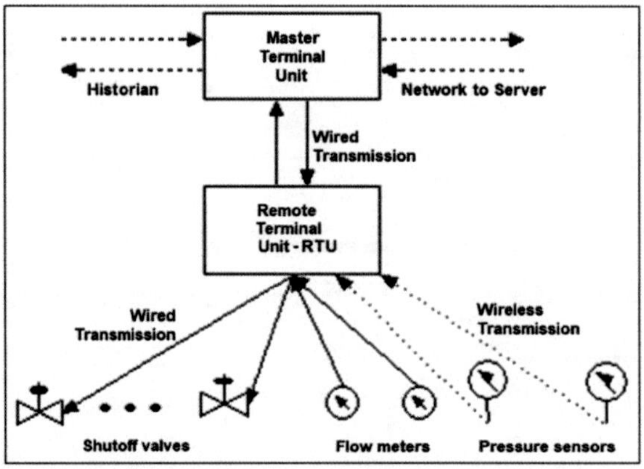

Fig. 9. The laboratory SCADA architecture used in the project.

With this architecture and the corresponding attack surface interfacing the Master to the external components, the following security services are to be addressed to model respective protection mechanisms:

- one of the SCADA security protocols to cover transmissions to the RTU;
- data confidentiality, integrity and availability for the Historian interface;
- port vulnerabilities to communicate with the Server, and
- authentication and authorization services for the Human Machine Interface.

The next subsection outlines the integration of modeling related mechanisms through the use of MP tools and Shodan within the context of the NFR.

4.2 Integration of the Simulation and Pentesting with the NFR

The ultimate goal of deriving different security models for SCADA, one based on simulation with Monterey Phoenix and another based on experimentation with Shodan is to merge them with the NFR approach to improve the reasoning about security. This subsection presents a respective diagram for the SCADA case study.

In simplest terms, the top level softgoal of SCADA security is achieved when meeting the subgoals of confidentiality, integrity and availability, possibly others. Each of these subgoals can be achieved by applying respective protection mechanisms, named here countermeasures, addressing vulnerabilities and mitigating potential attacks. Respective countermeasures may involve the most obvious, such as encryption and access control, complemented by others, which is illustrated at level 3.

Potential negative contributions from the participating devices are shown at the lowest level, with the intermediate inclusion of Masters and RTU's between the two lowest levels, which is not shown here for simplicity. It is assumed that not all devices, or their respective Masters and RTU's, would contribute vulnerabilities to each subgoal, so they would not require corresponding countermeasure. For example, measuring devices (meters, sensors, etc.) would not be subject to violations of availability, which is illustrated by the lack of respective contributions in Fig. 10.

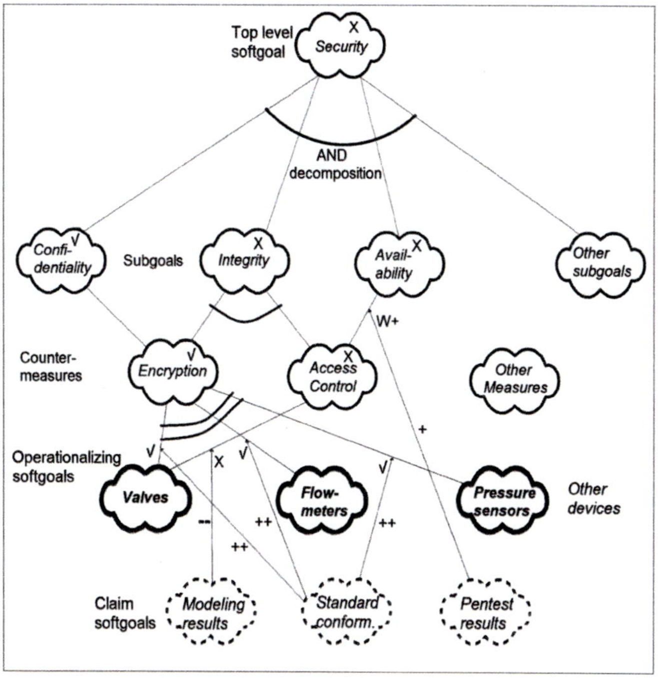

Fig. 10. Building the NFR SIG model for a laboratory SCADA example.

The next step in the NFR approach is to include the claim softgoals, marked with dotted clouds in Fig. 10. Claim softgoals are softgoals capturing the design decisions. For example, conformance to a certain professional standard can be a claim softgoal, which is reflected in the middle claim softgoal on the diagram.

Names of claim softgoals in clouds from Fig. 10 are kept generic to show the applicability of the concept to a wide array of design decisions. For example, "Standard conformance" can be viewed as compliance with any of the underlying standards, in particular with IEEE 1711.2 [10] or ISA 62443-4-2 [12] or both. Likewise, "Pentest results" may concern the operation of any pentesting tool(s), including Shodan, but also Wireshark, Nmap, etc. The cloud labeled "Modeling results" can be related to the use of simulation tools, an example of which is Monterey Phoenix used in this project, but there are several more that are applicable.

Sample reasoning about the SCADA security using the NFR approach for the SCADA architecture outlined in Fig. 9, with respective SIG shown in Fig. 10, can be done as follows. Going bottom-up, claim softgoals are considered first. For "Standard conformance" softgoal, since sensor data, as transmitted from both the flowmeters and pressure sensors, follow the respective standards for encryption, both offer positive contributions to the Encryption softgoal, which is reflected by "++" and Satisficed label in Fig. 10. So is the contribution from the valves, since valve commands are also encrypted as per the standard. Thus, the Encryption softgoal is fully satisficed with all three positive contributions, even though it only requires one such, as per the OR decomposition marked with double arc. This is marked by a "V" sign in the Encryption softgoal cloud.

However, the "Modeling results" claim softgoal, represented by running Monterey Phoenix simulations and contributing to the Access Control softgoal, is making a negative contribution marked by a double negative sign "—", because the simulation detected a flaw in the authentication process, which is a part of Access Control. This has a very negative impact on the entire process of providing security, because the Access Control sofgoal makes a negative contribution to the Integrity softgoal, as well as to the Availability softgoal. In case of Integrity, it does not matter that the Encryption softgoal offers a positive contribution, because it is combined via AND decomposition with Access Control, thus making the result negative, as reflected in Fig. 10.

Additionally, one has to consider the "Pentesting results" claim softgoal, which in this project is represented by the results of the Shodan penetration tests. It contributes to the Availability softgoal and it is a weakly positive contribution, which in NFR is marked as W+. The reason for a positive weak contribution from pentesting is that, although Shodan scan did not discover any vulnerabilities, it may have not scanned all involved domains, so chances are that there may be some vulnerabilities discovered in the future, which would have a negative effect on Availability. In case of this study, however, this weak contribution does not actually matter, because of the negative contribution for the Availability softgoal from Access Control, as discussed above.

All this discussion is illustrated in Fig. 10, which shows that the Security, being the top-level softgoal, is not met due to the requirement of having an AND decomposition, which needs positive contributions from all constituent softgoals, that is, Confidentiality, which is positive, and Integrity and Availability, which are both negative. In summary, the example illustrates how security modeling and pentesting can assist in overall determination of security with the NFR approach, thus enhancing the method and extending its applicability.

5 Conclusion

This project concerned a study on security modeling of cyberphysical systems. The adopted model assumed identifying three essential components: the attack surface, security services and protection mechanisms. For such model, the objective was to integrate the theoretical approach, realized here through the Non-Functional Requirements method, with simulation and experimentation, implemented with Monterey Phoenix and Shodan tools, respectively.

A model built for simulation followed the IEEE Std 1711.2-2019, for which a state transition diagram for the SCADA slave has been adopted for use with the Monterey Phoenix, and the traces have been collected for a limited number of transition events. For pentesting experiments, the Shodan search engine has been applied. Results from both models, generated by Monterey Phoenix and Shodan, have been incorporated next into a broader NFR model for SCADA security for the laboratory example with three kinds of devices (valves, flowmeters and sensors), in terms of architectural properties of the SCADA system. The resulting NFR's Softgoal Interdependency Graph showed that it is beneficial to reveal both weaknesses and strengths of SCADA system security, and adding both types of modeling enhances the reasoning process as well as contributes to better understanding and enriching the protection mechanisms.

Further work should include some alternative views for comparison, for example, the ISA-95 architectural model, viewing the human interface as a Human System Interface (HSI), etc.

Acknowledgments. Part of this work has been done during the author's fellowship at the U.S. Air Force Academy. Professor Mikhail Auguston of Naval Postgraduate School is gratefully acknowledged for his guidance through the intricacies of MP. The author is grateful to anonymous reviewers for useful remarks improving the quality of the paper.

References

1. International Telecommunication Union, Recommendation X.800. Data Communication Networks: Open Systems Interconnection. Security, Structure and Applications. ITU, Geneva (1991)
2. Stallings, W.: Network Security Essentials. Applications and Standards. 6th edn. Pearson, New York (2017)
3. Sanz, R., Zalewski, J.: Pattern-based control systems engineering. IEEE Control. Syst. **23**(3), 43–60 (2003)
4. Glimm, J., Sharp, D.H.: Complex fluid mixing flows: simulation vs. theory vs. experiment. SIAM News. **39**(5) (2006)
5. Subramanian, N., Zalewski, J.: Quantitative assessment of safety and security of system architectures for cyberphysical systems using the NFR approach. IEEE Syst. J. **10**(2), 397–409 (2016)

6. Monterey Phoenix Behavior Modeling Tool. https://wiki.nps.edu/display/MP/Monterey+Phoenix+Home. Accessed 18 April 2022
7. Shodan: The Search Engine for Internet of Everything. https://www.shodan.io/. Accessed 18 April 2022
8. Subramanian, N., Zalewski, J.: Safety and security integrated SIL evaluation using the NFR approach. In: Jarzabek, S., Poniszewska-Marańda, A., Madeyski, L. (eds.) Integrating Research and Practice in Software Engineering. SCI, vol. 851, pp. 53–68. Springer, Cham (2020). https://doi.org/10.1007/978-3-030-26574-8_5
9. Simon, H.A.: Rational choice and the structure of the environment. Psychol. Rev. **63**(2), 129–138 (1956)
10. IEEE Std 1711.2-2019. Secure SCADA Communications Protocol (SSCP). IEEE, New York (2020)
11. Leverett, E.P.: Quantitatively Assessing and Visualising Industrial System Attack Surfaces. M.Phil. Dissertation, University of Cambridge Computer Laboratory (2011)
12. ISA/IEC-62443-4-2: Security for Industrial Automation and Control Systems: Technical Security Requirements for IACS Components. International Society of Automation, Research Triangle Park, NC (2018)

Artificial Neural Networks

Training of Deep Learning Models Using Synthetic Datasets

Zdzisław Kowalczuk[1](✉) and Jan Glinko[1,2]

[1] Gdansk University of Technology, Gdansk, Poland
{kova,janglink}@pg.edu.pl
[2] Avena Technologie, Gdansk, Poland

Abstract. In order to solve increasingly complex problems, the complexity of Deep Neural Networks also needs to be constantly increased, and therefore training such networks requires more and more data. Unfortunately, obtaining such massive real world training data to optimize neural networks parameters is a challenging and time-consuming task. To solve this problem, we propose an easy-to-use and general approach to training deep learning models for object detection and instance segmentation without being involved in the generation of real world datasets. In principle, we generate and annotate images with open-source software and 3D models that mimic real life objects. This approach allows us significantly reduce the effort required to gather pictures as well as automatize data tagging. It is worth noting that such synthetic datasets can be easily manipulated, e.g. to reduce the texture bias that often occurs in the resulting trained convolutional networks. Using the Mask R-CNN instance segmentation model as an example, we demonstrate that a network trained on the synthetic dataset of kitchen facilities shows remarkable performance on the validation dataset of real-world human-annotated photos. We show that our approach helps to bridge the domain gap between pre-trained models and their specific applications. In summary, such synthetic datasets help overcome the problem of acquiring and tagging thousands of images, while reducing the time and labor costs associated with the preparation of an appropriate real dataset.

Keywords: Deep learning · Instance segmentation · Synthetic dataset

1 Introduction

A single-layer dense network with a sufficiently large number of neurons is able to approximate any continuous function [1], thus becoming a strong competitor to classical modeling methods. The growing popularity of deep neural networks makes them the first choice in many fields, like computer vision [2], natural language processing [3], recommendation systems [4] or financial applications [5]. Each field of application adopts a different architecture: for example, Convolutional Neural Networks are most popular in the computer vision community and are used for tasks such as 6D pose estimation [6, 7], keypoint detection [8, 9], stereo disparity estimation [10, 11] or instance segmentation [8].

Z. Kowalczuk (Ed.): DPS 2022, LNNS 545, pp. 141–152, 2023.
https://doi.org/10.1007/978-3-031-16159-9_12

Following the history of the competitions in the field of computer vision, where convolutional neural networks are used, it can be observed that each successive winning solution is built on increasingly deeper neural models. For example, the winner of the 2012 ImageNet Large Scale Visual Recognition Challenge (ILSVRC) [12], ALEXNET [13], has 61 Milion trainable parameters, and the 2018's best solution, PNASNET-5 [14], has 86 Milion of them. Novel image recognition networks based on a 'self-attention' architecture have up to 632 Milion trainable parameters [15]. This trend of increasing the complexity of neural networks follows the rule: the more parameters to train, the more data is needed to train. Consequently, it is necessary to obtain a large amount of data, which requires a lot of human work.

Currently, the main approach to essentially reducing the problem of generating big data from scratch and building and training large networks is the transfer learning technique [16]. The deep learning community provides parameters for models trained on complex datasets from various domains, such as Common Objects in Context (COCO) [17] for instance segmentation and keypoints detection, ShapeNet [18] dataset for 6D object pose estimation or ImageNet dataset [12] for object classification. These networks are used as a starting point for training a new model to work in a specific domain, greatly reducing the amount of data and time required to train this model. In computer vision, where mainly Convolutional Neural Networks (CNN) are applied, the transfer learning technique is extremely effective [19, 20]. This efficiency is due to the two-tier model architecture in which CNN acts as a backbone (feature generator) on top of which specialized heads make inferences. As backbones are only feature generators, they are independent of heads and thus are freely switchable between different architectures and applications [21–23]. Feature generators learn robust generic features that are 'independent' of the individual objects in the training dataset [24]. In transfer learning via feature extraction, the backbone parameters are kept constant and only the head layers are trained. In transfer learning through fine-tuning, only the first few layers of the feature generator are frozen while the rest (including the heads) are trained. Unfortunately, any additional training requires a large set of labeled data, as training the network on too small a data set often leads to unsatisfactory performance.

Extra complexity is introduced when a neural network is trained for a more specific task. For example, when estimating stereo disparity, the annotation process requires additional measurements of the distance between the camera and the scene, which is usually difficult for a human to estimate. In cases such as image classification or object detection, data can be tagged by a human operator. However, the preparation of a complete representative dataset requires a great deal of work and is prone to human error.

In summary, learning data acquisition is an initial step in the learning process. It is also its bottleneck that you must go through before you start training the neural network. The usual approach is to scrap (download content) the Internet or take own photos and annotate them manually or semi-automatically using software tools such as paid V7, free CVAT or other. Moreover, it is imperative to verify the quality of the images and labels, as any shortcomings or errors in this regard may lead to inadequate training. Finally, insufficient diversity in the dataset can lead to over-fitting [25].

One of the ways to facilitate the acquisition of training data is to prepare synthetic datasets. Initially, such datasets were generated by cropping objects from real photos and overlaying them on appropriate canvases [26]. Instead, complex scenes are built up of primitive scenes. Thus, the ground truth is preserved, as the position of each localized object results from the known composition of the scene [27]. By automating and making the process of stage preparation more flexible, this approach greatly simplify the whole process of obtaining a large amount of appropriate training data.

Some work has already focused on building huge synthetic datasets for various computer vision tasks, namely: KITTI-CARLA [28], SYNTHIA [29] or ShapeNet [18]. New solutions use generative adversarial networks to generate data that mimics real photos [2]. However, these networks require real world data for training. Natural extensions of synthetic datasets are simulation environments such as CARLA [30]. The general rules for the preparation of a synthetic dataset are presented in [31].

In this work we propose the automated generation of synthetic datasets for training neural networks on a customized set of objects. Remarkably, our method is an efficient, fast and cheap alternative for manual training data acquisition that requires very limited 3D modelling skills. We demonstrate that our proposed process of generating such dataset is simple and straightforward. At the same time, the automatization of data annotation eliminates the bottleneck of costly, time consuming tagging of training data. The availability of cost-free modifications of dataset parameters, such as lighting, background, object placing and others makes the data more flexible, thus training more effective. Using the Mask-RCNN model and real-world images as an example, we show that a network trained on a synthetic dataset achieves surprisingly high performance in the instance segmentation task.

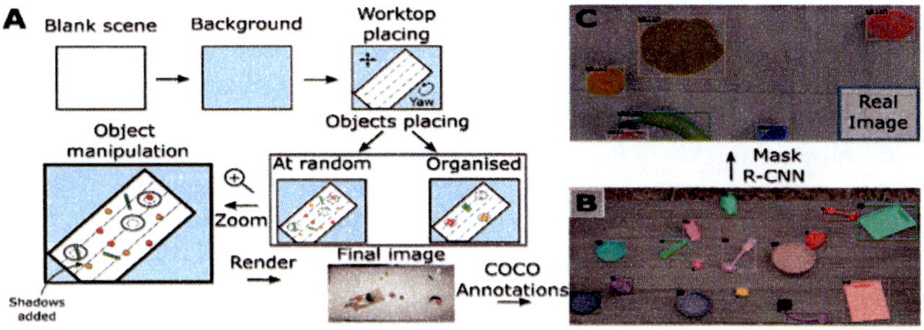

Fig. 1. (A) Scene preparation sequence: starting with a blank screen, the background is randomized (step 1), the worktop texture and camera perspective are set (step 2), objects are placed on the worktop (step 3), object properties and camera zoom are selected (step 4), and finally the image is rendered. (B) Representative depiction of fundamental truths in a rendered image. (C) Example of a prediction (bounding boxes and instance masks) in a real photo.

We provide with a complete pipeline (sequence of operations) to train the neural network of instance segmentation on a given limited set of custom objects. We show how to systematically optimize scene composition and details (Fig. 1) and network hyperparameters. We test several variations of each step of our pipeline for the best network performance. The results are described in detail in Sect. 3.

2 Applied Methods and Techniques

In this section we present technical details of our proposal and show how to proceed and reproduce our results (scripts are at https://github.com/avena-robotics/items3).

2.1 Technical Details

A PC running Ubuntu 20.04, an Intel Core i9-9900K processor and an Nvidia RTX A5000 graphic processor are used both to generate synthetic datasets and to train the network. The equipment was selected so that parallel data generation and network training could be implemented.

2.2 Collecting 3D Models

3D models of real kitchen appliances and consumables are sourced from open-source databases such as *ShapeNet* [18] and websites. We model the missing elements ourselves using *Blender*.

2.3 Synthetic Dataset Generation

BlenderProc2 [32], which wraps the *Blender* API into easy-to-use functions, is used for image preparation and annotation. The process is as follows: a) setting the scene, b) setting the position and parameters of the camera, c) rendering the image, d) annotating the image.

The basis of each scene is a worktop model. The model is made on the basis of a real worktop in our laboratory. As part of the scene, items (Fig. 2) are placed only on the worktop. The placement of objects is limited by two rules: containers cannot be oriented upside down and the pose of each object is stabilized by a simulated gravitational field. Additionally, two variants of the scene are prepared: in the first variant, consumables are placed either in containers or on cutting boards (organized way), and in the second, all items are placed randomly (unorganized way).

Fig. 2. Representative snapshots of synthetic dataset objects: consumables, containers, cutting boards and tools.

Then the position and parameters of the camera are set. In particular, the camera targets a point of interest in the scene, defined as the average position value of all objects in the scene. A field of view of the camera is set to 1 degree. The camera's distance from the scene is fixed either at 138 m (far) or 78 m (close) from the center of the worktop. The scene is illuminated by evenly distributed area light, the intensity of which is set to 10 according to Blender units. The light sources are placed 0.9 m above the table top. The image is then rendered in the camera view with the image resolution set to 1000 × 720 pixels. Finally, the rendered image is annotated based on the scene's ground truth stored in the scene state in the internal *Blender* files. The annotations are exported to the COCO standard for subsequent network training.

2.4 Validation Dataset Generation

100 photos (1920 × 1080 RGB, using Intel D415 camera) of actual kitchen appliances and consumables are prepared as a validation dataset. Ground truths are defined by manual annotation using the V7 web tool. The scenes are designed according to the preliminary rules established for synthetic data sets: consumables are placed on the table top, either randomly (50 photos) or in containers and cutting boards (50 photos).

2.5 Neural Network Architecture

The Mask R-CNN serves as the basic structure of our neural network. The building blocks of this architecture consist of fixed heads, a region proposal network, and a switchable backbone.

Detectron2 [33] is used as the start framework of the neural network in the adaptation process. EfficientNetB0 [34] or Feature Pyramid Network with ResNet [35] of various sizes (50, 101, 152) is used as the backbone.

2.6 Transfer Learning via Fine-Tuning

Model trained on ImageNet dataset [12] is used as a starting point for transfer learning via fine-tuning. In the case of ResNet backbone, the number of frozen segments varies from 2 to 5 for the training procedure. In the case of EfficientNetB0 all backbone layers are trained.

Table 1. Learning hyperparameters

Hyperparameter	Value
Learning Ratio (LR)	0.01
Number of Iterations	20000
Images per Batch	2
LR Decay Value	0.9
LR Decay Steps	12000, 17000
Warm-up Iterations	1000

The learning hyperparameters are set according to the rules proposed by Goyal et al. [36] and are kept constant during the optimization process. The values of these parameters are presented in Table 1.

2.7 Scene Parameters Optimization

The simulation world background is randomized using HDRI from the Haven Texture Pack to influence the light and differentiate the background (non-object) learning samples. Scene parameters such as worktop textures and object light absorption/emissivity and texture visibility are optimized to maximize the performance of the trained neural network in the validation dataset. The worktop textures are modified by changing the texture image node with the textures from the Haven Texture Pack. The light absorption/emissivity of the object ranged from –10 (absorption) to 10 (emission) *Blender* units. The visibility of the texture is modified in two ways: in the first one, objects are randomly covered with dust, where the amount of dust varies from 0 to 1; in the second case, the textures of the objects are turned off randomly with a probability of 0.6.

2.8 Network Parameters and Architecture Optimization

In the process of additional optimization of the Mask R-CNN architecture (to select ROI) appropriate thresholds were set for (I) initial classification and (II) filtration of regions of interest (RoI) generated by a sub-system called region proposal network (RPN). This classification is made by calculating the IoU (Intersection over Union) index according to the training data for each of these proposed areas. Based on the threshold (I), the area is classified as an object (positive) or as a background (negative). In the experiment, the threshold (I) was used with the value: 0.2 or 0.3 for the classification of the area as a background and 0.7 or 0.8 for the classification of the area as an object. Threshold (II) controls the extent to which the ROIs generated by the RPN overlap. Its value was set to the level of 0.6 or 0.7. Regions of interest with a common IoU greater than the threshold (II) are subject to the non-maximum suppression (NMS) algorithm. In additional experiments, the L1 regularization of weights was replaced with the L2 (milder) regularization, and additional layers of Group Normalization (GN) [37] were added.

2.9 Validation

The performance of the neural networks is verified using our validation dataset consisting of real images (Sect. 2.4). We use averaged precision (AP) as an index of neural network performance in the segmentation task. To evaluate the effectiveness of the training process (indications of overfitting or underfitting), we use losses in training and validation processes, defined as the sum of losses in classification, masking, and box-bound region restriction.

Moreover, our synthetic datasets are additionally tested using the PointRend network, the authors of which report better mask accuracy than that obtained with the Mask R-CNN network [38].

3 Results

Our main goal was to generate a synthetic dataset that allows us to train the neural network to work efficiently on real photos of kitchen facilities. Initially, we started by collecting 120 models from 29 different object classes. Our scene consisted of a worktop on which we placed 6 to 20 kitchen items. An upper limit has been set on the number of objects to prevent the scene from overflowing. Unless otherwise noted, the camera has been set to the far position. In this step, we have kept the object properties, worktop texture and background image unchanged. As an example of a neural network, we chose the Mask R-CNN with the ResNet101 backbone. Initially, the detailed architecture hyperparameters were set to: non-maximum suppression threshold for RPN of 0.7, intersection over union threshold for proposals of 0.3 and 0.7 for negatives and positives respectively, weights regularization to L1 and no additional normalization layers. We used transfer learning to initialize the network weights and froze the first 2 backbone layers. Each synthetic dataset used for learning consisted of 10,000 images.

3.1 The Role of Scene Organization in the Learning Process

At first, we just wanted to see how the organization of the scene influences the learning process. For this purpose, we have prepared two sets of synthetic data. In the first one, items (consumables, cutting boards, containers, etc.) were randomly placed on the worktop (Fig. 3, left). In the second set, consumables were placed in containers or on cutting boards to mimic real-world scenarios for organizing a typical kitchen worktop (Fig. 3, right).

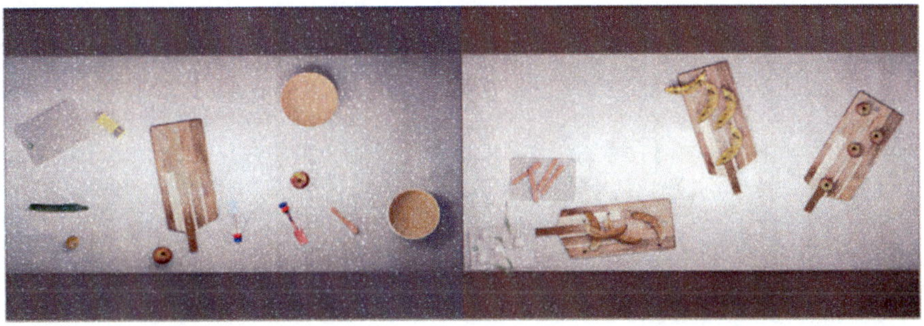

Fig. 3. Representative snapshots of scene organization: unorganized (left) and organized (right).

Note that for both synthetic datasets, the adopted set of hyperparameters allowed for appropriate convergence of network weights. Both networks (trained on data describing organized and random collections of items) showed AP of 14.72 and 17.85 (for real validating photos). Thus, the performance of the networks trained on our (preliminary) synthetic datasets immediately turned out to be relatively high - as it showed half the quality of the original Mask R-CNN [8] network which achieved a validation efficiency (AP) of 36.4 (admittedly on another popular *Cityscapes* dataset). The validation loss for our Mask R-CNN networks (tuned on both our sets) turned out to be over 5 times higher (worse) than the learning loss (Table 2), which clearly shows that the model is overfitted and (consequently) generalizes poorly.

Table 2. First experiment results

Dataset	Validation loss	Training loss
Organized	1.07	0.37
Unorganized	0.926	0.183

Our first approach to reducing overfitting was to diversify background training samples (non-subject or non-object) by randomizing the world background (using HDRI) and tabletop textures. The models trained on the new datasets showed a significant improvement in network performance compared to the previous models (AP increases of 2.46 and 5.45 for disorganized and organized datasets, respectively). It is noteworthy that the difference in terms of AP between networks trained on unorganized and organized datasets has been reduced from 3.13 to 0.14, showing that background randomization reduces the impact of scene organization on the learning process. Overall, our results show that the randomization of background and worktop textures significantly improves the quality of instance segmentation.

3.2 Impact of Object Texture Properties on the Accuracy of a Neural Network

Since randomization of both background images and object placement has improved network performance, we will also investigate whether additional randomization of object properties will improve training results. Namely, to obtain such a modification of objects, we use 3 different parameters: dust opacity, complete removal of texture and light absorption (giving the effect of shadows). Given that network performance has proven to be independent of how objects are arranged, we now use the unorganized method of preparing a dataset.

The datasets were prepared as follows: 70% of the images were from a unorganized dataset with a random background image and worktop texture; the remaining 30% were prepared as a randomized dataset with one of the following modifications: dust coverage; no texture; with shadows (3 datasets in total).

The network trained on these three datasets achieved APs of 19.32, 17.07, and 30.23 for dust, texture, and shadow modifications, respectively. Our results show that enriching information about contours, including by shadows, has a positive effect on performance, and blurring the contours of objects by lack of texture or dust reduces the effectiveness of the network. So you can see that strongly emphasizing the objects on their background helps the segmentation head to better distinguish objects.

3.3 Impact of Camera Position on the Accuracy of a Neural Network

Since the training dataset of unorganized data with random background and additional shadow data turned out to be the best of the tested varieties, we used this dataset to investigate how camera positioning affects the training process. We compared the datasets with the near and far camera position. Interestingly, a network trained on a dataset with a close camera configuration showed a significant performance improvement of 7.29 (total AP was 37.52) compared to further camera configuration. Our results show that positioning the camera up close can help distinguish objects due to a fine-grained representation of objects (more pixels per object and clearer texture).

3.4 Network Architecture and Hyperparameters Optimization

So far, we have studied the effect of dataset attributes on the learning process. We independently checked how changes in architecture affect the training process. We used close camera setup, 70% unorganized data with random background, and 30% data with shadow to test the impact of different backbones (ResNet50, ResNet152, EfficientNetB0). We obtained APs of 35.28, 23.81, and 4.35 for ResNet50, ResNet152, and EfficientNetB0, respectively. Since we have observed overfitting even when using a simple ResNet101 network as the backbone, the inferior performance of ResNet152 is most likely due to the enhancement of this phenomenon (even more overtrained parameters). On the other hand, in the case of Efficient-NetB0 and ResNet50, the number of trainable parameters is too small to discriminate effectively between objects. So we come to the conclusion that ResNet101 represents the best compromise between good generalization and overfitting.

Next we investigated the effect of the number of frozen layers on transfer learning by appropriate additional tuning of ResNet101 as the best architecture among the tested backbones. We obtained AP values of 40.8, 35.24, and 28.82 on the validation dataset at 3, 4, and 5 frozen layers, respectively. The results suggest that the initial 3 layers of ResNet101 pre-trained in ImageNet are sufficient to recognize primitive features. Layers 4 and 5, on the other hand, seem too suited to ImageNet.

Finally, we tested how the more detailed hyperparameters affect the performance of the instance segmentation network using our best-performing architecture. We considered individual cases separately: network without normalization and with group normalization (GN), threshold values for IoU in ROI (of conducted classification) $\{0.3, 0.2\}$ and $\{0.7, 0.8\}$ respectively for negatives and positives, the threshold for NMS in ROI $\{0.7, 0.6\}$, and regularization of weights according to L1 and L2.

Among these changes, L2-regulation of weights and more restrictive ROI classification (narrower IoU threshold range) improved network performance AP by 1.42 and 0.74, respectively; changing the NMS threshold had little or no effect on network performance (AP change of 0.02); while the introduction of GN significantly worsened AP (by 7.11). It is worth noting that taking into account all positive changes (in terms of IoU, NMS and L2 normalization) at once gave a synergistic effect on the network performance AP (by 2.55). Overall, the resulting 'synergistic' structure achieved the highest efficiency AP of 43.35 in the validation data set.

3.5 PointRend Network

After identifying best practices and principles for both training dataset development and network architecture, we assessed the PointRend network. We trained this network (with ResNet101 as the backbone) on a dataset of which 70% were random worktop and background textures, and 30% were shadows. We got an AP of 43.02 on the validation data set. This result is similar (0.33 lower) to the one achieved by our best network, Mask R-CNN. However, we have observed that the PointRend network validation loss is greater than that of Mask R-CNN. Probably an additional adaptation of this network should ensure its better operation (in relation to the R-CNN Mask).

4 Conclusion

In this article, we propose a cheap and simple method to synthetically generate training datasets for deep neural networks. We show that open source tools allow for efficient generation of training datasets and their automatic annotations. Nowadays, acquiring 3D models to generate specific synthetic datasets is possible, which can be seen in the example of kitchen facilities and consumables shown. Importantly, we also show that a network trained on a synthetic dataset and verified from real photos of a sample kitchen environment shows a peak performance AP of 43.35, which is higher than that of networks trained on real photos (e.g. the original/*vanilla* network Mask R-CNN reaches AP of 36.4 for the validation dataset of *Cityscapes*). Overall, the best results were achieved with the ResNet101 backbone, a random shadow dataset, altered worktop texture images, and a closer camera.

Such good results of the network trained on a synthetic set are counterintuitive, but this success may come from the flexibility of such a dataset and the unlimited possibilities of randomizing image parameters. Synthetic datasets can be easily adapted to training purposes. They also easily ensure the desired randomness of the scene organization. Importantly, we also show that rational modifications of objects and backgrounds significantly improve the precision of the trained network. Thus, we have clearly shown that an appropriate adaptation of the training dataset can substantially contribute to improving the effectiveness of network learning. On the other hand, however, such a procedure should be approached with due caution, as it has a large impact on the learning outcomes and the quality of network operation, which are very sensitive in this respect (apart from the problem of selecting the size of the training dataset to the complexity of the structure).

Acknowledgements. This research was founded by The National Centre for Research and Development, grant no.: POIR.01.01.01-00-0833/18. We would like to thank Łukasz Nierzwicki for valuable suggestions.

References

1. Cybenko, G.: Approximation by superpositions of a sigmoidal function (1989). https://doi.org/10.1007/BF02551274
2. Goodfellow, I.J., et al.: Generative Adversarial Networks. arXiv:1406.2661 [cs, stat] (2014)
3. Arik, S., et al.: Deep Voice 2: Multi-Speaker Neural Text-to-Speech (2017)
4. Rendle, S., Zhang, L., Koren, Y.: On the Difficulty of Evaluating Baselines: A Study on Recommender Systems. arXiv:1905.01395 [cs] (2019)
5. Calvo-Pardo, H.F., Mancini, T., Olmo, J.: Neural Network Models for Empirical Finance (2020). https://doi.org/10.3390/jrfm13110265

6. Xiang, Y., Schmidt, T., Narayanan, V., Fox, D.: PoseCNN: A Convolutional Neural Network for 6D Object Pose Estimation in Cluttered Scenes. arXiv:1711.00199 [cs] (2018)
7. Wang, C., et al.: DenseFusion: 6D Object Pose Estimation by Iterative Dense Fusion (2019). https://doi.org/10.1109/CVPR.2019.00346
8. He, K., Gkioxari, G., Dollár, P., Girshick, R.: Mask R-CNN (2017). https://doi.org/10.1109/ICCV.2017.322
9. Zhou, X., Wang, D., Krähenbühl, P.: Objects as Points. arXiv:1904.07850 [cs] (2019)
10. Lipson, L., Teed, Z., Deng, J.: RAFT-Stereo: Multilevel Recurrent Field Transforms for Stereo Matching. arXiv:2109.07547 [cs] (2021)
11. Tankovich, V., Häne, C., Zhang, Y., Kowdle, A., Fanello, S., Bouaziz, S.: HITNet: Hierarchical Iterative Tile Refinement Network for Real-time Stereo Matching (2021). https://doi.org/10.1109/CVPR46437.2021.01413
12. Russakovsky, O., et al.: ImageNet large scale visual recognition challenge. Int. J. Comput. Vis. **115**(3), 211–252 (2015). https://doi.org/10.1007/s11263-015-0816-y
13. Krizhevsky, A.: One weird trick for parallelizing convolutional neural networks. arXiv:1404.5997 [cs] (2014)
14. Liu, C., et al..: Progressive neural architecture search. In: Ferrari, V., Hebert, M., Sminchisescu, C., Weiss, Y. (eds.) ECCV 2018. LNCS, vol. 11205, pp. 19–35. Springer, Cham (2018). https://doi.org/10.1007/978-3-030-01246-5_2
15. Dosovitskiy, A., et al.: An Image is Worth 16x16 Words: Transformers for Image Recognition at Scale. arXiv:2010.11929 [cs] (2021)
16. Zhuang, F., et al.: A Comprehensive Survey on Transfer Learning (2020). https://doi.org/10.1109/JPROC.2020.3004555
17. Lin, T.-Y., et al.: Microsoft COCO: common objects in context. In: Fleet, D., Pajdla, T., Schiele, B., Tuytelaars, T. (eds.) ECCV 2014. LNCS, vol. 8693, pp. 740–755. Springer, Cham (2014). https://doi.org/10.1007/978-3-319-10602-1_48
18. Chang, A.X., et al.: ShapeNet: An Information-Rich 3D Model Repository. arXiv:1512.03012 [cs] (2015)
19. Yin, X., Chen, W., Wu, X., Yue, H.: Fine-tuning and visualization of convolutional neural networks (2017). https://doi.org/10.1109/ICIEA.2017.8283041
20. Sonntag, D., et al.: Fine-tuning deep CNN models on specific MS COCO categories. arXiv:1709.01476 [cs] (2017)
21. Liu, W., et al.: SSD: single shot multibox detector. In: Leibe, B., Matas, J., Sebe, N., Welling, M. (eds.) ECCV 2016. LNCS, vol. 9905, pp. 21–37. Springer, Cham (2016). https://doi.org/10.1007/978-3-319-46448-0_2
22. Redmon, J., Divvala, S., Girshick, R., Farhadi, A.: You Only Look Once: Unified, Real-Time Object Detection (2016). https://doi.org/10.1109/CVPR.2016.91
23. Girshick, R.: Fast R-CNN. arXiv:1504.08083 [cs]. (2015)
24. Zeiler, M.D., Fergus, R.: Visualizing and understanding convolutional networks. In: Fleet, D., Pajdla, T., Schiele, B., Tuytelaars, T. (eds.) ECCV 2014. LNCS, vol. 8689, pp. 818–833. Springer, Cham (2014). https://doi.org/10.1007/978-3-319-10590-1_53
25. Li, Z., Kamnitsas, K., Glocker, B.: Overfitting of neural nets under class imbalance: analysis and improvements for segmentation. In: Shen, D., et al. (eds.) MICCAI 2019. LNCS, vol. 11766, pp. 402–410. Springer, Cham (2019). https://doi.org/10.1007/978-3-030-32248-9_45
26. McCane, B., Novins, K., Crannitch, D., Galvin, B.: On Benchmarking Optical Flow (2001). https://doi.org/10.1006/cviu.2001.0930
27. Meister, S., Kondermann, D.: Real versus realistically rendered scenes for optical flow evaluation. In: 14th ITG Conference on Electronic Media Technology, pp. 1–6 (2011)
28. Deschaud, J.-E.: KITTI-CARLA: a KITTI-like dataset generated by CARLA Simulator. arXiv:2109.00892 [cs] (2021)

29. Ros, G., Sellart, L., Materzynska, J., Vazquez, D., Lopez, A.M.: The SYNTHIA Dataset: A Large Collection of Synthetic Images for Semantic Segmentation of Urban Scenes (2016). https://doi.org/10.1109/CVPR.2016.352
30. Dosovitskiy, A., Ros, G., Codevilla, F., Lopez, A., Koltun, V.: CARLA: An Open Urban Driving Simulator. arXiv:1711.03938 [cs] (2017)
31. Zhang, Y., et al.: Physically-Based Rendering for Indoor Scene Understanding Using Convolutional Neural Networks (2017). https://doi.org/10.1109/CVPR.2017.537
32. Denninger, M., et al.: BlenderProc. arXiv:1911.01911 [cs] (2019)
33. Wu, Y., Kirilov, A., Massa, F., Lo, W.-Y., Girshick, R.: Detectron2 (2019)
34. Tan, M., Le, Q.V.: EfficientNet: Rethinking Model Scaling for Convolutional Neural Networks. arXiv:1905.11946 [cs, stat] (2020)
35. He, K., Zhang, X., Ren, S., Sun, J.: Deep Residual Learning for Image Recognition (2016). https://doi.org/10.1109/CVPR.2016.90
36. Goyal, P., et al.: Accurate, Large Minibatch SGD: Training ImageNet in 1 Hour. arXiv:1706.02677 [cs] (2018)
37. Wu, Y., He, K.: Group normalization. Int. J. Comput. Vision **128**(3), 742–755 (2019). https://doi.org/10.1007/s11263-019-01198-w
38. Kirillov, A., Wu, Y., He, K., Girshick, R.: PointRend: Image Segmentation as Rendering. arXiv:1912.08193 [cs] (2020)

Autonomous Perception and Grasp Generation Based on Multiple 3D Sensors and Deep Learning

Zdzisław Kowalczuk[1](✉) and Jan Glinko[1,2]

[1] Gdansk University of Technology, Gdansk, Poland
{kova,janglink}@pg.edu.pl
[2] Avena Technologie, Gdansk, Poland

Abstract. Grasping objects and manipulating them is the main way the robot interacts with its environment. However, for robots to operate in a dynamic environment, a system for determining the gripping position for objects in the scene is also required. For this purpose, neural networks segmenting the point cloud are usually applied. However, training such networks is very complex and their results are unsatisfactory. Therefore, we propose an innovative and end-to-end approach to generating the grip position that replaces (3D) point cloud segmentation with 2D image segmentation. For this purpose, we create an OrthoView module that acts as an adapter between 3D space (point cloud) and 2D space (image). The 2D object mask created in it serves as basic information in the process of selecting the final grip from among the grips generated for the entire scene using Contact GraspNet. An unquestionable conceptual advantage of our solution (OrthoView) is the fact that only one resulting 2D image is created from the point cloud, which can be the result of merging (integrating) 3D images from many cameras. Therefore, it allows for the fusion of information from any number of cameras, without the need to implement solutions for identifying the same objects seen from different perspectives (cameras). In order to test our solution stream, we created 6 scenes of different complexity, on the basis of which we present the effectiveness of our solution. Thus, eliminating the need for 3D point cloud segmentation and reducing the image segmentation problem to infer from only one 2D photo can significantly improve the performance of the position estimation system for grasping objects.

Keywords: Pick-and-place · Deep learning · Robotics · Computer vision

1 Introduction

Vision systems are used in many areas of automation and robotics. Cameras are present in autonomous cars [1], drones [2] and industrial robots [3]. The semantic understanding of the environment that can be achieved through the use of information contained in images is essential for the autonomous operation of systems such as vehicles and robots. In robotic systems, more and more often the combination of image information with spatial information obtained from stereovision or LIDAR system is observed [4, 5].

Z. Kowalczuk (Ed.): DPS 2022, LNNS 545, pp. 153–166, 2023.
https://doi.org/10.1007/978-3-031-16159-9_13

Such data fusion allows for obtaining in-depth, technically more precise information about the environment.

The analysis of spatial information, represented in the form of a point cloud, has recently been the subject of much research.

In such preliminary deep neural network solutions, it is assumed that the entire 3D point cloud is the best representation of a given object. On the other hand, by rendering views of objects from different perspectives, we reduce the problem of classification in 3D space to classification in 2D space [6, 7]. In parallel, networks of inference were developed on the basis of some ordered (*voxelized*) space [8, 9]. Later, solutions operating directly on the point cloud were introduced [10, 11]. Such networks have high performance in detecting 3D-objects. Their disadvantage is the complicated learning process, usually requiring CAD models of the analyzed objects.

The depth image is the structure containing information equivalent to the point cloud. The combination of such (*spatially ordered*) representation with the RGB image is used in various solutions presented in [12–14], in order to obtain a semantic segmentation of the point cloud.

In this work, we are interested in using only information about the color of points in the cloud to determine the space occupied by an object. For this purpose, we use the image obtained by the orthogonal projection of the point cloud. Solutions similar to ours were presented in [15–20], as well as in [21], where the LIDAR technique supports automatic analysis of the environment. In particular, in the works [16, 17, 20] the projection (the so-called bird-eye-view) of the point cloud obtained from LIDAR onto a plane was used. The inference takes place on the thus created occupancy grid. In [15, 19], apart from projection onto a plane, camera image data are also used. The solution presented by us stands out from the above-mentioned works in that in the inference process we do not use any additional information about the point cloud, but only the image created by projecting this cloud onto a selected plane associated with the base camera (we do not use here, for example, images directly from this camera). It is worth noting that in this case, training the network for image segmentation is much easier and less time-consuming than training the network for 3D point cloud segmentation. For the segmentation of images in 2D, we chose the Mask R-CNN network [22] with the ResNet101 backbone [23].

We show the effectiveness of our solution, which is based on the operation of a neural network, on the example of a system generating positions for gripping objects for the needs of a robot manipulator. The main division of such networks in literature concerns model-based solutions [24] and generic solutions [25, 26]. Model-based solutions use predefined grips assigned to an object and transform them according to the transformation of the object itself. Generic solutions learn universal grasping positions, independent of a specific object (object-agnostic). To avoid training the network generating grips for objects in our set, we decided to use the generic *Contact GraspNet* architecture, trained on the ACRONYM set [27].

On the basis of separately created test scenes, we prove that we are able to determine the area of the object and its position on the basis of the image segmentation created by the orthogonal projection of the point cloud onto a certain base plane. Note that in this way we will be able to select the appropriate grip from all generated for a given scene.

We also present the impact of various methods of limiting the 2D mask area on the effectiveness of filtering out the positions of grips belonging to a given object. Finally, we provide a complete pipeline to generate a robotic grasp of an object of unknown 3D representation, shown in the summary diagram in Fig. 1. Our simple approach only requires training an instance segmentation network on a customized dataset.

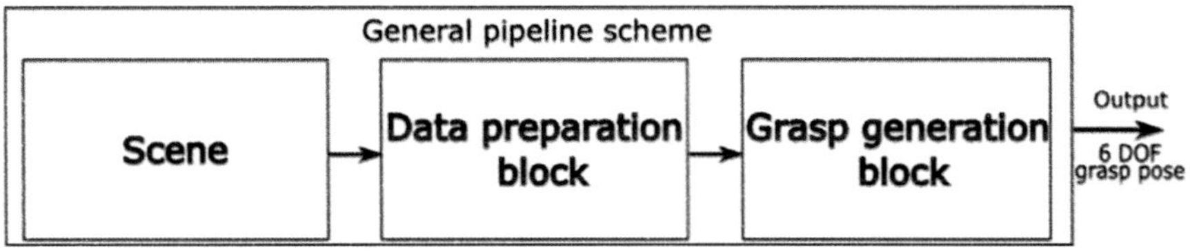

Fig. 1. General diagram of the proposed pipeline. First, 3D images of the analyzed scene are taken; then, in the perception block, the data is prepared by merging point clouds; finally, the grasp positions are determined in the Grasp generation block.

The basic assumption is the horizontal alignment of the base camera plane. Thus, objects are also placed on the appropriate horizontal frame. Otherwise, an additional transformation would be required. Another important assumption is the spatial distribution of objects so that the objects do not significantly obscure each other.

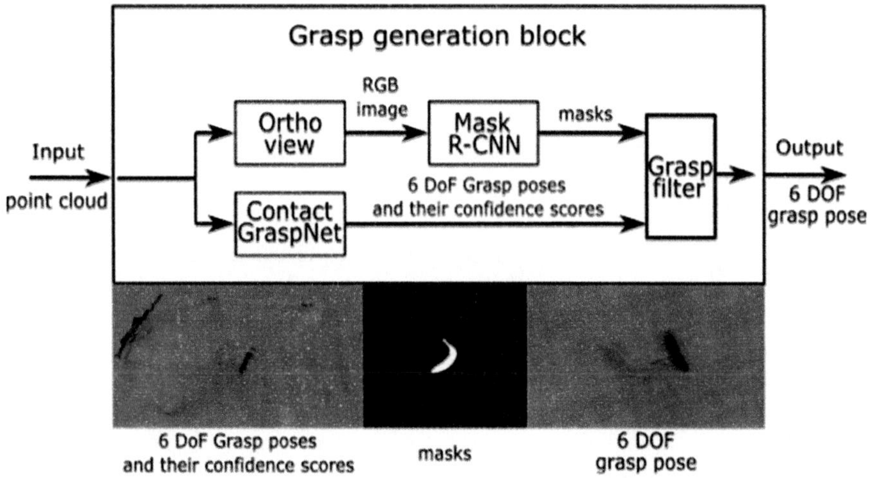

Fig. 2. Block generating gripping positions. We predict all the grasps for the scene using Contant-GraspNet and at the same time we run the "upper" branch of (2D) image processing (OrthoView block and Mask-R-CNN block which segment instances). Finally, in the decision block/filter, we select the final grip based on all the predicted grasps for the scene, the mask of the selected object, and the final filtering policy (GraspFilter).

Note that the discussed robotic perception includes object detection, which is essential in industrial robotic systems based on a video channel with (3D) cameras and used in industrial processes of bin picking or picking up objects from a conveyor belt. Detailed scheme of our perception block is presented in Fig. 2.

The rest of this article is organized as follows: Section 2 introduces the main parts of the proposed pipeline, especially the *OrthoView* and *GraspFilter* modules. In Sect. 3 we present the results of the individual modules of our pipeline and the final results of the processing. Finally, in Sect. 4, we summarize our research and indicate possible ways to extend this approach.

2 Methods

In this section, we will cover the following topics: camera setup and calibration, point cloud merge, cloud to RGB conversion, instance segmentation, robot grip (grasp) generation and filtering.

2.1 Camera Setup

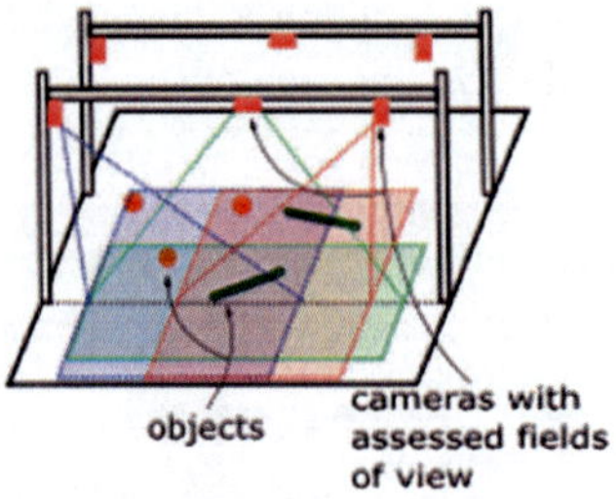

Fig. 3. Setup 6 Intel Realsense D415 cameras above the stage. The colored regions represent the camera fields of view.

As a solution, we suggest using 6 Intel RealSense D415 cameras to cover our entire working area. Above the stage, three cameras are placed on two parallel frames, in the center and on the sides, as shown in Fig. 3. Each camera provides RGB and depth images, working at 30 FPS (all the necessary calculations are done on-board). We take into account the built-in temporal and spatial filtering of the depth image for better results. The depth image is aligned with the RGB image using an internal transformation matrix and camera model (inverse Brown-Conrady), so that each pixel of the depth image can be enriched with additional information about the color. Such enhanced image data will hereinafter be referred to as RGB-D. It is worth recalling here that the pixel depth value is some measure of the distance of the point from the parallel base plane of the source camera (representing the Z value), not the absolute distance of the object from the camera.

2.2 Camera Calibration

Cameras need to be calibrated before fusing data from multiple devices. We selected one camera as the source and computed a transformation matrix between each camera and the source one. We have divided our calibration process into two parts. We start with a global calibration, and then refine the obtained results with local calibration.

Global Calibration. We use a ChAruco board (9×12, 60 mm, 5×5 Aruco dictionary) for global calibration. A particular advantage of the ChAruco mat is the unique code of each square, which facilitates correct calibration, even if the mat is not fully visible in the camera. We have prepared a set of 10 photos of the calibration mat for each pair (target, origin). Based on these photos, we calculate the transformation matrix for each pair of cameras.

Local Calibration. To refine the effects of global calibration, we use an Iterative Closest Point (ICP) algorithm that compares the common features of two point clouds and compute a transformation matrix based on the comparison of these features. The algorithm must have information about the global transformation to work properly. Because, in particular, an inaccurate global transformation matrix can lead to divergent results. Thus, we use the transformation matrix calculated in the global calibration procedure as the initial transform for the local refinement. The output of the ICP algorithm is the final refined value of a given transformation matrix.

2.3 Point Cloud Merging

Each RGB-D image from the camera can be converted to a point cloud based on the camera's intrinsic parameters, namely the focal length and optical center of the camera. A point cloud is an unorganized collection of points, each of which is represented by six variables: three spatial coordinates (X, Y, Z) and three colors (R, G, B values).

We use the Truncated Signed Distance Function (TSDF) algorithm to filter and merge point clouds from all cameras. The input are 6 point clouds and the output is one, merged point cloud, as presented in Fig. 4.

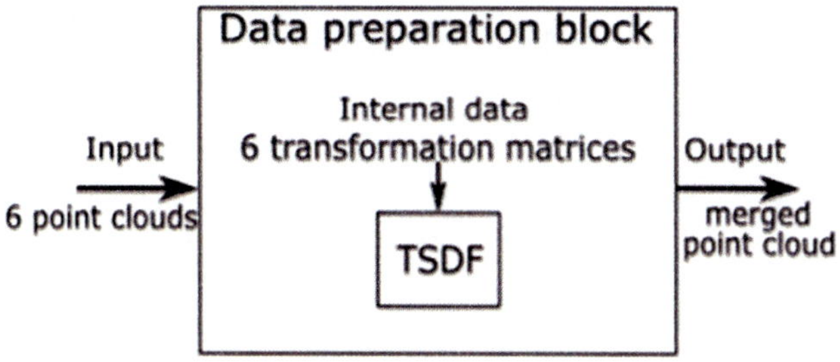

Fig. 4. Perception - data preparation. Using transformation matrices (taken from the calibration process) we merge 6 point clouds from Intel Realsense D415 cameras into one point cloud.

For convenience, we use the Open3D [28] framework to perform the required calculations on point clouds. The Open3D TSDF implementation uses the GPU to speed up the computation, which is critical as point cloud merging is the computing bottleneck of our system. TSDF accepts RGB-D, camera intrinsic parameters (to covert depth to point cloud) and a transformation matrix as input. Additional parameters are set as follows: voxel size to 0.001 m and signed distance function truncation value to 0.005 m.

Each point cloud is merged with the source/base camera point cloud, and we cut off all points with a depth value (Z) greater than 1.2 m in order to limit the working space. It is worth noting that the final point cloud (the only one) is represented in the coordinate frame of the source camera.

2.4 Converting a Point Cloud to an RGB Image

To extract an RGB 2D image (a three-dimensional array of size *i* x *j* x *RGB*) from a merged 3D point cloud (an array of points, where each point is described by 6 values, i.e. X, Y, Z coordinates in the camera coordinate system and RGB color), we perform the following operations, called the ***OrthoView*** ***pipeline***: **(a)** quantization (*voxelization*) of the 3D space in which the cloud is located (the size of such a *voxel cube* is 1 mm); **(b)** assigning a color to each *voxel cube*, whereby if many points fall into one *cube*, the voxel color is the average of the colors of all points in this *voxel* [28]; **(c)** projection of the resulting *voxel grid* on the XY plane (eliminating the depth/Z), and the color conflict is resolved according to the equation below

$$color(i,j,0)|_{\substack{1 \le i \le m \\ 1 \le j \le n}} = color(i,j,\mathrm{argmin}k_{1 \le k \le p} : color(i,j,k) \neq \varnothing) \quad (1)$$

where $color(\bullet) = (\mathrm{R,G,B})$ is a function that returns or assign the RGB colors of a *voxel cube* (i, j, k), and m, n, and p limit the sizes of the *voxel grid* X, Y, Z, respectively. Since different colors can appear in one column of the voxel grid (along the Z axis), there is (above) ambiguity in flattening the voxel grid that we solve by taking the RGB color of the point closest to the XY plane (indicated by the lowest Z depth coordinate with any 'non-zero' color). Thus, as **(d)** we realize a simple conversion of such a flattened voxel grid into a 2D RGB image according to the equation

$$I(i,j)|_{\substack{1 \le i \le m \\ 1 \le j \le n}} = color(i,j,0)|color(i,j,0) \neq \varnothing \quad (2)$$

where I is resulting image of sizes same as flattened voxel grid. Conclusively, the last OrthoView step is **(e)** filling the vacant image values with black color with fixed parameters (0, 0, 0).

2.5 Instance Segmentation

To process the image prepared in this way (point 2.4), we use the Mask R-CNN [22] architecture with the ResNet101 [23] backbone, trained on a certain synthetic data set, in order to obtain semantic information about the scene content. There are 29 kitchen and dining facilities in this training dataset. Note that this set consists of 30,000 images synthesized with different object parameters (texture image, texture visibility, light absorption attribute) and scene parameters (camera position, number of objects in the scene, and different object relationships). Details of generating such a synthetic data set and training the network based on it are described in our other work [29].

2.6 Generating the Robotic Grips

For the initial generation of robotic grasps, we use the Contact GraspNet neural network [30], which provides gripping according to the point of contact, preventing the change of the initial position of the grasped objects. The input for this network is a point cloud, and its output is a list of grasp positions (X, Y, Z, roll, pitch, yaw) with a corresponding confidence score. This network is trained on the ACRONYM [27] data set for the FE gripper on the PANDA robotic arm (which limits the maximum grip width), however, it can be easily adapted to any other parallel gripper with the appropriate scaling of the input point cloud. In the case of more specific robotic grippers, this network must be trained separately.

2.7 Grasp Filtration Using GraspFilter

Grasps predicted by Contact GraspNet should be filtered out to get a feasible version of grasping operations. Namely, gripping away from the center of gravity of the object may require greater forces for a smooth/stable grasp. Also, it is more likely that the object will dynamically change its position in the gripper as the robot moves. On the other hand, tools such as a knife, spatula or ladle need to be picked on the right end in order to obtain the desired result (handling them appropriately). For these reasons, we created software for filtering grasps, *GraspFilter*. Filtration is divided into two separate stages: coarse and fine (granular) filtering.

Coarse Filtering. In this procedure, we use the mask generated by the Mask R-CNN network to filter out all the grasps not belonging to the selected object. The applied bounding area of each mask is generated in three ways: (a) an axis-aligned bounding box (AABB); (b) a minimum area boundary rectangle (MABR); (c) the contours detected. We choose one of these bounding areas and discard all grips, positioned outside of this area.

Fine Filtering. To filter the remaining grasps, we use two different indicators: (a) confidence score and (b) distance from the center of the bounding area. As the resulting grasp of the robotic gripper, the one with the highest score (relative to the profit policy) is selected.

3 Results

Our main goal was to build a system for generating object gripping positions based on a colored point cloud. To avoid using instance segmentation in 3D space, we created the OrthoView module (Sect. 2.4) that converts the point cloud into a properly positioned RGB image (preserving spatial information in 3D). As a result of the OrthoView module, we obtain properly crafted RGB 2D images that can be processed with the Mask R-CNN network for instance segmentation of objects. The resulting object mask (another image preparation) allows us to filter the optimal grasp from the set of grips previously generated by the Contact GraspNet network from the initial point cloud. For such a fusion of data from two processing paths, we have created the GraspFilter software package described in Sect. 2.7.

3.1 Point Cloud Merging

We have prepared 6 scenes with different arrangement of items and their mutual relations in order to assess the quality of the (final) point cloud. Having a point cloud from each camera and its transformation matrix (relative to the base/source camera), we used the TSDF algorithm (with the parameters given in Sect. 2.3) to merge the points into a single cloud, located in the base camera coordinate system. The results that we obtained after the merge are shown in Fig. 5.

It is worth emphasizing here that the basic factor influencing the quality of the merged point cloud is camera calibration. It must be properly carried out and its results thoroughly verified. If the results of the calibration are unsatisfactory, the calibration process should be repeated.

Fig. 5. Point clouds of our test scenes. Note that we have specially changed the view perspective to show the true value of the data, as a point cloud viewed directly from the camera perspective is always visually better (no holes are visible).

The obtained point clouds are of high quality. It is worth noting that even the structures of light-reflecting objects (knife, ladle, plate), although often a problem for depth estimation algorithms, have been correctly reproduced. Interestingly, by placing cameras on both sides of the stage and above the stage, we reduced the number and size of holes (missing points) in the point cloud. Point clouds of such high quality should positively influence the process of generating the grips and the results of the OrthoView module (and thus the segmentation results).

3.2 OrthoView

The point clouds presented in Sect. 3.1 served as input to our OrthoView module (Sect. 2.4). The results are shown in Fig. 6.

We expected object gaps due to the imprecision of the point cloud and occlusion, but they are minor. In addition, blackline artifacts appeared, possibly due to quantization. However, we believe that the RGB images obtained in this way are of good quality, and minor errors should not affect the subsequent segmentation of the instance.

Fig. 6. RGB images of point clouds created by our OrthoView module.

Since the optical plane of the base camera is parallel to the scene, the obtained images are projected orthogonally (they are not loaded with perspective and distance), which is a remarkable advantage of OrthoView. Applying instance segmentation to such an image will result in the creation of object masks with a minimal (projected) area (assuming a collision-free location and a stable position of objects). It is worth noting that an additional advantage of such conversion is that the pixels of the image are organized with respect to the voxel grid, so that information about the X and Y dimensions can be obtained directly from such an image (while a conventional RGB image loses information about actual dimensions).

3.3 Instance Segmentation

The images obtained from the OrthoView module serve as input to the instance segmentation network. The results of such segmentation of instances using the Mask R-CNN network with the ResNet101 backbone, trained on the synthetic training set, are shown in Fig. 7. We noticed classification errors (red circles) or no masks (orange circles). Significant mask defects may prevent proper filtering of the position of grips belonging to a given object, while the lack of a mask completely eliminates the possibility of selecting grips. Overall, however, the network has shown great accuracy. 28 out of 38 objects in the test scenes are correctly classified and have appropriate masks. As can be seen (Fig. 7), the segmentation of containers (plates and bowls) is the most problematic case.

In this work, apart from the problem of instance segmentation we consider identification of grasp positions. For this we need to avoid an incomplete mask (with its incorrect classification), which is used to filter the position of the grasp.

Better segmentation results can be expected by selecting the training set for the scenarios used to collect data for the OrthoView module (e.g. non-perspective images, artifacts in the image, or blurred textures).

Fig. 7. Instance segmentation results obtained in OrthoView images. Examples of misclassified objects are marked with red circles, and objects with missing masks - with orange circles.

3.4 Initial-Grasps Generation

The gripping positions provided by Contact GraspNet are shown in Fig. 8. It was observed that the network assigned the correct grasps to most of the objects. The analysis of incorrect grip positions or the lack of assignment (for a given object) led us to the following conclusions: (a) Defects in the point cloud cause that some of the grasp positions proposed by Contact GraspNet cause a collision with the object (Fig. 8D, milk). This is evident in the case of tall and rectangular objects. There are deficiencies in the cloud of points on the sidewall which result in incorrect suggestions of the grip position. (b) The problem is also low objects, the tips of which do not stick out clearly from the table surface (Fig. 8D, small plate, knife). The network was unable to generate any gripping positions for these objects. (c) Grasps unrelated to any specific object were generated. They result from disturbances in the point cloud (Fig. 8A–E). (d) The number of proposed grasps for items in containers is limited (Fig. 8E). (e) Fewer grips have been generated for objects in the crowded scene (Fig. 8F), which is due to the gripper colliding with other objects. (f) The lack of the generated grasps for the orange is understandable as its diameter exceeds the maximum opening of the gripper fingers (Fig. 8B, E).

The use of Contact GraspNet gives good results, especially in uncrowded scenes where objects are loosely scattered and clearly distinguishable from the table surface. A solution to the problem with generating grips may be to cut a fragment of the point cloud with the use of a segment mask before generating the position of the grip. Performing such an operation is possible thanks to instance segmentation for photos obtained from OrthoView. It seems to be a promising direction for further development of this system, which will be included in future work.

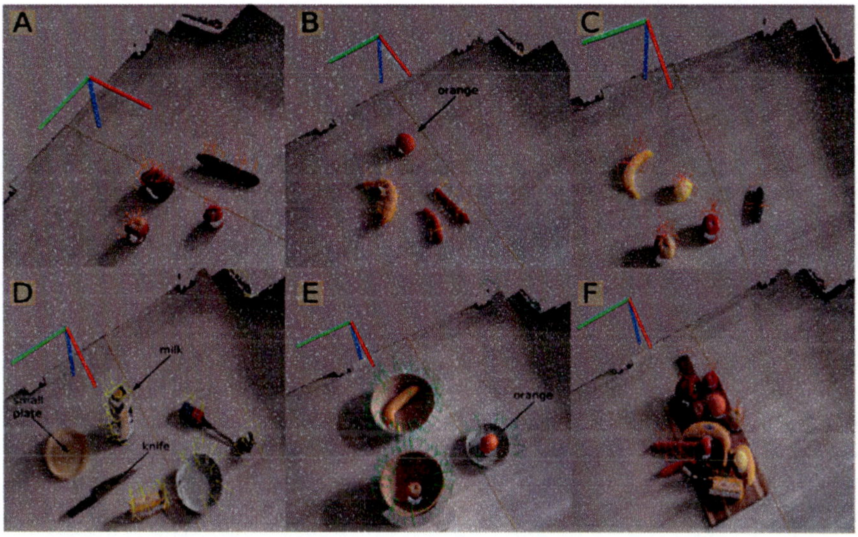

Fig. 8. Visualization of grasps predicted by Contact GraspNet on 6 different scenes.

3.5 Initial-Grasps Filtration by GraspFilter

We have tested how various filtering policies implemented in GraspFilter affect the final choice of object grasping. In uncrowded scenes (Figs. 9A, B), coarse filtering has no effect on the final grasp. We observed the same results for all 3 policies (AABB, MABR, contours). The last scene in Fig. 9A, where a banana is placed in the bowl, is an exception here. In this situation, only contour selection leads to a successful gripping of the bowl. In general, the contour policy should be applied whenever one object is the background of another object.

Fine filtering has a significant impact on the final grasp position. It is worth noting that the use of confidence score assessment promotes object grasping at its boundaries, which is optimal for non-tool objects. Conversely, the grips closest to the tip or center of the masks are usually more suitable for tools.

In crowded scenes, we suggest using contours to coarsely filter out the grips. In the case of non-rectangular objects, AABB and MABR cover a large portion of the non-object area, and thus there is a risk that grasps of adjacent objects will be included in fine filtering.

We have observed that tool objects (e.g., a ladle) should have a special filtering policy. We propose to consider detecting tool holders of unique colors to identify the tools. In this case, the grips will be limited to the tool holder, which solves the problem of (special) filtering.

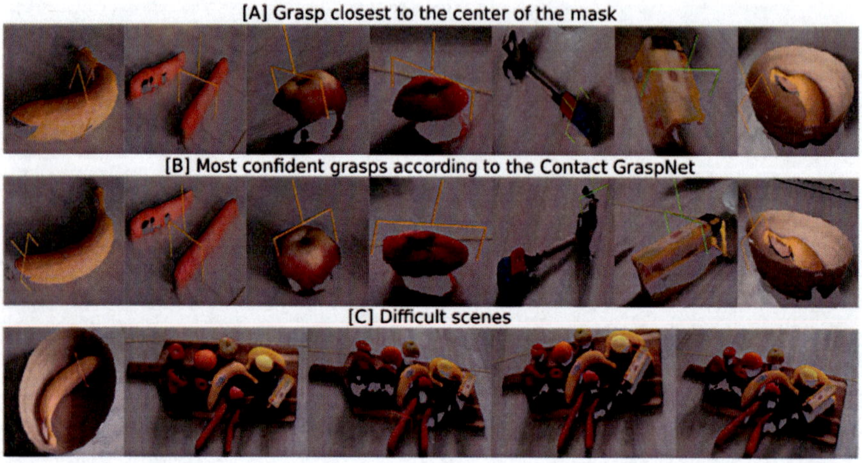

Fig. 9. Final grasps of selected objects: (A) Grips closest to the center of the object segmentation mask in an uncrowded scene. (B) The grasps with the highest confidence score of all the grips of the same object in an uncrowded scene. (C) Difficult scenes for grasping objects. In such crowded scenes, the strictest rule (contour selection) to limit the mask area should be used to avoid taking into account the grips of adjacent objects during fine filtering.

4 Discussion

In this article, we presented how to generate robot gripping methods for a selected object without knowing about its 3D representation. We introduced an OrthoView module which serves as an adapter between 3D and 2D space. This solution allows us to filter 3D positions based on an enhanced 2D image and thus avoid the need for instance segmentation in 3D, which is the greatest advantage of our approach.

An important role in the presented system is played by the appropriate arrangement of the cameras and their two-stage calibration. Thanks to the developed precise transformation matrices, the use of the TSDF algorithm allows to obtain a point cloud of a quality comparable to the quality obtained with the use of expensive cameras based on the structured light technology. Such a point cloud enables (by means of the Contact GraspNet) to predict the variety of grips for further filtration.

We prepared 6 scenes with different crowdedness and mutual relations between objects. For each scene, we tested different grasp filtering rules and presented their advantages and disadvantages depending on the relationship between the objects.

At this stage, we omitted the problem of the kinematic feasibility of grasping and the collision of the robot with the scene. Therefore, the integration of our system with a real robot is another challenge and a welcome extension of this work.

Note also that, regardless of the perception problem, the grasping itself does not have to be vertical (as shown in the bowl example).

Acknowledgements. This research was founded by The National Centre for Research and Development, grant no.: POIR.01.01.01-00-0833/18.

References

1. Kocić, J., Jovičić, N., Drndarević, V.: Sensors and sensor fusion in autonomous vehicles. In: 2018 26th Telecommunications Forum (TELFOR), pp. 420–425 (2018). https://doi.org/10.1109/TELFOR.2018.8612054
2. Zhu, P., Wen, L., Bian, X., Ling, H., Hu, Q.: Vision Meets Drones: A Challenge. arXiv:1804.07437 [cs] (2018)
3. Ebert, D.M., Henrich, D.D.: Safe human-robot-cooperation: image-based collision detection for industrial robots. In: IEEE/RSJ International Conference on Intelligent Robots and Systems, vol. 2, pp. 1826–1831 (2002). https://doi.org/10.1109/IRDS.2002.1044021
4. Caltagirone, L., Bellone, M., Svensson, L., Wahde, M.: LIDAR–camera fusion for road detection using fully convolutional neural networks. Robot. Auton. Syst. **111**, 125–131 (2019). https://doi.org/10.1016/j.robot.2018.11.002
5. Banerjee, K., Notz, D., Windelen, J., Gavarraju, S., He, M.: Online camera LiDAR fusion and object detection on hybrid data for autonomous driving. In: 2018 IEEE Intelligent Vehicles Symposium (IV), pp. 1632–1638 (2018). https://doi.org/10.1109/IVS.2018.8500699
6. Qi, C.R., Su, H., Niessner, M., Dai, A., Yan, M., Guibas, L.J.: Volumetric and Multi-View CNNs for Object Classification on 3D Data. arXiv:1604.03265 [cs] (2016)
7. Su, H., Maji, S., Kalogerakis, E., Learned-Miller, E.: Multi-view Convolutional Neural Networks for 3D Shape Recognition. arXiv:1505.00880 [cs] (2015)
8. Maturana, D., Scherer, S.: VoxNet: A 3D Convolutional Neural Network for real-time object recognition. In: 2015 IEEE/RSJ International Conference on Intelligent Robots and Systems (IROS), pp. 922–928 (2015). https://doi.org/10.1109/IROS.2015.7353481
9. Zhou, Y., Tuzel, O.: VoxelNet: End-to-End Learning for Point Cloud Based 3D Object Detection. arXiv:1711.06396 [cs] (2017)
10. Qi, C.R., Su, H., Mo, K., Guibas, L.J.: PointNet: Deep Learning on Point Sets for 3D Classification and Segmentation. arXiv:1612.00593 [cs] (2017)
11. Qi, C.R., Yi, L., Su, H., Guibas, L.J.: PointNet++: Deep Hierarchical Feature Learning on Point Sets in a Metric Space. arXiv:1706.02413 [cs] (2017)
12. Fooladgar, F., Kasaei, S.: Multi-Modal Attention-based Fusion Model for Semantic Segmentation of RGB-Depth Images. arXiv:1912.11691 [cs] (2019)
13. Chen, X., et al..: Bi-directional cross-modality feature propagation with separation-and-aggregation gate for RGB-D semantic segmentation. In: Vedaldi, A., Bischof, H., Brox, T., Frahm, J.-M. (eds.) ECCV 2020. LNCS, vol. 12356, pp. 561–577. Springer, Cham (2020). https://doi.org/10.1007/978-3-030-58621-8_33
14. Hazirbas, C., Ma, L., Domokos, C., Cremers, D.: FuseNet: incorporating depth into semantic segmentation via fusion-based cnn architecture. In: Lai, S.-H., Lepetit, V., Nishino, K., Sato, Y. (eds.) ACCV 2016. LNCS, vol. 10111, pp. 213–228. Springer, Cham (2017). https://doi.org/10.1007/978-3-319-54181-5_14
15. Chen, X., Ma, H., Wan, J., Li, B., Xia, T.: Multi-View 3D Object Detection Network for Autonomous Driving. arXiv:1611.07759 [cs] (2017)
16. Li, B., Zhang, T., Xia, T.: Vehicle Detection from 3D Lidar Using Fully Convolutional Network. arXiv:1608.07916 [cs] (2016)
17. Beltran, J., Guindel, C., Moreno, F.M., Cruzado, D., Garcia, F., de la Escalera, A.: BirdNet: a 3D Object Detection Framework from LiDAR information. arXiv:1805.01195 [cs] (2018)
18. Zhou, Y., et al.: End-to-End Multi-View Fusion for 3D Object Detection in LiDAR Point Clouds. arXiv:1910.06528 [cs] (2019)
19. Yang, B., Liang, M., Urtasun, R.: HDNET: Exploiting HD Maps for 3D Object Detection. arXiv:2012.11704 [cs] (2020)

20. Wirges, S., Fischer, T., Frias, J.B., Stiller, C.: Object Detection and Classification in Occupancy Grid Maps using Deep Convolutional Networks. arXiv:1805.08689 [cs] (2018)
21. Kowalczuk, Z., Szymański, K.: Classification of objects in the LIDAR point clouds using Deep Neural Networks based on the PointNet model. IFAC-PapersOnLine. **52**, 416–421 (2019). https://doi.org/10.1016/j.ifacol.2019.08.099
22. He, K., Gkioxari, G., Dollár, P., Girshick, R.: Mask R-CNN. arXiv:1703.06870 [cs] (2018)
23. He, K., Zhang, X., Ren, S., Sun, J.: Deep Residual Learning for Image Recognition. arXiv:1512.03385 [cs] (2015)
24. Deng, X., Xiang, Y., Mousavian, A., Eppner, C., Bretl, T., Fox, D.: Self-supervised 6D object pose estimation for robot manipulation. In: 2020 IEEE International Conference on Robotics and Automation (ICRA), pp. 3665–3671 (2020). https://doi.org/10.1109/ICRA40945.2020.9196714
25. Pas, A. ten, Gualtieri, M., Saenko, K., Platt, R.: Grasp Pose Detection in Point Clouds. arXiv:1706.09911 [cs] (2017)
26. Mousavian, A., Eppner, C., Fox, D.: 6-DOF GraspNet: Variational Grasp Generation for Object Manipulation. arXiv:1905.10520 [cs] (2019)
27. Eppner, C., Mousavian, A., Fox, D.: ACRONYM: A Large-Scale Grasp Dataset Based on Simulation. arXiv:2011.09584 [cs] (2020)
28. Zhou, Q.-Y., Park, J., Koltun, V.: Open3D: A Modern Library for 3D Data Processing. arXiv:1801.09847 [cs] (2018)
29. Kowalczuk, Z., Glinko, J.: Training of deep learning models using synthetic datasets. In: 2022 15th International Conference on Diagnostics of Processes and Systems (DPS). (POLSPAR,PAN), pp. 1–12 (2022)
30. Sundermeyer, M., Mousavian, A., Triebel, R., Fox, D.: Contact-GraspNet: Efficient 6-DoF Grasp Generation in Cluttered Scenes. arXiv:2103.14127 [cs] (2021)

Open-Set Speaker Identification Using Closed-Set Pretrained Embeddings

Michał Affek and Marek S. Tatara(✉)

Department of Robotics and Decision Systems, Faculty of Electronics, Telecommunications, and Informatics, Gdańsk Tech, Gdańsk, Poland
s165541@student.pg.edu.pl, martatar@pg.edu.pl

Abstract. The paper proposes an approach for extending deep neural networks-based solutions to closed-set speaker identification toward the open-set problem. The idea is built on the characteristics of deep neural networks trained for the classification tasks, where there is a layer consisting of a set of deep features extracted from the analyzed inputs. By extracting this vector and performing anomaly detection against the set of known speakers, new speakers can be detected and modeled for further re-identification. The approach is tested on the basis of NeMo toolkit with SpeakerNet architecture. The algorithm is shown to be working with multiple new speakers introduced.

Keywords: Speaker identification · Open-set identification · Speaker recognition · Feature extraction · Anomaly detection

1 Introduction

Throughout the years the solutions to the task of identifying a person became more accurate and robust. One of the aspects that plays a major role here is the identification of a person by his/her voice, on a basis of short utterances. The problem can be divided into a few subcategories: (i) closed-set identification, where the recording must be assigned to one of the already known speakers, (ii) open-set problem, where apart from the already known speakers there are lectors previously unheard, and (iii) speaker diarization, where subsequent parts of a recording are attributed to particular speakers [2]. Although the closed-set problem is the object of research for quite some time, the higher accuracy for this problem with numerous speakers was achieved just recently [1]. The open-set problem, on the other hand, is still not sufficiently addressed and more scientific efforts are still needed.

Early approaches to address the open-set problem were using Gaussian Mixture Models (GMM) [13], which were further extended by introducing Universal Background Model (UBM) [14] to include speaker-independent background in the model, commonly used across different application of speaker recognition [10,11,15]. Later, on top of GMM-UBM, the i-vectors (identity vectors) were introduced [4,6]. These vectors used to extract fixed-length feature vectors from

Z. Kowalczuk (Ed.): DPS 2022, LNNS 545, pp. 167–177, 2023.
https://doi.org/10.1007/978-3-031-16159-9_14

utterances using UBM combined with Baum-Welch statistics, which were used in numerous applications [5]. Instead of relying on a predefined, usually statistics-based set of features, deep models map utterances to a set of fixed-length deep features, which are further used for the classification task, usually, with an associated neural model [1,3]. The approach proposed in this paper, which can be seen as the original contribution of the Authors, takes advantage of the deep neural models trained with an aim of speaker identification in a closed-set, and extend them toward the generalized open-set problems.

2 Proposed Approach

The authors propose to build the proposed solution on the deep neural networks approach used for a closed-set identification with numerous speakers and extend it toward an open-set problem without retraining. A vector of features is extracted by probing the last layer (of a neural network) that is prior to the classifier. If the network is adequately trained, this layer should represent low-level deep features differentiating the speakers. Further, customized algorithms for classifying speakers can be implemented. In the simplest form, it can be implemented as thresholding of the cosine distance between two embeddings. If the distance is smaller than the assumed threshold, then two embeddings belong to the same speaker, otherwise - they do not. If a test feature vector does not belong to any known speaker, a new class (representing a speaker) is created.

2.1 Rationale

The rationale behind the proposed approach is the fact, that each new training requires new data and resources, and might even not return a viable solution. Instead, it is proposed to use an already trained neural network and cut off the part responsible for classification. In that way, the output of the network will be a vector containing features that represent the characteristics of the speaker. Such vectors will be called embedding.

The success of the authors' reasoning depends on whether the embeddings are rich enough and discriminative in a way allowing for the successful classification of new speakers. For a sufficiently large and heterogeneous dataset, the deep neural networks should learn the representation of such discriminative features to make further classification possible. By assuming, that the human voice can be characterized by a finite number of features, it should be sufficient to use these features and perform the identification even for previously unheard speakers.

2.2 Data Flow

The proposed solution attempts to take advantage of the pretrained models (especially those trained with a high volume of data) without the need to retrain them or collect new data. The proposed high-level data flow for the speaker identification task is presented in Fig. 1. The utterances to be analyzed are provided to the input of a neural network. Then, the weights on one of the last layers are read, forming a feature vector (embedding). The feature vector is compared with all the models of known speakers and if a match according to specified criteria is found, then the utterance is classified as the known speaker that was matched. Otherwise, the utterance is assigned to a new speaker, and based on the extracted embedding a new model is created and added to the speakers' models database.

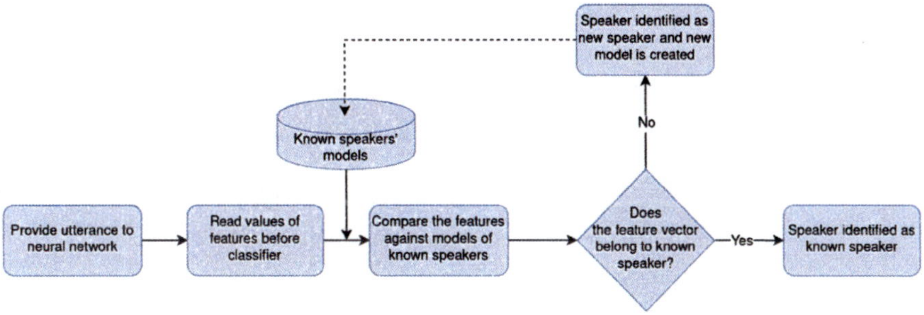

Fig. 1. Chart presenting the data flow in the proposed method.

Note that it is advisable to analyze utterances belonging to a single person in a batch so that the model representing a speaker can be built on a set of features containing representative statistical characteristics.

2.3 Speaker Recognition Backbone

NeMo (Neural Modules) is a toolkit developed by the international company NVIDIA and published under Apache 2.0 license in 2019 [8]. Since its very first publication, it is still under development and major bug fixes are constantly being introduced. The representative idea behind the system is to show the capacity of NVIDIA's products in Machine Learning (ML) applications to the public.

Both training and inference functions from NeMo were used to establish and test the selected model. The exact structure of the neural model is entirely based on SpeakerNet - available in pretrained collection in NeMo [7]. The model, with the help of structural scripts from NeMo, was fine-tuned using a selected part of LibriSpeech [12]. This establishes the neural backbone for the proposed solution. Theoretically, any other backbone can be used but the latter part (classifier and anomaly detection) must be adapted to a specifically chosen approach. The architecture of SpeakerNet with the indicated part where embeddings are extracted is shown in Fig. 2. Note that the SpeakerNet is used in the context of the proposed approach only as a feature extractor, and the open-set classification scheme is implemented by the Authors.

In summary, the NeMo backbone provides utilities for training the model and produces embeddings for each utterance, serving as inputs to the anomaly detection algorithm, followed by the speaker identification part. The embeddings are

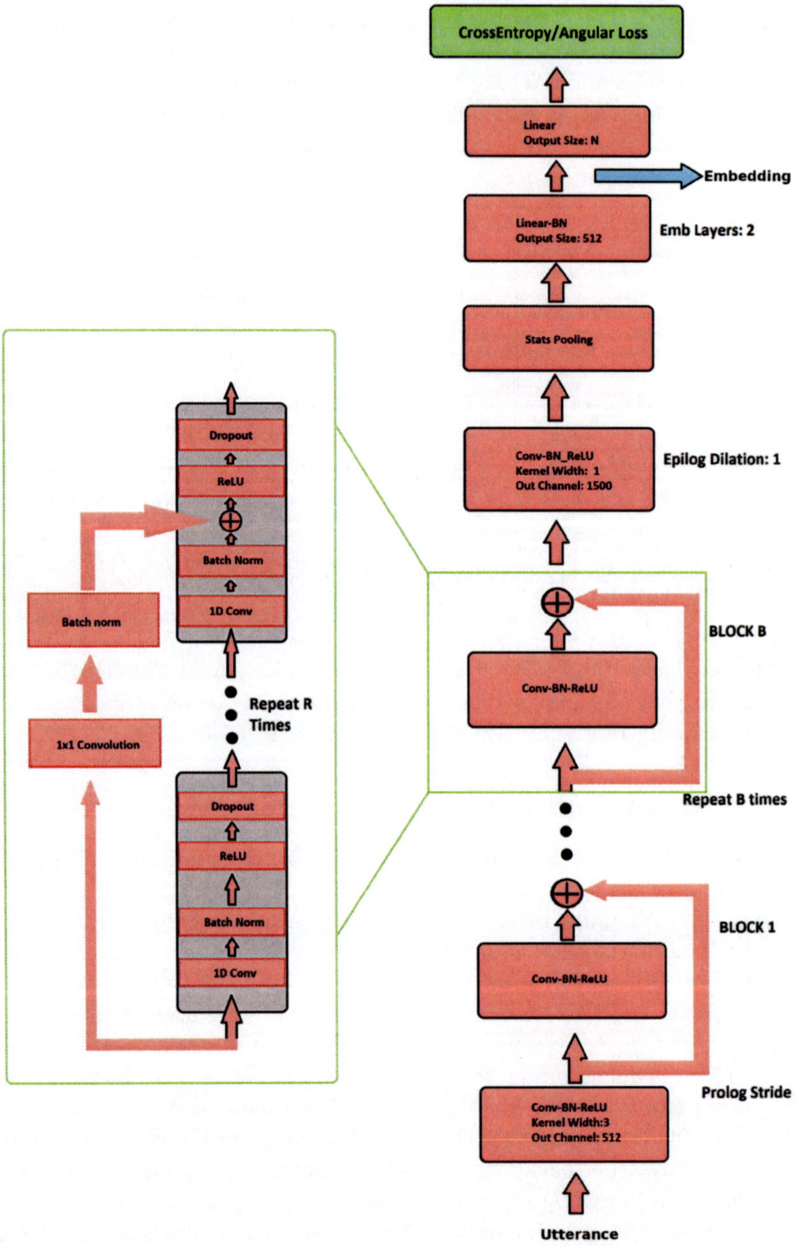

Fig. 2. SpeakerNet architecture with the embedding extraction point indicated (teal arrow). The architecture is taken from [7].

represented as vectors with 512 elements each (*i.e.*, containing one-dimensional numeric data of size 512). The values inside mentioned embedding vectors are ranging between -0.2 and 0.2 with the mean value close to 0. For example, the embedding of the speaker with ID number 1455 has the mean value equal to 0.001, maximal value in the vector equal to 0.143 and minimal equal to -0.079. This is part of characteristic of LibriSpeech dataset - it shows that the data is quite consistent and correctly cleaned from noise or significant outliers. For clarification, these characteristics are extracted by loading the trained model and using NeMo's inference functions. Then the results are saved inside *Pickle* (*.pkl*) file (Python package which provides the user with the possibility to transform Python object into a stream of bytes). Thanks to the byte representation of the data a full portability of embeddings is assured, which is important in case of evaluation in various environments (*i.e.* different operating systems).

3 Embedding-Based Classification of Speakers

Embeddings (*i.e.* feature vectors extracted from audio by SpeakerNet consisting of 512 values each) described previously are the main output of the model's inference. General usage of such embeddings which provides the information on the model's sufficiency is based on a trial approach. Users with the help of scripts generate trial files that include two utterances with the ground truth information (same class - 1, mismatch - 0). After preparing such files, the evaluation of cases starts. Program loads data from embeddings *.pkl* file pointed by the provided trial text file. With both vectors loaded the script evaluates the similarity between them. The similarity function used is the basic cosine similarity function. Since it produces values in the range $(-1, 1)$ the decision to scale them to the range $(0, 1)$ was made. By obtaining the new file with scores, it is possible to see how the model behaves in each case. Greater value indicates that vectors are more similar, while lower values can impede that the utterances are not coming from the same speaker.

After collecting the scores on the test utterances it was essential to calculate Equal Error Rate (EER) for this trial. EER is an error measure associated with a threshold, for which the probability of false rejection (false negative) is equal to the probability of false acceptance (false positive). The measurement is equal to 0.83% of EER, which was estimated with a threshold equal to 0.71. With increasing the number of trials the EER went down to 0.58% and the threshold changed insignificantly to 0.72. Further tests showed that the convergence on EER is going to around 0.35% and the threshold stabilizes at 0.7. Note that these values, in general, can change with new data, as it is specific to a particular set of utterances. Nevertheless, due to low error and approximately stable threshold value, the threshold equal to 0.7 was used for further analysis. Table 1 presents the exemplary output of the test which evaluates classification capability by performing a verification task (same class examples are not using the same utterances for comparison).

Note that in case of applying the solution fitted to the closed-set problem, the detection of unseen speakers can be referred to as anomaly detection, as this speaker is previously unseen (anomaly) to the model.

Table 1. Exemplary scores produced by cosine similarity function on two feature vectors (IDs, rescaled cosine similarity value, ground truth.

Indexes	Cosine similarity	Ground truth
211 and 211	0.830	1
7402 and 7402	0.890	1
233 and 233	0.812	1
3242 and 3242	0.718	1
6848 and 6848	0.917	1
211 and 211	0.868	1
4340 and 4340	0.893	1
1069 and 226	0.493	0
2136 and 2911	0.524	0
8324 and 2436	0.518	0
4014 and 730	0.468	0
6415 and 5561	0.599	0
4640 and 1235	0.603	0
2836 and 27	0.368	0

To visualize the data, the t-distributed stochastic neighbor embedding (t-SNE) algorithm is introduced [9]. It performs the projection of multidimensional data from embeddings to two- or three-dimensional space. It is worth noting that in this method there is a random initialization of points in the target projection. This is causing the results to diverge between runs. The projected space contains datapoints representing embeddings for particular utterances in a human-comprehensible way.

The main observation from analyzing the distribution of embedding in these spaces is that additional unseen points most of the time are isolated in the space. The model which did not "heard" this speaker's utterances in the training stage is still able to produce embeddings, yet they are distinct enough to not fall close to a known class. In addition, in the case of more than one unseen speaker the datapoints are organized in clusters which proves that the model is recognizing similar characteristics in data from an unknown source, and can be applied to the open-set problem.

The anomaly detection algorithm is implemented in its simplest form and brings down to the calculation of cosine distance between available datapoints. The distance threshold is used to categorize the results. The exact threshold level is indicated empirically (here 0.7 was assumed), so that the ratio of correctly classified new speakers is maximized, and at the same time the number of misclassification of known speakers is minimized. Tested behavior covers two-dimensional projections of embeddings.

4 Results

To check the correctness of the trained model it is advisable to visualize the embeddings in an identifiable manner. It is possible to represent feature vectors in reduced-dimensionality form. Again, to make it human-readable and possible to visually evaluate, reduction to two latent dimensions (with no identified physical meaning) is performed. In Fig. 3 groups of samely colored points hint that embeddings of speakers accurately portray the speaker's voice characteristics. In addition, standard inference (created by NeMo authors) tests are performed. These with usage of cosine similarity on embeddings yielded more than 99% of accuracy on the test set (known speaker, unseen utterances). It should be noted that every dot in the mentioned figure is from the test subset of the LibriSpeech dataset. They were not previously seen by the trained model. Also it is worth mentioning general characteristics of used LibriSpeech dataset subset. In training 251 speakers were used (each speaker has portion of utterances saved for test subset), 2 speakers from completely new LibriSpeech subset (for evaluating anomaly detection methods). Duration of utterances is not exceeding 20 s; they are preprocessed by dataset authors - cleaned from noise. In following figures all 251 seen speakers are showed (plus unseen ones) except Fig. 3 where the speaker's count was reduced to emphasize NeMo identification accuracy.

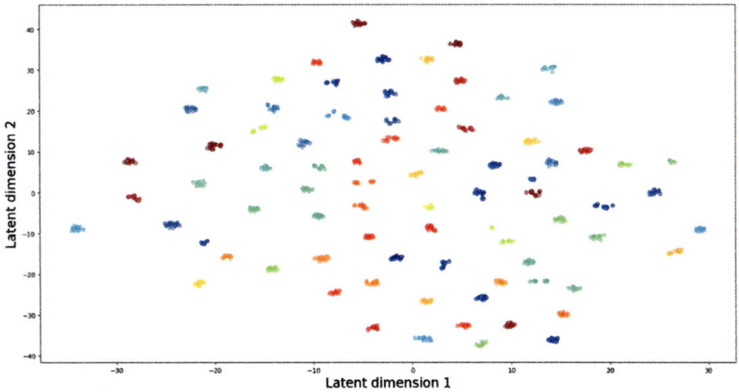

Fig. 3. Classification of generated embeddings by membership to speaker class.

Another test's results can be observed in Fig. 4. They present visualization fn mentioned dimensionality principles with the division between male/female speakers. Such differentiation and self-organization of males and females is a premise that the trained model is suitable for further open-set evaluation.

Unseen speaker is a lector which was not present during the training process of the model. However, the model is used in inference for evaluating utterances and producing embeddings for this unseen speaker. Further anomaly detection algorithm decides which utterances provided in testing phase were from outside

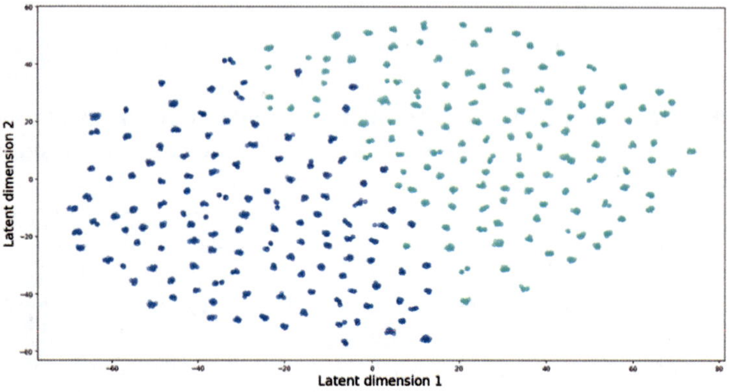

Fig. 4. Visualization of male (blue) and female (cyan) speakers' embeddings.

of training set. This method can be tuned for specific datasets (or parts of it) with tunable threshold. The parameter is chosen empirically. The exact value in example presented in Fig. 5 converges to 0.7. Multiple trials with different setups did not disclosed noticeable deviation from this value (oscillations were ± 0.1).

Setting too high threshold increases the number of False Negatives - the embeddings which are really close to particular classes (set of similar characteristics in voice is substantial), but are not aggregated into a single cluster. The effect of too high threshold can be observed in Fig. 6. On the other hand, too small value leads to too high number of False Positives and embeddings are assigned wrongfully to other classes.

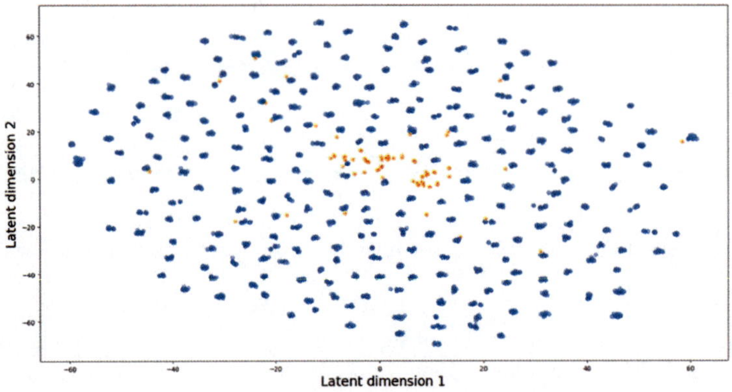

Fig. 5. Visualization of anomaly detection algorithm with threshold equal to 0.7. Blue stands for seen speakers, yellow for unseen utterances in the training stage. Small red dots constitute the results of anomaly detection method (red dot - anomalous data).

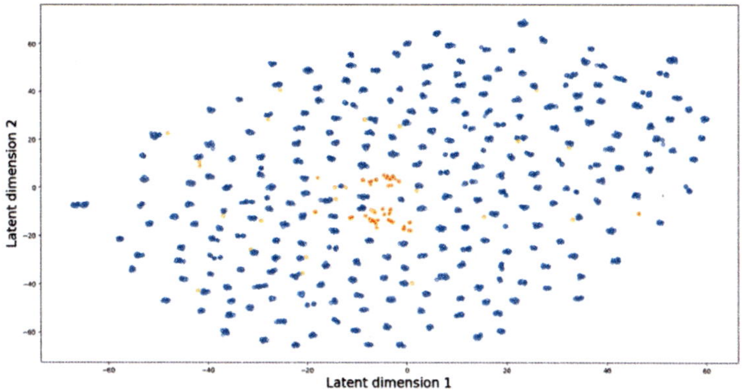

Fig. 6. Visualization of anomaly detection algorithm with too high threshold. Blue stands for seen speakers, yellow for unseen utterances in the training stage. Small red dots constitute the results of anomaly detection method (red dot - anomalous data).

Additionally, another unseen speaker was added to evaluate how anomaly detection algorithm behaves with more than one unseen speaker. The visualizing results Fig. 7 contains two new unseen speakers. The t-SNE algorithm detects that these embeddings are highly different from the rest of the data points. The anomaly detection algorithm does not have much problems with solving such cases (the threshold can be even enlarged).

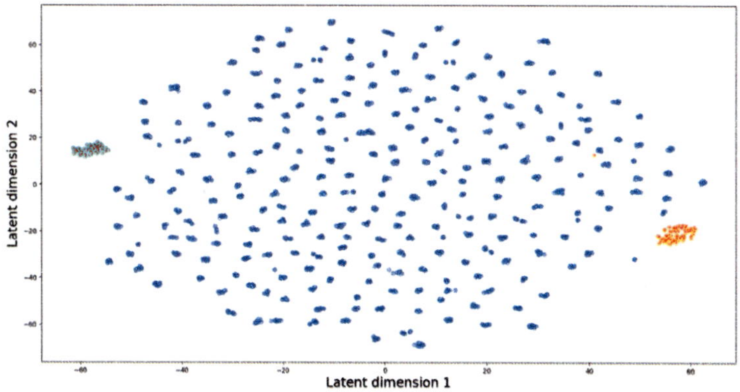

Fig. 7. Visualization of anomaly detection algorithm with threshold equal to 0.7. Blue stands for seen speakers, yellow and teal (accordingly label 464 and 7314 from train_clean_360 subset of LibriSpeech) for not seen in training stage utterances. Small red dots constitute the results of anomaly detection method (red dot - anomalous data).

5 Summary

The paper has shown a methodology how to build a functional speaker open-set recognition system. The system's operation has been tested with functional tests. These preliminary study shows the correctness of the proposed approach.

The speaker recognition backbone can classify the provided utterances into a set of speakers seen during the training with high recognition accuracy exceeding 99% for known speakers. It is almost infallible in the new speaker detection task, even for multiple new speaker, yet the accuracy for associating new speakers with the models has to be determined on a more diverse dataset. Created tests have confirmed the usability and accuracy of the constructed system. Performed checks were taking data (recordings) from different LibriSpeech collections. Interesting task would be evaluating the approach on noisy data (for example on the SITW datasets).

The proposed anomaly detection algorithm is not entirely universal and has some limitations. Firstly, the t-SNE projection is changing every run, which makes the visual analysis more difficult. Also the threshold is now predefined, but ideally, it should be adaptive and calculated on the basis of the already collected database of speakers. This would prevent the incorrect anomaly detection in case of providing embeddings from retrained model. Another challenge is the time of computation. The distance is calculated between all possible points. In case of huge number of inputs, it is time consuming.

The focus should also be directed on the way of creating the model of the detected anomaly, which may use more sophisticated method like, for instance, Gaussian Mixture Model. That would make the system less prone to misclassification and can increase the overall accuracy in the open-set problem.

References

1. Bai, Z., Zhang, X.L.: Speaker recognition based on deep learning: an overview. Neural Netw. **140**, 65–99 (2021). https://doi.org/10.1016/j.neunet.2021.03.004. https://www.sciencedirect.com/science/article/pii/S0893608021000848
2. Brew, A., Cunningham, P.: Combining cohort and UBM models in open set speaker identification. In: 2009 Seventh International Workshop on Content-Based Multimedia Indexing, pp. 62–67 (2009). https://doi.org/10.1109/CBMI.2009.30
3. Chung, J.S., Nagrani, A., Zisserman, A.: VoxCeleb2: deep speaker recognition. In: Proceedings Interspeech 2018, pp. 1086–1090 (2018). https://doi.org/10.21437/Interspeech.2018-1929
4. Dehak, N., Kenny, P.J., Dehak, R., Dumouchel, P., Ouellet, P.: Front-end factor analysis for speaker verification. IEEE Trans. Audio Speech Lang. Process. **19**(4), 788–798 (2011). https://doi.org/10.1109/TASL.2010.2064307
5. Ibrahim, N.S., Ramli, D.A.: I-vector extraction for speaker recognition based on dimensionality reduction. Procedia Comput. Sci. **126**, 1534–1540 (2018). https://doi.org/10.1016/j.procs.2018.08.126. http://www.sciencedirect.com/science/article/pii/S1877050918314042. Knowledge-Based and Intelligent Information & Engineering Systems: Proceedings of the 22nd International Conference, KES-2018, Belgrade, Serbia

6. Kenny, P., Boulianne, G., Ouellet, P., Dumouchel, P.: Joint factor analysis versus eigenchannels in speaker recognition. IEEE Trans. Audio Speech Lang. Process. **15**(4), 1435–1447 (2007). https://doi.org/10.1109/TASL.2006.881693
7. Koluguri, N.R., Li, J., Lavrukhin, V., Ginsburg, B.: SpeakerNet: 1D depth-wise separable convolutional network for text-independent speaker recognition and verification (2020). https://doi.org/10.48550/ARXIV.2010.12653. http://arxiv.org/abs/2010.12653
8. Kuchaiev, O., et al.: NeMo: a toolkit for building AI applications using neural modules. arXiv preprint arXiv:1909.09577 (2019)
9. Linderman, G.C., Steinerberger, S.: Clustering with t-SNE, provably. SIAM J. Math. Data Sci. **1**(2), 313–332 (2019)
10. Liu, M., Dai, B., Xie, Y., Yao, Z.: Improved GMM-UBM/SVM for speaker verification. In: 2006 IEEE International Conference on Acoustics Speech and Signal Processing Proceedings, vol. 1, p. I. IEEE (2006)
11. McLaughlin, J., Reynolds, D.A., Gleason, T.: A study of computation speed-ups of the GMM-UBM speaker recognition system. In: Sixth European Conference on Speech Communication and Technology. Citeseer (1999)
12. Panayotov, V., Chen, G., Povey, D., Khudanpur, S.: LibriSpeech: an ASR corpus based on public domain audio books. In: 2015 IEEE International Conference on Acoustics, Speech and Signal Processing (ICASSP), pp. 5206–5210 (2015). https://doi.org/10.1109/ICASSP.2015.7178964
13. Reynolds, D.A.: Speaker identification and verification using Gaussian mixture speaker models. Speech Commun. **17**(1), 91–108 (1995). https://doi.org/10.1016/0167-6393(95)00009-D. https://www.sciencedirect.com/science/article/pii/016763939500009D
14. Reynolds, D.A.: Comparison of background normalization methods for text-independent speaker verification. In: EUROSPEECH (1997)
15. Zheng, R., Zhang, S., Xu, B.: Text-independent speaker identification using GMM-UBM and frame level likelihood normalization. In: 2004 International Symposium on Chinese Spoken Language Processing, pp. 289–292. IEEE (2004)

Condition-Based Monitoring of DC Motors Performed with Autoencoders

Krzysztof Włódarczak[1], Łukasz Grzymkowski[2], and Tomasz P. Stefański[1(✉)]

[1] Faculty of Electronics, Telecommunications and Informatics, Gdansk University of Technology, 80-233 Gdansk, Poland
tomasz.stefanski@pg.edu.pl
[2] Arrow Electronics, 80-309 Gdansk, Poland

Abstract. This paper describes a condition-based monitoring system estimating DC motor degradation with the use of an autoencoder. Two methods of training the autoencoder are evaluated, namely backpropagation and extreme learning machines. The root mean square (RMS) error in the reconstruction of successive fragments of the measured DC motor angular-frequency signal, which is fed to the input of autoencoder, is used to determine the health indicator (HI). A complete test bench is built using a Raspberry Pi system (i.e., motor driver controlling angular frequency) and Jetson Nano (i.e., embedded compute node to estimate motor degradation) to perform exploratory analysis of autoencoders for condition-based monitoring and comparison of several classical artificial intelligence algorithms. The experiments include detection of degradation of DC motor working in both constant and variable work points. Results indicate that the HI obtained with the autoencoders trained with the use of either training method is suitable for both work points. Next, an experiment with multiple autoencoders trained on each specific work point and running in parallel is reviewed. It is shown that, in this case, the minimum value of RMS error among all autoencoders should be taken as HI. Furthermore, it has been shown that there is a near-linear relationship between HI and the difference between measured and reconstructed angular-frequency waveforms.

Keywords: Autoencoders · Condition-based monitoring · DC motors

1 Introduction

Automating processes and substituting human operators in order to obtain semi- or fully-autonomous machines are becoming more and more common. Machines are able to work faster, for a longer time without breaks and often with higher precision than humans, and so reduce the total manufacturing cost. The benefits of continuous operation are significant. However, the devices will eventually degrade and wear-down. Degradation can have various origins, e.g., unbalanced

Z. Kowalczuk (Ed.): DPS 2022, LNNS 545, pp. 178–189, 2023.
https://doi.org/10.1007/978-3-031-16159-9_15

bearing, damaged components or thermal degradation, all leading to some deterioration of performance and throughput. If no maintenance is performed in time, machines may break down completely and stop the production, which can be extremely costly. On the other hand, unnecessary maintenance also results in costs and makes it necessary to stop the machines. The goal of condition-based monitoring is to provide a solution to the above-mentioned problem, as well as to ensure that maintenance is performed in time and when it is needed, to avoid damage in the future.

In recent years, there has been significant progress with regard to machine learning and artificial intelligence in nearly every area, including autonomous cars, healthcare, warehouses, agriculture, etc. In many cases, the current results were impossible to achieve prior to the application of deep learning methods and neural networks, and their introduction allowed for entirely new use cases to be implemented, and for improving the accuracy of existing ones by orders of magnitude. It is therefore no surprise that these modern methods have also been applied in the field of condition-based monitoring, where classifying the system condition and predicting its expected life-time until breakdown are suitable for machine learning models. These include methods which estimate the condition of devices and inform users of their degradation, in some cases also including the data on the extent and location of the wear-down or damage if it has already occurred [1].

The purpose of this paper is to compare modern deep learning autoencoders with classical solutions in various experimental settings and draw conclusions on their applicability in condition-based monitoring setting. Experiments are explorative in nature, with multiple datasets and test settings evaluated. The used autoencoders are trained using backpropagation (BP) or extreme learning machines (ELMs), and then compared with support vector machines (SVMs) and random forests (RFs). This paper is organized as follows: Sect. 2 reviews related works and other condition-based monitoring systems. Section 3 gives an overview of autoencoders and training methods. Section 4 describes condition-based monitoring used in this work for DC motors. The experimental results for various cases estimating degradation of DC motors are presented in Sect. 5, with conclusions and a short discussion following in Sect. 6.

2 Related Works

Various types of learning can be used for condition-based monitoring. If one has access to a labeled dataset, ideally from both healthy and degraded systems, then supervised learning may be applied. The number of distinctive fault types which may occur is large, and so it is not an easy task to collect samples covering all such cases [2]. Another approach, instead of identifying a specific fault, focuses on calculating a general degree of degradation, which is indicative of either degradation or health of a device, and shows whether it requires maintenance immediately, or in near future. To solve this problem, one can employ unsupervised learning using unlabeled data collected from systems. In most cases, only healthy data points are sampled and the task becomes focused on detection of anomalies in a properly behaving device. The benefit here is undoubtedly the lack of necessity to collect and label data from production environment, which may be difficult or, from the equipment owner's perspective, nearly impossible. That is, the data of erroneous or faulty situations should never occur. Therefore, various methods, like clustering [3], or neural networks used in unsupervised learning, including generative adversarial networks [4], autoencoders [5,6], convolutional networks [7], and deep belief networks, are used for prognostics and health management applications [2,8].

In this paper, the focus is on autoencoders, described in more detail in the following section. This type of neural network, built from two subnetworks, compresses (i.e., encodes) the input vector (or more generally a tensor) into a reduced representation vector in latent space, and then decompresses (i.e., decodes) it into a reconstructed input vector. The reconstruction error is calculated as a distance between the input vector and the reconstructed vector in metric space. If the error is small, then the input vector is similar to the samples used during training and the autoencoder is able to reconstruct it effectively from its latent space representation. However, if the error is significant, it contains an anomaly, as the encoder will amplify any deviation from the learned samples. It is therefore important to ensure that the training set includes only samples from a healthy system, and not a deteriorated one. Autoencoders are used in numerous existing condition-based monitoring systems. Most applications compare the measured signal and the reconstructed signal. In [2,9], a system consisting of stacked autoencoders and a binary classifier, called a hierarchical extreme learning machine (HELM), is presented. The confidence of classification indicates the health of the system and, consequently, also the level of degradation. The HELM structure provides good results, although only suitable when the monitored machine works in a single and known work point [9]. In [10], a binary classifier is tasked with determining if the measured signal is either an impulse or a non-impulse response signal. Then, the signals classified as impulse responses are processed further to extract the dynamic parameters of the monitored machine, and are later used to determine the health indicator (HI). In [11], a condition-based monitoring system of a ball screw using a variational autoencoder is presented, where the RMS error of reconstruction is used as HI. The probability distribution of HI samples using a dynamic sliding window is obtained by using the kernel density estimation. Next, the probability-distribution points which exceed a predefined threshold are used to evaluate the degradation degree.

3 Overview

This section gives an overview of algorithms and approaches used during experiments, with a short introduction to autoencoders and a presentation of the setup of the condition-based monitoring system. The system architecture and devices used to built the test bench are also presented with the processing pipeline.

3.1 Autoencoders

Autoencoders are neural networks built from two coupled subnetworks, i.e., an encoder which performs dimensionality reduction (i.e., compression) into a latent space representation, and a decoder which reconstructs (i.e., decompresses) the signal, both individually parametrized. A standard topology of an autoencoder is presented in Fig. 1. The latent space is a lower dimensionality representation of data from the feature space (i.e., inputs) in which data points resembling each other are distributed more closely. The training objective is to minimize the reconstruction error between the decoder output and the true output

$$L(x, x') = \|x - x'\|^2 \tag{1}$$

where L is the loss function, x is the input in feature space, x' is the reconstructed input in feature space (i.e., the output of decoder). The encoder subnetwork and the latent space are partially similar to applying the principal component analysis to map the feature space into a lower number of dimensions with the use of eigenvectors, while maximizing the retained information. The advantage of autoencoders lays in non-linearities in activation functions evaluated after each hidden layer, which allows them to map the entire feature space more accurately [12]. Probabilistic counterparts of autoencoders are called variational autoencoders (VAE) with different mathematical formulations. In the case of a VAE, the data is sampled from parametrized distribution (prior) as the input to the probabilistic encoder. The training goal is to minimize a reconstruction error based on the Kullback–Leibler divergence between the parametric posterior (i.e., decoder output) and the true posterior [13]. This paper focuses exclusively on non-probabilistic autoencoders.

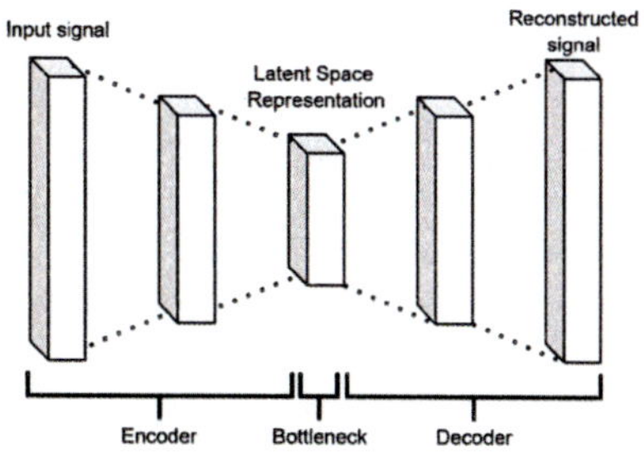

Fig. 1. Classic autoencoder network topology with encoder, bottleneck (latent space representation) and decoder.

Autoencoders are most commonly trained using the BP method by iteratively updating the weights of each layer in order to minimize the selected loss function. Alternatively, the ELM algorithm [14] may be used. This method updates only the weights between the hidden and output layers using the Moore-Penrose inverse, also called the pseudo-inverse [15], while the weights between input and hidden layers are randomly generated and never updated. The training with the ELM method is much faster (by multiple orders of magnitude), while giving good generalization results, and the optimization of weight values produces a unique solution [2,14].

4 Experiment Setup

In this section, the hardware and software setup is presented.

4.1 Hardware and Software

The device under test is a traditional feedback-based closed-loop control system, shown in a simplified diagram in Fig. 2. The required angular frequency of the DC motor shaft ω_{ref} is the reference signal. The control of the DC motor is implemented using the Raspberry Pi microcomputer with software implementation of the proportional-integral (PI) controller with the anti-windup system to limit the input voltage to the DC motor. The measured angular frequency of the DC motor ω_m is the system output.

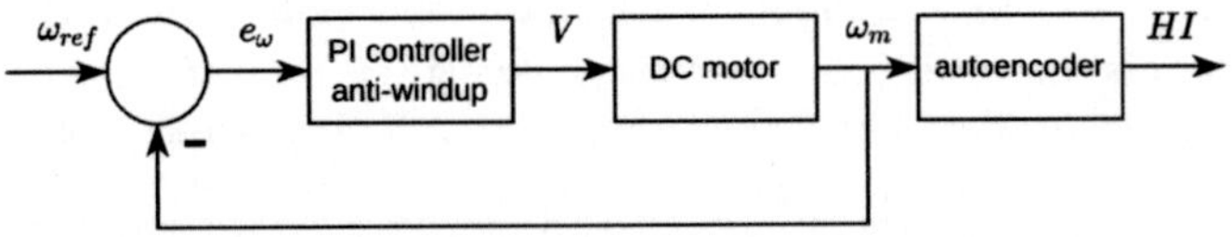

Fig. 2. Simplified diagram of condition-based monitoring system for DC motor. Additional components (current control loop, nonlinearities) are not included for sake of brevity.

The condition-based monitoring system consists of two main controllers: a) Raspberry Pi 4B microcomputer which, together with the L298N driver, acts as the DC motor driver controlling the Pololu 37D DC motor, b) NVIDIA Jetson Nano B01 microcomputer as the compute node with condition-based monitoring algorithm. In addition, a standard development computer is used to orchestrate the experiments and manage the controllers listed above. Raspberry Pi generates the pulse-width modulation (PWM) signal which is passed to the DC motor input. The current angular frequency of the DC motor is calculated from impulses sent by the DC motor encoder, delivered to the compute node, i.e., Jetson Nano, and subsequently to the autoencoder. The RMS error between the reconstructed and measured signals is computed and considered as HI. The lower the HI value, the higher the degree of degradation, which is the difference between healthy-system and current HI values.

The software is implemented using Python, and runs on embedded Linux on Raspberry Pi. The DC motor is controlled using the RPi.GPIO library, which uses standard Linux IO control. Autoencoders and models are built and trained using the Tensorflow/Keras framework.

4.2 Application

Measurements of the angular-frequency signal, obtained from a healthy DC motor, are collected by the condition-based monitoring system and split into windows of equal length. Next, the windows are used to train the autoencoder in an offline mode. The trained models are then used in the online mode with fixed weights to avoid model weight drift after deployment and testing. Samples of the measured signal are passed to the autoencoder, which returns the reconstructed signal. When the input signal is similar to the samples used for training (i.e., healthy samples), then the reconstruction error is low, indicating a healthy system. When the input signal is different from the training samples, the reconstruction error increases. The reconstruction error is calculated as RMS of the difference between the input and the reconstructed signal vectors. The pipeline is presented in Fig. 3.

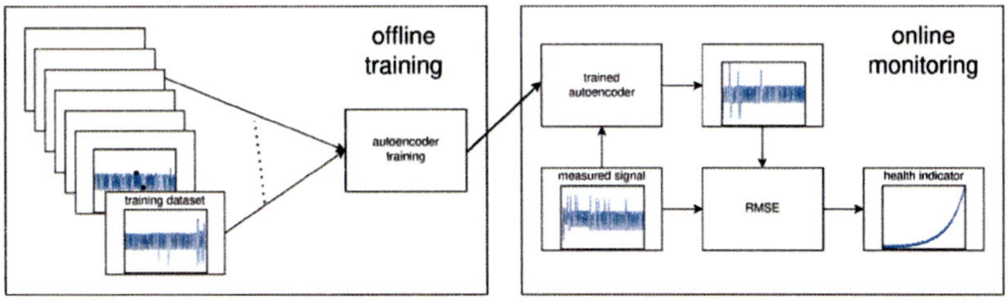

Fig. 3. Autoencoder-based algorithm to determine condition of DC motor.

5 Results

Results of experiments are presented and discussed in this section. The evaluation is performed under various conditions and is explorative in nature to cover multiple experiment scenarios and use findings for future research, focused on specific areas. The evaluation of condition-based monitoring system uses constant and variable reference signals (i.e., set or work points) to test robustness of settings with single and multiple autoencoders, each trained on the data collected from a single work point. Tests are also performed for a harmonic reference signal. Further tests evaluate the system behaviour when the disturbances are introduced to the spinning DC motor. The disturbances are introduced by physically obstructing rotation of the DC motor shaft with a piece of material, which results in an unevenly spinning motor. The correlation between HI and the measured angular-frequency waveform is evaluated. Finally, a comparison between autoencoders and other artificial intelligence classification algorithms, i.e., SVM and RF, is performed.

5.1 Parameters

Autoencoders are used with the dense layers of shape [250, 60, 250]. The feature space is of the size 250 (same as the number of samples in data windows), and latent space dimension is equal to 60. For BP-trained autoencoders, ReLU and *tanh* activation functions are used for hidden and output layers, respectively. The model is trained for 100 epochs, using the Adam optimizer [16], with the learning rate of 10^{-3}, the MSE loss function, and the batch size equal to 128. The ELM autoencoders have the same layers as the BP autoencoders, but with the *sigmoid* activation function on the hidden layer and without activation function on the output layer. The parameters of PI controller with the anti-windup system are as follows: $K_p = 1$ Vs/rad, $T_i = 0.8$ rad/V, $anti-windup-saturation = 12$ V, $T_a = 0.3$ s. The angular-frequency signal is sampled with the sampling frequency equal to $F_s = 25$ Hz.

5.2 Datasets

Several datasets are prepared to run the experiments. They are built from windows, each consisting of 250 samples. The windows are classified as windows from either healthy or degraded DC motors. However, this information is not used during training of autoencoders, but only for evaluation purposes in tests. All the datasets prepared and used are listed in Table 1.

Table 1. Summary of datasets prepared for experiments. [1] used only for testing, [2] indicates number of healthy samples in dataset, 50% indicates that half of samples are obtained from degraded motor.

Name	Work point [rad/s]	Size [samples]	Healthy[2] [%]	Experiment
D_{C10}	10	16 000	100%	Sects. 5.3, 5.4, 5.5
D_{C20}	20	16 000	100%	Sect. 5.4
D_{C30}	30	16 000	100%	Sect. 5.4
D_V	$2sin(0.08t) + 10$	16 000	100%	Sects. 5.3, 5.5
D_{C10CH}	10 + sin. disturbance	32 000	50%	Sect. 5.6
D_{VCH}	$2sin(0.08t) + 10$ + sin. disturbance	32 000	50%	Sect. 5.6
D_{C10CD}	10 + saw disturbance	2000[1]	50%	Sect. 5.6
D_{VCD}	$2sin(0.08t) + 10$ + saw disturbance	2000[1]	50%	Sect. 5.6

5.3 Single Autoencoder, Single Work Point

HI is calculated in the case of a single autoencoder for perfectly constant and disturbed work points. The results are presented in Figs. 4 and 5.

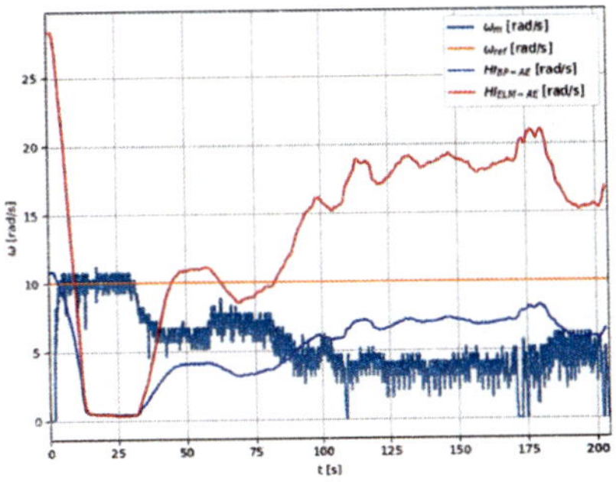

Fig. 4. Reconstruction error of autoencoders working at constant work point.

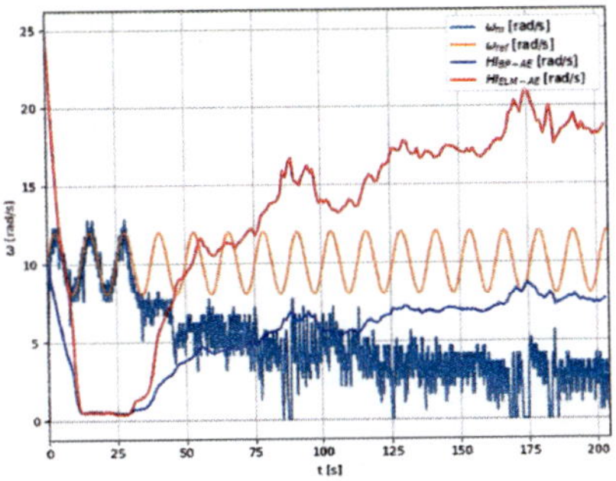

Fig. 5. Reconstruction error of autoencoders working at disturbed work point.

The decrease in the initial high values of HIs is the result of the input buffer of autoencoder filled with non-zero samples after its early initialization with zeros. Once the input buffer is completely filled with valid samples, the HI values are correct and indicate a healthy system. After about 30 s, the disturbances are introduced into the DC motor shaft, causing increase in the HI values. Both autoencoders, trained using BP and ELM methods, produce HI values with similar curves. The ELM method is significantly faster when it comes to training time, while also yielding a valid HI curve. Note that the HI values should be normalized in accordance with the selected algorithm.

5.4 Multiple Autoencoders, Multiple Work Points

The scenario where the DC motor operates with several constant work points is investigated in this section. The work points are 10, 20 and 30 rad/s. The reference signal starts as one of the work points used to generate the above datasets, e.g., 30 rad/s. It is kept constant for 200 s and then moves linearly to the next work point, e.g., 20 rad/s, and so on. The measured angular frequency is processed by all of the autoencoders working in parallel. All autoencoders were trained prior to the experiment on separate datasets, each on the data for a single work point. The results are shown in Fig. 6. It can be seen that each autoencoder correctly detects its respective work point and provides a low value of HI. In the case of DC motor working in several work points, the lowest HI signal among all the autoencoders can be used to identify the state of health of a DC motor. In a setting where various work points are expected, it may be useful to train and tune multiple encoders for each point. Using the ELM method may provide a simple and fast way to retrain autoencoders, even in a highly embedded setting, where compute capabilities are limited.

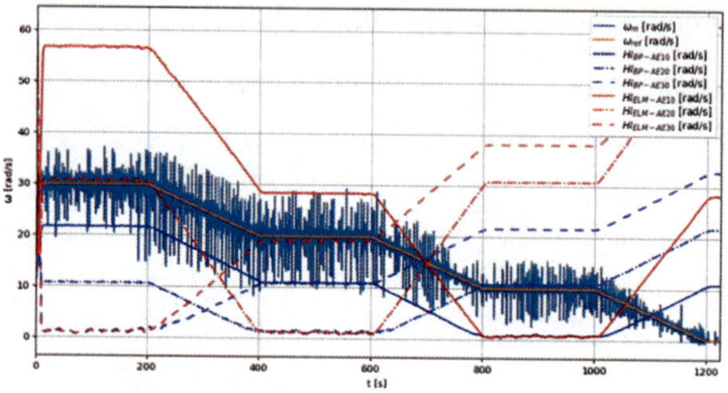

Fig. 6. HI results for multiple autoencoders trained on individual work points.

5.5 Health Indicator and Signal Correlation

In Figs. 4, 5 and 6, one can see that the HI signals are similar to the scaled difference between the reference and measured signals. It suggests that there is a near-linear relationship between the HI and difference signals. This is verified through computing their correlation by utilizing curves from the experiments introduced in Sect. 5.3. As the measured angular-frequency signal is noisy, an averaging window with the length N_{avg} is applied to the signal. The results are presented in Table 2.

Table 2. Correlation values for autoencoders trained on D_{C10} and D_V.

Dataset	N_{avg}	1	25	50	100	200
D_{C10}	$HI_{BP}(t)$	−0.736	−0.786	−0.822	−0.857	−0.835
	$HI_{ELM}(t)$	−0.783	−0.836	−0.872	−0.901	−0.878
D_V	$HI_{BP}(t)$	−0.765	−0.810	−0.82	−0.818	−0.844
	$HI_{ELM}(t)$	−0.797	−0.844	−0.852	−0.851	−0.875

5.6 Comparison with Classical Methods

The autoencoder is a neural network often used as an alternative to classical artificial intelligence methods for detecting degradation and anomalies. Comparison of autoencoders is performed against the SVM [17] and RF [18] algorithms. Two versions of models are tested - using sampled data and extracted features. The SVM and RF methods are implemented with the Python library scikit-learn [19]. For SVM, the regularization parameter is set to 1 and γ is set to 1 with linear kernel. The tested RF uses 50 estimators. Both methods are trained on the full datasets D_{C10CH} and D_{VCH}. The autoencoders are trained using data from a healthy DC motor.

For comparison purposes, the task is to classify sample windows as either deteriorated or healthy, and not to compare the HI curves. The autoencoders cannot be used for classification tasks in a straightforward manner, as the reconstruction error does not give a clear indication whether the input signal is obtained from a healthy DC motor. Instead, they produce information about the difference between the input signal and the samples used for training, i.e., the degradation degree. Therefore, a threshold value is applied to RMS of the difference signal in order to classify the DC motor as either healthy or degraded. The value of threshold is chosen empirically. For the BP autoencoder, the threshold is taken as $Thr = 0.8$, and for the ELM autoencoder as $Thr = 0.65$.

The metrics used are accuracy, precision, recall and f_1 score. The accuracy is the ratio of correct predictions to all predictions. The precision is a ratio of correctly classified healthy samples to samples classified as healthy both correctly and erroneously. The recall is a ratio of correctly classified healthy samples to correctly classified healthy samples and those marked incorrectly as degraded. The f_1 score uses recall and precision to calculate a harmonic mean for a given classifier [20].

Table 3. Results of comparison of methods.

Dataset	Algorithm	TPR	TNR	FNR	FPR	Accuracy	Precision	Recall	f1 score
D_{C10CH}	BP-AE	0.5	0.5	0.0	0.0	1.0	1.0	1.0	1.0
	ELM-AE	0.5	0.5	0.0	0.0	1.0	1.0	1.0	1.0
	RF	0.5	0.5	0.0	0.0	1.0	1.0	1.0	1.0
	SVM	0.188	0.5	0.313	0.0	0.688	1.0	0.375	0.545
D_{VCH}	BP-AE	0.5	0.5	0.0	0.0	1.0	1.0	1.0	1.0
	ELM-AE	0.5	0.5	0.0	0.0	1.0	1.0	1.0	1.0
	RF	0.5	0.5	0.0	0.0	1.0	1.0	1.0	1.0
	SVM	0.188	0.5	0.313	0.0	0.688	1.0	0.375	0.545
D_{C10CH} D_{C10CD}	BP-AE	0.5	0.499	0.0	0.001	0.999	0.999	1.0	0.999
	ELM-AE	0.5	0.499	0.0	0.001	0.999	0.999	1.0	0.999
	RF	0.488	0.5	0.013	0.0	0.988	1.0	1.0	0.987
	SVM	0.0	0.5	0.5	0.0	0.5	NaN	0.0	0.0
D_{VCH} D_{VCD}	BP-AE	0.5	0.479	0.0	0.022	0.979	0.959	1.0	0.979
	ELM-AE	0.5	0.499	0.0	0.002	0.999	0.997	1.0	0.999
	RF	0.227	0.5	0.273	0.0	0.727	1.0	0.454	0.624
	SVM	0.0	0.5	0.5	0.0	0.5	NaN	0.0	0.0

The results in Table 3 show that autoencoders and RF models trained and evaluated either on datasets D_{C10CH} and D_{VCH} both give accurate results. The SVM method correctly detects all samples from a degraded DC motor. However, it fails to classify samples from a healthy motor. When trained on the dataset D_{C10CH} and evaluated on D_{C10CD}, both autoencoders produce slightly worse results, seldom failing to classify samples as degraded. Nevertheless, autoencoders are more accurate than the results of RF and SVM. Using these datasets, the RF correctly classifies, as before, the samples from a degraded DC motor.

Both RF and SVM method misclassify healthy samples. When trained on dataset D_{VCH} and evaluated on D_{VCD}, the ELM autoencoder yields improved results compared to other algorithms, correctly classifying windows from a healthy DC motor, with slightly lower precision for a faulty motor. This is more pronounced when using a BP-trained autoencoder, which more often fails for faulty data, while being more accurate for healthy data. Classical methods fail for a healthy DC motor, with the RF method incorrectly classifying almost half of the windows, and the SVM unable to recognize any window at all. These experiments indicate that autoencoders are more suitable for this application than RF and SVM methods, and can adaptively detect degradations of monitored systems.

6 Conclusion

This work presents a condition-based monitoring system using an autoencoder for DC motors with explorative research under various conditions. It is shown that autoencoders are valuable tools to measure the degradation of DC motors working in a single constant or variable work points. It is also proved for a DC motor working on several work points, where multiple autoencoders are evaluated simultaneously.

The ELM autoencoders produce similar results to BP-based autoencoders, but have different amplification, with a need for normalization of output values. Then, their training process is faster by orders of magnitude. Future research will evaluate whether the ELM autoencoders could be applied in areas where quick retraining of a model may prove valuable, for instance to fine-tune it for a specific device or work point.

Autoencoders are compared with different artificial intelligence algorithms. In most cases, autoencoders give better results or very closely match the accuracy of classical methods like SVM and RF. However, to train the autoencoders, only the signals measured from a healthy DC motor are necessary, contrary to SVM and RF, where a full labeled dataset is needed, including both healthy and degraded samples. Another advantage is the fact that, in addition to the HI value, autoencoders can also measure the degree of degradation. They do not only perform a binary classification task.

References

1. Ran, Y., Zhou, X., Lin, P., Wen, Y., Deng, R.: A survey of predictive maintenance: systems, purposes and approaches. arXiv:1912.07383 (2019)
2. Michau, G., Hu, Y., Palmé, T., Fink, O.: Feature learning for fault detection in high-dimensional condition-monitoring signals. Proc. Inst. Mech. Eng. Part O J. Risk Reliab. **234**(1), 104–115 (2020)
3. Bengio, Y., Courville, A., Vincent, P.: Representation learning: a review and new perspectives. IEEE Trans. Pattern Anal. Mach. Intell. **35**(8), 1798–1828 (2013)
4. Creswell, A., White, T., Dumoulin, V., Arulkumaran, K., Sengupta, B., Bharath, A.A.: Generative adversarial networks: an overview. IEEE Sig. Process. Mag. **35**(1), 53–65 (2018)

5. Wang, W., Huang, Y., Wang, Y., Wang, L.: Generalized autoencoder: a neural network framework for dimensionality reduction. In: 2014 IEEE Conference on Computer Vision and Pattern Recognition Workshops, pp. 496–503 (2014). https://doi.org/10.1109/CVPRW.2014.79
6. Zhai, J., Zhang, S., Chen, J., He, Q.: Autoencoder and its various variants. In: 2018 IEEE International Conference on Systems, Man, and Cybernetics (SMC), pp. 415–419 (2018). https://doi.org/10.1109/SMC.2018.00080
7. LeCun, Y.: Deep learning & convolutional networks. In: 2015 IEEE Hot Chips 27 Symposium (HCS), pp. 1–95 (2015). https://doi.org/10.1109/HOTCHIPS.2015.7477328
8. Chen, Z., Li, W.: Multisensor feature fusion for bearing fault diagnosis using sparse autoencoder and deep belief network. IEEE Trans. Instrum. Meas. **66**(7), 1693–1702 (2017)
9. Michau, G.: From data to physics: signal processing for measurement head degradation detection. In: European Conference of the PHM Society 2018, Utrecht, Netherlands (2018)
10. Luo, B., Wang, H., Liu, H., Li, B., Peng, F.: Early fault detection of machine tools based on deep learning and dynamic identification. IEEE Trans. Ind. Electron. **66**(1), 509–518 (2019)
11. Wen, J., Gao, H.: Degradation assessment for the ball screw with variational autoencoder and kernel density estimation. Adv. Mech. Eng. **10**(9), 1–12 (2018)
12. Almotiri, J., Elleithy, K., Elleithy, A.: Comparison of autoencoder and principal component analysis followed by neural network for e-learning using handwritten recognition. In: 2017 IEEE Long Island Systems, Applications and Technology Conference (LISAT), pp. 1–5 (2017). https://doi.org/10.1109/LISAT.2017.8001963
13. An, J., Cho, S.: Variational autoencoder based anomaly detection using reconstruction probability. Special Lecture on IE, SNU Data Mining Centre **2**(1), 1–18 (2015)
14. Huang, G.-B., Zhu, Q.-Y., Siew, C.-K.: Extreme learning machine: theory and applications. Neurocomputing **70**(1), 489–501 (2006)
15. Strang, G.: Linear Algebra and Its Applications, 2nd edn., pp. 139–142. Academic Press Inc., Orlando (1980)
16. Kingma, D., Ba, J.: Adam: a method for stochastic optimization. In: 2015 International Conference on Learning Representations (2015). arXiv:1412.6980v9
17. Srivastava, D., Bhambhu, L.: Data classification using support vector machine. J. Theor. Appl. Inf. Technol. **12**(1), 1–7 (2010)
18. Breiman, L.: Random forests. Mach. Learn. **45**(1), 5–32 (2001). https://doi.org/10.1023/A:1010933404324
19. Pedregosa, F., et al.: Scikit-learn: machine learning in Python. J. Mach. Learn. Res. **12**(1), 2825–2830 (2011)
20. Grandini, M., Bagli, E., Visani, G.: Metrics for multi-class classification: an overview. arXiv:2008.05756v1 (2020)

Estimation of Mass Flow Rates of Two-Phase Flow Using Convolutional Neural Networks

M. F. Rocha-Mancera[1], S. Arce-Benítez[1], L. Torres[2(✉)], and J. E. G. Vázquez[2]

[1] Facultad de Ingeniería, Universidad Nacional Autónoma de México, Alcaldía Coyoacán, 04510 Mexico City, Mexico

[2] Instituto de Ingeniería, Universidad Nacional Autónoma de México, Alcaldía Coyoacán, 04510 Mexico City, Mexico

{ftorreso,jguzmanv}@iingen.unam.mx

Abstract. We present the design and training process of a couple of convolutional neural networks (CNNs) to predict the mass flow rates of a glycerin-air mixture injected into an looped horizontal pressurized pipeline. The CNNs were trained with spectrogram images generated from pressure differentials, which were calculated from pressure measurements taken at different points along the pipeline. To obtain enough data for the conception of the CNNs, we performed a series of experiments in which several combinations of glycerin-air flow mass rates were supplied. To program the CNNs of this work, we used Anaconda tools for Python developers.

Keywords: Convolutional neural network · Two-phase flow · Pipeline safety · Python

1 Introduction

For years, considerable effort has been invested in the study of two-phase flow because its relevance for the analysis and design of refrigeration cycles, power plants, compressors, condensers, evaporators, oil production, boilers, nuclear reactors and the development of space technology. Two-phase flow in pipelines is dynamically complex and develops different behaviors depending on the size of the pipeline, flow velocity, fluid properties, and pipe slope. Due to this situation, it is not easy to develop techniques and technologies that help to solve typical problems caused by two-phase flow, such as high pressure drops, undesirable flow patterns, as well as erosion due to velocities. To address these problems, it is essential to have real-time information about the two-phase flow. However, instrumentation to measure two-phase flow variables is extremely expensive, fact that increase

This contribution is a result of Project 4730 financed by CONACYT (Atención a Problemas Nacionales, Convocatoria 2017).

Z. Kowalczuk (Ed.): DPS 2022, LNNS 545, pp. 190–201, 2023.
https://doi.org/10.1007/978-3-031-16159-9_16

the operation cost of transport pipelines. Hence, new and cheap alternatives are always desirable, especially, those used for safeguarding the operation of pipelines.

In this work, we propose the use of pressure transducers, together with a couple of CNNs, for the indirect estimation (prediction) of the mass flow rates injected into a pipeline. There are two reasons for this proposal: (1) pressure transducers are cheap and easy to install, (2) CNNs have been extensively researched; therefore there is a lot of information that allow their easy implementation.

The use of CNNs as tools for applications involving two-phase flow in pipes has already been explored. For example, Du et al. (2018) proposed in [2] a CNN to identify the flow patterns of an oil-water mixture transported in a standpipe. For the CNN training, images were acquired by a high-speed camera placed in front of a transparent pipeline section. To improve the image acquisition, LED backlighting was used. A very similar contribution to the one proposed in this work was presented in [9], by Xu et al. (2020), who designed a CNN to predict flow rates, flow patterns and gas void fraction of an oil-gas flow. To train the network, the authors collected electrical capacitance tomographies of the two-phase flow at different oil and gas flow rates. The instruments used to acquire the tomographies were placed before and after of a venturi tube. From the data provided by these instruments, images were obtained using an algorithm based on local binary patterns. According to the authors, the prediction results using images of the flow pattern as CNN inputs were much better than the prediction results using the original capacitance data as input. In [6], Torisaki and Miwa (2020) proposed a CNN to extract bubbling multiphase flow features in real time. From the coordinate information of each individually detected bubble, characteristics such as void fraction, bubble number density, and average bubble size can be instantly acquired. Zhang et al. (2020) presented in [10] the results provided by a CNN trained to identify water-air flow patterns in horizontal pipes. The CNN inputs were contour lines of flow velocity measured by Doppler ultrasonic velocimetry. Urbina-Salas et al. (2021) presented in [8] results of the performance of the training process, validation and classification tests of images of two-phase flow patterns using a CNN. To carry out these tasks, a series of experimental video frames were extracted from a working heat exchanger taken by a FASTCAM-PCI R2 camera at 400 fps. Subsequently, a catalog of images with samples of representative flow patterns with slug, annular, semi-annular and disperse annular flow types was defined. The CNN proposed by the authors was of 9 layers and, with this network, they obtained 90% effectiveness in identifying the flow patterns developed in a small vertical channel.

As can be seen, the works mentioned are relatively recent. This means that the use of CNNs for the study and measurement of parameters that characterize two-phase flow is an ongoing line of research, which can still contribute much knowledge to the area of fluid mechanics. In this work, Sect. 2 presents the experimental work done for the training, testing and validation of the proposed CNNs. Section 3 presents the overall conception of the proposed CNNs. Section 4 describes the results and Sect. 5 draws some conclusions and future work.

2 Experimental Work

2.1 Experimental Setup

The experiments were carried out in the flow loop shown in Fig. 1. In this loop, the liquid phase is supplied to the test section through a progressive cavity pump (Seepex Mod. BN35-24). This pump is capable of delivering constant mass flow rates in the range of 0.0 [kg/s] to 6.1 [kg/s]. The outlet of the test section is connected to a separate tank with an internal capacity of 1.5 [m^3]. On the other hand, a Kaeser Aircenter SK.2 compressor supplies a constant mass of dry air at room temperature at a pressure in the range of 0.0 [Pa] to 1.6×10^6 [Pa]. The mass flow rate is carefully set with a regulator and globe valves. Mixing occurs in the 3-way connection shown in Fig. 1. Mass flow rates can be measured at the inlet with an Endress-Hauser coriolis flowmeter. All pressures are measured with an array of conventional MEAS U5300 transducers. The selected measurement ranges for these instruments are 0.0 [Pa] to 1.03×10^5 [Pa] and 0.0 [Pa] to 3.45×10^5 [Pa].

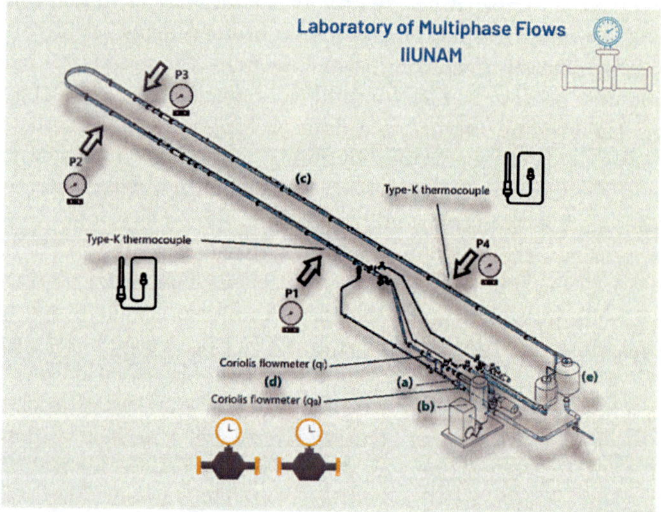

Fig. 1. Test apparatus. a) pumping subsystem, b) pressurized gas subsystem, c) flow loop, d) flow-meters and e) separator and storage tanks. The flow loop consists of a 54m long pipe with pressure transducers placed along.

2.2 Methodology

Six experiments were executed to obtain the required data to conceive the CNNs. In each experiment, 15 air-glycerin mass flow combinations were injected into the pipeline. These combinations were obtained by pairing each mass flow rate of air with each mass flow rate of glycerin as indicated in Table 1.

Table 1. Matrix of experiments. Glycerine flow rate (q_l) and air flow rate (q_g) injected into the test section of the pipeline.

q_g [kg/s]	q_l [kg/s]
	1.3
0.005	2.5
0.01	3.7
0.015	4.9
	6.1

The pressure transducers were placed in four different points (coordinates) distributed along the pipeline space domain. These points were labeled in this work as: x_{P1}, x_{P2}, x_{P3}, x_{P4}. The distance between these points were labeled as Δx_1, Δx_2, Δx_3. The size of these distances are summarized in Table 2. The pressures measured at the four different points were labeled as P_1, P_2, P_3 and P_4, respectively (check Fig 1). The pressure differential were calculated as: $\Delta P_{12} = P_1 - P_2$, $\Delta P_{23} = P_2 - P_3$ and $\Delta P_{34} = P_3 - P_4$. All this information is summarized in Table 2.

Table 2. Information about pressure differentials

Item	Definition	Value	Units
Δx_1	$x_{P2} - x_{P1}$	18	[m]
Δx_2	$x_{P3} - x_{P2}$	4	[m]
Δx_3	$x_{P4} - x_{P3}$	21	[m]
ΔP_{12}	$P_1 - P_2$	–	[Pa]
ΔP_{23}	$P_2 - P_3$	–	[Pa]
ΔP_{34}	$P_3 - P_4$	–	[Pa]

Given that 3 pressure differentials can be calculated with 4 pressure measurements acquired at 4 different pipeline points, and given that 15 different glycerin-air flow combinations were injected, then at each sampling time, 3 data matrices were obtained with 15 pressure differential values, i.e., 45 pressure measurements at each sampling instant (check Fig. 2). The sample period was set $\Delta t = 0.2$ [s] and every performed experiment had a duration of $t_f = 120$ [s]. This means that for each experiment we obtained $N_s = 900 \times 45 = 40500$ pressure samples to construct 45 spectrograms.

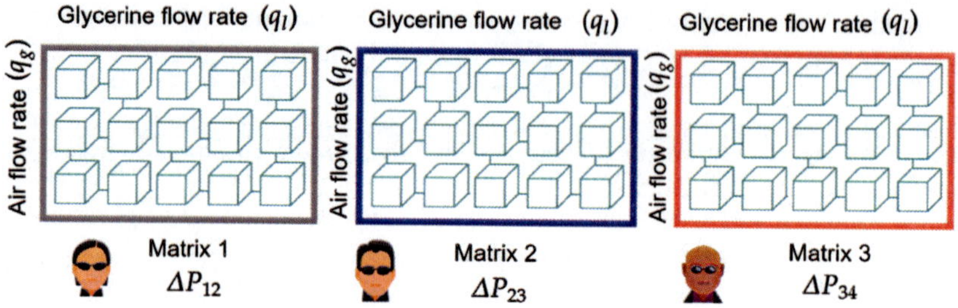

Fig. 2. Data arrays at each instant of time.

3 Convolutional Neural Networks for Estimation

The origin of the name of CNN comes from the mathematical concept of convolution, which is a linear transformation of two functions into a third. Because the convolution operations can be performed between two-dimensional arrays, these neural networks are very effective for computer vision tasks, such as image classification. The basis of CNNs were given by Fukushima and Miyake (1982) in [3], and later improved by LeCun et al. (1998) in [5], who introduced a backpropagation-based learning method for a correct training.

In practice, a CNN is designed, programmed, coded, and tested in the programming environment of choice. The two most popular frameworks to do it are `Anaconda` by Anaconda Inc and `Google Collab` by Google. CNN created in these environments are coded in the Python programming language. To design the couple of CNNs of this work, we used `Anaconda`.

The goal of the CNNs presented in this work is the discrete estimation of the mass flow rates of glycerin and air simultaneously injected into a pipeline. The aim of the first CNN is the estimation of the the glycerin mass flow rate and the purpose of the second one is the estimation of the air flow rate. For this purpose, spectrograms[1] of pressure differentials were used as inputs of the CNN. To construct the spectrograms, pressure measurements were obtained from four pressure transducers installed along a two-phase flow pipeline housed at the Engineering Institute of UNAM[2]. The recorded pressure were transformed into pressure differentials. Then, by using the short-time Fourier transform (STFT), spectrograms were constructed from the pressured differentials. It is worth to say that the STFT algorithm was programmed in MATLAB. Finally, the spectrograms were converted into image format files, which were stored in appropriately labeled folders, to later serve as CNN feeds. A scheme of this methodology is presented in Fig. 3.

[1] Images that show the time evolution of the spectrum of a signal.

[2] Further details of the experimentation are given in Sect. 2.

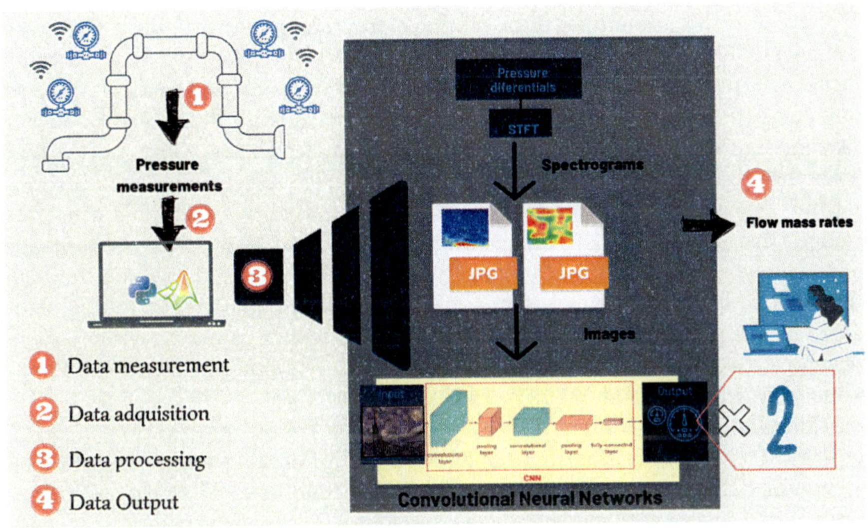

Fig. 3. Schema of the estimation of flow mass rates using CNNs

3.1 Image Classification

For training, a CNN needs a set of images properly classified. The classification is quite simple. To classify images, it is enough that they are inside a folder with the name of the class. We classified the spectrograms according to (a) the mass flow rate of glycerin (labeled as: `quantity+[kg/s]`) and (b) the mass flow rate of air (labeled as: `quantity+[g/s]`).

- Spectrograms of air mass flow rate:
  `5g/s` → 🗁 `10g/s` → 🗁 `15g/s` → 🗁
- Spectrograms of glycerine mass flow rate:
  `1.3kg/s` → 🗁 `2.5kg/s` → 🗁 `3.7kg/s` → 🗁 `4.9kg/s` → 🗁
  `6.1kg/s` → 🗁

In short, all images were manually separated into 3 folders for the conception and use of a first CNN, and into 5 folders for the conception and use of a second CNN. At the end, 8 folders were created to feed both CNNs.

3.2 Data Augmentation

The images that were available at the beginning of the CNN were 279. This number of images is very small in contrast to the number regularly used in other neural networks. To address this issue, a technique called data augmentation was used [7]. Data augmentation is an external process whose objective is to increase the amount of data available to train a network when these are insufficient. This process is done prior to the CNN training and it is not part of the CNN program. Image augmentation refers to modifying multiple aspects of an image in order to multiply them and increase the total image pool. For this contribution, `ImageDataGenerator` was used, which is a `Keras`[3] class that allows the augmentation of the number of images. There are many modifications that this `Keras` class can perform, but the ones used were the following: (1) Change Channel. Makes the image whiter.(2) Vertical Mirror. Flips the image about the vertical axis. (3) Random Rotation. Rotate the image in a random range. It can be clockwise or counterclockwise rotation. (4) Highlight. Highlight figures in the image. (5) Random Zoom. Make a certain zoom on some part of the image. (6) Obscured. Darkens the image. (7) Movement. Moves the image in one direction: a) Right, b) Left, c) Up, and d) Down.

Figure 4 shows an original image and the use of data augmentation to multiply it by 10. To obtain even more images, the original images were transformed by changing their color palette using MATLAB functions. The original images were colored with the default palette of MATLAB: the JET palette; then, for recoloring them, we chose to use the HSV palette. Once the images with the new colors were obtained, the number of images was increased. Examples of the recolored images are shown in Fig. 5.

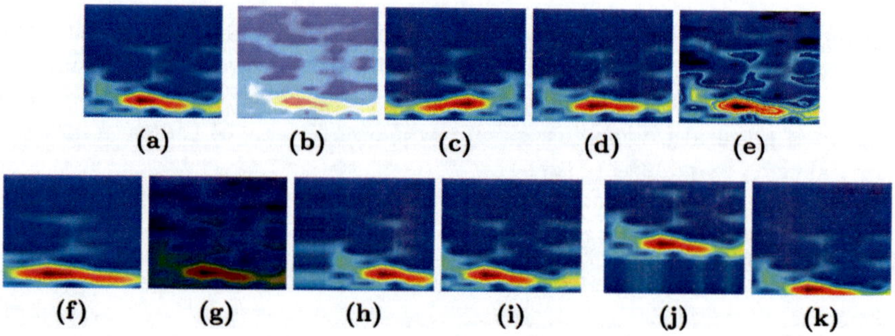

Fig. 4. Data augmentation, MATLAB JET palette colors: **(a)** Original image. **(b)** Lightened image. **(c)** Image flipped on the vertical axis. **(d)** Image slightly rotated. **(e)** Highlighted image. **(f)** Zoomed image. **(g)** Darkened image. **(h)** Image dragged to the right. **(i)** Image dragged to the left. **(j)** Image dragged up. **(k)** Image dragged down.

3.3 Training, Validation and Testing of the CNN

The CNNs for this project were programmed in `Anaconda` and was based on a CNN architecture proposed in [1]. Figure 6 shows the steps followed for training,

[3] It is a high level application programming interface (API) designed for the creation and training of *deep learning models.*

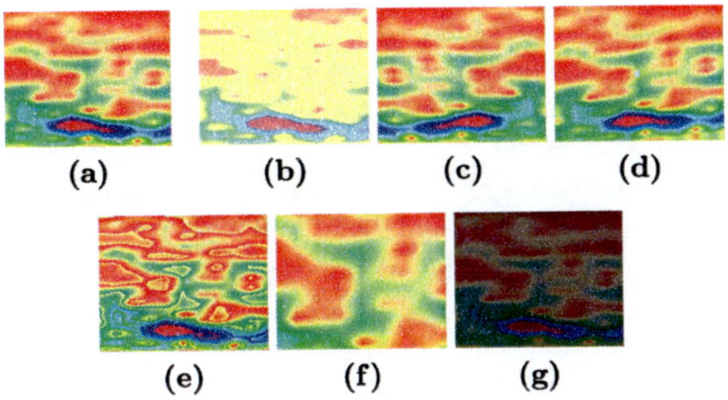

Fig. 5. Data augmentation, MATLAB HSV palette colors: **(a)** Original image. **(b)** Lightened image. **(c)** Image flipped on the vertical axis. **(d)** Image slightly rotated. **(e)** Highlighted image. **(f)** Zoomed image. **(g)** Darkened image.

validating and testing both CNNs. Each stage of this process will be following explained.

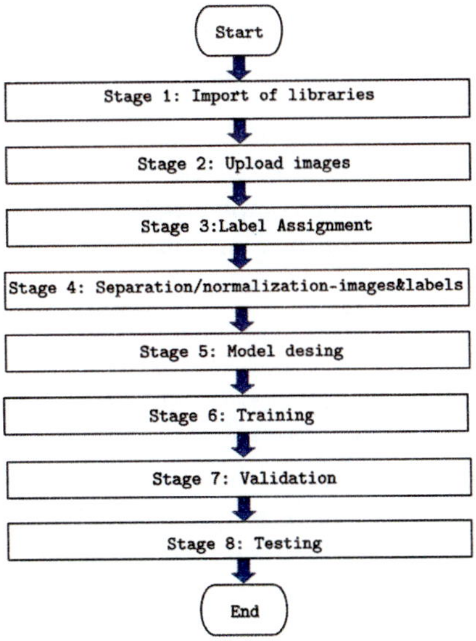

Fig. 6. Overall workflow of the CNNs training, testing and validation

Import of Libraries. The different libraries used for the proper functioning of the CNNs were: (1) `TensorFlow`: open source library that can simultaneously relate multiple data to build and train neural networks. For our network, it was used as the *backend*[4] of `Keras`. (2) `Keras`: it is open-source software that serves as the interface between `Tensorflow` and the programmer, which offers an environment more friendly and a code less extensive. `Keras` also serves as *backend* of `Microsoft Cognitive Toolkit`, `Theano`, and `PlaidML`. (3) `Numpy`: it means `Python Numeric`. It is a library that allows powerful calculations with multidimensional arrays and contains tools to work with these arrays. (4) `Matplotlib.pyplot`: it provides a way of graphing similar to MATLAB. It is mainly designed to generate interactive graphs. (5) `Os` (module): This module provides a portable way of using operating system dependent functionality. It is used to open, close or read files, manipulate routes, create temporary files and addresses among other utilities. Particularly, in our networks, it is used for getting the path of the folder in which the images are located. (6) `Re` (module): this module is used to search for matches with a search pattern. Particularly, in our networks, it was used to find the images that were used for training. (7) `Scikit-learn`: it includes several popular machine learning algorithms, such as regression and classification. (8) `Scikit-image`: it contains a collection of algorithms for image processing.

Upload Images. Once the libraries have been loaded, the CNN training program looks for the folders where the images are held. Each folder contains each class that we want to predict. These images are uploaded to the network. In this stage, the images were resized. For our networks, the original size of the images was 276×217 pixels, and they then were resized to 56×56 pixels. If the images were left at their real size, many more neurons would be used and the learning process would become much more complex, which would occupy resources that are not available. The resizing of each image is done immediately it is loaded.

Label Assignment. Each image was labeled with the name of the folder in which is located. Each image was converted to pixel arrays using `numpy` library. In a type-matrix variable, the pixel matrices of all the images were placed, and in a type-array variable, the labels of each image were stored. A class was assigned to each label. For the classes related to the air mass flow rate, a number from 0 to 2 was given to each folder, because we only worked with 3 classes of air mass flow rate. This labeling was done for when the network is asked to identify an image, identify it with the number that corresponds to it. For the classes related to the glycerin mass flow rate, a number from 0 to 4 was given to each folder, because we worked with 5 classes of glycerin mass flow rate. In Table 3 and Table 4, information about the classes, labels and number of images is summarized.

Separation. In this stage, the number of images to be used in the training, validation and testing was assigned. The division was done as follows: 72.25% of the total images were assigned to training, 15% of the total images to validation, and the remaining 12.75% to testing. The training arrays were stored in the

[4] It is the part of the software that processes data and perform math operations.

Table 3. Number of images per label (Glycerin mass flow rate)

Label	Class	Number of images
0	`1.3 kg/s`	1126
1	`2.5 kg/s`	786
2	`3.7 kg/s`	1115
3	`4.9 kg/s`	786
4	`6.1 kg/s`	1110

Table 4. Number of images per label (air mass flow rate)

Label	Class	Number of images
0	`5 g/s`	1274
1	`10 g/s`	1176
2	`15 g/s`	1176

variable `train_image` (which contains the images) and `train_label` (which contains the labels). The test arrays were stored in `test_image` and `test_label`.

Model Design. Our networks were built with the following layers:

- A convolutional layer made up of 32 *kernels* with size 3×3. The layer includes a padding set in *same* configuration, i.e., it is made up of zeros. The *stride* of this layer was 2×2. The activation function used in this layer was ReLU. All these metaparameters were chosen heuristically by comparing the results of multiple network configurations.
- A max pooling layer with size 3×3 and padding in *same* configuration.
- A dropout layer, which disables 25% of the parameters that pass through it.
- A batch normalization layer.
- A flatten layer.
- A first densely connected layer with 32 neurons. Softmax was chosen as its activation function.
- A dropout layer was placed again to disable 30% of the parameters that pass through it.
- A dense layer, with a number of neurons to the number of classes. The activation function of this final layer was the Softmax function.

In the compilation of the model, we used as loss function (a measure of how well a CNN perform a prediction) the *categorical cross entropy*. To reduce the overall loss provided by the loss function, we used the Adam optimizer, which is an algorithm that modifies the attributes of a CNN, such as weights and learning rate. The Adam optimizer is commonly used instead of the classical stochastic gradient descent procedure. The advantages of this algorithm were listed in [4]. After repeated tests, we concluded that the best learning rate for the Adam algorithm was $\alpha = 0.0002$ (Table 5).

Table 5. Model architecture

Type of layer	Output size	Parameters
Convolutional	28 × 28 × 32	896
Max pooling	10 × 10 × 32	0
Dropout	10 × 10 × 32	0
Batch normalization	10 × 10 × 32	128
Flatten	3200	0
Dense	32	102432
Dropout	32	0
Dense	3	99

Training. In this stage, the networks were trained. For this purpose the meta-parameter *epochs*, which is the number of iterations that a CNN executes, was defined. In order to analyze the behavior of the training, this variable was setting from 10 to 200. At the end, *epochs* was fixed at 100, because better results were obtained with this value.

Validation. In this stage, the CNNs were examined by using the images that were reserved exclusively for validating. Since these images were not used in training, CNNs did not know about them, so they are suitable to validate if the CNNs recognizes them with what they learned. The validation yields the accuracy and loss of the model.

Testing. For this stage a new program was generated. Its goal was to load the trained model and then load a completely new image. This image was not in the Training and Validation subsets, and is intended to show that the network works, and that it is capable of classifying it in some label.

4 Results

In the testing stage, the CNN for predicting the mass flow rate of glycerin produced results of 68% accuracy and a loss of 0.9982. From the set of test images (739 in total), the network was able to correctly label 503 images and miss 236. The CNN for predicting the mass flow rate of air gave results, for the test stage, of 78% accuracy and a loss of 0.6185. Among the set of test images (544 in total), the network was able to correctly match the label of 428 images and missed 116.

5 Conclusions a Future Work

This work presented the design of two CNNs to predict the mass flow rates of glycerin and air simultaneously injected into a pipeline. The accuracy of the

CNN trained to predict the glycerin flow rate was 68%. The accuracy of the CNN trained to predict the air flow rate was 78%. These results, which seem modest, were obtained with a limited number of images thanks to the use of the Data Augmentation technique, which consists in modifying the original images to obtain more. It is fair to mention that the estimation of the mass flows performed by the CNNs is discrete, i.e., a discrete value is estimated for the mass flow of glycerin, and another discrete value for the mass flow of air. These discrete values were predefined during the conception of the CNNs and were labeled to categorize them. Therefore, as future work, we plan to design a network in which the estimation is continuous and not discrete using interpolation techniques. We are also envisioning incorporating the temperature variable in the design of the networks, since the viscosity of glycerin is sensitive to it. In addition, we will look at how to train a single network that provides both values of mass flow, although for this, we may have to consider performing more experiments to obtain more images.

References

1. Bagnato, J.I.: Clasificación de Imágenes en Python | Aprende Machine Learning (2018). http://www.aprendemachinelearning.com/clasificacion-de-imagenes-en-python/
2. Du, M., Yin, H., Chen, X., Wang, X.: Oil-in-water two-phase flow pattern identification from experimental snapshots using convolutional neural network. IEEE Access **7**, 6219–6225 (2018)
3. Fukushima, K., Miyake, S.: Neocognitron: a self-organizing neural network model for a mechanism of visual pattern recognition. In: Amari, Si., Arbib, M.A. (eds.) Competition and Cooperation in Neural Nets. LNBM, vol. 45, pp. 267–285. Springer, Heidelberg (1982). https://doi.org/10.1007/978-3-642-46466-9_18
4. Kingma, D.P., Ba, J.: Adam: A method for stochastic optimization. arXiv preprint arXiv:1412.6980 (2014)
5. LeCun, Y., Bottou, L., Bengio, Y., Haffner, P.: Gradient-based learning applied to document recognition. Proc. IEEE **86**(11), 2278–2324 (1998)
6. Torisaki, S., Miwa, S.: Robust bubble feature extraction in gas-liquid two-phase flow using object detection technique. J. Nuclear Sci. Technol. **57**, 1–14 (2020)
7. Torres, J.: Deep Learning Introducción práictica con Keras. Primera parte, Watch This Space (2018)
8. Urbina-Salas, I., et al.: Application of convolutional neural networks for the classification of two-phase flow patterns. In: 2021 IEEE International Autumn Meeting on Power, Electronics and Computing (ROPEC), vol. 5, pp. 1–6. IEEE (2021)
9. Xu, Z., Wu, F., Yang, X., Li, Y.: Measurement of gas-oil two-phase flow patterns by using CNN algorithm based on dual ECT sensors with venturi tube. Sensors **20**(4), 1200 (2020)
10. Zhang, Y., Azman, A.N., Xu, K.-W., Kang, C., Kim, H.-B.: Two-phase flow regime identification based on the liquid-phase velocity information and machine learning. Expe. Fluids **61**(10), 1–16 (2020). https://doi.org/10.1007/s00348-020-03046-x

Recurrent Neural Network Based Adaptive Variable-Order Fractional PID Controller for Small Modular Reactor Thermal Power Control

Bartosz Puchalski[1,2](✉), Tomasz Adam Rutkowski[1,2], Jarosław Tarnawski[1,2], and Tomasz Karla[1,2]

[1] Digital Technologies Center, Gdańsk University of Technology, 11/12 Gabriela Narutowicza Street, 80-233 Gdańsk, Poland
{bartosz.puchalski,tomasz.adam.rutkowski,jaroslaw.tarnawski, tomasz.karla}@pg.edu.pl

[2] Department of Intelligent Control and Decision Support Systems, Faculty of Electrical and Control Engineering, Gdańsk University of Technology, 11/12 Gabriela Narutowicza Street, 80-233 Gdańsk, Poland

Abstract. This paper presents the synthesis of an adaptive PID type controller in which the variable-order fractional operators are used. Due to the implementation difficulties of fractional order operators, both with a fixed and variable order, on digital control platforms caused by the requirement of infinite memory resources, the fractional operators that are part of the discussed controller were approximated by recurrent neural networks based on Gated Recurrent Unit cells. The study compares the performance of the proposed neural controller with other solutions, which are based on definitional fractional-order operators exploiting an infinite memory buffer and a classical adaptive PID controller. The proposed neural approximations of variable-order fractional operators applied to a PID-type controller provide a viable solution that can be successfully implemented on present-day digital control platforms. The research presented here focuses on the aspects of accuracy of approximators in simulated operating conditions within the thermal power control system of the challenging plant such as Small Modular Nuclear Reactor.

Keywords: Fractional calculus · Recurrent neural networks · Approximation methods · Control system synthesis · Adaptive control

1 Introduction

First mentions of Fractional Order (FO) calculus appeared at the end of the 17th century. This calculus generalizes the classical differential and integral calculus into orders, which can take real or complex values. Assuming that the order of the general integro-differential operator denoted as $\mathscr{D}$ is given as α, then the fundamental FO operator can be formalized as [17]

Z. Kowalczuk (Ed.): DPS 2022, LNNS 545, pp. 202–214, 2023.
https://doi.org/10.1007/978-3-031-16159-9_17

$$
{}_{t_0}\mathscr{D}_t^{\alpha} f(t) = \begin{cases} d^{\alpha} f(t)/dt^{\alpha}, & \text{if } \alpha > 0, \\ f(t), & \text{if } \alpha = 0, \\ \int_{t_0}^{t} f(\tau) d\tau^{-\alpha} & \text{if } \alpha < 0, \end{cases} \tag{1}
$$

where $\alpha \in \mathbb{R}$ or in general case $\alpha \in \mathbb{C}$. If in definition (1) the order α is introduced as a function of arbitrary independent variable, e.g., time t, then the fundamental operator (1) will be given as ${}_{t_0}\mathscr{D}_t^{\alpha(t)}$. This modification represents a Variable-Order Fractional Calculus (VO-FC) operator. The introduction of order variation denoted as $\alpha(t)$ to the above-mentioned fundamental FO operator (1) undoubtedly extends its usability. Although VO-FC operators have been formalized in recent years, the research community has been intensively exploring this relatively new branch of integro-differential calculus. Current applications of VO-FC operators are presented in a broad review [5].

One field that FO operators have significant influence is control engineering. They are used in such applications as: modeling [2], sensing and filtering [4], linear control [14], and many other categories. With FO operators, the most common type of controller used in industry, i.e. PID controller has evolved to Fractional PID (FPID) or also called $PI^{\lambda}D^{\mu}$ controller [6]. In FPID controller, two new degrees of freedom were introduced, and they are associated with the order of integration – λ and differentiation – μ operations accordingly. This evolution has allowed for a more flexible process of FPID controller synthesis, which directly improves the control quality metrics in comparison to the classical PID controller solutions [10]. The mentioned FPID control also is available for nonlinear plants, for example, by applying fuzzy switching mechanisms [8,11,12].

Fuzzy switched FPID controllers used to control nonlinear plants are characterized by a complex synthesis process mainly because of the problem caused by a large number of parameters to be adjusted. These include the parameters of the local FPID controllers and parameters related to the shape of the associated membership functions used in the fuzzy part of the controller.

The introduction of VO-FC operators simplifies the synthesis problem, as it becomes possible to directly implement an adaptation technique based for instance on the well-known gain-scheduling mechanism into the structure of the FPID controllers. In this paper the FPID controller with a mechanism for adaptation of parameters and orders of integration and differentiation operators will be called VO-FPID controller.

Depending on the VO-FC definition, operators are characterized by different properties that mainly address issues related to two memory effects. The first memory effect is called fading memory [3], and it is associated with both fixed-order and variable-order fractional operators. The second effect is associated only with VO-FC operators, and it addresses the memory of the varying order [15]. This effect may occur with different intensities, i.e. no order memory, weak order memory, and strong order memory. The use of FO operators and models in practical real-time applications which involve implementation on digital platforms is problematic due to the aforementioned fading memory effect. Depending on which definition of the FO operator is used, there arises the singular integral problem (Riemann-Liouville or Caputo definitions) [13] or the problem related to processing an infinite number of signal samples (Grünwald-Letnikov definition). Regardless of the nature of the fading memory problem, the very core of this problem causes significant difficulties in implementing FO or VO-FC operators on digital computing platforms, resulting explicitly in need for infinite memory resources. Therefore, there is a need to develop techniques for approximation and modeling FO and VO-FC operators. These approximations should retain the properties of definitional operators (exploiting infinite memory), using finite memory. Typical approximations found in the state-of-the-art literature for FO operators include: standard and refined Oustaloup filters, frequency response fitting approach, continued fraction-based approximations, and many others as described in [17]. These typical solutions listed above perform efficiently for fractional operators involving fixed order. Unfortunately, they are not practical when dealing with VO-FC operators because they require real-time re-calculation of their parameters, which depend directly on order.

Given the issues outlined above, the authors focused on an approximation technique for VO-FC operators based on the application of modern recurrent neural network architectures. The main advantage of using neural approximators is that they don't need an infinite memory buffer and also the order of the neural fractional operator can be changed without recalculating its parameters. However, the neural form of the approximator requires a learning process with the usage of appropriate data that leads to satisfactory operator responses. Pre-trained neural networks were used in the research presented in this paper. Broad investigations along with a detailed presentation of the learning process on such neural approximations for both FO and VO-FC operators has been done previously with satisfactory results, and the outcomes of these studies were presented in [7,9].

The paper is organized as follows. In Sect. 2 the considered problem is described. In Sect. 3 methodology used in the study is presented. Section 4 includes and discusses the quantitative and qualitative results that were obtained from the simulation studies. Finally, Sect. 5 concludes the paper.

2 Problem Statement

Nuclear reactors are complex non-stationary plants characterized by non-linear processes dynamics with various time scales. A nuclear reactor in the Small Modular Reactor (SMR) technology, considered in this paper, is a modern, developmental technology important from the point of view of the energy industry with particular emphasis on economic aspects. Due to the continuous increase in energy production from renewable energy sources (e.g., photovoltaics, hydroelectric power, and off-shore wind farms), it is necessary to adopt nuclear power plants to operate in load-following mode. Therefore, nuclear power plant control

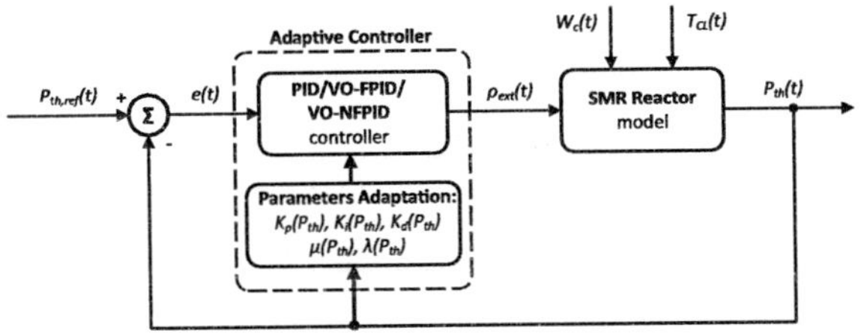

Fig. 1. Considered control system structure, where: $P_{th,ref}(t)$ and $P_{th}(t)$ are set point, $SP(t)$ and process value, $PV(t)$, $e(t)$ - error signal, $\rho(t)$ - external reactivity form control rods movement (control signal, $CV(t)$), $W_c(t)$ and $T_c(t)$ are disturbance inputs in the form of coolant mass flow rate and coolant temperature at the inlet to the SMR reactor, respectively.

systems are required to generate control signals enabling stable and effective system operation considering a widely varying operating point related to thermal power of nuclear reactor and electric power fed into the power grid. One technique that makes this adaptation possible is adding to classically used solutions, such as PID controllers, mechanisms for adapting their parameters depending on the current system operating point related to nuclear reactor thermal power. Furthermore, this adaptation can be combined with extension of the classical PID control algorithm's capabilities by introduction of the VO-FC operators. Therefore, the SMR nuclear reactor was chosen as a case study in the presented research. It should be noted that the goal of the authors in this case was to verify the performance and accuracy of neural approximators of VO-FC operators in a practical and complex approach, i.e. in the SMR rector power control system. It is also known that the design of control systems for critical infrastructure plants, to which the SRM reactor undoubtedly belongs, should include an extensive analysis of safety, reliability, and robustness. However, these elements at the current stage of the authors' investigations on various possibilities of approximation of VO-FC operators go beyond the scope of this paper. The analysis of the mentioned issues will undoubtedly be the subject of further research by the authors when satisfactory results on neural approximators are obtained.

General structure of the SMR reactor power control system considered in the study with adaptive controller is shown in Fig. 1. It consists of the main feedback loop and internal information loop. The controller's parameters used in the control system structure are adapted based on the current system operating point related to reactor thermal power output $P_{th}(t)$. This paper proposes, exploiting formerly prepared neural approximations, an implementation of the VO-NFPID controller, i.e., the Variable Order Neural Fractional PID controller with adaptive parameters and adaptive integration and differentiation operators orders. The main contribution of the paper is represented by original research results that address: 1) implementation of VO-NFPID controller with recurrent neural networks used for approximation of VO-FC operators, 2) demonstration of the methodology used for the synthesis of the VO-NFPID controller based on the case study of SMR type non-linear nuclear reactor power control, 3) comparison of the behavior of the classical adaptive PID controller with the adaptive VO-NFPID controllers, 4) comparison of the obtained results with approach based on the definition-based infinite memory implementation of the VO-FPID controller.

2.1 Mathematical Model of SMR Nuclear Reactor

The physics of the reactor core is modeled considering simplifications like one-speed neutron diffusion model, utilization of six groups of neutron precursor concentrations, and the definitions of the averaged fuel and moderator temperatures feedback coefficients. The neutron kinetics model is defined as follows

$$\frac{dP_{th}(t)}{dt} = \frac{(\rho_{tot}(t) - \beta)P_{th}(t)}{\Lambda} + \sum_{i=1}^{6} \lambda_i C_i(t), \tag{2}$$

$$\frac{dC_i(t)}{dt} = \frac{\beta_i}{\Lambda} P_{th}(t) - \lambda_i C_i(t), \text{ for } i = 1, ..., 6. \tag{3}$$

where: $P_{th}(t)$ is reactor thermal power, Λ is mean neutron generation time, $\rho_{tot}(t)$ is total reactivity, λ_i is decay constant of the ith delayed neutron precursor, β_i is i^{th} delayed neutron fraction, $\beta = \sum_i \beta_i$ is total fraction of fission neutrons which are delayed, C_i is i^{th} delayed neutron precursor concentration. Data for the model parameters mentioned above can be found in [1]. The energy balance for the heat transfer from the nuclear fuel to the reactor coolant is modelled using Mann's lumped node approach [1]. Differential equations related to core heat transfer are presented below

$$\frac{dT_f(t)}{dt} = \frac{fP_0}{m_f c_{p_f}} P_{th}(t) + \frac{U_{fc}A_{fc}}{m_f c_{p_f}} (\theta_1(t) - T_f(t)) \tag{4}$$

$$\frac{d\theta_1(t)}{dt} = (1-f)\frac{P_0}{m_{c_1} c_{p_c}} P_{th}(t) + \frac{U_{fc}A_{f_{c_1}}}{m_{c_1} c_{p_c}} (T_f(t) - \theta_1(t)) + \frac{W_c}{m_{c_1}} (T_{CL} - \theta_1(t)) \tag{5}$$

$$\frac{d\theta_2(t)}{dt} = \left(\frac{1-f}{2}\right) \frac{P_0}{m_{c_2} c_{p_c}} P_{th}(t) + \frac{U_{fc}A_{f_{c_2}}}{m_{c_2} c_{p_c}} (T_f(t) - \theta_1(t)) + \frac{W_c}{m_{c_2}} (\theta_1(t) - \theta_2(t)) \tag{6}$$

$$\frac{dT_{HL}(t)}{dt} = \frac{W_c(\theta_2(t) - T_{HL}(t))}{m_c} \tag{7}$$

where: T_f is the temperature of the fuel node, $\theta_{1,2}$ are coolant temperatures in respective nodes, m_f is the fuel mass, $m_{c_{1,2}}$ are the coolant masses in respective nodes, c_{pf} is the fuel specific heat capacity, c_{pc} is the coolant specific heat capacity, f is the fraction of the total power generated in the reactor fuel rods,

$A_{fc_{1,2}}$ are the overall areas of the effective heat transfer from fuel to coolant in respective nodes, U_{fc} is the average overall heat transfer coefficient, W_c is the mass flow rate of the coolant within the reactor core, T_{HL} and T_{CL} are temperatures of coolant in hot and cold leg respectively. Reactivity balance, which includes temperature feedbacks from fuel, and coolant nodes and also from reactivity introduced via control rods is given as

$$\rho_{tot}(t) = \rho_{ext}\beta + \alpha_f(T_f(t) - T_{f_0}) + \frac{1}{2}\alpha_c\Big((\theta_1 + \theta_2) - (\theta_{1_0} + \theta_{2_0})\Big) \tag{8}$$

where: α_f is the reactivity coefficient of the fuel, T_{f0} is the nominal (initial) fuel temperature, α_c is the reactivity coefficient of the coolant and $\theta_{1,2_0}$ are the nominal (initial) coolant temperatures. The most important state variable of the presented model from the investigated control perspective is the reactor thermal power P_{th}. The model inputs treated as disturbances but not used in the study are related to the temperature of the coolant T_{CL} and the mass flow rate W_c of the coolant. The external reactivity ρ_{ext} is treated as a control input with which it is possible to change the state of the nuclear reactor.

3 Research Method

3.1 Considered Controller Types

In the paper, the effectiveness of controllers with the following discrete forms are verified:

- basic, discrete PID controller in non-interacting form:

$$E[k] = SP[k] - PV[k], \tag{9}$$
$$iE[k] = iE[k-1] + E[k] \cdot \Delta t, \tag{10}$$
$$dE[k] = (E[k] - E[k-1])/\Delta t, \tag{11}$$
$$CV_{PID}[k] = CV[0] + K_p \cdot E[k] + K_i \cdot iE[k] + K_d \cdot dE[k], \tag{12}$$

  where: k denotes discrete time instant, SP is the set point, PV is the process value, E is the error signal, Δt is sampling time, iE is a discrete approximation of integral part of PID controller, dE is a discrete approximation of derivative part of PID controller, CV_{PID} is the control signal, $CV[0]$ is an initial control signal value, and K_p, K_i, K_d are PID controller parameters,
- discrete version of Variable-Order Fractional PID (VO-FPID) controller based on the discrete definition of the variable order fractional Grünwald-Letnikov operator [7]:

$$E[k] = SP[k] - PV[k], \tag{13}$$

$$_0\mathscr{D}_{\infty}^{\alpha[k]} E[k] \approx \Delta t^{-\alpha[k]} \cdot \sum_{i=0}^{\infty} \big(w_i[k] \cdot E[k - i \cdot \Delta t]\big), \tag{14}$$

$$w_0[k] = 1 \ , \ w_i[k] = \big(1 - \frac{\alpha[k] + 1}{i}\big) \cdot w_{i-1}[k], \text{ for } i = 1, 2, \dots \ , \tag{15}$$

$$CV_{FPID}[k] = CV[0] + K_p \cdot E[k] + K_i \cdot \mathscr{D}_{int}^{-\lambda[k]} E[k] + K_d \cdot \mathscr{D}_{diff}^{\mu[k]} E[k], \tag{16}$$

where: $_0\mathscr{D}_{\infty}^{\alpha[k]}$ denotes discrete version of the variable order fractional Grünwald-Letnikov operator with fractional order $\alpha[k]$ at discrete time instant k (operator with infinite memory [7]), w_i is recursively calculated coefficients of the fractional operator, $\mathscr{D}_{int}^{-\lambda[k]}$ and $\mathscr{D}_{diff}^{\mu[k]}$ are discrete versions of fractional order integral ($\lambda \geq 0$) and differential operators ($\mu \geq 0$) based on the Grünwald-Letnikov definition [7], and CV_{FPID} is controller control signal,

- discrete version of Variable-Order Fractional PID controller with recurrent neural networks (RNN) based on GRU cells to approximate fractional operators presented as VO-NFPID controller [7]:

$$E[k] = SP[k] - PV[k], \tag{17}$$

$$CV_{NFPID}[k] = CV[0] + K_p \cdot E[k] + K_i \cdot \mathscr{N}_{int}^{-\lambda[k]} E[k] + K_d \cdot \mathscr{N}_{diff}^{\mu[k]} E[k] \tag{18}$$

where: $\mathscr{N}_{int}^{-\lambda[k]}$ and $\mathscr{N}_{diff}^{\mu[k]}$ denotes the recurrent neural networks approximations of the fractional integral and derivative operators [7], and CV_{NFPID} is VO-NFPID controller control signal.

3.2 Adaptation Mechanism

An idea of the adaptation mechanism is shown in Fig. 1. Based on the information about current thermal power P_{th}, the parameters of the controller (PID, FPID, or NFPID) are continuously adapted. It is proposed that the parameter adaptation is based on the gain-schedule approach. This ensures smooth controller parameter adjusting according to the system operating point changes. The following three-step procedure was used to determine the structure of the appropriate formulas for parameters adaptation of the designed controller: **Step 1**: the mathematical model of the SMR reactor presented in Sect. 2.1 was linearized using classical approach with the Taylor series expansion in eight a priori selected operating points within the range 30–100% with the step of 10% of its nominal normalized thermal power, **Step 2**: for each linearized model of the SMR reactor the optimal parameters for the local controllers of the different types PID, FPID, and NFPID were determined, optimized parameters of the local PID (Equations (19)–(21)), FPID and NFPID controllers (Equations (22)–(24)) were determined by solving suitably defined optimization tasks using a simulation model of the entire control system on the fixed time interval, **Step 3**: the relations between the changes in the parameters of each considered controller (PID, FPID, and NFPID) and the changes in the reactor's thermal power P_{th} were determined. Model of the control system is represented by $F(\Omega, t, \circ)$ operator ((20) and (23)), where '$\circ$' symbol denotes all process variables.

$$\underset{\Omega \in \{K_p, K_i, K_d\}}{\arg\min} \quad J(\Omega) = \gamma_1 \cdot \sum_{k=0}^{n \cdot \Delta t} E[k]^2 + \gamma_2 \cdot \sum_{j=1}^{|\Omega|} (\Omega\{j\})^2 \tag{19}$$

subject to :

$$F(\Omega, t, \circ) = 0, \tag{20}$$

$$[K_p, K_i, K_d] \geq [0, 0, 0]. \tag{21}$$

$$\underset{\Omega \in \{K_p, K_i, K_d, \lambda, \mu\}}{\arg\min} \quad J(\Omega) = \gamma_1 \cdot \sum_{k=0}^{n \cdot \Delta t} E[k]^2 + \gamma_2 \cdot \sum_{j=1}^{|\Omega|} (\Omega\{j\})^2 \tag{22}$$

subject to :

$$F(\Omega, t, \circ) = 0, \tag{23}$$

$$[K_p, K_i, K_d, \lambda, \mu] \geq [0, 0, 0, 0, 0], \tag{24}$$

In the numerical simulations, set point changes at 10 s were applied as step changes by +5% of the maximum normalized thermal power from a given operating point. Sampling time was set to 0.01 sec, and an overall numerical simulation time horizon was set to 160 sec. In the objective function $J(\Omega)$ (Equations (19), (22) accordingly), in addition to the metric term related to the square of the error, the $L2$ regularization term was used to guarantee a stable and unique solution. Weights γ_1 and γ_2, for all the calculations carried out at this stage, were chosen experimentally to maintain a compromise between the observed transients' dynamics and the values of the parameters of the individual controllers.

Finally, in step 3 of the mentioned procedure, using the method of least squares, interpolation formulas were determined to approximate the variations of parameters of particular controllers (K_p, K_i and K_d for PID controller, and K_p, K_i, K_d, λ and μ for FPID and NFPID controllers). Interpolation formulas were treated as the functions of thermal power P_{th}, in the form of appropriate polynomials of the 2nd-degree as stated in Eq. (25). Evaluated parameters of these polynomials are listed in Table 1, respectively.

$$\Omega(P_{th}[k]) = a \cdot (P_{th}[k])^2 + b \cdot P_{th}[k] + c \tag{25}$$

where $\Omega \in \{K_p, K_i, K_d\}$ for PID,

and $\Omega \in \{K_p, K_i, K_d, \lambda, \mu\}$ for $FPID$ and $NFPID$.

Table 1. Coefficients of 2nd-degree polynomials (25) in adaptation rules for PID, FPID, and FNPID controller parameters.

Controller	PID			FPID			NFPID		
Parameters	a	b	c	a	b	c	a	b	c
K_p	0.003	−0.85	1.26	3.00E-3	−0.81	1.02	−6.00E-3	−1.09	1.02
K_i	0.008	−1.61	2.12	8.00E-3	−1.52	1.99	3.00E-3	−0.93	1.53
K_d	0.0004	−0.07	0.03	3.00E-3	−0.81	1.02	7.00E-3	−1.39	2.00
λ	–	–	–	4.00E-4	−0.14	−0.96	5.00E-5	0.05	0.90
μ	–	–	–	9.00E-19	−1.00E-16	3.00E-17	3.00E-5	3.00E-3	7.00E-4

4 Simulation Results

This section presents the results of the simulation experiments of the load-following control scenario of the nonlinear SMR reactor. The synthesis of the adaptation mechanism for individual controllers and the simulation experiments presented in this section were conducted in Python simulation environment with additional packages including Scipy, Numpy and PyTorch.

In relation to the procedure used to determine the particular controller parameter adaptation mechanism, a numerical optimization based on SLSQP algorithm [16], initialized from a single starting point, was used for the synthesis of the PID (Eqs. (9)–(12)) and FOPID (Eqs. (13)–(16)) adaptive controllers. While in the case of the NFPID controller synthesis (Eqs. (17)–(18)), where the fractional-order operators were approximated by GRU recurrent neural networks, it was observed that the solutions of optimization problem got stuck in local minima. Therefore, in this case, a hybrid population algorithm based on stochastic Differential Evolution was used. The algorithm was set to the 'best1bin' strategy with 60 individuals in the population. The found solution was further polished using the L-BFGS-B algorithm [16]. Optimization process for each controller under consideration is performed only once in the offline mode.

Obtained simulation results presented in Figs. 2 and 3 demonstrate appropriate controller implementation in the presence of rapid and wide range of set point changes. The simulation experiment of the power load-following scenario was conducted by changing the reference thermal power trajectory in the range of 55–105% in the form of several steps changes, increasing and decreasing the reference power trajectory. It should be noted here that the stepwise scenario of the set point power trajectory is rare in the valid operation of a nuclear power plant. Power changes typically take place with ramp-type trajectories. The responses of the control system with selected adaptive controllers were also tested during the study for various set point power trajectories including ramp. However, from the point of view of efficiency analysis of the proposed neural approximators in this paper the most challenging stepwise scenario without the control rods actuator is presented. The reference trajectory SP ($P_{th,ref}(t)$), as well as its realization PV ($P_{th}(t)$), are presented in each case in the upper waveform, while the lower waveform shows the control signal CV ($\rho_{ext}(t)$) generated by the appropriate

Table 2. Quality metrics for the controllers – stepwise power trajectory P_{th}.

Step trajectory	Type	$\sum_k (E[k])^2$	$\sum_k (\Delta CV[k])^2$
PID	Single, tuned for 100% P_th	7.620	0.322
FPID	Single, tuned for 100% P_th	9.095	1.238
PID	Adaptive	6.391	0.682
VO-FPID	Adaptive	4.644	0.333
VO-NFPID	Adaptive	10.966	0.114
VO-NFPID/VO-FPID	Adaptive, with VO-FPID parameters	6.999	0.186

controller. Table 2 presents values of the quality metrics of the control system operation, defined as the sum of squared error signal $\sum_k (E[k])^2$ and the sum of squared changes in the control signal values $\sum_k (\Delta CV[k])^2$.

In Fig. 2 it can be seen that control signal CV ($\rho_{ext}(t)$) calculated by the VO-FPID controller is less aggressive compared to classical single and adaptive PID controllers. It is characterized by smaller peaks at time instants of step changes in the set point. This can also be observed in the values of quality indicators characterizing the control system quality performance presented in Table 2. The sum of squares of errors and sum of squares of changes in the control signal has been significantly reduced in comparison to all other tested controllers. As shown in Fig. 3 and also in Table 2, the VO-NFPID controller, in which the corresponding GRU neural networks approximate the FO operators, showed the worst performance in the terms of tracking errors and the control signal quality performance. However, it should be noted that in this case, the proposed VO-NFPID controller with neural fractional-order operators, unlike the VO-FPID controller, is characterized by finite memory. On the other hand, usage of the adaptation mechanism from infinite-memory VO-FPID controller to finite-memory VO-NFPID/VO-FPID controller improved overall performance of the control system. Although the tracking error performance indicator $\sum_k (E[k])^2$ of the control system, in this case, is worse than for adaptive PID and VO-FPID controllers (Table 2), the waveform of the control signal is the smoothest (Fig. 3). This can also be seen in the control signal metric i.e. $\sum_k (\Delta CV[k])^2$ for VO-NFPID/VO-FPID controller. The presented results showed that the VO-NFPID/VO-FPID controller, in which the neural approximation of VO-FC operators was used, ranked third in terms of tracking error metric $\sum_k (E[k])^2$. Referring directly to the result obtained by the adaptive PID and the VO-FPID controller with infinite memory, there undoubtedly is space for improvement. However, the result achieved by the VO-NFPID/VO-FPID controller is satisfactory and exceeded the authors' initial expectations for the use of this type of neural approximators in an adaptive control system.

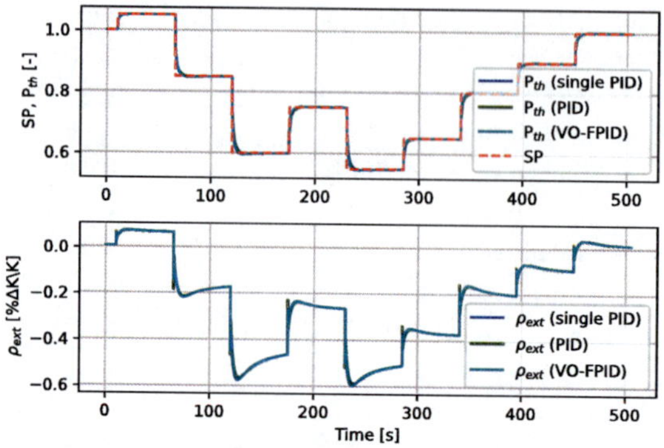

Fig. 2. Responses of the control system to the given stepwise power trajectory P_{th} together with the control signal transient for selected controllers part 1.

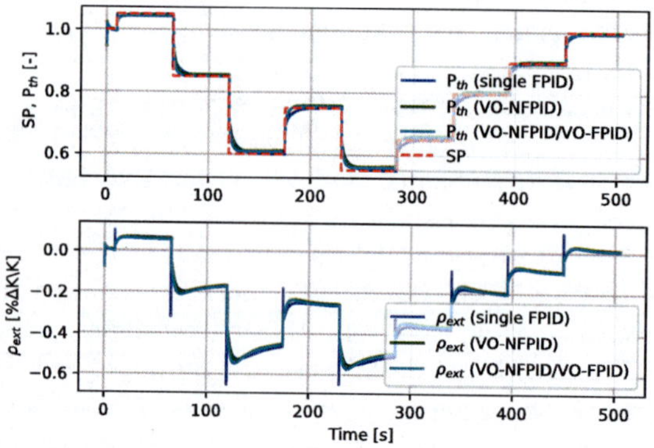

Fig. 3. Responses of the control system to the given stepwise power trajectory P_{th} together with the control signal transient for selected controllers part 2.

5 Conclusions

The Variable Order Neural Fractional PID controller presented in this paper is an original approach to avoiding the problem associated with the implementation of fractional order operators on digital platforms with special emphasis on fractional operators with variable order. Although it is not as efficient as its definition counterpart, it does not require the infinite memory buffer. Presented research shows that the neural approach to approximation of variable-order fractional operators is a successful direction. However, it still needs much development to approximate the accuracy of the definitional operators. Further development of neural approximators is an open problem and may concern: the use of more complex network architectures, the appropriate tuning of the hyperparameters used in the learning process, the selection of appropriate learning data, and the implementation of variable order operations in a single neural operator. Future research efforts by the authors will include a formal analysis of the applicability of the neural adaptive controller presented in this paper, considering the uncertainty and reliability of the proposed solution. In the context of the latter, it is also planned in the future to implement presented neural controllers on industrial digital control platforms such as: PLCs, FPGAs, or industrial computers.

Acknowledgements. Financial support of these studies from Gdańsk University of Technology by the DEC-33/2020/IDUB/I.3.3 grant under the ARGENTUM - 'Excellence Initiative - Research University' program is gratefully acknowledged.

References

1. Kapernick, J.R.: Dynamic modeling of a small modular reactor for control and monitoring (2015)
2. Kothari, K., Mehta, U.V., Prasad, R.: Fractional-order system modeling and its applications. J. Eng. Sci. Technol. Rev. **12**(6), 1–10 (2019)
3. Lorenzo, C.F., Hartley, T.T.: Variable order and distributed order fractional operators. Nonlinear Dyn. **29**(1), 57–98 (2002). https://doi.org/10.1023/A:1016586905654
4. Muresan, C.I., Birs, I.R., Dulf, E.H., Copot, D., Miclea, L.: A review of recent advances in fractional-order sensing and filtering techniques. Sensors **21**(17), 5920 (2021)
5. Patnaik, S., Hollkamp, J.P., Semperlotti, F.: Applications of variable-order fractional operators: a review. Proc. R. Soc. Math. Phys. Eng. Sci. **476**(2234), 20190498 (2020)
6. Podlubny, I.: Fractional Differential Equations: An Introduction to Fractional Derivatives, Fractional Differential Equations, to Methods of Their Solution and Some of Their Applications. Elsevier, San Diego, Boston, New York, London, Tokyo, Toronto (1999)
7. Puchalski, B.: Neural approximators for variable-order fractional calculus operators (VO-FC). IEEE Access **10**, 7989–8004 (2022)
8. Puchalski, B., Duzinkiewicz, K., Rutkowski, T.: Multi-region fuzzy logic controller with local PID controllers for U-tube steam generator in nuclear power plant. Arch. Control Sci. **25**(4), 429–444 (2015)
9. Puchalski, B., Rutkowski, T.A.: Approximation of fractional order dynamic systems using Elman, GRU and LSTM neural networks. In: Rutkowski, L., Scherer, R., Korytkowski, M., Pedrycz, W., Tadeusiewicz, R., Zurada, J.M. (eds.) ICAISC 2020, Part I. LNCS (LNAI), vol. 12415, pp. 215–230. Springer, Cham (2020). https://doi.org/10.1007/978-3-030-61401-0_21
10. Puchalski, B., Rutkowski, T.A., Duzinkiewicz, K.: Implementation of the Fopid algorithm in the PLC controller - PWR thermal power control case study. In: 2018 23rd International Conference on Methods Models in Automation Robotics (MMAR), pp. 229–234 (2018)
11. Puchalski, B., Rutkowski, T.A., Duzinkiewicz, K.: Fuzzy multi-regional fractional PID controller for pressurized water nuclear reactor. ISA Trans. **103**, 86–102 (2020)
12. Puchalski, B., Rutkowski, T.A., Tarnawski, J., Duzinkiewicz, K.: Programowo-sprzetowa platforma symulacyjna - Hardware In the Loop - zaawansowanego ukladu sterowania poziomem wody w pionowej wytwornicy pary elektrowni jadrowej. Aktualne problemy automatyki i robotyki pod red. K. Malinowski, J. Jözefczyk, J. Swiatek, Oficyna Wydawnicza EXIT, pp. 570–580 (2014)

13. Sabatier, J.: Fractional order models are doubly infinite dimensional models and thus of infinite memory: consequences on initialization and some solutions. Symmetry **13**(6), 1099 (2021)
14. Shah, P., Agashe, S.: Review of fractional PID controller. Mechatronics **38**, 29–41 (2016)
15. Sun, H., Chen, W., Wei, H., Chen, Y.: A comparative study of constant-order and variable-order fractional models in characterizing memory property of systems. Eur. Phys. J. Spec. Top. **193**(1), 185–192 (2011). https://doi.org/10.1140/epjst/e2011-01390-6
16. Virtanen, P., et al.: SciPY 1.0: fundamental algorithms for scientific computing in python. Nat. Methods **17**(3), 261–272 (2020)
17. Xue, D.: Fractional-Order Control Systems: Fundamentals and Numerical Implementations. De Gruyter, June 2017

Fault Detection

LSTM Model-Based Fault Detection for Electric Vehicle's Battery Packs

Grzegorz Wójcik[1,2,3](✉) and Piotr Przystałka[1]

[1] Department of Fundamentals of Machinery Design, Silesian University of Technology, 18a Konarskiego Street, 44-100 Gliwice, Poland
[2] Doctoral School, Silesian University of Technology, 2a Akademicka Street, 44-100 Gliwice, Poland
[3] DIP Draexlmaier Engineering Polska Sp. z o.o, Gliwice, Poland
grzegorz.wojcik@polsl.pl

Abstract. An LSTM model-based fault detection method dedicated for a liquid leakage and liquid intrusion detection system was proposed and described. The method utilized two individual residual evaluation approaches. The first based on statistical analysis and the second based on model error modelling methodology. Both approaches were trained and tested using collected datasets of a prototyped laboratory stand, simulating liquid intrusion and liquid leakage faults of an electric vehicle's battery pack with direct liquid cooling battery thermal management system. Obtained results were compared and discussed in details and have revealed notably higher robustness of one of the proposed approaches.

Keywords: Electric vehicle · Battery pack · Model-based fault detection · Liquid leakage · Liquid intrusion · Long short-term memory model

1 Introduction

Electric vehicle's (EVs) lithium ion battery packs (LIBs) are known as devices of high power, high energy density, and low self-discharge performance. However, their safety operating window is relatively small when compared to the conditions that EVs are susceptible to. LIB performance and lifetime are highly dependent on environmental factors and operation modes. Faults usually come from extreme operating conditions, manufacturing flaws, or battery ageing [1]. A component playing a critical role in a battery pack system is a battery management system (BMS), which main purpose is to maintain all cells and modules within their operating limits, to supervise their states, and to provide appropriate mechanisms in case of uncontrolled conditions or emergency. LIBs are susceptible to a number of factors that affect their operation and lifetime. Collision and shock, vibration, deformation, metallic lithium plating, formation of a solid electrolyte interphase (SEI), and formation of dendrites are among the causes leading to

Z. Kowalczuk (Ed.): DPS 2022, LNNS 545, pp. 217–229, 2023.
https://doi.org/10.1007/978-3-031-16159-9_18

multiple faults of the battery pack system and ultimately to a system failure [2]. LIB pack faults are categorized into external and internal. External include faults of sensors, cell connection, and cooling system. Internal faults consist of overcharge, overdischarge, internal and external short circuits, overheating, accelerated degradation, and thermal runaway. An accelerated degradation and thermal runaway are supposed to be the most severe faults for an EV's battery pack. The accelerated degradation is an abnormal condition that shortens the battery pack lifetime and can also lead, on a battery cell level, to surface layer formation and contact deterioration [3], which results in electrode and material disintegration, and loss of lithium. These phenomena can result in penetration of the separator, cause an internal short circuit, and ultimately, thermal runaway. Aging and self-discharging mechanisms, higher frequency of cycle, state of charge (SoC) change, and voltage rates are among the reasons of accelerated degradation. It can be accelerated even further due to elevated temperatures, that might occur due to a cooling system fault caused, e.g., by a coolant leakage [4]. A thermal runaway is an exothermic reaction that occurs as the generated heat (caused, e.g., by internal short-circuit) can not be dissipated as rapidly as it formed. Such an event is a very dangerous situation for EV battery packs, as the failure might propagate further to other battery cells and structures. In such cases, the excessive heat damages the insulating separators between cell electrodes, damaging them further and causing consecutive short circuits. As the heat increases, flammable and toxic gases are released and there is a substantial rise in pressure of the battery cell. The consequences of such cascaded events in large LIB packs are severe and the failure propagation could even lead to explosion [5,6].

EV battery packs can be discharged with currents in the order of hundreds of amperes upon car acceleration or charged with similar currents upon fast charging. Currents of this magnitude lead to excessive heat generation and decreased performance of the whole system. The need for higher performance drives the research for new Battery Thermal Management (BTM) designs. Some of the already developed methods are based on: air, liquid, direct refrigerant, phase change, thermoelectric, and heat pipe cooling [7]. From the perspective of process diagnostics, one of the most interesting BTM designs is the one based on direct liquid cooling (or immersive cooling). This approach utilizes battery cells fully immersed in non-conductive liquids for lower thermal resistances and reduced risk of thermal runaway. Thanks to an electric pump, the coolant circulates through the battery modules and heat exchanger, which finally releases the heat into the ambient environment via radiators. Although one of the most efficient in terms of cooling and the most safe in terms of thermal runaway causes [8], this method presents placement limitations, increased maintenance costs, and strict sealing requirements [9]. Risks associated with the operation of a LIB packs can be minimized by the safety functions of the BMS. Since external and internal short-circuits are the most common causes for thermal runaway accidents [10], the BMS shall ensure that any coolant leakage or liquid intrusion would be detected as soon as possible. Because LIBs are complex systems undergoing various degradation mechanisms due to their electrochemical properties, fault symptoms are even harder to be extracted because of hysteresis and inconsistency among cells [11]. For that reason, these safety measures are often not adequate and fault diagnostic algorithms are required. Model-based methods utilize battery pack or a particular process models to generate residuals. Used models depend on the application and purpose. In general, models such as electrochemical, electrical, thermal, and combinations such as electro-thermal may be used. This kind of diagnosis is often used in fault diagnosis for its simplicity and cost efficiency [2].

2 Research Methodology

The aim of this paper is to develop a fault detection method dedicated for a liquid leakage and liquid intrusion detection system (in the form of electronic control unit, ECU), that is aimed to be used by an EV's battery pack, with the battery pack using a direct liquid cooling BTM approach. The ECU is connected to liquid-type sensitive sensors. Due to a proceeding patent application process, no details about the sensors can be disclaimed and the sensors are referred to as "data sources".

2.1 Liquid Leakage and Liquid Intrusion Detection Method

The proposed fault detection method is based on a process model trained with the use of an artificial neural network (ANN), more specifically a long short-term memory neural network (LSTM) is employed [12]. ANN is chosen due to its self-adaptability and learning ability, capable of capturing the dynamics of highly nonlinear systems, such as LIB. Generated residuals are evaluated with two individual methods. The first method is based on statistical analysis, and the second method is based on the model error modelling methodology in which a feed-forward input-delay backpropagation network (TDNN) is applied [13].

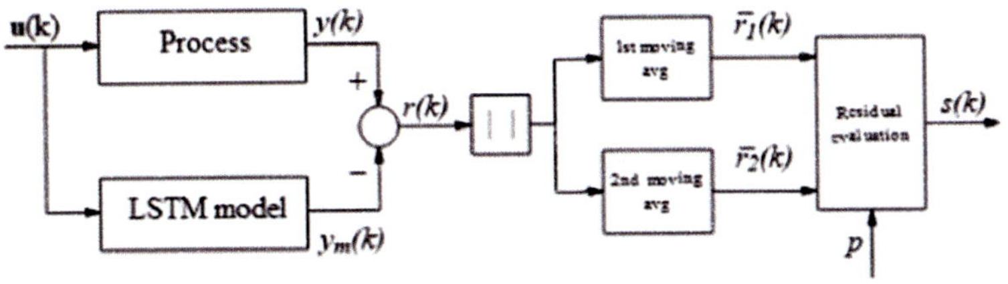

Fig. 1. Method 1: simple model-based fault detection scheme

The first approach is shown in Fig. 1. This is the typical model-based approach in which the model of a process is trained using the data collected for normal conditions. An LSTM network is applied herein because it is known in literature as recurrent neural network (RNN) that can learn long-term dependencies between time steps of sequence data. The main components of such network are a sequence input layer and an LSTM layer. The architecture of an LSTM layer is shown in Fig. 2. This diagram illustrates the flow of a time series $\mathbf{X}$ with $\mathbf{C}$ features (channels) of length $\mathbf{S}$ through an LSTM layer. In the diagram, $\mathbf{h}_t$, $\mathbf{c}_t$ denote the output (also known as the hidden state) and the cell state at time step t, respectively.

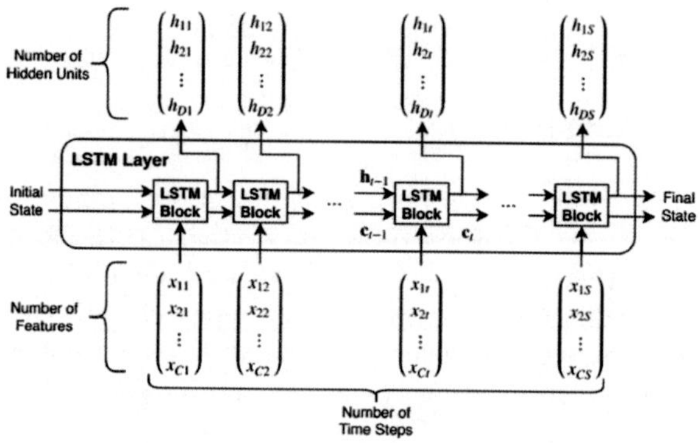

Fig. 2. LSTM layer architecture [12]

Initial network state and the sequence's first time step are provided to the first LSTM block to compute the first output and the updated cell state. For a given time step t, the cell state $\mathbf{c}_t$ is derived based on the current state of the network $(\mathbf{c}_{t-1}, \mathbf{h}_{t-1})$ and the next sequence's time step:

$$\mathbf{c_t} = \mathbf{f}_t \odot \mathbf{c_{t-1}} + \mathbf{i}_t \odot \mathbf{g}_t, \tag{1}$$

where $\odot$ denotes the element-wise multiplication of vectors. The hidden state $\mathbf{h}_t$ contains the output of the corresponding LSTM layer, and the cell state $\mathbf{c}_t$ contains information learned from the previous time steps:

$$\mathbf{h_t} = \mathbf{o}_t \odot \sigma_c\left(\mathbf{c_t}\right), \tag{2}$$

where $\sigma_c\left(\mathbf{c}_t\right)$ denotes the state activation function.

The learnable weights of an LSTM layer are the input weights $\mathbf{W}$, the recurrent weights $\mathbf{R}$, and the bias $\mathbf{b}$. The matrices $\mathbf{W}$, $\mathbf{R}$, and $\mathbf{b}$ are series of interconnected input weights, recurrent weights, and bias of each component, respectively:

$$\mathbf{W} = \begin{bmatrix} \mathbf{W}_i \\ \mathbf{W}_f \\ \mathbf{W}_g \\ \mathbf{W}_o \end{bmatrix}, \mathbf{R} = \begin{bmatrix} \mathbf{R}_i \\ \mathbf{R}_f \\ \mathbf{R}_g \\ \mathbf{R}_o \end{bmatrix}, \mathbf{b} = \begin{bmatrix} \mathbf{b}_i \\ \mathbf{b}_f \\ \mathbf{b}_g \\ \mathbf{b}_o \end{bmatrix}. \tag{3}$$

An LSTM layer adds or removes information from the cell state at each time step, which is done using so-called gates, namely input gate

$$\mathbf{i}_t = \sigma_g(\mathbf{W}_i\mathbf{x}_t + \mathbf{R}_i\mathbf{h}_{t-1} + \mathbf{b}_i), \tag{4}$$

forget gate

$$\mathbf{f}_t = \sigma_g(\mathbf{W}_f\mathbf{x}_t + \mathbf{R}_f\mathbf{h}_{t-1} + \mathbf{b}_f), \tag{5}$$

cell candidate

$$\mathbf{g}_t = \sigma_c(\mathbf{W}_g\mathbf{x}_t + \mathbf{R}_g\mathbf{h}_{t-1} + \mathbf{b}_g), \tag{6}$$

and output gate

$$\mathbf{o}_t = \sigma_g(\mathbf{W}_o\mathbf{x}_t + \mathbf{R}_o\mathbf{h}_{t-1} + \mathbf{b}_o), \tag{7}$$

where $\sigma_g(\mathbf{c}_t)$ denotes the gate activation function.

Such LSTM neural model corresponds to the faultless state and it is applied in order to generate the residual signal according to the formula:

$$r(k) = y(k) - y_m(k), \tag{8}$$

where $y(k)$ is the measured signal, $y_m(k)$ is the signal that is calculated using the aforementioned LSTM model.

The binary diagnostic signal $s(k) \in \{0,1\}$ is computed according to the proposed equation:

$$s(k) = \begin{cases} 0 \text{ for } \bar{r}_1(k) - \bar{r}_2(k) \leq p \\ 1 \text{ otherwise} \end{cases}, \tag{9}$$

where $\bar{r}_1(k)$ and $\bar{r}_2(k)$ denote the first and second moving average window of the residual signal with different sample sizes, p is the arbitrary threshold.

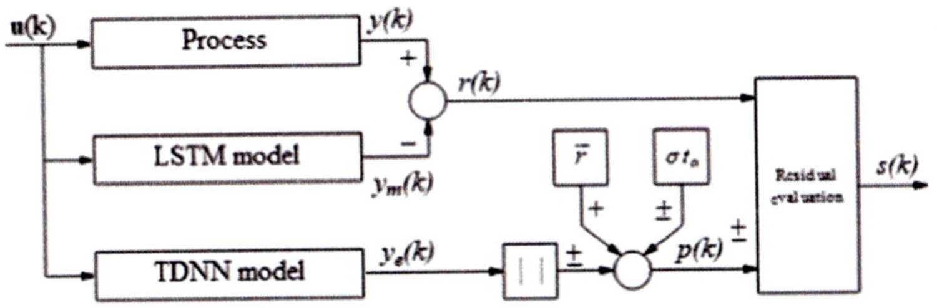

Fig. 3. Method 2: robust model-based fault detection scheme

On the other hand, the model-based approach with the use of model error modelling methodology proposed by [14] is adopted and extended in order to create a robust fault detection scheme. The scheme presented in Fig. 3 illustrates the main idea of the model-based approach with adaptive threshold evaluation of the residuals. In this approach the model of the process is created using training and test data collected for fault-free operations of the system. The output of this model (y_m) is used to calculate a residual signal (r). The test dataset is also applied to prepare additional training and test data subsets which are needed to create a model of the residual (model error). The estimation of the residual signal (y_e) is then used to compute adaptive thresholds ($\mathbf{p}$). In this way it is possible to obtain a robust decision block of the fault detection scheme. The robustness of the fault detection algorithm to uncertainties of different nature without losing sensitivity to faults is performed by evaluating adaptive thresholds which are given in the following form:

$$p^{\pm}(k) = \bar{r} \pm t^{\pm}_{\alpha_i}\sigma_r \pm \left|\mathbf{LW}\boldsymbol{f}^1\left(\mathbf{IW}\mathbf{u}'(k) + \mathbf{b}^1\right) + \mathbf{b}^2\right|, \qquad (10)$$

where $\bar{r}$ and σ_r represent the mean and standard deviation values of the residual (these statistics are calculated for the faultless condition), $t^{\pm}_{\alpha}$ denotes the critical value corresponding to a given significance level α. The second part of the Eq. (10) concerns the estimation of the residual behavior and it becomes realized with the help of the TDNN model (y_e), where $\mathbf{IW}$ is an input weight matrix, $\mathbf{LW}$ is an output weight matrix, $\mathbf{b}^1$, $\mathbf{b}^2$ are bias matrices, $\boldsymbol{f}^1$ is a non-linear activation function, $\mathbf{u}'(k) = [\mathbf{u}(k), \mathbf{u}(k-1), \mathbf{u}(k-2), \ldots, \mathbf{u}(k-n)]^T$.

The binary diagnostic signal $s(k) \in \{0, 1\}$ is originated as a result of two-value evaluation of residuals and, therefore it can be computed according to the following rule:

$$s(k) = \begin{cases} 0 \ \textbf{for} \ p^{+}(k) \geq r(k) \geq p^{-}(k) \\ 1 \ \textbf{otherwise} \end{cases}. \qquad (11)$$

2.2 Laboratory stand and experiment methodology

A test rig was constructed (Fig. 4) to simulate liquid leakage and liquid intrusion faults, and to obtain data during these events using a data logger. Main driver and peripheral devices were supplied by a bench power supply. The test rig also consists of two 500 ml containers. Only the first container was filled with liquid, i.e., mineral oil (300 ml). Liquid intrusion fault was simulated using distilled water and a syringe, while liquid leakage was simulated using a peristaltic pump, which was used to transport the coolant between containers. The data source was immersed in liquid to monitor its state. A reference data source was used for additional reference measurements. Both the reference data source as well as the main data source were provided with electrical PWM input signals: C02, and C05 respectively. Since water and oil are immiscible, any intruded amount of water will form an oil-water emulsion. If the formed emulsion is quiescent, then the particles will settle (based on their densities) and both liquids will separate over time. A magnetic stirrer was occasionally used to stir both liquids and simulate the EV's movement. To link the proposed fault detection method and signals of the real object, the following relationships can be provided $\mathbf{u}(k) = \left[C05, C02\right], y(k) = \left[C06\right]$.

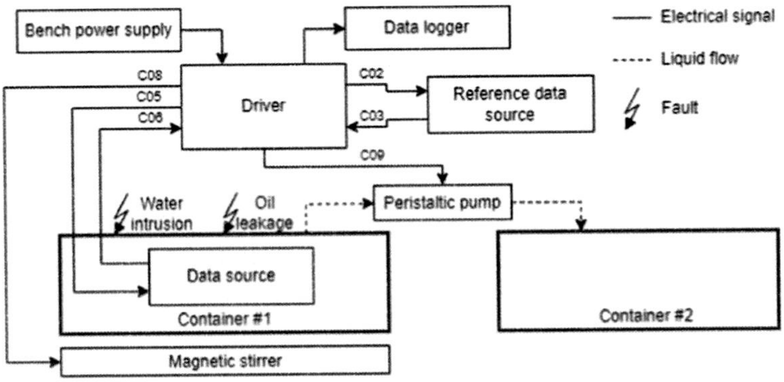

Fig. 4. Test rig block diagram

In total, 25 experiments were carried out: 10 experiments for faultless condition (F0), 5 experiments for oil leakage fault (F1), and 10 experiments for water intrusion faults (F2). The magnitude of simulated faults $Q1$ was variable and can be found with other parameters in Table 1.

Table 1. Experiment scenarios

Fault ID	Description	Magnitude	Q_1	t_1	t_2	t_3
F0	No fault 1	0%	-	-	-	-
F0	No fault 2	0%	-	-	-	-
F0	No fault 3	0%	-	-	-	-
F0	No fault 4	0%	-	-	-	-
F0	No fault 5	0%	-	-	-	-
F0	No fault stirred 1	0%	-	-	1243	5742
F0	No fault stirred 2	0%	-	-	1148	6747
F0	No fault stirred 3	0%	-	-	2322	6738
F0	No fault stirred 4	0%	-	-	-	4797
F0	No fault stirred 5	0%	-	-	668	2967
F1	Oil leakage 1	100%	300 ml	4452	-	-
F1	Oil leakage 2	100%	300 ml	730	-	-
F1	Oil leakage 3	50%	150 ml	4246	-	-
F1	Oil leakage 4	100%	300 ml	1658	-	-
F1	Oil leakage 5	100%	300 ml	2322	-	-
F2	Water intrusion 1	9%	30 ml	5460	1460	17523
F2	Water intrusion 2	14%	20 ml	1652	0	12151
F2	Water intrusion 3	14%	-	-	1687	10686
F2	Water intrusion 4	14%	-	-	0	8868
F2	Water intrusion 5	14%	-	-	408	7407
F2	Water intrusion 6	21%	30 ml	1784	0	8783
F2	Water intrusion 7	21%	-	-	1928	6427
F2	Water intrusion 8	21%	-	-	2353; 6853	4852; 9852
F2	Water intrusion 9	30%	50 ml	1054	0	9553
F2	Water intrusion 10	30%	130 ml	-	1896	6495

For the oil leakage, two variants were performed: a 100% leakage and a 50% leakage. More magnitude variants were introduced for water intrusion faults. Final water concentrations, i.e., after water intrusion, were: 9%, 14%, 21%, and 30%. Each experiment started with system initialization, followed by the start of data logging. Three experiment variants can be distinguished, including a faultless condition, a liquid leakage fault, and a water intrusion fault. For a water intrusion fault, the specified amount of water $Q1$ was intruded at the time $t1$. During this experiment, a magnetic stirrer was occassionally enabled at the time $t2$ and disable at the time $t3$. In case of oil leakage, the parameter $Q1$ corresponds to the amount of leaked oil. For a faultless condition, no liquid leakage or intrusion was simulated. The magnetic stirrer was periodically enabled as well.

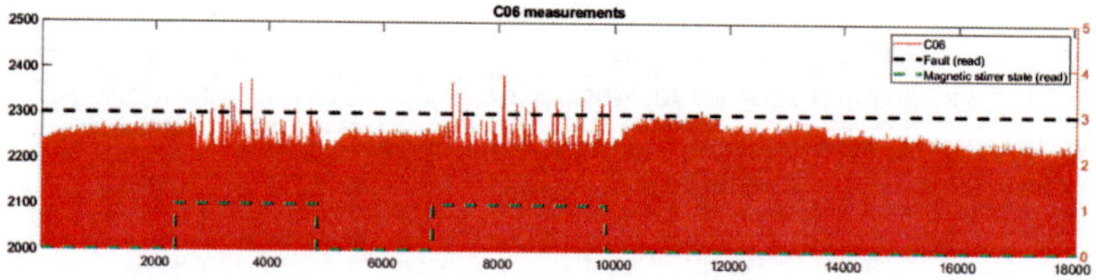

Fig. 5. Exemplary measurements for *Water intrusion 8* dataset

The dataset labeled as "Water intrusion 8" is the only dataset, where the magnetic stirrer was enabled and disabled twice. Raw measurements, together with known fault and magnetic stirrer states are presented in Fig.5. Raw signal variations are clearly visible at the moment of stirring the emulsion, and less visible, or not visible at all, when the emulsion is no longer stirred -depending on the fault magnitude. Known fault state is set to 3, which in this case corresponds to the fault magnitude of 21%.

3 Results and Discussion

The proposed LSTM model was trained using 80% of datasets collected under faultless condition. The testing phase was performed on the remaining 20% of datasets (faultless conditions), 100% of water intrusion datasets, and 100% of oil leakage datasets. The following layer order was used: sequence input layer (2), LSTM layer (55), dropout layer (0.1), LSTM layer (55), fully connected layer (1), regression output layer. Exemplary residuals and results of the Method I are shown in Fig. 6. This exemplary dataset illustrates an essential characteristic of water intrusion faults. If the fault's magnitude is small enough and the emulsion is quiescent, the water particles will settle over time and if the sensor's placement is not low enough, the generated residuals may not trigger the diagnostics signal, unless the emulsion is stirred again.

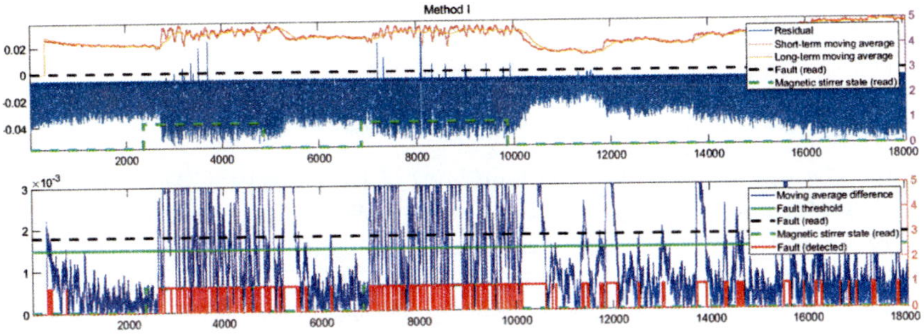

Fig. 6. Exemplary residuals generated for *Water intrusion 8*

In this case, sample sizes of the first moving average and the second moving average were chosen empirically and were 300 and 50 respectively. Although the fault detection is delayed, this computational undemanding method allows to detect both oil leakage as well as water intrusion faults. Figure 7 presents exemplary results of the Method II for the same dataset. Besides a noticeably higher performance, a large variation of the diagnostic signal is visible.

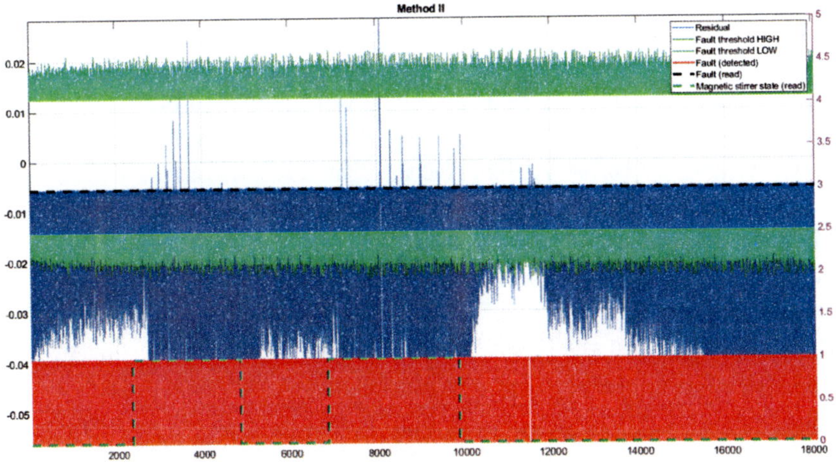

Fig. 7. Exemplary results for Method II and *Water intrusion 8* dataset

The performance of both methods was evaluated using a set of indices proposed by Bartys et al. (2006). First being a true detection rate r_{td}:

$$r_{fd} = \frac{\Sigma_n t_{fd}^n}{t_{from} - t_{on}} \tag{12}$$

where t_{fd}^n is the n_{th} period of the high level of the binary signal indicating the existence of a fault (F1 or F2) in the system between t_{on} and t_{from}.

Second being a false detection rate r_{fd}:

$$r_{td} = 1 - \frac{\Sigma_n t_{td}^n}{t_{hor} - t_{from}} \tag{13}$$

where t_{td}^n is the n_{th} period of the high level of the binary signal indicating the existence of a fault (F1 or F2) in the system between t_{from} and t_{hor}.

For the water intrusion fault, where under the stirring conditions the formed emulsion may become a nonstable suspension of small water droplets in oil, a third index r'_{td} was proposed. This index is derived similarly to r_{td}, but only under stirring conditions - to provide additional measure of method performance.

Table 2. Fault detection performance measures

Dataset description	r_{td}		r_{fd}		$r_{td'}$		Magnitude
	Met. I	Met. II	Met. I	Met. II	Met. I	Met. II	
No fault 5	0.000	0.000	0.000	0.019	0.000	0.000	0%
No fault stirred 5	0.000	0.000	0.000	0.000	0.000	0.000	0%
Water intrusion 1	0.034	0.007	0.000	0.002	0.034	0.007	9%
Water intrusion 2	0.124	0.010	0.000	0.000	0.148	0.014	14%
Water intrusion 3	0.143	0.198	0.000	0.000	0.206	0.152	14%
Water intrusion 4	0.154	0.603	0.000	0.000	0.224	0.553	14%
Water intrusion 5	0.125	0.573	0.000	0.000	0.219	0.467	21%
Water intrusion 6	0.388	0.753	0.000	0.000	0.509	0.751	21%
Water intrusion 7	0.154	0.632	0.000	0.000	0.541	0.728	21%
Water intrusion 8	0.291	0.736	0.000	0.000	0.568	0.763	21%
Water intrusion 9	0.556	0.689	0.000	0.000	0.649	0.708	30%
Water intrusion 10	0.150	0.784	0.000	0.000	0.452	0.790	30%
Oil leakage 1	0.375	0.405	0.000	0.001	-	-	100%
Oil leakage 2	0.372	0.466	0.000	0.003	-	-	100%
Oil leakage 3	0.047	0.305	0.000	0.013	-	-	50%
Oil leakage 4	0.411	0.551	0.011	0.096	-	-	100%
Oil leakage 5	0.362	0.614	0.013	0.059	-	-	100%

Aforementioned indices were calculated for Method I and Method II and listed in Table 2 together with fault magnitude. The obtained results indicate that the highest performance by means of r_{td} and r'_{td} was achieved using a model error modelling methodology. In the case of water intrusion fault, Method II's true detection rate indices are significantly higher than Method I's indices. For datasets with fault magnitudes of 9%, 14%, 21%, and 30%, Method II's r_{td} indices were higher on average of 21%, 92%, 181%, and 208% respectively. In the case of true detection rate indices under stirring conditions, the difference is notably lower due to the higher performance of the first method. For datasets with fault magnitudes of 9%, 14%, 21%, and 30%, Method II's r'_{td} indices were higher on average of 21%, 24%, 47%, and 36% respectively. A significant positive correlation was found between the magnitude of faults and the performance of both methods, especially under stirring conditions. Oil leakage detection based on Method II provided higher true detection rate indices on average by 34% for 50% leakage and as much as 548% for 100% leakage. The only false detection rate indices for Method I were measured for oil leakage faults and were merely about 1%. It should be noted that for water intrusion faults there was only the first dataset, for which the false detection index could be measured. The oil for the remaining F1 datasets was intruded with water and the formed emulsion had at least 9% water concentration. The highest false detection rate indices r_{fd} were reported for Method II, in particular for F2 faults, as well as the faultless condition F0. The highest r_{fd} of Method II was measured for *Oil leakage 4* and was almost 10%.

4 Conclusions

Two residual evaluation methods were developed and tested for collected datasets. Both approaches were compared by means of performance under simulated water intrusion and oil leakage faults. The obtained results emphasized the importance of the quality of data that are feed to neural models. For EV battery packs that utilize a BTM system based on direct liquid-cooling strategy, it is important to apply a correct sensor placement to obtain useful data. For oil leakage, the sensors shall be placed in an upper part of the battery pack so the sensor can detect a change in fluid type as quickly as possible. On the other hand, additional sensors shall be placed in the battery pack's lower part to effectively sense water intrusion, as the water droplets will settle there when the formed emulsion is quiescent. Due to characteristics of these faults, they can be categorized as abrupt (oil leakage) and intermittent (water intrusion). If the amount of intruded water is not enough to cover sensing devices during quiescent conditions, the fault will appear mainly during EV's movement, or in the case of this research paper - during stirring conditions. Both of the proposed approaches had the same LSTM process model and differed in the residual evaluation. Method I utilized a statistical analysis, based on short term and long term moving averages. This method presented limited performance for faults of lower magnitudes, but increased performance for abrupt changes and stirring conditions. Method II utilized a robust residual evaluation approach by means of TDNN model. Although more complicated and computational demanding, this method performed notably better for all cases, and the difference of its performance between quiescent and stirring conditions was significantly lower. However, the generated diagnostic signal presented signal variations of frequency similar to the frequency of provided input signals (C03, C05). To increase the method's performance even further, an additional diagnostic signal state timeout should have been used to filter out low states of the C06 signal. Presented detection method was provided with datasets that cover only a fraction of factors that affect system measurements. On a daily basis, EVs are exposed to elevated operating temperatures, vibrations, component ageing, etc. To obtain datasets affected by those factors, expensive and time consuming tests are required and are a subject of further research.

Acknowledgements. This research was financed by the Ministry of Education and Science of Poland under grant No DWD/33/33/2019. This publication is partially supported from the statutory funds of Department of Fundamentals of Machinery Design.

References

1. Lu, L., Han, X., Jianqiu, L., Hua, J., Ouyang, M.: A review on the key issues for lithium-ion battery management in electric vehicles. J. Power Sour. **226**, 272–288 (2013)
2. Tran, M.-K., Fowler, M.: A review of lithium-ion battery fault diagnostic algorithms: current progress and future challenges. Algorithms **13**(3) (2020)
3. Kanevskii, L.S., Dubasova, V.S.: Degradation of Lithium-Ion batteries and how to fight it: a review. Russ. J. Electrochem. **41**, 1–16 (2005)
4. Xu, J., Deshpande, R., Pan, J., Cheng, Y.-T., Battaglia, V.: Electrode side reactions, capacity loss and mechanical degradation in lithium-ion batteries. J. Electrochem. Soc. **162**, 2026–2035 (2015)
5. Mao B., Chen H., Cui Z., Tang qin, W., Wang, Q.: Failure mechanism of the lithium ion battery during nail penetration. Int. J. Heat Mass Transfer **122**, 1103–1115 (2018)
6. Galushkin, N., Yazvinskaya, N., Galushkin, D.: Mechanism of thermal runaway in lithium-ion cells. J. Electrochem. Soc. **165** (2018)
7. Asef, P., Milan, M., Lapthorn, A., Sanjeevikumar, P.: Future trends and aging analysis of battery energy storage systems for electric vehicles. Sustainability **13** (2021)
8. Yue, Q.L., He, C.X., Zhao, T.: Advances in thermal management systems for next-generation power batteries. Int. J. Heat Mass Transf. **181** (2021)
9. Patil, M., Seo, J.-H., Jianqiu, L., Hua, J., Ouyang, M.: A novel dielectric fluid immersion cooling technology for Li-ion battery thermal management. Energy Conver. Manag. **48** (2021)
10. Feng, X., Ouyang, M., Liu, X., Lu, L., Xia, Y., He, X..: Thermal runaway mechanism of lithium ion battery for electric vehicles: a review. Energy Stor. Mater. **10**, (2017)
11. Hendricks, C., Williard, N., Mathew, S., Pecht, M.: A failure modes, mechanisms, and effects analysis (FMMEA) of lithium-ion batteries. J. Power Sourc. **297**, 113–120 (2015)
12. Hochreiter, S., Schmidhuber, J.: Long short-term memory. Neural Comput. **9**(8), 1735–1780 (1997)
13. Shao, Y.E., Lin, S.-C.: Using a Time Delay Neural Network Approach to Diagnose the Out-of-Control Signals for a Multivariate Normal Process with Variance Shifts. Mathematics **7**, 959 (2019)
14. Patan, K.: Artificial Neural Networks for the Modelling and Fault Diagnosis of Technical Processes, Lecture Notes in Control and Information Sciences, Springer, Heidelberg (2008). https://doi.org/10.1007/978-3-540-79872-9

15. Bartyś, M, Patton, R., Syfert, M., de las Heras, S., Quevedo, J.: Introduction to the DAMADICS actuator FDI benchmark study, control engineering practice, (Invited Special Issue Paper). Control Eng. Pract. **14**, 577–596 (2006)
16. Przystałka, P.: Performance optimization of a leak detection scheme for water distribution networks. IFAC-PapersOnLine **51**, 914–921 (2018)

Remaining Useful Life Prediction of the Li-Ion Batteries

Bogdan Lipiec(✉), Marcin Mrugalski, and Marcin Witczak

Institute of Control and Computation Engineering, Faculty of Computer, Electrical and Control Engineering, University of Zielona Góra, ul. Szafrana 2, 65-246 Zielona Góra, Poland
{b.lipiec,m.mrugalski,m.witczak}@issi.uz.zgora.pl

Abstract. Many different prediction methods have been developed in recent years and data-driven methods are often used. The aim of this paper is to present the new method of prediction remaining useful life of components. The proposed soft computing approach bridges the fuzzy-logic and data-driven health prognostic approaches. The result of this combination is the practical method for determining the remaining useful life. Proposed method is based on a Takagi-Sugeno multiple-based framework. Compared to other data-driven methods, the proposed algorithm differs in the use of historical data in order to improve the quality of prediction and to create a flexible scheme. The entire method is used to predict remaining useful life of batteries. Finally, the validation of the proposed algorithm is made with NASA PCoE data set of Li-Ion battery. This benchmark consist run-to-failure tests.

Keywords: Fuzzy logic · Remaining useful life · Battery capacity

1 Introduction

Numerous and important practical applications of batteries and their great importance for the safety of other systems make it necessary to develop new Remaining Useful Life (RUL) prediction methods of such components. It should be noted that in particular a Li-Ion batteries are increasingly used in many fields of customer and industry applications i.e.: laptops [15], military [10], aerospace [2] and Automated Guided Vehicles (AGVs) [9]. Their advantage is low weight and long exploitation time which makes them eagerly used in the automotive industry. In addition, at the end of their life, they are used in renewable energy storage systems [3]. The continuous development of these batteries makes them more and more popular and they replace a old type such us a lead-acid batteries [13].

Due to the growing popularity of the battery-based systems, the importance of diagnostics of batteries has also increased [14]. In the literature, there is a

The work was supported by the National Science Centre of Poland under Grant: UMO-2017/27/B/ST7/00620.

Z. Kowalczuk (Ed.): DPS 2022, LNNS 545, pp. 230–241, 2023.
https://doi.org/10.1007/978-3-031-16159-9_19

large group of diagnostic methods that can be used in battery fault detection tasks [5,12,18,20]. These methods are mature and do not seem to require further development. On the other hand, these methods do not predict the remaining useful life of battery. It should be emphasized that this knowledge is needed for better organization of maintenance and reducing costs caused by downtime. The application of a methods that allows for the prediction of the remaining useful life of technical systems significantly increase its effectiveness [8,17,21].

The paper is organized as follows. Section 2 presents the motivation to undertake research and briefly define the class of RUL prediction methods. Section 3 presents the a new fuzzy logic based approach to modelling of Li-Ion batteries. Section 5 shows the results of the efficiency evaluation of the Takagi-Sugeno based prediction framework and Sect. 6 concludes the paper.

2 RUL Prediction Methods

The life cycle of a battery goes from the fully functional stage, through the period of reduced efficiency to the unusable stage. According to the diagram presented in Fig. 1, the battery start operating in reduced efficiency stage when the capacity crosses the fault-threshold. This situation occurs at time t_d. This fault may be caused by a natural ageing process or by the destructive influence of various external factors [23]. The occurrence of a fault does not have to mean that the system is completely unable to work, but it may mean a decrease in its effectiveness. If factors that caused the fault still affect the system, it may lead to its complete failure. This means that the system is unable to work. Such a state is usually called as End of Life (EoL). Thus, the process of RUL can be perceived as a time from fault detection t_d to EoL at time t_f.

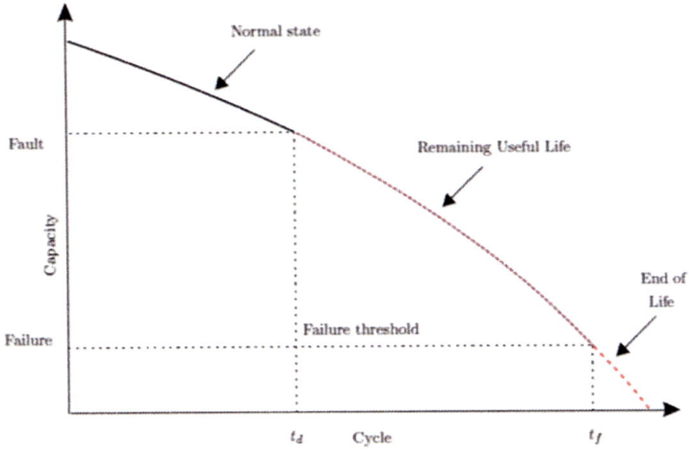

Fig. 1. Concept of RUL prediction.

The aim of the paper is to propose a methodology enabling RUL prediction of the batteries. In particular, the determination of the so-called Time-To-Failure (TTF) being an estimate of the time remaining until the system fails. Figure 2 presents the concept of the RUL prediction. The TTF is calculated from time of fault detection t_d to EoL at time t_f. During the RUL prediction process, the accuracy of the RUL increases as the TTF decreases. This phenomenon is illustrated by decreasing the uncertainty interval of the RUL while approaching t_f.

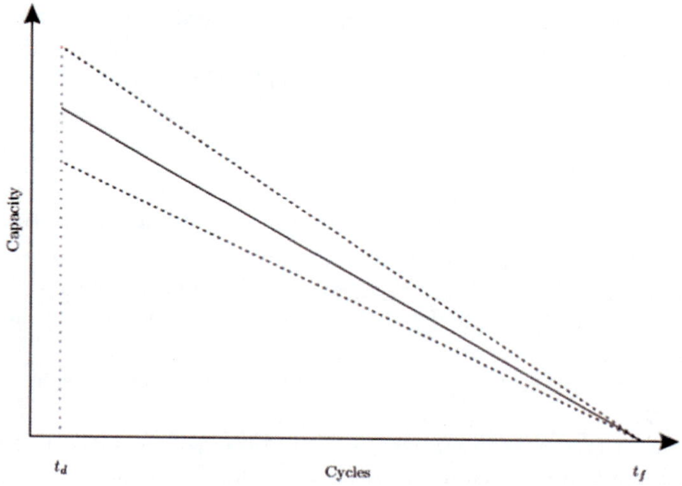

Fig. 2. TTF prediction.

In the Table 1 a set of different RUL prediction methods of Li-On batteries are presented. The most popular methods are based on Artificial Neural networks [4, 6,7,11,22]. Secondly methods are based on Filters (Kalman Filter, Extended Kalman Filter and Particle Filter) [6,11,16,22]. Moreover, a vector machine methods are commonly used too [7,11,22]. Due to the variety of data, classical fuzzy logic methods are rarely used [6].

3 Fuzzy Logic Degradation Modeling Framework

This section describes the theory of fuzzy logic modeling of the degradation process. It is assume that: $m_k = x_k$ and m_k represents a the degradation signal. The discussed signal must be bound upper and lower as:

$$\underline{m} \leq m_k \leq \bar{m}, \tag{1}$$

where $\underline{m} \geq 0$ describes situation when the degradation m_k has a minimum value. Thus, $\underline{m} = 0$ represents the situation when the component is not degraded yet. A

Table 1. Classification of approaches presented in literature.

Author	Ref.	Classification
H. Meng, Y. F. Li	[11]	Physical Models, Artificial Neural Network, Machine Learning, Vector Machines, Particle Filter, Kalman Filter
L. Wu, X. Fu, Y, Guan	[22]	AR Model, Relevance Vector Machine, Support Vector Machine, Artificial Neural Network, Kalman Filters,Particle Filter, Gausian Processes
B. Saha et al.	[16]	ARIMA, RMV, EKF, Particle Filter
M. S. Hosen et al.,	[4]	Gaussian Process Regression, Artificial Neural Network, Nonlinear Autoregressive Exogenous
Y. Li et al.	[7]	AR Model, Support Vector Machine, Machine Learning, Artificial Neural Network, Gaussian Process Regression, Relevance Vector Machine
N. Khayat, N. Karami	[6]	Kalman Filter, Artificial Neural Network, Fuzzy Logic

discussed component is called a healthy one. What is more, $\bar{m} > 0$ describes the maximum degradation level. The maximum value can be called EOL and this is the maximum allowable degradation value. TTF is presented as a difference between current m_k time and $\bar{m}$ time. The different levels of degradation are presented by n classes. Thus, the degradation signal must be divided into classes. This is possible due to the boundaries of the signal (1). Due to the division of the signal into classes, they should be described as follows:

$$z_j = \underline{m} + (j-1)\frac{\bar{m} - \underline{m}}{n-1}, \quad j = 1, \ldots, n. \tag{2}$$

Foundation of fuzzy logic are the membership functions, which are correlated with classes, which were proposed above.

$$\begin{aligned} &a_j = b_{i-1}, \quad b_j = z_i, \quad c_i = b_{j+1}, \quad j = 2, \ldots, n-1, \\ &a_1 = b_1, \quad c_n = b_n, \end{aligned} \tag{3}$$

The parameters a_j, b_j, c_j are defining the i-th membership function.

The theory of Tagaki-Sugeno [19] enables to describe the membership functions into sub-models in the given form:

$$\text{IF } m_k \in M_{m,i} \text{ THEN } m_k = r_k^T p^j + v_k, \quad j = 1, \ldots, n. \tag{4}$$

where $M_{m,i}$ is a set of fuzzy sub-models related to membership functions, which were defined by (3), $r_k = [1, t_c]^T$, where t_c is the load-discharge cycle number. What is more, uncertainty factor of modelling and measurement modelling is presented as v_k. Thus, the condensed form of the system (4) can be described as:

$$m_k = \sum_{j=1}^{n} \mu_j(m_k) \left(r_k^T p^j + v_k\right), \tag{5}$$

$$\sum_{j=1}^{n} \mu_j(m_k) = 1, \quad \mu_j(m_k) \geq 0,$$

where the normalised jth is presented as: $\mu_j(m_k)$ ($j = 1, \ldots, n$). Thus, the triangular shape (3) of the function makes it possible to calculate the strength of the rule. Otherwise, the system (5) can be defined as:

$$m_k = \bar{r}_k^T \bar{p} + v_k, \tag{6}$$

where

$$\bar{r} = [\mu_1(m_k) r_k^T, \mu_2(m_k) r_k^T, \ldots, \mu_n(m_k) r_k^T]^T, \tag{7}$$

$$\bar{p} = [(p^1)^T, \ldots, (p^n)^T]^T. \tag{8}$$

Should be noted that some elements $\mu_l(m_k) = 0$ until the given class will be reached by the degradation signal.

Triangular membership functions are used to avoid the situation when some regressor element are equal to zero. Thanks to this, it is possible to eliminate the problem that could prevent the parameter estimation. Thus, almost always two functions are active for each x_k. This leads to a reduced form of the system (5):

$$m_k = \mu_j(m_k) r_k^T p^j + \mu_{j+1}(m_k) r_k^T p^{j+1} + v_k, \tag{9}$$

$$\mu_j(m_k) + \mu_{j+1}(m_k) = 1, \quad \mu_i(m_k) \geq 0, \quad j = 1, \ldots, n-1.$$

The vector of parameters $\bar{p}_h = [(p_h^1)^T, \ldots, (p_h^n)^T]^T$, which describes the degradation is stored by the proposed approach. The algorithm for estimation $\bar{p}_h$ is the another objective of this paper. Various methods based on so-called bounded error methods [1] enable to tackle this problem. The Quasi-Outer Bounding Ellipsoid [1] algorithm was chosen as the satisfying solution. The structure of this algorithm is close to Recursive Least-Square (RLS), which is one of the most common solution. The number of computation during the execution of the MOBE algorithm is reduced because the parameters are not updated when the following condition (absolute error value) is met:

$$|e_k^i| > \bar{v} \tag{10}$$

Finally, a set of models is received by the proposed algorithm. This corresponds to run-to-failure tests of n_h batteries. A set of $n_h \times n$ degradation model parameters, which were possible to prepare by sufficient amount of tested batteries.

4 Battery Remaining Useful Life Prediction

This section describes the algorithm to predict the remaining useful life of Li-ion batteries. The proposed algorithm uses a set of historical models n_h, which are run-to-failure fuzzy models. Construction of fuzzy models was presented in Sect. 3. The proposed method enables the modeling of degradation of the Li-On batteries and allows to predict RUL of them. The algorithm is presented in the following way:

Algorithm 1: Battery remaining useful life prediction

S_0: Let us assume:

$$p_0^j = 0 \tag{11}$$

$$P_0^i = \rho \mathbf{I}, \tag{12}$$

for $i = 1, \ldots, n-1 j = 1, \ldots, n$ where $\rho > 0$ and $k = 1$.
S_1: A set of currently operating fuzzy sub-models must be specified for $j = 1, \ldots, n-1$ as follows:

$$i = \{j : \mu_j(m_k) + \mu_{j+1}(m_k) = 1\}. \tag{13}$$

S_2: Temporary vector and parameter vector must be assumed:

$$q_k^i = [\mu_i(m_k) q_k^T, \mu_{i+1}(m_k) q_k^T]^T, \tag{14}$$

$$w_k^i = [(p_k^i)^T, (p_k^{i+1})^T]^T. \tag{15}$$

S_3: Calculate

$$e_k^i = m_k - (q_k^i)^T w_{k-1}^i, \tag{16}$$

S_4: If $|e_k^i| > \bar{v}$ then

$$h_k^i = (q_k^i)^T R_k^i q_k^i, \qquad \lambda_k^i = \frac{h_k^i}{\frac{|e_k^i|}{\bar{v}} - 1}, \tag{17}$$

$$R_k^i = (\lambda_k^i)^{-1} \left(R_{k-1}^i - R_{k-1}^i q_k^i (q_k^i)^T R_{k-1}^i (h_k^i)^{-1} \right), \tag{18}$$

$$w_k^i = w_{k-1}^i + R_k^i q_k^i e_k^i, \tag{19}$$

else

$$R_k^i = R_{k-1}^i, \qquad w_k^i = w_{k-1}^i. \tag{20}$$

S_5: Search for the closest historical model

$$h = \arg \min_{h=1,\ldots,n_h} Q_l^i. \tag{21}$$

S_6: Predict time-to-failure $t_{c,f}$:

$$t_{c,f} = \frac{\bar{d} - p_{1,l}^n}{p_{2,l}^n} \tag{22}$$

S_7: Calculate RUL

$$RUL_k = t_{c,f} - k \tag{23}$$

S_7: Set $k = k + 1$ and go to S_1.

It should be explained that in Step 5 the given i class is determined in Step 1 while a set of n_h models for class i is described as follows:

$$\begin{aligned} z_{m,k} &= \mu_i(m_k) q_k^T p_h^i + \mu_{i+1}(m_k) q_k^T p_h^{i+1}, \\ \mu_i(m_k) &+ \mu_{i+1}(m_k) = 1, \quad \mu_i(m_k) \geq 0, \quad h = 1, \ldots, n_h. \end{aligned} \tag{24}$$

The below formulation describes the normalized difference between the current model and the historical one for $h = 1, \ldots, n_h$:

$$Q_h^i = \left(\mu_i(m_k) q_k^T (p_h^i - p^i) + \mu_{i+1}(m_k) q_k^T (p_h^{i+1} - p^{i+1})\right)^2. \tag{25}$$

The most accurate historical model to the currently operating one can be easily find using (21). The h-th model should specify the Eq. 26, this situation describes the maximum degradation value.

$$\begin{aligned} \bar{m} &= \mu_j(\bar{m}) q_f^T p_h^j + \mu_{j+1}(\bar{m}) q_f^T p_h^{j+1}, \\ \mu_j(\bar{m}) &+ \mu_{j+1}(\bar{m}) = 1, \quad \mu_j(\bar{m}) \geq 0, \end{aligned} \tag{26}$$

for a given $i \leq j \leq n - 1$ and $f > k$. According to (2) it should be noted that $j = n - 1$ and must be clarified that $\mu_{j-1}(\bar{m}) = 0$ and $\mu_j(\bar{m}) = 1$ thus it is necessary to obtain $f > k$ for which (26) holds, i.e.:

$$\bar{m} = q_f^T p_h^n. \tag{27}$$

According to (8) it yields $r_f = [1, t_{c,f}]^T$. Finally, Eq. (27) enables obtaining the time of failure presented by (22).

5 Validation of Remaining Useful Life Prediction

Benchmark provided by NASA PCoE was chosen to verify the correctness of the presented approach. Tests were conducted at 25 °C (room temperature). This made it possible to gather reliable data, about the capacity degradation, from the tested batteries. The batteries were designed with the nominal capacity 2 Ah.

- The battery was charged when value 4.2 V was reached and the charging current was set to 1.5 A.
- The battery was discharged when the value 2.5 V was reached and the discharging current was set to 2 A.

As an indicator of the health status of this type of battery, the capacity has been chosen. Degradation of battery capacity to about 75% of the nominal value is determined as the End of Life. A 25% drop in capacity significantly affects battery performance. The nominal capacity of this type of battery is set to $2Ah$ and the critical drop in capacity is determined when the battery capacity is 1.5 Ah. The degradation in Li-Ion battery performance is due to the loss of capacity.

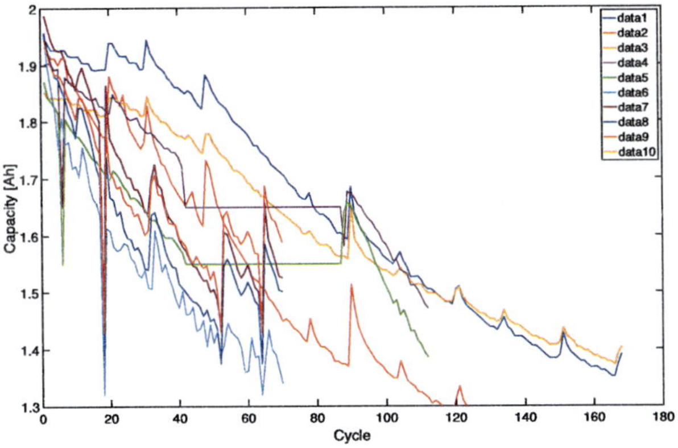

Fig. 3. Capacity of 10 tested batteries

Modeling of degradation of ten batteries is presented in Fig. 3. This models are stored in historical dataset, which are used to predict remaining useful life. The End of Life was determined as 1.5 Ah, which is the last common value of capacity for all batteries.

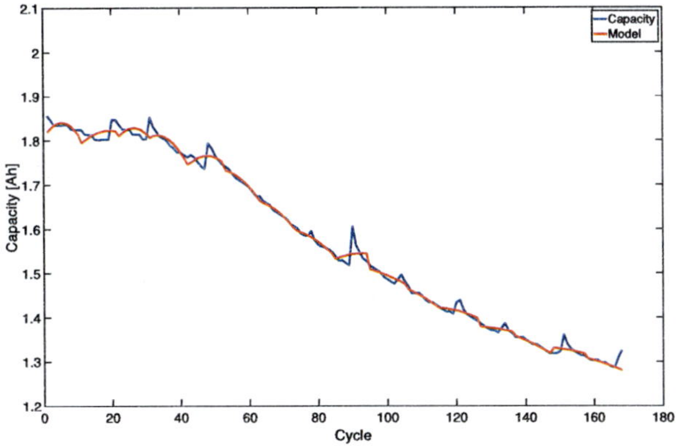

Fig. 4. Learning process

Modeling of the degradation is presented in Fig. 4. The historical dataset contains this model, which is used to make a prediction of the remaining useful life. The discussed model contains sub-models, which are satisfying estimate the degradation. What is more, the effect of self-healing does not degrade the quality of modeling. The described example is divided into 15 sub-models, which are based on the classes.

The prediction of RUL after 25% of cycles for the tested battery is presented in Fig. 5. The proposed algorithm chooses two historical models to predict the remaining useful life. The first one (green line) shows the End of Life in the 97*th* cycle, and the second (orange line) shows EoL in the 75*th* cycle. The figure 6 shows the comparison in the prediction of the remaining life of the selected models. To better predict Time-to-Failure second prediction was made after 40% of cycles.

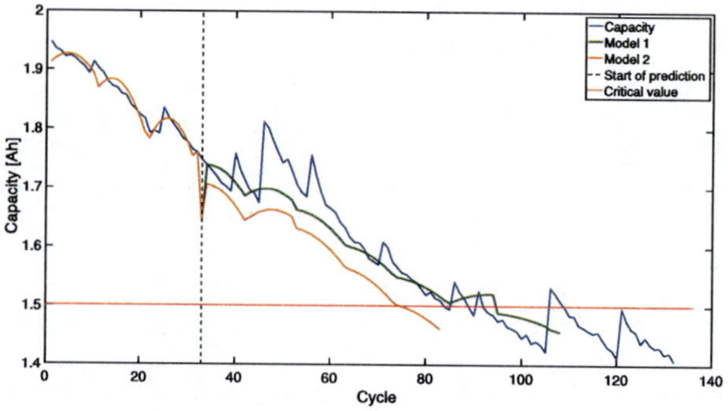

Fig. 5. Estimation after 25% of cycles

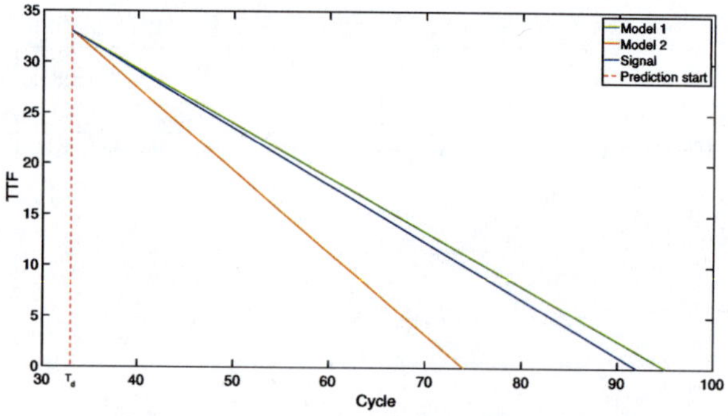

Fig. 6. TTF prediction after 25% of cycles

The prediction after 40% was shown in Fig. 7. As in the previous case, the first model (green line) was chosen and shows EoL in the 97*th* cycle. The second model presents EoL in the 100*th* cycle. The comparison in prediction between the two selected models is shown in the Fig. 8. In both cases the algorithm indicated the first model, therefore it was chosen as the most approximate prediction of the remaining useful life.

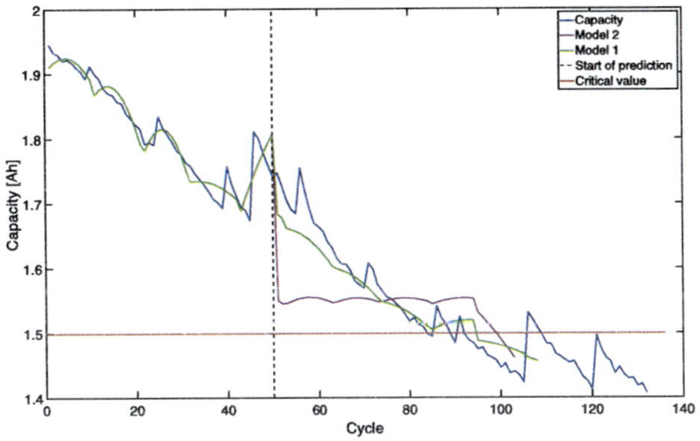

Fig. 7. Estimation after 40% of cycles

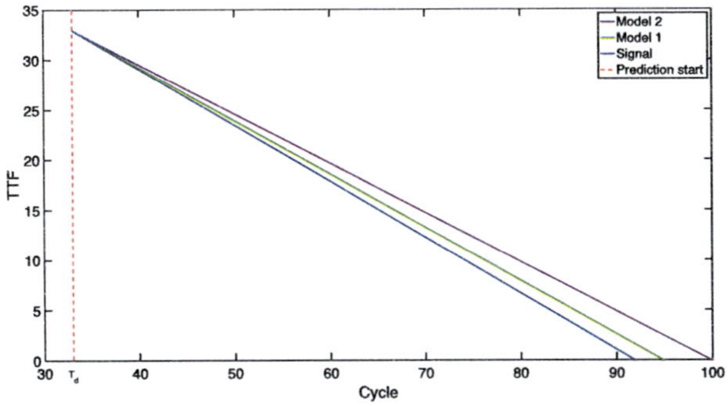

Fig. 8. TTF prediction after 40% of cycles

6 Conclusion Remarks

The aim of this paper was to present the new method of prediction remaining useful life of components. The proposed soft computing approach combines the fuzzy-logic and data-driven health prognostic approaches. The result of this combination wass the practical method for determining the remaining useful life. Proposed method was based on a Takagi-Sugeno multiple-based framework. Compared to other data-driven methods, the proposed algorithm differs in the use of historical data in order to improve the quality of prediction and to create a flexible scheme. The entire method was used to predict remaining useful life of batteries. Finally, the validation of the proposed algorithm was made with NASA PCoE data set of Li-Ion battery. This benchmark consist run-to-failure tests. The presented results correctly describe the predictions of the remaining useful life and confirmed the effectiveness of the solution.

Acknowledgements. The work was supported by the National Science Centre of Poland under Grant: UMO-2017/27/B/ST7/00620.

References

1. Arablouei, R., Doğançay, K.: Modified quasi-OBE algorithm with improved numerical properties. Signal Process. **93**(4), 797–803 (2013)
2. Damiano, A., et al.: Batteries for aerospace: a brief review. In: 2018 AEIT International Annual Conference, pp. 1–6. IEEE (2018)
3. Diouf, B., Pode, R.: Potential of lithium-ion batteries in renewable energy. Renewable Energy **76**, 375–380 (2015)
4. Hosen, M.S., Jaguemont, J., Van Mierlo, J., Berecibar, M.: Battery lifetime prediction and performance assessment of different modeling approaches. IScience **24**(2) (2021)
5. Iaremko, I., Senkerik, R., Jasek, R., Lukastik, P.: An effective data reduction model for machine emergency state detection from big data tree topology structures. Int. J. Appl. Math. Comput. Sci. **31**(4) (2021)
6. Khayat, N., Karami, N.: Adaptive techniques used for lifetime estimation of lithium-ion batteries. In: 2016 Third International Conference on Electrical, Electronics, Computer Engineering and their Applications (EECEA), pp. 98–103. IEEE (2016)
7. Li, Y., et al.: Data-driven health estimation and lifetime prediction of lithium-ion batteries: a review. Renew. Sustain. Energy Rev. **113** (2019)
8. Lipiec, B., Mrugalski, M., Witczak, M.: Health-aware fault-tolerant control of multiple cooperating autonoumous vehicles. In: 2021 IEEE International Conference on Fuzzy Systems (FUZZ-IEEE), pp. 1–7. IEEE (2021)
9. Liu, X., Li, W., Zhou, A.: PNGV equivalent circuit model and SOC estimation algorithm for lithium battery pack adopted in AGV vehicle. IEEE Access **6**, 23639–23647 (2018)
10. Mamun, A.A., Liu, Z., Rizzo, D.M., Onori, S.: An integrated design and control optimization framework for hybrid military vehicle using lithium-ion battery and supercapacitor as energy storage devices. IEEE Trans. Transp. Electr.. **5**(1), 239–251 (2018)
11. Meng, H., Li, Y.F.: A review on prognostics and health management (PHM) methods of lithium-ion batteries. Renew. Sustain. Energy Rev. **116** (2019)
12. Mrugalski, M., Korbicz, J.: Least mean square vs. outer bounding ellipsoid algorithm in confidence estimation of the GMDH neural networks. In: Beliczynski, B., Dzielinski, A., Iwanowski, M., Ribeiro, B. (eds.) ICANNGA 2007. LNCS, vol. 4432, pp. 19–26. Springer, Heidelberg (2007). https://doi.org/10.1007/978-3-540-71629-7_3
13. Nayak, P.K., Yang, L., Brehm, W., Adelhelm, P.: From lithium-ion to sodium-ion batteries: advantages, challenges, and surprises. Angewandte Chemie Int. Edn. **57**(1), 102–120 (2018)

14. Rauh, A., Butt, S.S., Aschemann, H.: Nonlinear state observers and extended Kalman filters for battery systems. Int. J. Appl. Math. Comput. Sci. **23**(3) (2013)
15. Sabbaghi, M., Esmaeilian, B., Raihanian Mashhadi, A., Cade, W., Behdad, S.: Reusability assessment of lithium-ion laptop batteries based on consumers actual usage behavior. J. Mech. Des. **137**(12) (2015)
16. Saha, B., Goebel, K., Christophersen, J.: Comparison of prognostic algorithms for estimating remaining useful life of batteries. Trans. Inst. Measur. Control **31**(3–4), 293–308 (2009)
17. Stetter, R., Till, M., Witczak, M., Lipiec, B.: Health aware fault-tolerant forklift design and control in industry 4.0. In: 2021 5th International Conference on Control and Fault-Tolerant Systems (SysTol), pp. 255–260. IEEE (2021)
18. Straka, O., Čochář, I.P.: Decentralized and distributed active fault diagnosis: multiple model estimation algorithms. Int. J. Appl. Math. Comput. Sci. **30**(2), 239–249 (2020)
19. Tanaka, K., Sugeno, M.: Stability analysis and design of fuzzy control systems. Fuzzy Sets Syst. **45**(2), 135–156 (1992)
20. Witczak, M., Mrugalski, M., Lipiec, B.: Remaining useful life prediction of MOSFETs via the takagi-sugeno framework. Energies **14**(8), 2135 (2021)
21. Witczak, M., Seybold, L., Bocewicz, G., Mrugalski, M., Gola, A., Banaszak, Z.: A fuzzy logic approach to remaining useful life control and scheduling of cooperating forklifts. In: 2021 IEEE International Conference on Fuzzy Systems (FUZZ-IEEE), pp. 1–8. IEEE (2021)
22. Wu, L., Fu, X., Guan, Y.: Review of the remaining useful life prognostics of vehicle lithium-ion batteries using data-driven methodologies. Appl. Sci. **6**(6), 166 (2016)
23. Zhu, J., et al.: Investigation of lithium-ion battery degradation mechanisms by combining differential voltage analysis and alternating current impedance. J. Power Sour. **448** (2020)

Detection of Multiple Leaks in Liquid Transmission Pipelines Using Static Flow Model

Pawel Ostapkowicz[1(✉)] and Andrzej Bratek[2]

[1] Bialystok University of Technology, Wiejska Street 45A, 15-351 Białystok, Poland
p.ostapkowicz@pb.edu.pl

[2] ŁUKASIEWICZ Research Network - Industrial Research Institute for Automation and Measurements PIAP, Al. Jerozolimskie Street 202, 02-486 Warsaw, Poland
andrzej.bratek@piap.lukasiewicz.gov.pl

Abstract. This work presents two methods for double leaks detection in liquid transmission pipelines. They may be used to detect both two simultaneous leaks and non-simultaneous ones. The first method is based on the pipeline's static model with two leaks and minimization of the objective function, which is determined as the square deviation between modeled pressure and pressure measurements taken from the pipeline. The second method uses the pipeline's static model, gradient index functions for leakage detection and algorithms to determine leaks' location and size. Both methods were verified with measurements, gathered from the experimental pipeline to transmit water. The study results confirmed the highly satisfactory effectiveness of both proposed diagnostic methods.

Keywords: Pipelines · Leak detection · Double leakages · Static flow model

1 Introduction

The leak detection systems (LDS) are an indispensable element to ensure safe operation of liquid transmission pipelines. These systems usually use so called analytical methods, which are focused on measuring of the flow internal parameters: flow rate, pressure and temperature [1–3].

Considering pipelines' operators expectations and requirements, LDS, besides single leak detection, should also ensure detection, localization and size estimation of multiple leaks. Due to its highest probability and frequent occurrence, the most obvious cases of multiple leakages are double leakages.

Double leakages, taking into account the time of their occurrence, could be divided into: *non-concurrent* or *concurrent*. Examples of internal methods aimed at double leaks diagnosis might be reviewed in the following papers [2, 4–8].

It is worth mentioning that it is easier to diagnose non-simultaneous leaks than simultaneous or almost simultaneous ones, when transient states provoked by pipeline's leakages overlap in time. In the first case, known methods might be applied, which enable single leaks detection and localization, e.g. wave pressure or gradient methods [9, 10]. It is possible to do only when leaks are delayed in time to such extent that each leakage

Z. Kowalczuk (Ed.): DPS 2022, LNNS 545, pp. 242–253, 2023.
https://doi.org/10.1007/978-3-031-16159-9_20

might be treated separately as a single leak. In the second case, straight forward applying of diagnostic methods targeted at single leaks will no longer be efficient. This could result in obtaining erroneous localization results for both leaks. Therefore, for simultaneous leaks diagnosis more advanced solutions are being applied, which often use dynamic flow models [2, 4–8]. Dynamic models of two leaks in steady state conditions correspond to a case of a single leak, which was observed by [2]. By using of such models, it is possible to detect that two simultaneous leakages occurred, however, their identification (isolation) is not possible without analyzing of changes in their dynamics.

The use of methods based on dynamic models in practice involves resolving many issues. Mainly they relate to aspects such as: elaborating and finding model's solutions, when it may be difficult to determine boundary conditions. Another significant problem consist in the existence of insufficient measurement systems' dynamics, which may result in incorrect determination of real changes and measured values of process variables. Erroneous measurement data introduced into a model impede obtaining a correct, final diagnosis.

The improvement of the knowledge of the boundary conditions could be achieved with tests involving sudden opening of the valve located at the extreme pipe section [5]. However, such a solution is not welcomed by pipelines' operators, as it provokes potential danger resulting from the possibility of uncontrolled pressure increase in a pipeline as well as distorts desirable, stable flow conditions.

Taking into account the above-mentioned conditions, the intention of the authors of this work is to elaborate simplified diagnostic methods without the use of pipeline's dynamic models.

It is worth mentioning that authors of this study previously successfully applied simplified methods to diagnose single leakages of significantly low intensity, i.e. from 0.1 to 1.0% of the nominal flow rate, which occurred under the conditions of stable flow in the pipeline [10]. Besides, the authors proposed solutions to simplified diagnostic methods that proved to be effective in diagnosing single leakages occurring in a transient state. The analyzed transient state was a consequence of the change of the pipeline's operating point, consisting in increasing the pump's rotation velocity [11].

Here, these simplified methods are aimed at detecting, localizing and determining the intensity of two simultaneous leaks in steady state conditions as well as two non-simultaneous leaks. The second leakage occurs already in a transient state, which is a result of the first leak occurrence.

2 General Characteristics of Diagnostic Methods

The proposed diagnostic methods have the following initial assumptions:

- availability of pressure measurements $p_1, \ldots, p_N$ along a pipeline, at $N - 1$ pipeline's segments extremities established by N pressure sensors' location, and flow rate measurements q_1, q_2 at the pipeline's inlet and outlet,
- determination of parameters of both leaks: z_{u1}, z_{u2} - position coordinates, q_{u1}, q_{u2} - leaks' size (volume flows).

2.1 Method I

The proposed solution is based on using a static pipeline model.

The pipeline with two leaks might be described with a static model, according to the schema shown in Model's data flow schema Fig. 1.

Fig. 1. Model's data flow schema.

The pressure and flow volume measurements at the pipeline's inlet and outlet are the model's solution boundary conditions, while a drop leak location is its parameter. The model determines p_{mi} pressure in pressure measurement points $i = 2, \ldots, N - 1$ and leaks parameters.

In the pipeline's model with two leakages a static model of a liquid flow was used. The p pressure at any pipeline's point might be calculated with the following formula:

$$p(z) = p_0 - \lambda \frac{\rho \cdot q^2}{2 \cdot A^2} \frac{z}{d} \tag{1}$$

where p_0 is initial point's pressure ($z = 0$), z is pipeline's length coordinate, d is pipeline's diameter, A is cross sectional area of a pipe, q is liquid volume flow rate in a pipeline, ρ is liquid's density, λ is linear loss coefficient.

The approach applied in the model consisted in defining of pressure at the inlet segment of the pipeline till the position of the first leak z_{u1}, on the basis of p_1, q_1 measurements. Pressure at the outlet part of the pipeline, i.e. starting from the location of the second leak until the pipeline's end, is determined according to the linear function stretched on the p_{u1}, p_{u2} pressures, while the flow rate q_1' in pipeline's segment between the leak points - on the basis of q_1 and the pressure gradient at this segment. The following model's equations might be considered:

$$p_{mi} = p_1 - \lambda \frac{\rho \cdot q_1^2}{2 \cdot A^2} \frac{(z_i - z_1)}{d}, \text{ when } z_i <= z_{u1} \tag{2}$$

$$p_{u1} = p_1 - \lambda \frac{\rho \cdot q_1^2}{2 \cdot A^2} \frac{(z_{u1} - z_1)}{d} \tag{3}$$

$$p_{mi} = p_N + \lambda \frac{\rho \cdot q_2^2}{2 \cdot A^2} \frac{(z_N - z_i)}{d}, \text{ when } z_i >= z_{u2} \tag{4}$$

$$p_{u2} = p_N + \lambda \frac{\rho \cdot q_2^2}{2 \cdot A^2} \frac{(z_N - z_{u2})}{d} \tag{5}$$

$$q_1' = \sqrt{\frac{2 \cdot d \cdot A^2}{\lambda \cdot \rho} \cdot \frac{p_{u1} - p_{u2}}{z_{u2} - z_{u1}}} \tag{6}$$

$$p_{mi} = p_{u1} - \lambda \frac{\rho \cdot q_1'^2}{2 \cdot A^2} \frac{(z_{u2} - z_{u1})}{d}, \text{ when } z_{u1} < z_i < z_{u2} \tag{7}$$

$$\begin{aligned} q_{u1} &= q_1 - q_1' \\ q_{u2} &= q_1' - q_2 \end{aligned} \tag{8}$$

The discussed model should fulfill the tasks such as determination of leaks' location and their size. While identifying an issue, that in a pipeline a leak occurred, the presented model might enable defining double leak parameters. For this purpose, we might apply an objective function F_C, dependent from z_{u1} and z_{u2}, and determined as a square deviation of modeled pressure from measured pressure (9), and identify its minimum, which will ensure the best model fitting to measurements gathered during leaks at z_{u1} and z_{u2} points (10).

$$F_C = \sum_{i=2}^{N-1} (p_{mi} - p_i)^2 \tag{9}$$

$$\min F_C(z_{i(1)}, z_{i(2)}) \Rightarrow \{z_{u1}, z_{u2}\}$$

$$\text{for } z_1 < z_{u1} < z_N, \ z_{u1} < z_{u2} < z_N \tag{10}$$

2.2 Method II

The proposed solution is also based on application of a static pipeline's model.

The pressure drop in the L length segment might be calculated as follows:

$$p_0 - p_L = \lambda \frac{\rho \cdot q^2}{2 \cdot A^2} \frac{L}{d} \tag{11}$$

where p_0 is initial point's pressure ($z = 0$), p_L is pressure at the segment's end,

L is pipeline's segment length, d is pipeline's diameter, A is cross sectional area of a pipe, q is liquid's flow rate, ρ is liquid's density, λ is linear loss coefficient; or by using pressure gradient gr:

$$\begin{aligned} gr &= \frac{p_0 - p_L}{L} \\ gr &= \lambda \frac{\rho \cdot q^2}{2 \cdot A^2} \frac{1}{d} \end{aligned} \tag{12}$$

Once a leakage occurs at a point with z_u coordinate, located at the i-th pipeline's segment, liquid flow distribution take the form such as the one presented in Fig. 2. Figure 3 presents the distribution of the liquid flow rate, corresponding to this situation, which is perceived through the prism of pressure measurement along the pipeline. The volume flow rate q_1' at the i-th pipeline's segment achieves the intermediate value between the pipeline's inbound flow rate q_1 and outbound flow rate q_2, according to the formula (11).

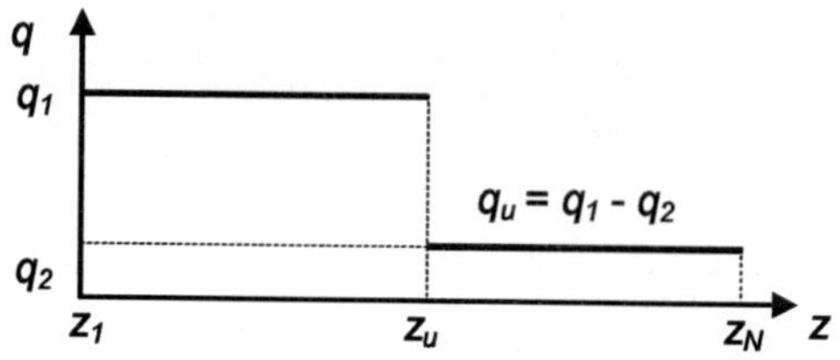

Fig. 2. Distribution of q volume flow rate in the pipeline with a leak of z_u coordinate in static conditions.

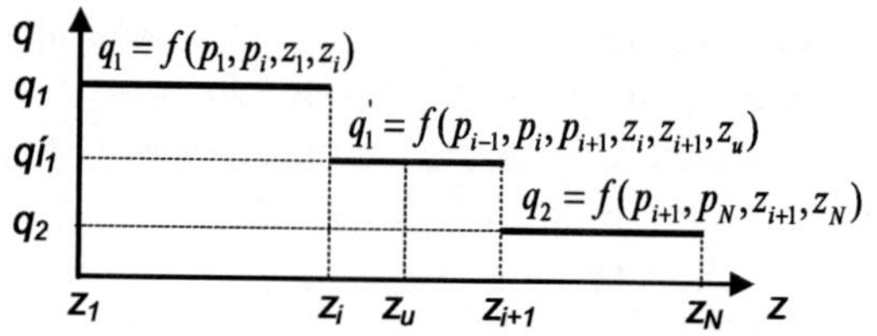

Fig. 3. Distribution of q volume flow rate, based on pressure measurements, in the pipeline with a leak of z_u coordinate found on the i-th sensor's segment.

After transformations the dependency (11) leads to a simple solution for the leak location:

$$z_u = z_i + (z_{i+1} - z_i) \cdot \frac{q_i'^2 - q_{i+1}^2}{q_{i-1}^2 - q_{i+1}^2} \tag{13}$$

By applying the pressure gradients the z_u coordinate may be determined as follows:

$$z_u = z_i + (z_{i+1} - z_i) \cdot \frac{\lambda_{i-1} \cdot (\lambda_{i+1} \cdot gr_i - \lambda_i \cdot gr_{i+1})}{\lambda_i \cdot (\lambda_{i+1} \cdot gr_{i-1} - \lambda_{i-1} \cdot gr_{i+1})} \tag{14}$$

where, assuming k denotes the pipeline's segment, $k \in \langle i-1, i+1 \rangle$,

$$gr_k = \frac{p_k - p_{k+1}}{z_{k+1} - z_k} \tag{15}$$

while λ_k is its linear loss coefficient and leakage intensity q_u from the relationship:

$$q_u = \sqrt{\frac{2 \cdot d \cdot A^2}{\lambda_{i-1} \cdot \rho} \cdot gr_{i-1}} - \sqrt{\frac{2 \cdot d \cdot A^2}{\lambda_{i+1} \cdot \rho} \cdot gr_{i+1}} \tag{16}$$

In order to define leak parameters z_u, q_u with the presented method, it is crucial to understand the pressure at two pipeline's points before the leak's location and at two points behind the leak as well as to know the pipeline's parameters.

The above mentioned method could be directly applied in case of double leakage, unless the leaks appear in exactly the same pipeline's segment or in two adjacent segments. We might distinguish two distinct pipeline's segments with a single leak and apply the Eqs. (14)–(16) to each of them. If leaks occur in the adjacent pipeline's segments, then the solution require to consider additional equations.

The crucial factor of performing such diagnostics is the ability to determine that leaks occurred at two different locations in a pipeline.

To facilitate that $IG_{i\text{-}j}$ index functions were introduced, expressed as a difference between the pressure gradient in the segment $j - 1$ (i.e. between pressure sensors numbered $i - 1$ and i) and the j segment (i.e. between pressure sensors numbered j and $j + 1$), while $i \in \, < 2, N - 1 >, j \in \, < i, N - 1 >$.

The index functions $IG_{i\text{-}j}$ are compared with threshold values in order to detect a leak. It is mandatory then to determine the thresholds' margins, below which index functions achieve values in no leak conditions. Overpassing the assumed threshold by a defined $IG_{i\text{-}j}$ function, indicates that a leakage occurred between i and j sensors.

In a single leak conditions, $IG_{i\text{-}j}$ indicator, i.e. for the adjacent pipeline's segments through pressure sensor i, achieves residual values only in case of two consecutive i indices. The segment with a leak is then either the initial or final segment in residual area. If there are two leaks (occurring at two different segments), the indicators achieve standing out values for from two to four indices i. A simple logical analysis of these values enables recognizing in which segments leaks occurred. Consequently, if these segments are not adjacent, a pipeline's decomposition into two single leak segments might be performed.

3 Experimental Data Acquired from the Laboratory Pipeline

The verification of both methods presented above was carried out utilizing measurement data collected from the laboratory pipeline pumping water.

3.1 Pipeline Stand

The main pipeline's segment is 380 m long (between coordinates $z = 0$ and $z = 380$) and is made of polyethylene (PEHD) pipes which are 34 mm in internal and 40 mm in external diameter. The pipeline is equipped with measuring devices - two electromagnetic flow meters (located at the inlet and outlet), nine pressure transducers (at the inlet and outlet, and also in several points along the pipeline) and two thermometers. Sensors are connected to a PC provided with the 16 bit A/D converter (see Table 1).

3.2 Conditions of Experiments

The experiments performed on a pipeline consisted in simulating two leaks using electromagnetic valves. Leakages are provoked at the point about coordinates $z_{leak\text{-}L1} = 155$ m and $z_{leak\text{-}L2} = 315$ m. Two cases were considered i.e. simultaneous and separate leak occurrences. In case of simulating both leaks not in exactly the same time, the second

Table 1. Measurement system characteristics.

Devices	Pressure sensors	Flow rate sensors
Location [m]	$p_1 = -7.2$ $p_2 = 1$ $p_3 = 61$ $p_4 = 141$ $p_5 = 201$ $p_6 = 281$ $p_7 = 341$ $p_8 = 378$ $p_9 = 381.5$	$q_{1(in)} = -6.5$ $q_{2(out)} = 382.2$
Range	$0 \div 10$ [bar]	$0 \div 200$ [l/min]
Accuracy	0.1% of range	0.2% of range
Uncertainty (sensor + converter)	±0.012 [bar]	± 0.44 [l/min]

leak occurred at the moment when the pressure wave provoked by the first one, propagating with the sound speed, reached its position (see Table 2). The leaks were simulated by sudden (step) aperture of the electromagnetic valve. Before the leak simulation, the pipeline was operating in steady state conditions, with a nominal flow rate of about $q_{nom} \approx 140$ l/min. The intensity of both leaks was about $0.9 \div 1.5$ l/min. They were estimated by measuring with the use of a measuring cylinder and a stopwatch. The sensors were scanned with the frequency $f_P = 100$ Hz. Two sample experiments were considered for each simulated leakages configuration.

Table 2. Parameters of leakages simulated in experiments.

No.	Simulated leaks		Coordinate z_{leak} [m]	Occurrence time [s]	q_{leak} [l/min]	[%] q_{nom}
1	Concurrent	L1	155	180.00	0.96	0.69
2		L2	315	180.00	1.08	0.77
3	Non-concurrent	L1	155	180.00	0.85	0.71
4		L2	315	180.50	1.07	0.71
5	Concurrent	L1	155	180.00	1.49	1.06
6		L2	315	180.00	1.21	0.86
7	Non-concurrent	L1	155	180.00	1.49	1.06
8		L2	315	180.50	1.19	0.85

4 Results of Verification

The pipeline's model and diagnostic algorithms discussed in above methods were implemented in MATLAB program.

To carry out their verification, MATLAB's program was enriched with data, gathered from measurement data files registered during experiments.

4.1 Method I

Initial Testing Phase. To start with, on the basis of p_i ($i \in <1, N>$), q_1, q_2 measurements, using time windows with 100 signal samples, their corresponding ps_i, qs_1, qs_2 variables were obtained. For a pipeline in steady state conditions, i.e. before having a leak simulated, also variables' statistics were determined for leak detection procedures and leak detection thresholds for leak identification procedures. The loss coefficient λ, based on the relation (1), was determined as well.

Leak Detection. With the view to detect an emergency state (a leak) the same approach was adopted as the one used in a single leak diagnosis. The algorithm based on PPA method (pressure point analysis), considering calculation of index functions IF_i based on ps_i variables, where $i \in <2, 8>$, was exploit in this case.

Table 3 presents T_w leak detection times (in relation to the moment of the first leak occurrence) through IF_3 to IF_7 index functions, obtained for every single data set.

Table 3. Response times of detection functions used for Method I and Method II, observed in carried out experiments (the shortest one in each experiment is bolded).

No.	Response time T_w [s]						
	Method I					Method II	
	IF3	IF_4	IF_5	IF_6	IF_7	*IG4–5*	*IG6–7*
1	1.25	0.72	0.72	**0.66**	0.69	1.1	0.7
2	0.82	**0.57**	0.61	0.58	0.60	1.1	0.8
3	1.28	**0.90**	0.95	0.99	1.02	1.8	1.1
4	1.51	**0.88**	0.95	0.89	0.96	2.2	1.1
5	1.10	0.60	0.60	**0.54**	0.55	1.1	0.6
6	0.81	**0.34**	0.49	0.55	0.63	0.8	0.7
7	0.95	**0.59**	0.67	0.76	0.83	0.6	1.0
8	0.84	**0.47**	0.60	0.76	0.86	0.6	1.0

By analyzing the achieved detection times, we might state that IF_4, IF_6 indicators were the most useful, as they were the quickest to inform about the leak. These indicators corresponded to p_4 or p_6 pressure measurement points, i.e. positioned directly close to the leak. The shortest times of leakage detection amounted from 0.34 s to 0.90 s. These values correspond to the transient state of a pumping process - detection succeeded still during propagation of pressure waves, provoked by the leak event, in the pipeline.

The IF_i index functions' values, corresponding to different sensors, however didn't give foundation to state that two distinct leaks occurred. This considered both experiments with simultaneous leaks (samples: 1, 2, 5, 6) and the ones with leaks not made at the same time (samples: 3, 4, 7, 8).

Nevertheless, we might notice a regularity, which consist in shortening of T_w detection times in case of enlarged leakages. Besides, T_w detection times were getting longer, in comparison to samples with simultaneous leaks, which could be explained by the postponed effect of pressure waves accumulation, provoked by both leaks.

Leak Location and Size Determination. Within 4.5 s after having realized that there is a leak in a pipeline, i.e. yet in new steady state conditions, the leak location began. This was carried out in 50 s timeframe by sequentially launching leak pipeline's model computations and the objective function minimization algorithm (9). Such calculations were fed with measurement data every one second. Table 4 presents localization results and leak size estimations, expressed as average values over the relevant 50-s time span. It should be explained here that these values are given in a set with an accuracy of 0.1 l/min, which corresponds to the metrological characteristics of measuring devices (electromagnetic volume flow meters) applied on the examined pipeline.

On the basis of the obtained results for the registered measurement data, we might ascertain that errors in determining both leak locations z_{u1}, z_{u2}, amount from −9.8 m to −20.4 m and respectively from 7.7 m to 15.7 m, while in case of q_{u1}, q_{u2} from $-0.1 \div 0.1$ l/min and $0.1 \div 0.4$ l/min. The obtained leak position and leakage size estimation values for both leak points should be considered to be enough satisfactory.

Table 4. Results of the leak localization and leak intensity estimation.

No.	Method I				Method II					
	Leak localization [m]		Leak intensity [l/min]		Leak localization [m]		Leak intensity [l/min]			
	$zu1$	z_{u2}	$qu1$	$qu2$	$zu1$	z_{u2}	$qu1$	$qu2$	qu	Δqu
1	142.9	327.6	0.9	1.4	151.1	321.3	0.8	1.2	2.2	0.2
2	142.8	326.3	1.0	1.2	153.0	321.4	0.9	1.2	2.2	0.1
3	134.6	323.8	0.9	1.2	149.9	320.0	0.8	1.1	2.1	0.2
4	139.7	322.7	0.8	1.2	148.7	318.4	0.7	1.1	1.9	0.1
5	145.2	327.3	1.4	1.5	148.7	318.4	1.3	1.4	2.9	0.2
6	140.9	326.0	1.4	1.4	151.3	320.9	1.3	1.3	2.8	0.2
7	145.1	330.7	1.4	1.6	151.8	323.1	1.3	1.4	3.0	0.3
8	137.4	324.5	1.5	1.4	150.7	321.1	1.3	1.4	2.8	0.1

4.2 Method II

Initial Testing Phase. This phase was exactly the same as in case of the method I. On the basis of the relation (1) λ_k the loss coefficient was determined for all pipeline's segments.

Leak Detection. Similar approach was adopted as in the method I for the initial stage. Moreover, $IG_{i\text{-}i}$ index functions analysis was performed, where $i \in <2, 7>$. The index functions were expressed as gradient differences for the adjacent segments. The analysis's results were similar for all experiment, so that they are presented as a single row (not marked with a sample number) in Table 5. They indicate a leak occurrence in segments $k = 4$ and $k = 6$, in segments that are not close to each other. Due to this fact these segments were selected to perform the exact leak localization in the next diagnostic stage. In Table 5 the 'segment' row indicates pipeline's segments numbering, which are being tested with the index function. The "+" sign in the 'result' row means overpassing the agreed threshold by the index function, while "–" sign means beneath threshold index function value. As the test results were the same in case of all samples (samples 1–8), the 'results' row corresponds to all of them.

To ascertain the right choice $IG_{i\text{-}j}$ index functions were analyzed as well, where $i \in$ <2, 6> and $j = i + 1$, which define gradients' differences for the segments located on both sides of the segment $k = i$. This process ended the identification of segments with leaks. In Table 3 the times of this identification are presented. The determination of the segment with a leak came slightly later than the identification of the leak event, i.e. 0.8 ÷ 1.1 s for simultaneous leaks and 1.0 ÷ 2.2 s in case of leaks not occurring at the same time, counting the time starting from the first leak occurrence.

Table 5. Results of leak detection with $IG_{i\text{-}i}$ index functions corresponding to adjoining pipeline's segments.

Indicator	$IG2\text{–}2$	$IG_{3\text{–}3}$	$IG_{4\text{–}4}$	$IG_{5\text{–}5}$	$IG_{6\text{–}6}$	$IG_{7\text{–}7}$	$IG_{8\text{–}8}$
Segment	1–2	2–3	3–4	4–5	5–6	6–7	7–8
Result	–	–	+	–	+	+	–

Localization of the Leak and Determination of its Size. The leak localization was carried out in two modes: from the steady state standpoint and considering the period of a transient state in a pipeline.

Mode 1 – Steady State. Identically as in case of the method I, within 4.5 s from the distinguishing that a leak occurred in the pipeline, i.e. in conditions of a new steady state, the leak location process began. This was carried out in 50-s timeframe on the initially identified leak segments $k = 4$ and $k = 6$, sequentially initiating localization procedures according to relationships (14), (15), (16). The procedures were enriched with measurement data every single second. Table 4 presents corresponding results expressed as averaged values of the whole considered 50-s timeframe. In the penultimate table's column a summarized sluice loss (18) is presented, which results from flow balance at the pipeline's inlet and outlet, whereas the leak flow rate determination error (19) is given in the last column.

$$ERx = x_u - x_{leak} \tag{17}$$

$$qu = q_1 - q_2 \tag{18}$$

$$\Delta qu = qu - (q_u 1 + q_u 2) \tag{19}$$

Considering the results obtained for the collected measurement data, we might state that the errors calculated according to the general relationship (17), in case of leak location z_{u1}, z_{u2} varied from −2.0 m to −6.3 respectively and from 3.4 m to 8.1 m. In reference to leak flows, the estimation error q_{u1} amounts to −0.1 ÷ −0.2 l/min, while q_{u2} amounts to 0.0 ÷ 0.2 l/min. The leak flow rate sum error defined on the basis of liquid flow balance is from 0.1 l/min to 0.3 l/min. The obtained result values of leak location and its intensity estimation for both leaks were even better, than in case of the method I.

Mode 2 – Transient State. The process of leak localization was initiated just after having identified the leaks with $IG_{4\text{–}5}$, $IG_{6\text{–}7}$ functions, considering the starting point for a given sample the moment corresponding to the greater of two times presented in Table 3. It was carried out on initially determined leaks' segments $k = 4$ and $k = 6$ in the 8-s time windows. Calculation procedures were fed with measurement data every 50 ms.

It may be noticed that localization procedure's results and the estimated leaks' intensity become useful only within a bit time after having detected a leak and identified the pipeline's segments with a leakage. Useless time interval from the leak localization standpoint oscillated between around 1.2 and 1.8 s for the leak at 155 m, and around 1.8 and 2.1 s for the leak at 315 m.

Besides, it was observed that for the experiments with not simultaneous leaks (experiments: 3, 4, 7, 8) reaching a stabilization of the leak position z_{u1} was lengthened in time in comparison to the samples with simultaneous leaks (experiments: 1, 2, 5, 6), while in case of positioning z_{u2} the length of the instable period in both cases was much similar. It results from the shift in time of the impact on the pipeline's pressure parameters of the pressure wave, being a consequence of the second leak in the simulations of not simultaneous leaks.

5 Conclusion

This paper presents two methods to diagnose double leakages, which were verified in experiments with the use of measurement data, obtained on the laboratory pipeline.

The effectiveness of the algorithms, dedicated to both methods, was confirmed with examples of small leaks diagnosing, with the intensity of 0.6 ÷ 1% of nominal flow rate, which were simulated at two different pipeline's points.

The acquired results of the performed experiments allow to assume that both methods should manage double leaks' events provoked by the changes in real pipelines' operating conditions.

In the future work it would be worth examining the performance of both proposed methods in relation to smaller leakages (e.g. even 0.2 l/min) and taking into account not only their constant, but also increasing nature.

Acknowledgement. Research were partially carried out within project No. N N504 494439 supported by the National Science Centre – Poland in the years 2010–2015.

Research work was funded by the Bialystok University of Technology – project No. WZ/WM-IIM/2/2022.

References

1. Colombo, A.F., Lee, P., Karney, B.W.: A selective literature review of transient-based leak detection methods. J. Hydroenviron. Res. **2**(4), 212–227 (2009)
2. Kowalczuk, Z., Gunawickrama, K.: Detecting and locating leaks in transmission pipelines. In: Korbicz, K. J., Koscielny, J. M., Kowalczuk, Z., Cholewa, W. (eds.) Fault Diagnosis: Models, Artificial Intelligence, Applications, pp. 821–864. Springer, Heidelberg (2004). https://doi.org/10.1007/978-3-642-18615-8_21
3. Turkowski, M., Bratek, A., Slowikowski, M.: Methods and systems of leak detection in long range pipelines. J. Autom. Mobile Robot. Intell. Syst. **1**, 39–46 (2007)
4. Delgado-Aguiñaga, J.A., Besançon, G., Begovich, O., Carvajal, J.E.: Multi-leak diagnosis in pipelines based on Extended Kalman Filter. Control. Eng. Pract. **49**, 139–148 (2016)
5. Lazar, A., Hadj-Taïeb, L., Hadj-Taïeb, E.: Two leaks detection in viscoelastic pipeline systems by means of transient. J. Loss Prev. Process Ind. **26**(6), 1341–1351 (2013)
6. Torres, L., Besanon, G., Georges, D.: Multi-leak estimator for pipelines based on an orthogonal collocation model. In: Proceedings of the 48th IEEE Conference on Decision and Control, pp. 410–415. Shanghai, China (2009)
7. Verde, C.: Multi-leak detection and isolation in fluid pipelines. Control. Eng. Pract. **9**(6), 673–682 (2001)
8. Verde, C., Molina, L., Torres, L.: Parameterized transient model of a pipeline for multiple leaks location. J. Loss Prev. Process Ind. **29**, 177–185 (2014)
9. Ostapkowicz, P.: Leak detection in liquid transmission pipelines using simplified pressure analysis techniques employing a minimum of standard and non-standard measuring devices. Eng. Struct. **113**, 194–205 (2016)
10. Ostapkowicz, P., Bratek, A.: Possible leakage detection level in transmission pipelines using improved simplified methods. Eksploatacja i Niezawodność – Mainten. Reliab. **18**(3), 469–480 (2016)
11. Ostapkowicz, P.; Bratek, A.: Leak detection in liquid transmission pipelines during transient state related to a change of operating point. In: Kowalczuk, Z.: Advanced and Intelligent Computations in Diagnosis and Control, AISC, vol. 386 pp. 253–265. Springer, Cham, Switzerland (2016). https://doi.org/10.1007/978-3-319-23180-8_18

Application of Bayesian Functional Gaussian Mixture Model Classifier for Cable Fault Isolation

Jerzy Baranowski(✉)

AGH University of Science and Technology, Kraków, Poland
jb@agh.edu.pl

Abstract. Fault isolation is an important problem allowing to addres faulty solutions appropriately. In this paper we consider using Bayesian methods represented by functional Gaussian Mixture Model to classify time series representing faults to appropriate categories. We use spline representation of time series and perform Markov Chain Monte Carlo computation to estimate probability of class membership. We show results and supplement them with sensitivity analysis with respect to data quality.

Keywords: Bayesian statistics · Gaussian mixture models · Fuctional data analysis · Cable diagnostics

1 Introduction

Effective and reliable monitoring and diagnostics of energy installations is of utmost importance, as they are important part of world's economy. Algorithms for fault detection and isolation allow extension of system lifetime, reduction in operation interruption and can lead to significant savings. The main difficulty in their development is that power installations have a high level of complexity, are usually nonlinear and influenced by stochastic disturbances and parameter variations. Therefore, approaches based on first principles models are difficult or even impossible to use on a wider scale. That is why methods based on statistical models or machine learning are those most researched.

Functional data analysis (FDA) is a group of methods for analysis of data in form of functions, with a special focus on time series data. The main idea is to create a model of the signal using certain function basis, and coefficients of that basis representation can be considered as reduced dimensionality. FDA is a matured field in the area of statistics, with focus on bases in functional spaces such as polynomials, wavelets and others. Maturity of the field can be observed

This research was funded by AGH's Research University Excellence Initiative under project "Interpretable methods of process diagnosis using statistics and machine learning" and by Polish National Science Centre project "Process Fault Prediction and Detection" contract no. UMO-2021/41/B/ST7/03851.

Z. Kowalczuk (Ed.): DPS 2022, LNNS 545, pp. 254–265, 2023.
https://doi.org/10.1007/978-3-031-16159-9_21

by recent review papers [9] or special issues [1] in prestigious journals covering the field of statistics. We join FDA with Bayesian approach in order to obtain probability distributions of basis coefficients, get generative models and model uncertainty.

Typically, machine learning, data-driven models are providing complicated 'black-box' models, which are not transparent and hard to interpret. This is flaws are less prevalent in statistical approaches which is a cause of their dominance of statistical approaches in the field. Both of those groups however suffer from the typical situation that the real data for system faults is extremely rare, and even if present is often incomplete. That is why it is crucial to develop methods that can handle issues of non-representative or missing data. Bayesian methods (especially in technical sciences) are an emerging set of tools for solving many kinds of diagnostic problems [3,4]. One interesting solution that can be applied for the field of classification are Gaussian Mixture Model Classifiers [8].

In case of cable fault monitoring popular approaches are based on first principle model fitting. We have provided an extensive review of those methods in our earlier paper [2]. Important aspect shared by all those methods is that they are not focused on uncertainty. They are often considering point estimates or least squares fits. In this paper we want to provide a certain proof of concept for using statistical methods for fault detection and distinguishing types of faults. For this purpose we focus on a real problem of distinguishing between pole-to-ground and pole-to-pole short circuits. Pole-to-pole and pole-to-ground short circuits are the typical DC cable faults. These generally result in fast discharge of the DC-link capacitor through the DC circuit, leading to transient overcurrent, which can damage system components.

In this paper we propose a functional extension to Bayesian Mixture Model classifier via using spline signal representation and Hamiltonian Monte Carlo computation to provide model learning. Moreover we present a model structure allowing use of labeled and unlabeled data to learn the model and perform classification.

This paper is organized as follows. First we cover the theory behind Bayesian Functional Gaussian Mixture modelling using splines and normal mixtures. Then we explain how we are considering multiple kinds of data and describe how to reconstruct probabilities of class membership. Then we describe considered use case of VSC DC cable diagnostics and present detailed coverage of the considered case. Following we provide statistically oriented evaluation of classifier quality and sensitivity to data quality. Paper ends with conclusions.

2 Bayesian Functional Gaussian Mixture Model

In this section we will cover main theoretical concepts that are needed for construction of functional Gaussian Mixture Model Classifier. We will not cover main principles of Bayesian statistics, for more details we refer reader to Gelman's book [6]. Briefly, we will note that in the Bayesian setting all of the model parameters are random variables, i.e. they have their own distribution. The only thing that is fixed in the model is the available data. Both model parameters describing the underlying process (mean of normal distribution) and parameters describing the measurement process (standard deviation) are then random variables and we infer their joint posterior distribution from the data via sampling.

2.1 Spline Representation

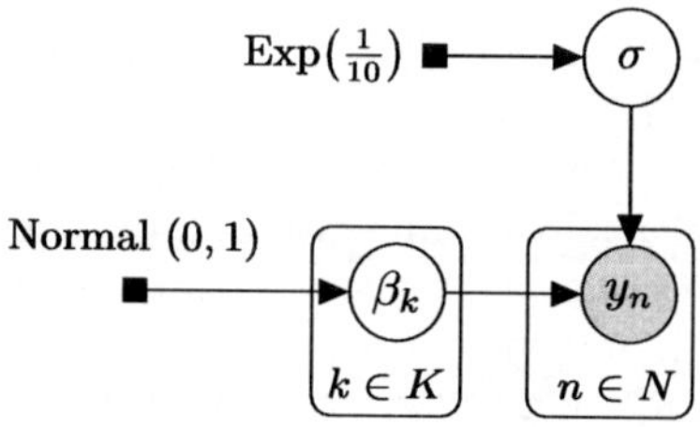

Fig. 1. Graphical representation of parameters in Bayesian model for functional representation of the signal. Spline weights β_k priors are standard normal distributions. Measurement process of y_n's is represented by a normal likelihood with standard deviation of σ. Standard deviation's prior is an exponential distribution with mean equal to 1/10.

In order to model a time series signal we will assume that it comes from a spline function. This allows sufficient flexibility to model relatively smooth functions and local phenomena. We consider that our data generating process is given by

$$\begin{aligned} y_n &\sim \text{Normal}(\mu_n, \sigma) \\ \mu_n &= \sum_{k=1}^{K} \beta_k \phi_k(t_n) \end{aligned} \tag{1}$$

where y_n are our sampled measurements, $n = 1, \ldots, N$, which with uncertainty of normal distribution given by parameter (random variable) σ. Functions $\phi_k(t)$ are B-splines on the assigned K knot grid and β_k are their coefficients, which are model parameters (also random variables). In other words we assume that the underlying process is an unknown function (represented by a finite spline series) from which we take measurements at t_k, which are disturbed normally with standard deviation of σ. Underlying process is then a random function, which distribution is completely determined by a joint distribution of β coefficients. Measurement process is a normal likelihood over that function with unknown parameter σ.

Spline coefficients β_k have a common prior distribution Normal(0, 1) and standard deviation σ has an exponential prior Exp $\left(\frac{1}{10}\right)$. Prior parameters were chosen to provide relative ignorance about parameters, but enough regularization so the Markov chains would behave properly (so called *weakly informative priors*).

While priors are independent of each other, we need to stress that in the joint posterior distribution of the parameters no independence occurs, as all parameters are deeply connected.

Quantities μ_n are transformed parameters of the model, as it corresponds to the mean of fitted distribution for individual measurements. It could be avoided, as actual model parameters are coefficients of B-splines combination β_k, but it improves formula clarity.

Relations of the entire model are presented using Bayesian network plate notation in the Fig. 1. Example of spline basis is presented in the Fig. 2.

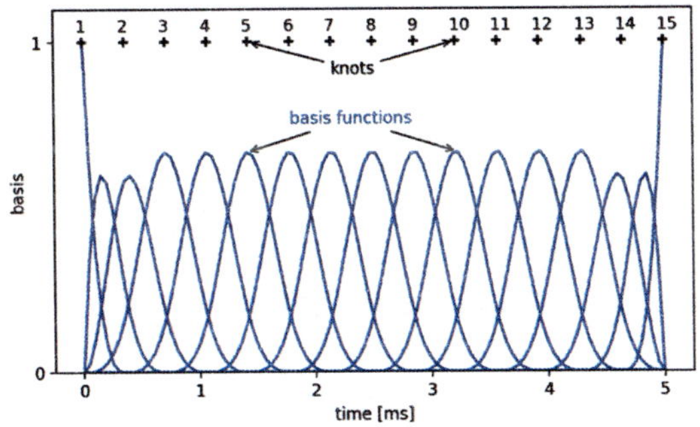

Fig. 2. Spline basis used for modelling of current time series in the considered VSC DC cable faults. We duplicate knots on the edges to obtain lower order splines for boundary behavior.

If we have assumed uniform priors on β and fixed value of σ the problem reduces to ordinary least squares. In our case are using MCMC, which in this case might be overreaching (only σ could not be infered by linear algebra), but if some other parameter priors are used (perhaps sparsity inducing ones) it becomes necessary anyway. However it is based on creation of design matrix from base spline function values at the measurement points and fitting them. Efficiency requires certain modifications presented with in the following code

Remark 1. Stan is a language that separates its code into blocks with different functionalities `data` - describes inputs to the program, `transformed data` are auxiliary variables and algebraic operations on the data, `parameters` are random variables declarations, `model` defines the posterior distribution using data and parameters, `generated quantities` are operations that can be done with obtained posterior samples (but outside of Markov Chain Monte Carlo).

2.2 Multiple Levels of Data in Diagnostics

Diagnostics of fault in processes is a tricky issue, as we often do not know the underlying phenomena that are responsible for fault occurrence. That is why data that is available can be often idealized or not representative. That is why in this paper we focus on a three level structure of data availability:

1. Labeled data. We have isolated the cases of faults in available data, we use cleaned, perhaps model based trajectories which we provide with labels of what case we are considering.
2. Non-labeled but available data. This is a situation that we have data available, which is not that well investigated, but is relevant and its analysis can be used to inform the model without time constraints.
3. Outside data, for which the classification has to be performed. Those data cannot be used to inform the model, but model has to provide inference for them. For example, these are the data that we got after learning the model.

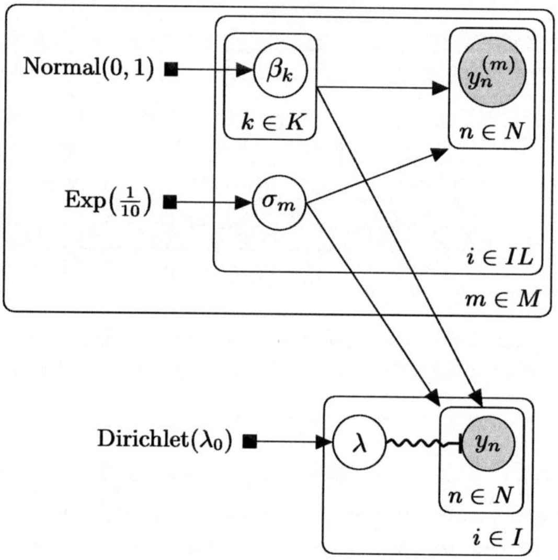

Fig. 3. Bayesian network representing mixture model that can be used for classification of faulty signals. Each of mixture components m is preinformed with labeled data $y^{(m)}$ which consists of total IL time series. Mixture model is assigned a prior λ_0 and then being informed (both spline and mixture parameters) via unlabeled data.

We can address those levels using mixture modelling. Classically unlabeled data assignment is the clusterization problem which is in a sense problem of finding a mixture model of all possible components. This is generally ill posed in Bayesian context, as it is naturally multimodal because of label switching and other issues. Here however comes the idea of using labeled data to inform initial locations of clusters, to combat multimodality. Our model will then determine the mixtures using both labeled data and then non-labeled adding necessary variance that can be lost in the idealized labeled data.

Here we are not addressing the issue of discrete vs continuous variables and we do not use categorical distribution to characterize mixture components $z_m \in 1, \ldots M$ as its out of the scope. Sufficient to say that discrete parameters are a problem for MCMC sampling, and marginalizing them out is sufficient for this case (for details see Stan manual[1]). For our purposes we can say that consider the following model is equivalent to the mixture of discrete components:

$$
\begin{aligned}
y_n &\sim \sum_{m=1}^{M} \lambda_m \mathrm{Normal}(\mu_{\mathrm{n}}^{(\mathrm{m})}, \sigma_{\mathrm{m}}) \\
y_n^{(m)} &\sim \mathrm{Normal}(\mu_{\mathrm{n}}^{(\mathrm{m})}, \sigma_{\mathrm{m}}) \\
\lambda &\sim \mathrm{Dirichlet}(\lambda_0)
\end{aligned}
\tag{2}
$$

In this model by y we denote unlabeled data, and by $y^{(m)}$ we consider one labelled as belonging to mixture component m. Obviously model (2) is logical for a single time series, but it can be clearly repeated for a number IL of labeled signals and I non-labeled. $\mu_n^{(m)}$ corresponds to (1) for each of components. Mixture parameters λ can be informed with the labeling by providing prior Dirichlet distribution with appropriate simplex λ_0. Entire model can be expressed most easily in a graphical form in the Fig. 3.

Remark 2. It should be explicitly stated that under model (2) measurement is not generated from all components of the mixture, but rather that the probability density function of measurements is a weighted sum of probability densities of individual components, where weighting coefficients are infered from the joint posterior model.

2.3 Class Probability Reconstruction

The main point of diagnostics is how to use the created model to determine in which class particular signal belongs. In this case, we need to return to the discrete parameters z_m that we have marginalized out of our model in the previous section. And that can be obtained by relatively simple formulas. Their main advantage is that they do not require MCMC to use.

The class membership probability is equivalent to the posterior of the mixture indicator conditional on data and model parameters (for convenience we denote y as an entire time series) $p(z \mid y_n, \mu, \sigma)$ over the $z \in 1 : M$. It means that $p(z = m \mid y, \mu, \sigma)$ is the posterior probability that that observation y was generated by mixture component m. The posterior can be computed via Bayes's rule,

$$
\Pr\left(z = m \mid y_n, \mu, \sigma, \lambda\right) \propto p\left(y_n \mid z = m, \mu, \sigma\right) p\left(z = m \mid \lambda\right) \tag{3}
$$

$$
= \mathrm{Normal}\left(y \mid \mu^{(m)}, \sigma_m\right) \cdot \lambda_m \tag{4}
$$

[1] https://mc-stan.org/docs/2_29/stan-users-guide/mixture-modeling.html.

We can compute the normalization constant via summation, as $z \in 1{:}K$ takes on finitely many values. In detail,

$$p(z = k \mid y_n, \mu, \sigma, \lambda) = \frac{\text{Normal}(y_n \mid \mu_k, \sigma_k) \cdot \lambda_k}{\sum_{m=1}^{M} \text{Normal}(y_n \mid \mu_m, \sigma_m) \cdot \lambda_m} \tag{5}$$

This formulation creates us the necessary classifier, or more precisely method of estimating the probability of time series y to each individual mixture component (class). To do this in the Bayesian framework we use the parameters sampled from the posterior distribution of the mixture model (as in the Fig. 3) to compute such probability. However because we have multiple samples of entire parameter set (usually 4000) we do it repeatedly and then infer anbout categories using appropriate statistics.

After formulating the entire process we can now consider its application.

3 Application to VSC DC Cable Diagnostics

In this section we present results of our case study applying our algorithm to determining in the VSC (voltage source controlled) DC cable the type of fault between pole-to-pole (PTP) and pole-to-ground (PTG) short circuits. First we provide brief description of our data. Then we describe computational system used. Finally we go in depth in to analysis of current and voltage signals for their diagnostic use.

For our data we have considered results from simulation models given by Mesas et al. [7]. Proposed models represent the initial 5ms period after the fault occurrence. For those signals we have simulated with randomized parameters 100 voltage-current pairs for both pole-to-pole and pole-to-ground faults. We have taken parameters normally distributed with means equals to parameters provided by Mesas et al. and standard deviations corresponding to 10% of those values. We have focused on current measurement. More detailed description of the problem is given in our earlier paper [2].

3.1 Computational Setup

For Bayesian computation, we have used Hamiltonian Monte Carlo (HMC) algorithm. Currently, most advanced HMC software is Stan [5]. HMC algorithm is a type of Markov Chain Monte Carlo (MCMC) methods. MCMC generates a Markov chain of samples in a way that makes their limiting distribution converge to the desired probability distribution. MCMC methods are especially useful for Bayesian computation, as sampling from the posterior distribution is difficult. Using generated samples, we can estimate expected values of desired functions of random variables. Because of that, we can answer practically all relevant statistical questions.

HMC is a variant of the Metropolis-Hastings algorithm. Traditionally MH algorithm uses Gaussian random walk proposal distribution. Algorithm accepts

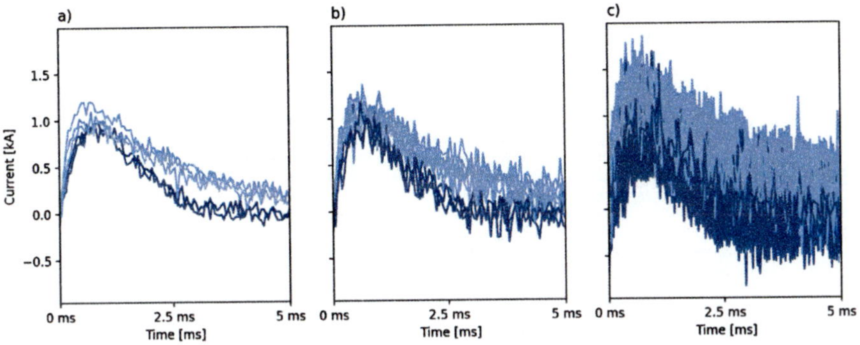

Fig. 4. Data used for model training. Lighter color corresponds to Pole-to-Pole fault and darker to Pole-to-Ground. a) Labelled data disturbed by 5% of max value. b) unlabeled data disturbed by 5% of max value. c) data for classification, disturbed by 15% of max value. Noise is significant enough that its hard to recognize the solution.

or rejects samples from the random walk depending on computed acceptance probability. In HMC, we generate proposals of random variables (system states) through a Hamiltonian dynamics evolution. This evolution is simulated using a time-reversible and volume-preserving numerical integrator (a symplectic integrator). HMC algorithm reduces the correlation between successive sampled states by proposing moves to distant states. Those states maintain a high probability of acceptance. This happens because symplectic integrators conserve energy of the simulated Hamiltonian dynamic. The reduced correlation in HMC means we need fewer Markov chain samples to get a desired level of Monte Carlo error when computing expectations.

Simulation of Hamiltonian dynamics might numerically destabilize for probability distributions with complicated geometry. This is an advantage of the method, because such complications usually mean problems in identifiability of parameters. Destabilization (known in the statistical field as "divergence") is a useful diagnostic that can suggest re-parametrization or other numerical adaptation of algorithm.

For model computation we have decided to use 10 labeled signals (5 of each type) and 10 unlabeled in order to increase the uncertainty and reduce averaging. For testing we have used 180 time series. In order enforce our assumption about degrading quality of data between labeled, available unlabeled and one for testing we introduced additional disturbances on each level. We added Gaussian random noise with zero mean and standard deviation equal to given percentage of maximum value.

3.2 Example of Use

In this section we will present application for one randomly selected dataset adding random disturbances on the level of 5%, 10% and 15% of max value on

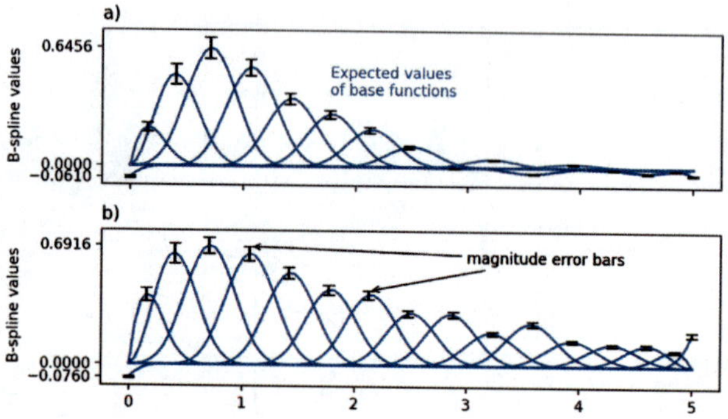

Fig. 5. Inferred spline representations of Pole-to-Ground (a) and Pole-to-Pole (b) fault waveforms. Errorbars represent marginal distributions of each β_k spline component. Errorbars are more noticeable in the initial part of the signal.

each current time series separately. Data used is illustrated in the Fig. 4. Our mixture has two components, which are represented via their spline fits presented in the Fig. 5.

Classification results are overly positive. In the Fig. 6 we have estimated probabilities of signal belonging to its proper class (PTG or PTP). Most cases were classified properly, and more dubious ones were provided with wide error bars, that for careful investigation also should be verified. We have only two noticeable false negatives, which can be connected with large noise magnitude. However it requires further investigation on what influences the classification quality, which is the focus of the next section.

4 Sensitivity Analysis

In this section we wanted to analyze how the classifier behaves in different scenarios and what problems it can encounter. Our methodology was as follows:

- Generate 100 datasets by randomly reshuffling of available signals and provide them with newly sampled random noises. We kept the composition of labeled data but rest was randomly distributed.
- We then trained the mixture model on each of those datasets.
- For each time series to be classified we evaluated its mean and if mean probability of correct classification was less than 0.5 we counted it as miss-classification. We then computed sensitivity for each class (which is a specificity for the another).
- Then we computed statistics for entire 100 runs.

We repeated this approach for three configurations of noise magnitudes. First one was modest disturbance i.e. 3%-7%-10% for each data level (labeled, unlabeled, to classify), second was identical to the one described in previous section i.e. 5%-10%-15% and the final one was the case of completely idealized, sanitized

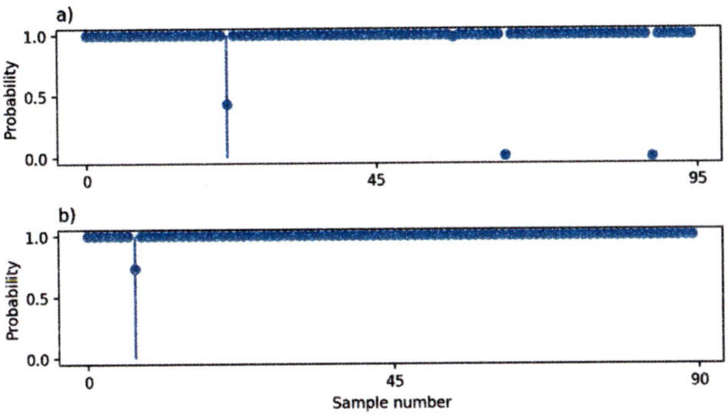

Fig. 6. Results of classification. Each subplots presents what was the probability of signal being assigned to the correct class. For PTG it was a) and for PTP it was b). Each probability is provided with errorbars corresponding to 93% highest density interval. In most cases the were not visible as they are indistinguishable from the point. In two cases probabilities had large errorbars which signifies unreliable estimate and suggest that the signal should be additionally verified. Two signals were completely wrongly classified.

Table 1. Statistics of classification over 100 datasets with disturbances structure: 3%-7%-10% for each data level (labeled, unlabeled, to classify)

	PTP sensitivity	PTG sensitivity
mean	0.995198	0.991173
std	0.014266	0.007466
25%	1.000000	0.988764
50%	1.000000	0.989071
75%	1.000000	1.000000

Table 2. Statistics of classification over 100 datasets with disturbances structure: 5%-10%-15% for each data level (labeled, unlabeled, to classify)

	PTP Sensitivity	PTG Sensitivity
mean	0.979150	0.982003
std	0.056188	0.028093
25%	0.988636	0.977778
50%	1.000000	0.988764
75%	1.000000	0.991848

data i.e. no disturbances on first level, 1% disturbance on the second one but 10% on the third.

First two cases shown very consistent behavior. Mean miss-classification was on the level below 1% in the first (see Table 1) and close to 1% in the second

Table 3. Statistics of classification over 100 datasets with disturbances structure: 0%-1%-10% for each data level (labeled, unlabeled, to classify). In this case very "clean" data for model learning impairs quality of classification.

	PTP Sensitivity	PTG Sensitivity
mean	0.969208	0.986643
std	0.070051	0.023383
25%	0.977778	0.988636
50%	1.000000	0.989011
75%	1.000000	1.000000

(Table 2). Of course mean and standard deviation for such asymetrical variables are obviously biased, that is why also median and quantiles were considered. The all show very solid behavior. However idealized case shown to be much more sensitive. As we can see in the Table 3, there is a possibility of significant errors visible in mean, showing that there were cases where classifier behaved very badly.

5 Conclusions

In this paper we have given a proof of concept solution for using Gaussian mixture models for classification of functional data. For considered case we have obtained good results and gained some intuition about learning data structure. The results show that the methodology can be used in more advanced scenarios. Because the paper's proof of concept character we have not used comparisons with popular methods, especially that results were rather good, and different dataset would be required for meaningful comparisons.

There are however some issues that require further investigation. There is a sporadic tendency of bad mixing of Markov chains during model fitting. This is a problem requiring investigation, as data inspection did not show anything unusual. This was however rare, and indentifiable by standard MCMC statistics such like $\hat{R}$. Also considered model is currently not optimal as sparsity structure of spline design matrix is completely unused. In case given here with only limited spline components (exactly 17) and total number of parameters less than 50 it is not an issue. But for cases with significant number of base functions, or noneven sampling between data series this can be an issue. Currently it is not solvable in stan, but there are other software solutions.

References

1. Aneiros, G., Cao, R., Fraiman, R., Genest, C., Vieu, P.: Recent advances in functional data analysis and high-dimensional statistics. J. Multivar. Anal. **170**, 3–9 (2019). Special Issue on Functional Data Analysis and Related Topics

2. Baranowski, J., Grobler-Dębska, K., Kucharska, E.: Recognizing VSC DC cable fault types using Bayesian functional data depth. Energies **14**(18) (2021). https://doi.org/10.3390/en14185893. https://www.mdpi.com/1996-1073/14/18/5893
3. Cai, B., et al.: Remaining useful life re-prediction methodology based on wiener process: subsea christmas tree system as a case study. Comput. Ind. Eng. **151**, 106983 (2021)
4. Cai, B., et al.: Fault detection and diagnostic method of diesel engine by combining rule-based algorithm and BNs/BPNNs. J. Manuf. Syst. **57**, 148–157 (2020)
5. Carpenter, B., et al.: Stan: a probabilistic programming language. J. Stat. Softw. **76**(1), 1–32 (2017)
6. Gelman, A., Carlin, J., Stern, H., Dunson, D., Vehtari, A., Rubin, D.: Bayesian Data Analysis, 3rd edn. Chapman & Hall/CRC Texts in Statistical Science, Taylor & Francis (2013)
7. Mesas, J., Monjo, L., Sainz, L., Pedra, J.: Cable fault characterization in VSC DC systems. In: 2016 International Symposium on Fundamentals of Electrical Engineering (ISFEE), pp. 1–5. IEEE (2016)
8. Wan, H., Wang, H., Scotney, B., Liu, J.: A novel gaussian mixture model for classification. In: 2019 IEEE International Conference on Systems, Man and Cybernetics (SMC), pp. 3298–3303 (2019). https://doi.org/10.1109/SMC.2019.8914215
9. Wang, J.L., Chiou, J.M., Müller, H.G.: Functional data analysis. Annu. Rev. Stat. Appl. **3**(1), 257–295 (2016)

Verification and Benchmarking in MPA Coprocessor Design Process

Tomasz P. Stefański[1(✉)], Kamil Rudnicki[2], and Wojciech Żebrowski[3]

[1] Faculty of Electronics, Telecommunications and Informatics, Gdansk University of Technology, 80-233 Gdansk, Poland
tomasz.stefanski@pg.edu.pl

[2] Department of Reconfigurable Systems, Brightelligence sp. z o.o., 90-562 Lodz, Poland
publications+DPS2022@brightelligence.com

[3] Aldec Inc., 80-288 Gdansk, Poland
wojtekz@aldec.com

Abstract. This paper presents verification and benchmarking required for the development of a coprocessor digital circuit for integer multiple-precision arithmetic (MPA). Its code is developed, with the use of very high speed integrated circuit hardware description language (VHDL), as an intellectual property core. Therefore, it can be used by a final user within their own computing system based on field-programmable gate arrays (FPGAs). The coprocessor is still under development and its open-source code is available on the Internet, based on the Mozilla Public License. Therefore, verification and benchmarking of the coprocessor code are vitally important issues as the sources are continually downloaded by users all over the world. In this contribution, we present software tools developed as a part of the system, allowing for detection of errors in the coprocessor code as well as for execution of its benchmarking tests. The research conclusion is that, without well-designed verification and benchmarking software tools, the development of any advanced digital circuit, such as a coprocessor, is actually impossible in realistic time. It stems from the fact that 60% of the project repository include hardware-description codes, whereas the rest of the codes support correct development of the project, i.e., verification and benchmarking in the design process.

Keywords: Verification · Benchmarking · Digital circuits · FPGAs · Coprocessors

1 Introduction

We have recently opened source codes of the coprocessor of multiple-precision arithmetic (MPA) [8]. Its purpose is to support a central processing unit (CPU) in computations which require precision higher than the standard 32/64 bits. The coprocessor makes it possible to offload CPU within a host machine by providing arithmetic with hardware support for addition and multiplication

Z. Kowalczuk (Ed.): DPS 2022, LNNS 545, pp. 266–277, 2023.
https://doi.org/10.1007/978-3-031-16159-9_22

operations up to 32 kbits. In this contribution, the design process of the MPA coprocessor is presented with the focus on its verification and benchmarking. That is, we present software tools allowing for detection of errors in the hardware as well as for benchmarking its throughput. The coprocessor is an intellectual property (IP) core implementable on field-programmable gate arrays (FPGAs) of various scales, e.g., on multi-processor system on chip (MPSoC) platforms consisting of CPU cores and FPGA, as well as on PCIe cards. The coprocessor code is written with the use of very high speed integrated circuit hardware description language (VHDL), which has already been implemented on Xilinx Zynq-7000 SoC (TySOM-1 board from Aldec) [9] and Xilinx Zynq UltraScale+ MPSoC (TySOM-3 board from Aldec) [12]. The unique features of our project stem from the release not only of selected VHDL codes, but also all the software developed for verification and benchmarking of the designed digital circuits. These software tools are vitally important for the project development because, without them, the detection of errors in hardware and its benchmarking would be impossible. It is due to a large number of logic elements inside a single coprocessor. For instance, the implementation on Xilinx Zynq-7000 SoC (Zynq UltraScale+ MPSoC) requires 8,745 slice look-up tables (LUTs), 16,249 slice registers, 482 F7 muxes, 66 F8 muxes, 18 block RAM tiles and 34 digital signal processing (DSP) modules (8,431 slice LUTs, 13,286 slice registers, 482 F7 muxes, 66 F8 muxes, 18 block RAM tiles and 34 DSP modules). Such a complex digital circuit of the coprocessor requires, simultaneously, the development of software tools which can ensure its correct operation. It is of vital importance, as the MPA coprocessor is continually downloaded by users all over the world. The purpose of this contribution is to present these verification and benchmarking software tools.

The paper is organized as follows: Sect. 2 reviews related works and presents verification methods for digital circuits. Section 3 and Sect. 4 presents, respectively, the developed MPA coprocessor and its design process. Section 5 provides data related to the programming languages employed in the design process and the ratio between hardware-description and verification codes. Finally, the conclusions are presented in Sect. 6.

2 Related Works

FPGA development includes several steps of synthesis and refinement. Hence, it also requires verifications (i.e., simulations) at each step [4]. These simulations are executed in order to check if high-level designs are correctly synthesized into low-level designs. Usually, designers develop test scenarios, then simulate each module in a testbench, and finally evaluate the simulation results against requirements. All these three verification activities are performed at each step individually and repetitively. Therefore, individual preparation for each verification (i.e., developing test scenarios and testbenches) is resource-consuming in terms of money and time. Hence, the topic of digital circuit verification has been present in literature for a number of years.

In [2], theoretical foundations of several approaches related to the logic-circuit verification are presented. These approaches employ the three-valued modelling capability found in most logic simulators, where the third-value X indicates a signal with unknown logic value. Although the circuit verification becomes significantly more complex when the circuit size is increased, several techniques can reduce the simulation complexity to a manageable level for numerous practical designs.

In [14], a method for generating design verification tests from behavioral-level VHDL programs is presented. The method generates stimuli to execute the desired control-flow paths in a given VHDL code. This method is based on path enumeration, constraint generation and constraint solving techniques which have traditionally been used for software testing.

A problem of introducing new verification methods into the design flow of a company is presented in [5]. The authors explore the difficulties resulting from the introduction of new verification methods into a *naturally grown* and well established design flow. The presented approach extends the capabilities of existing verification strategies by powerful new features, while keeping in mind the integration, reuse and applicability aspects. That is, it allows for reusing the existing verification components and test cases; hence, the running projects benefit from new techniques without the risk of losing their design efficiency or quality. As a result, the proposed approach results in maximum acceptance among the developers, which is essential for successful introduction of new methods.

In [18], the problems of quality are discussed for embedded systems based on FPGA. It is noticed that thorough verification of designs is currently a great challenge for designers. The authors provide the data that, during the whole process of FPGA design, verification takes up to 70%–80% of the time in the design cycle. Hence, the number of verification engineers may be twice as big as that of design engineers. In order to solve this problem, it may be concluded that the improvement of verification efficiency is required.

In [4], an integrated software testing framework is proposed for FPGA developments related to controllers in nuclear power plants. This solution allows for performing the three aforementioned activities (i.e., test development, module simulations, evaluation of simulation results against requirements) of the simulation-based verification in a single step and once only. That is, the framework generates common and meaningful test scenarios automatically, simulates all designs simultaneously and, finally, evaluates the simulation results against the expected ones. If any error is detected, the framework analyzes and compares the incorrect case in detail. Due to the fact that FPGA is applied in nuclear power plants, the proposed solution must meet strict requirements of critical-safety systems. Hence, it represents the state-of-art in the area of FPGA verification.

Unfortunately, none of the described verification and benchmarking solutions for FPGA provide a full repository of source codes, which is freely available to the public. Therefore, it is not straightforward, either to understand the software architecture of these tools. However, our integer MPA coprocessor is the solution presented in the literature, whose source codes are fully opened. Therefore, the presentation in this paper of our verification and benchmarking tools, developed for the MPA coprocessor, can fill in this gap, as well as support other researchers and engineers involved in the process of FPGA design.

3 MPA Coprocessor

In this section, we present the architecture of an MPA coprocessor, whose design process is analyzed throughout the paper. Because the coprocessor is already presented in several publications [6,7,9,12], we have limited this presentation to the required minimum.

The purpose of our open-source project [8] is to develop an MPA coprocessor which targets FPGAs of various scales (i.e., depending on the required processing power of final applications), in order to offload CPU in MPA computations. This coprocessor is a parametrizable IP core, and uses the sign-magnitude representation for integer numbers. Our fundamental design requirement is to ensure compatibility with the existing software codes and libraries employing the GMP library [3], which currently constitutes the standard software tool for CPU implementations of MPA computations. The coprocessor diagram is presented in Fig. 1. It exchanges data with the host CPU by using three 64-bit data buses (two of them are for the input data, whilst the remaining one is for the output data) and a single 8-bit program bus. All the buses are in AMBA (AXI Stream) standard. Data loaders transfer the data to the bank of registers, whilst the unloader transfers the data back to CPU. The coprocessor includes a set of 16 logical registers, which are mapped into 17 physical registers (the additional register is referred to as *shadow register*). That is, the logical representation of registers is not fixed and can change on-the-fly, i.e., during the code execution. The additional physical register and flexible association of physical and logical addresses accelerate the operations which do not immediately provide an unambiguous result. For instance, addition and subtraction operations are accelerated in this way (for detail see [7]). The registers store magnitudes of numbers of the maximal length equal to 32 kbits, whereas their signs and sizes are stored in the control unit. Instructions are fetched from the program bus, and then decoded in the instruction decoder. Control lines are set accordingly to allow for data transfers from the registers to the multiplication and adder units. The adder module provides the operation of number subtraction as well, because it also works with negative-sign MPA numbers. Finally, arithmetic operations are executed, and their results are sent back to registers from which they can be sent further to the host CPU. The instruction set of the MPA coprocessor consists of addition, subtraction, multiplication as well as data loading instructions. However, depending on the final user needs, it can be extended by other operations (e.g., bit manipulation instructions).

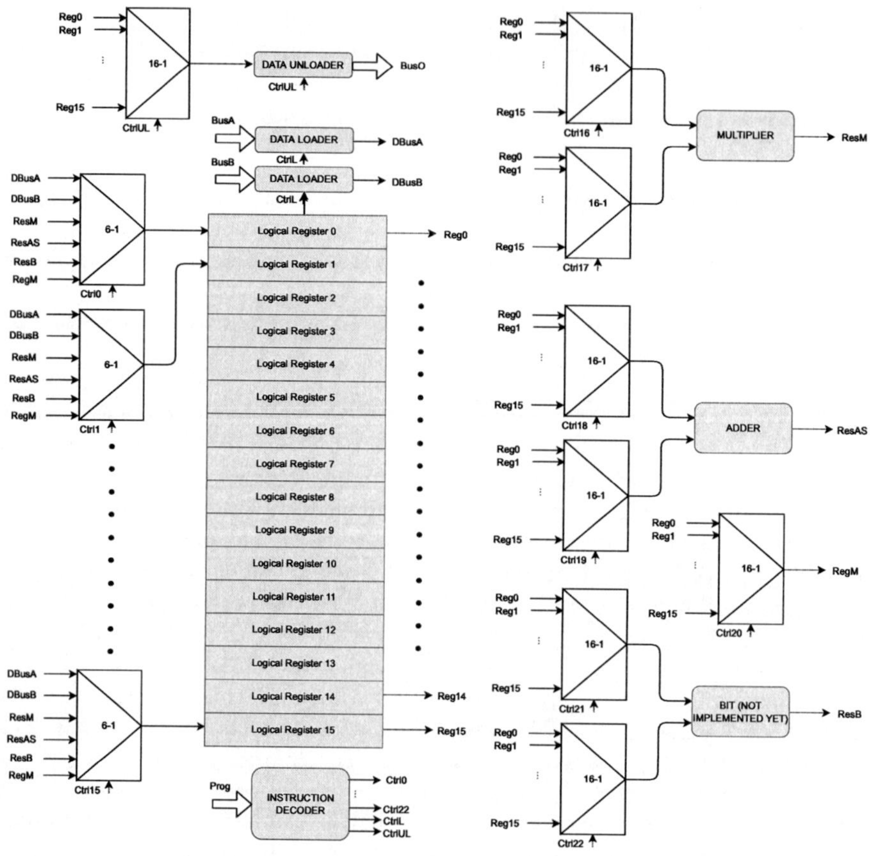

Fig. 1. Diagram of MPA coprocessor.

4 Design Process

Initially, the MPA coprocessor was verified and benchmarked in Xilinx Zynq-7000 SoC on TySOM-1 board. Then, the code was ported to Xilinx Zynq UltraScale+ MPSoC on TySOM-3 board. In this section, we describe the design process for the first aforementioned implementation of the MPA coprocessor.

The coprocessor is designed with the use of Vivado Design Suite from Xilinx [16]. It allows for the simulation of developed VHDL codes in terms of behavioral and functional operations. However, some simulation waveforms are taken from Aldec Riviera-PRO tool [1]. To be more precise, our behavioral simulations are aimed at obtaining waveforms of the MPA coprocessor under typical input conditions with the use of Vivado testbenches. Then, by way of visual inspection, the designer can detect differences between the obtained and required waveforms. In general, functional verification is the design step of verifying if the logic circuit conforms to specifications. Therefore, our functional simulations are aimed at obtaining results of arithmetic operations on the coprocessor with the use of an emulator. Differences between the results obtained by emulation and simulation of the MPA coprocessor in Vivado are automatically detected by a personal computer (PC). The design process of the MPA coprocessor is presented in Fig. 2. The development starts with specification of the requirements. As one can notice, two simultaneous paths of the development are performed, i.e., the design of digital circuits, as well as the verification and benchmarking path. That is, the coprocessor architecture and the test environment are simultaneously developed. Therefore, at each stage of the design process, feedback from the verification and benchmarking path allows us to modify the design to obtain the design requirements. At the beginning of the hardware development path, basic blocks such as adder, multiplier, data loader, register and control unit are designed. Along the path of verification and benchmarking, each of these basic blocks is tested with the use of behavioral simulations written in VHDL for Vivado. Additionally, for arithmetic modules of adder and multiplier, testbenches for functional simulations in Vivado are developed with the use of SystemVerilog (SV) Direct Programming Interface (DPI). Therefore, the modules of adder and multiplier are additionally tested in terms of correct execution of arithmetic operations. For this purpose, results of MPA operations on the designed modules are compared with reference values obtained with the use of SV/C codes and GMP library executed on a PC. These automatic tests are performed for large number of random integer operands with the size of up to 32 kbits. If either the obtained simulation results are wrong, or the performance of digital circuits is unsatisfactory (i.e., the maximum frequency of operation is too low), the design process is repeated. Further along the hardware development path, basic blocks are integrated into the MPA coprocessor. For the purpose of its verification and benchmarking, various behavioral testbenches are developed, as well as emulation software which exactly replicates the operation of the MPA coprocessor on a PC. Therefore, one can easily obtain reference results of MPA coprocessor operations in terms of its output data streams by executing coprocessor codes on a PC. Furthermore, one can compile the emulator in the mode allowing for its direct communication with the Vivado tool and execution of the same code step-by-step in a digital circuit simulator on a PC. Then, if differences occur between the output of the simulation tool and the emulator, the error flag is raised and the simulation is terminated, making it possible to find a *bug* in the coprocessor. The DPI functions written for the emulation of the coprocessor are summarized in Table 1.

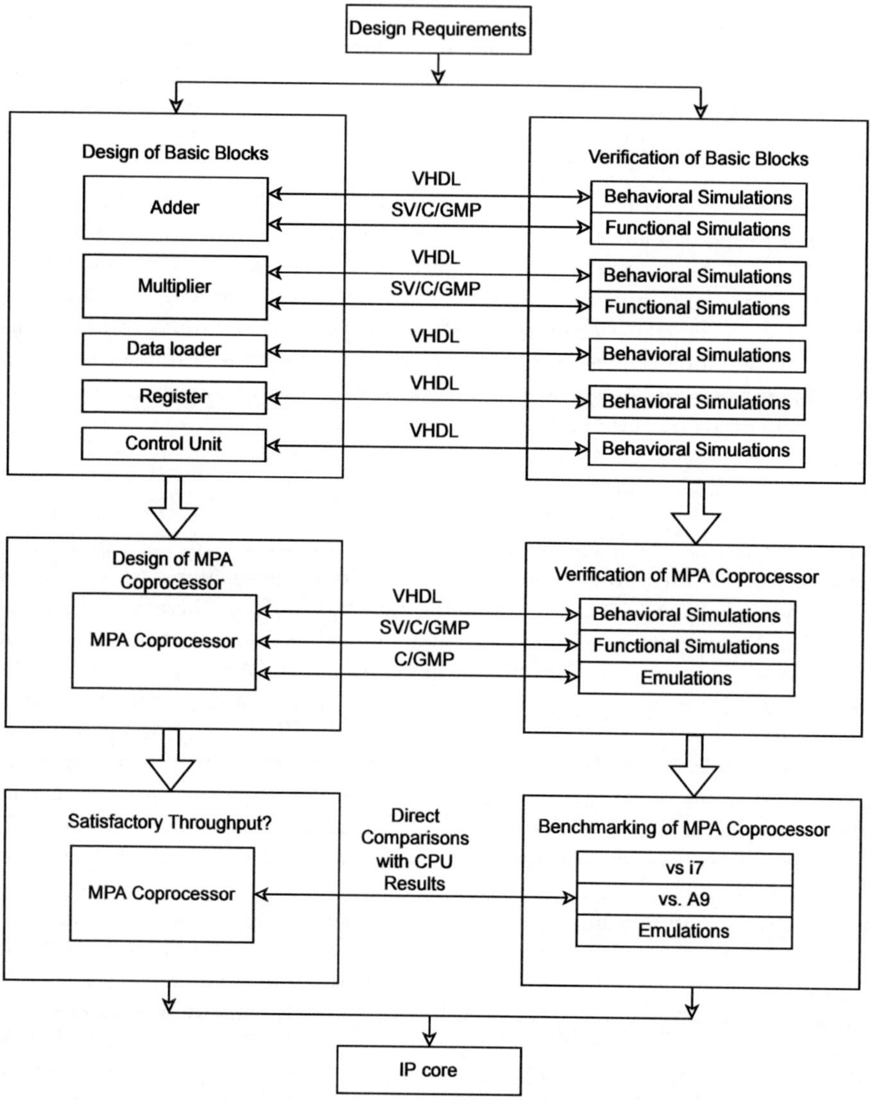

Fig. 2. Design flow of MPA coprocessor.

These functions allow for the execution of coprocessor codes in the background on a PC, i.e., using a server working as *daemon tool*. When the function tbEmusrupStart is called from a DPI testbench, arrays for storing handle and semaphore must be allocated at the SV side. It stems from the fact that SV does not provide (**char) type allowing for return of a pointer to the allocated

Table 1. SV testbench functions.

Name	Function
tbEmusrupStart	Initializes emulator for DPI debugging
tbEmusrupProceed	Emulator proceeds single step of code
tbEmusrupCheckLogic	Returns values of logical registers in MPA coprocessor
tbEmusrupCheckShadow	Returns value of shadow register in MPA coprocessor
tbEmusrupStop	Finalizes execution of emulator for DPI debugging

memory. However, equivalent data types to the C string (i.e., char*) are available; hence, memory for the string must be allocated in the SV testbench. Then, the simulation tool can execute step-by-step lines of code for the coprocessor on a PC by using the function tbEmusrupProceed. The status of logical and shadow registers of the coprocessor can be checked with the use of functions tbEmusrupCheckLogic and tbEmusrupCheckShadow, respectively. If any difference between the expected and obtained numbers stored in registers is detected, the simulation is terminated. Finally, the emulation of the coprocessor is stopped with the use of function tbEmusrupStop.

For the purpose of coprocessor verification as well as of its benchmarking, the codes are developed on PC, which allow for generating files with input streams for MPA computations of factorial ($n!$), exponentiation (n^n) and discrete Green's function (DGF) [10,11,13]. This set of basic computations is additionally implemented in the C language by codes executable on CPUs. Therefore, one can easily obtain benchmark results which are the speedup factors for the developed coprocessor against the reference CPU.

Finally, the coprocessor performance is compared to the ARM Cortex A9 core executing MPA computations with the use of GMP library. That is, the MPA coprocessor is verified and benchmarked on hardware with the setup presented in Fig. 3. It is worth mentioning that FPGA with the coprocessor and the CPU core are within the same chip; therefore, the comparison seems to be fair. The benchmarking codes (i.e., factorial $n!$, exponentiation n^n and DGF) are executed either on FPGA or reference CPU with an external PC used for benchmark management and acquisition of runtimes. The TySOM-1 board, used in our benchmarks, operates under the control of Linux (i.e., PetaLinux), loaded from an SD card. The system clock is employed for runtime measurements on CPU. However, for the MPA coprocessor in FPGA, the external PC communicates with the integrated logic analyzer (ILA) [15] within FPGA, using JTAG interface. ILA is an IP core implemented within FPGA in order to monitor internal signals and registers of a design. The code-execution time is measured on the MPA coprocessor with the use of a timer, which is triggered by the rising edge of the clock at the beginning of the execution of the first code instruction, whilst the end of computations is triggered by the first transfer of the resulting data. ILA acquires values in the timer and transfers them to the host computer, which is managed from the level of Xilinx Vivado tool (i.e., Program & Debug tab). The execution times are transferred to the PC with the use of JTAG server working as *daemon tool*. Finally, they are placed on time charts of the signal-state analyzer.

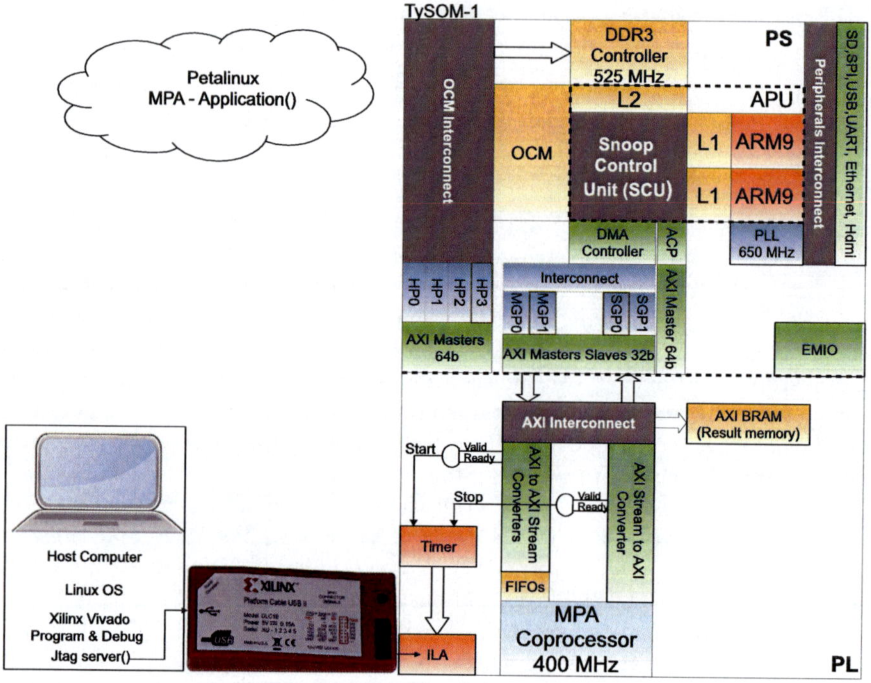

Fig. 3. Benchmarking setup for MPA coprocessor.

The developed software allows for efficient emulation of the MPA coprocessor on a PC as well as execution of benchmarks on hardware platforms, based on the Zynq processor. An alternative verification model can be obtained by using a virtual platform running in QEMU (i.e., quick emulator [17]) instead of physical platforms with Zynq processors. In both cases, using either hardware or a virtual platform, one works on the same PetaLinux system files and the same benchmarks written in the C language. The alternative virtual platform can be connected to any HDL simulator (with SystemC interface) in order to verify the MPA coprocessor. In such a solution, the simulator is connected to the IP core of the MPA coprocessor and proceeds in the same way as it is done on a hardware platform based on the Zynq processor. Hence, this solution can achieve compliance for the transaction-level model, but using the virtual platform, just as by using the hardware platform.

5 Verification and Benchmarking Software

The structure of project repository [8] is described in detail in Appendix. Statistics of the programming languages used in the repository is presented in Table 2. Although some VHDL codes are also related to verification of the MPA coprocessor at the behavioral level, one can notice that around 60% of the codes describe hardware and 40% are the supporting codes. Within these 40%, most of the codes verify and benchmark the MPA coprocessor. This ratio shows how important the verification and benchmarking are for the design of advanced digital circuits such as coprocessors. Our experience suggests that development of the MPA coprocessor is impossible without such tools developed in parallel. It occurs that the development of supporting codes requires similar time resources as the development of hardware in VHDL.

Table 2. Statistics of programming languages used in project of MPA coprocessor.

Name	Percentage
VHDL	60.7%
C	24.7%
SV	6.2%
C++	4.3%
Tcl	2.6%
Makefile	0.9%
Other	0.6%

6 Conclusions

A behavioral simulation in Vivado is a standard part of any digital-circuit design for Xilinx FPGAs. Due to the complexity of our design, for MPA units such as adder and multiplier, we decided to develop additional testbenches with the use of DPI, allowing us to verify the correct execution of MPA computations. Then, due to the additional complexity of our design, stemming from the integration of basic blocks into the coprocessor, we developed an emulator for the coprocessor, which exactly replicates its operation on a PC. Therefore, we are able to detect various design errors in reasonable time, as well as verify correct operation of the coprocessor in computational benchmarks. It occurs that 60% of the project repository include VHDL codes, whereas the remaining codes support correct development of the project, using other languages such as C/C++/SV/Tcl/Makefile. In our opinion, lack of well-designed verification and benchmarking software tools unnecessarily prolongs the time-to-market of any advanced digital circuit (such as a coprocessor).

Acknowledgments. Tomasz P. Stefański is grateful to Cathal McCabe at Xilinx Inc. for arranging the donation of design software tools.

Appendix: Project Repository

The project repository [8] consists of documentation (**doc**), **firmware** and **software** directories. Then, **firmware** directory consists of **zedboard**, **scripts**, **sim** and **src** directories. In **software** directory, the following directories are present: **DPI**, **common**, **data_apps**, **emusrup**, **helpers**, **runtimes**, **vhdl_gens**. Additionally, one can find there the scripts **clean_all_programs.sh** and **compile_all_programs.sh** which make it possible to clean and compile all the source files in this directory. The directory **DPI** includes the codes necessary to execute DPI simulations in Vivado for the MPA core and its modules alone. Then, the directory **common** includes various C functions used by many different codes in the project repository. In the directory **data_apps**, one can find codes allowing for the generation of input files for benchmarking of the MPA coprocessor. In the directory **emusrup**, one can find the code of the MPA coprocessor emulator. In the directory **helpers**, one can find some useful codes, e.g., allowing us to change the format of binary files. In the directory **runtimes**, one can find some useful codes allowing for evaluation of the CPU throughput in benchmarking codes. In the directory **vhdl_gens**, one can find useful codes allowing us to generate, automatically, some VHDL codes, e.g., for the multiplexers used in the coprocessor design.

References

1. Aldec Inc.: Riviera-PRO Manual (2017). www.aldec.com. Accessed 08 Aug 2019
2. Bryant, R.E.: A methodology for hardware verification based on logic simulation. J. ACM **38**(2), 299–328 (1991). https://doi.org/10.1145/103516.103519
3. Granlund, T.: GMP Development Team: The GNU multiple precision arithmetic library (Edition 6.1.2) (2016). www.gmplib.org. Accessed 08 Aug 2019
4. Kim, J., Kim, E.S., Yoo, J., Lee, Y.J., Choi, J.G.: An integrated software testing framework for FPGA-based controllers in nuclear power plants. Nucl. Eng. Technol. **48**(2), 470–481 (2016). https://doi.org/10.1016/j.net.2015.12.008
5. Lissel, R., Gerlach, J.: Introducing new verification methods into a company's design flow: an industrial user's point of view. In: 2007 Design, Automation and Test in Europe Conference and Exhibition, pp. 1–6 (2007). https://doi.org/10.1109/DATE.2007.364675
6. Rudnicki, K., Stefanski, T.P.: IP core of coprocessor for multiple-precision-arithmetic computations. In: 2018 25th International Conference Mixed Design of Integrated Circuits and System (MIXDES), pp. 416–419 (2018). https://doi.org/10.23919/MIXDES.2018.8436868
7. Rudnicki, K., Stefański, T.P.: Implementation of addition and subtraction operations in multiple precision arithmetic. In: 2019 MIXDES - 26th International Conference Mixed Design of Integrated Circuits and Systems, pp. 231–235 (2019). https://doi.org/10.23919/MIXDES.2019.8787156
8. Rudnicki, K., Stefański, T.P., Żebrowski, W.: Integer-MPA-coprocessor (2020). https://github.com/stafan26/integer-MPA-coprocessor
9. Rudnicki, K., Stefański, T.P., Żebrowski, W.: Open-source coprocessor for integer multiple precision arithmetic. Electronics **9**(7), 1141 (2020). https://doi.org/10.3390/electronics9071141

10. Stefański, T.: Applications of the discrete green's function in the finite-difference time-domain method. Prog. Electromagn. Res.-PIER **139**, 479–498 (2013). https://www.jpier.org/PIER/pier.php?paper=13032906
11. Stefański, T.P.: A new expression for the 3-D dyadic FDTD-compatible green's function based on multidimensional z-transform. IEEE Antennas Wirel. Propag. Lett. **14**, 1002–1005 (2015). https://doi.org/10.1109/LAWP.2015.2388955
12. Stefański, T.P., Rudnicki, K., Żebrowski, W.: Implementation of coprocessor for integer multiple precision arithmetic on Zynq Ultrascale+ MPSoC. In: 2021 28th International Conference on Mixed Design of Integrated Circuits and System, pp. 280–285 (2021). https://doi.org/10.23919/MIXDES52406.2021.9497554
13. Stefański, T.: Discrete green's function approach to disjoint domain simulations in 3D FDTD method. Electron. Lett. **49**, 597–598 (2013). https://digital-library.theiet.org/content/journals/10.1049/el.2012.4462
14. Vemuri, R., Kalyanaraman, R.: Generation of design verification tests from behavioral VHDL programs using path enumeration and constraint programming. IEEE Trans. Very Large Scale Integr. (VLSI) Syst. **3**(2), 201–214 (1995). https://doi.org/10.1109/92.386221
15. Xilinx Inc.: Integrated Logic Analyzer v6.2 - LogiCORE IP Product Guide, PG172 (2016). www.xilinx.com. Accessed 08 Aug 2019
16. Xilinx Inc.: Vivado Design Suite User Guide - Getting Started, UG910 (v2018.3) (2018). www.xilinx.com. Accessed 08 Aug 2019
17. Xilinx Inc.: Xilinx Quick Emulator User Guide, UG1169 (v2020.1) (2020). www.xilinx.com. Accessed 24 Mar 2022
18. Zheng, D., Wang, Y., Xueyi, Z.: The methods of FPGA software verification. In: 2011 IEEE International Conference on Computer Science and Automation Engineering, vol. 3, pp. 86–89 (2011). https://doi.org/10.1109/CSAE.2011.5952639

Sensor Fault Analysis of an Isolated Photovoltaic Generator

Ousmane W. Compaore[1,2](✉), Ghaleb Hoblos[1], and Zacharie Koalaga[2]

[1] Irseem, Esigelec, Unirouen, Normandy University, Rouen, France
{o.compaore,Ghaleb.hoblos}@esigelec.fr
[2] Lame, UFR-SEA, Université Joseph Ki-Zerbo, Ouagadougou, Burkina Faso

Abstract. This article deals with problems related to the isolation and identification of sensor faults in complex industrial processes. The idea is based on the quantitative analysis of residuals in the presence of sensor faults in order to establish binary fault signatures using the parity space method around the maximum power operating point (MPPT) of the generator. The residuals are generated thanks to the analytical redundancy relations (ARR) given by the system model and all the sensors used. The measured currents and voltages are used to perform fault detection and signature algorithms. The application of the method is done on a complex industrial system, consisting of a large number of photovoltaic panels organized in a field. Monitoring the state of health of this complex industrial process using this diagnostic approach will improve the reliability, performance and return on investment of the installation for sustainable development.

Keywords: Fault detection and isolation · Sensor fault · Parity space · Dynamic cumulative sum · Photovoltaic generator · Diagnosis · Analytical redundancy relation

Nomenclature

We start by defining some parameters used:

K_i :	the current/temperature coefficient of short-circuit
K_V :	the open circuit voltage/temperature coefficient
G *and* G_n	are the current and nominal illuminance
I_{phn} :	the nominal current of the PV cell, given under $T = 25$ °C, $G = 1000$ W/m^2
$I_m = I_c$:	current supplied by PV cell and the PV module
I_{stg} :	serial PV module current
V_{stg} :	voltage at the terminals of string
I_G :	generator current of PV field
V_G :	voltage at the terminals of generator or PV field
I_d :	the current in the diode
I_0 :	the saturation current of the diode
V_t :	the thermal voltage of the panel
V :	voltage at the terminals of PV cell
V_m :	voltage at the terminals of PV module.

Z. Kowalczuk (Ed.): DPS 2022, LNNS 545, pp. 278–290, 2023.
https://doi.org/10.1007/978-3-031-16159-9_23

Table 1. Used constants.

Constant	Value SI	Designation
a	1.3977	Coefficient of ideality of the diode
K_b	1.38e−23	Boltzmann's constant
q	1.602e−19	Charge of the electron
I_{sc}	3.56	The short-circuit current of the cell
V_{ocn}	0.6*N_{cell}	The open circuit voltage
R_s	0.015	The series resistance
R_{sh}	700000	The parallel resistance
k	3.5	Bishop coefficient
n	0.1	Bishop coefficient
V_b	−20	Breakdown voltage
T_n	273.15	The ambient temperature
T	298.15	The operating temperature
N_{cell}	36	Number of cells in series
N_s	4	Number of serial modules
N_p	4	Number of strings

1 Introduction

For the needs of process safety and reliability, monitoring strategies and fault detection are integrated into academic and industrial process applications [1]. Several shortcomings were also found with some certain methods, but this more or less led to enormous progress in fault detection techniques.

The photovoltaic generator is a sustainable, non-polluting and inexhaustible energy production system that represents a real alternative to fossil fuels. Defects due to design, installation, or operation affect photovoltaic production. In the field of research and development, fault diagnosis techniques consist in identifying the causes of faults, affecting the performance of such installations. In recent decades, various fault detection methods have been proposed in the literature [1–3]. Other research works has focused on measurement equipment, data acquisition and storage systems for measurement data, as well as data transmission methods and supervision software for monitoring systems [4, 5]. Other diagnostic systems adopted statistical analysis in the diagnosis of PV systems [6–8], while others has adopted methods based on artificial intelligence [6], in particular the neural networks.

The main contribution of this article can be considered as a verification test of the applicability of the analytical redundancy approach and the parity space for industrial purposes on a photovoltaic generator and to verify the efficiency of the detection of sensor faults in the presence of disturbances or measurement noise [3]. We will present in detail the method of the parity space in application on a nonlinear system, namely a photovoltaic generator [9, 10], based on model of Bishop [11] and on which we built a residual generator around the Maximum Power Point Tracking (MPPT). For this complex nonlinear process, we have successfully tested two sensor faults, one on the current sensor shown here and one on the voltage sensor.

After introduction, let us state the problem in Sect. 2. The application model, the PV generator (PVG) will presented in Sect. 3. Our contribution on the fault detection and isolation (FDI) method on the operating state of the studied system will be presented in Sect. 4. The discussion on our simulation results, in particular the signature of sensor faults, is summarized in Sect. 5. Finally, Sect. 6 will allow us to conclude on the proposed diagnostic method and to suggest some perspectives.

2 Problem Statement

The objective of this work is to implement a diagnostic method based on the parity space method, able to identify sensor faults in a photovoltaic generator. Application of the FDI method on an complex industrial system, which presents particularities of non-linearities and faults which can significantly affect its operating performance [11, 12]. The choice of sensor fault detection aims, on the one hand, to prove the effectiveness of the diagnostic method, when it is well established with instrumentation based on judiciously placed sensors; and on the other hand, to better monitor the state of health of the PVG in order to undertake corrective or predictive maintenance actions. The use of sensors is also non-invasive and non-intrusive in order to combine electronic efficiency and economy of the on-board instrumentation system, which must be as realistic as possible.

The major innovation of this work is because the application of this diagnosis technique is placed at the MPPT, in order to be able to generate the residuals necessary for the analysis of the diagnostic method. The simulation results from the detection and the signature make it possible to isolate any faults from the PVG with respect to the operating point considered.

3 Modeling of the PVG

The PV generator chosen for this work was built around the elementary cell of the so-called Bishop model [11], in Fig. 1. Its characterization allows us to place ourselves at the optimal operating point called most of the time (MPPT).

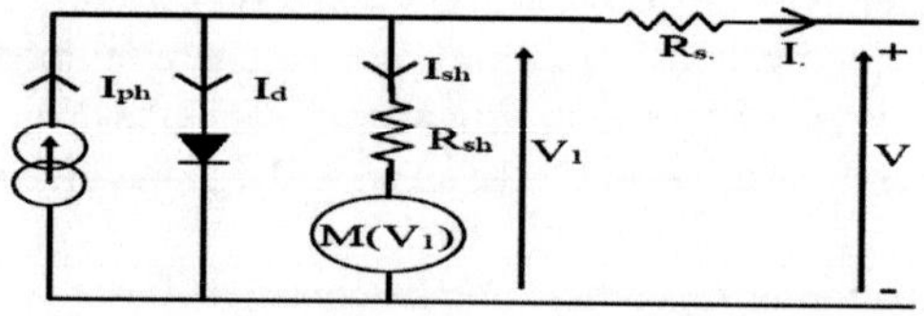

Fig. 1. Bishop model of a PV cell.

The following relation gives the output current:

$$I = I_{ph} - I_0\left[\exp\left(\frac{V + R_sI}{V_t}\right) - 1\right] - \frac{V + R_sI}{R_{sh}}\left[1 + k\left(1 - \frac{V + R_sI}{V_b}\right)^{-n}\right] \tag{1}$$

where $I_{ph} = (I_{phn} + K_i \Delta_T)\frac{G}{G_n}$ with $\Delta_T = T - T_n$.

We know the current in the diode is given by

$$I_d = I_0\left[\exp\left(\frac{V + R_s I}{V_t}\right) - 1\right] \tag{2}$$

where $I_0 = \frac{I_{scn} + K_i \Delta_T}{\left[\exp\left(\frac{V_{ocn} + K_V \Delta_T}{V_t}\right) - 1\right]}$

and the thermal voltage of the panel is

$$V_t = \frac{a k_b T}{q} \tag{3}$$

In a healthy operating mode, the identical N_{cell} cells are put in series, to reach the desired module voltage and protected by a bypass diode. The module thus created is subjected to the same temperature and sunlight conditions [4, 9]. The equivalent circuit for a PVG, comprising a number N_s of modules connected in series and a number N_p of branches (string) in parallel is shown in Fig. 2; relations (4) to (7) make it possible to establish the output current and voltage of the PVG created [3, 10].

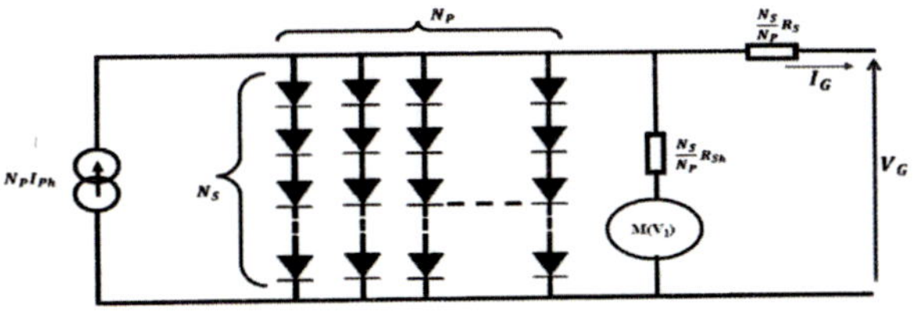

Fig. 2. Modeling of the PV generator.

In this way, we obtain the following dependencies in turn:

$$\text{from cell to module:} \quad \begin{cases} I_m = I_c = I \\ V_m = N_{cell} * V \end{cases} \tag{4}$$

$$\text{from module to string:} \quad \begin{cases} I_{stg} = I \\ V_{stg} = N_s * V_m \end{cases} \tag{5}$$

$$\text{from string to field:} \quad \begin{cases} I_G = N_p * I_{stg} \\ V_G = V_{stg} \end{cases} \tag{6}$$

$$\text{from the field to the generator:} \begin{cases} I_G = N_p * I \\ V_G = N_s * N_{cell} * V \end{cases} \tag{7}$$

4 Proposed Diagnostic Approach and Results

From the model of our system described in Sect. 3, we apply a diagnosis method based on the principle of parity space [8, 13, 14]. It based on the analysis of the analytical redundancy relations (ARR), which summarized in the diagram of Fig. 3.

The measured quantities then make it possible to feed the function for calculating the coefficients a_i and b_i of the operating point in which the generator is located. All this knowledge between theoretical and practical data in real time makes it possible to generate the residuals, then to apply the detection algorithm through the DCS. Isolation is done through the occurrence matrix in Table 2.

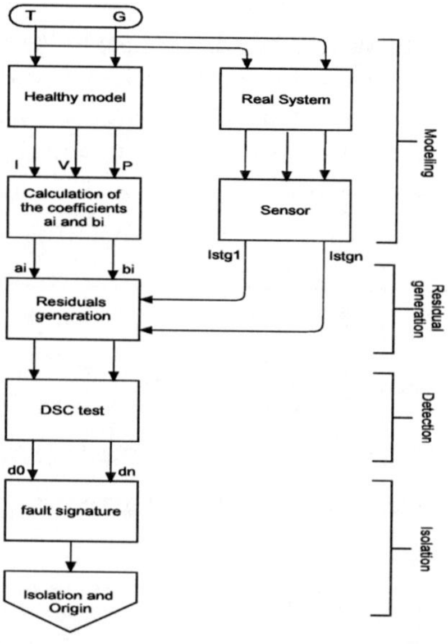

Fig. 3. Diagnostic strategy in the industrial process.

4.1 PVG Around the Operating Point

The application PVG of this study consists of sixteen (16) photovoltaic panels, organized in four chains of four modules placed in series.

Around the operating point (I_{sc}, V_{oc}), we can define the output current as

$$I = a_i * V + b_i \tag{8}$$

where a_i is the direction coefficient and b_i is the coefficient of the slope, defining the tangent line to the I-V characteristic of the system and corresponding to the MPPT

$$\text{with}\begin{cases} a_i = \left.\dfrac{\partial I\left(I_{stgi}, V_{stgi}\right)}{\partial V_{stgi}}\right|_{\substack{I = I_{sc} \\ V = V_{oc}}} \\ b_i = I_{stgi} - a_i V_{stgi}\Big|_{\substack{I = I_{sc} \\ V = V_{oc}}} \end{cases}$$

We obtain the following model of the system:

$$\begin{cases} I_G = I_{stg1} + I_{stg2} + I_{stg3} + I_{stg4} \\ V_G = V_{stg1} = V_{stg2} = V_{stg3} = V_{stg4} \\ V_G = V_{M1stg1} + V_{M2stg1} + V_{M3stg1} + V_{M4stg1} \end{cases} \quad (9)$$

Instrumentation measurements can be shown in the form of the following equations:

$$\begin{cases} \hat{I}_G = I_G + \varepsilon_G \\ \hat{I}_{stg1} = I_{stg1} + \varepsilon_{I_{stg1}} \\ \vdots \\ \hat{I}_{stgp} = I_{stgp} + \varepsilon_{I_{stgp}} \\ \hat{V}_G = V_G + \varepsilon_V \end{cases} \quad (10)$$

where:

- $\hat{I}_G$ *and* $\hat{V}_G$ are respectively the current and the voltage measured on the generator using a current sensor and a voltage sensor;
- $\hat{I}_{stg1}$ *to* $\hat{I}_{stgp}$ and $\varepsilon_{I_{stg1}}$ *to* $\varepsilon_{I_{stgp}}$ represent respectively the currents measured, and any noises on the outputs of strings i by current sensors;
- ε_V represent any noise on the outputs of strings by the voltage sensor.

The resulting PVG model is similar to the static system. But this model draws all its dynamics from the numerous fluctuations in the two main inputs, temperature and sunshine, and other major parameters such as R_s, R_{sh}, N_{cell}, and I_{ph}.

Figure 4 represents in order, the power, the voltage and the current of the PVG organized into four strings, of four series modules each one. The structural organization of the PV array of the PVG then makes it possible to observe structural anomalies related to other types of failures. This results in a voltage drop if one or more modules are removed from one or more strings. Or a drop in current which results in the loss of a string. These two situations are highly damaging for the PVG performances, whose the MPPT would be affected. And we now know that one or the other of the two causes mentioned above, can have several causes justifying a drop in voltage or output current.

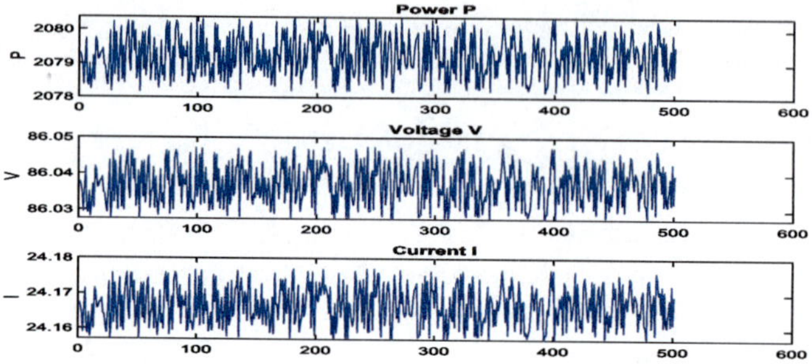

Fig. 4. Power, voltage and current signals in healthy mode.

4.2 Generation and Structuring

It is possible to simulate faults on all the sensors of the system. During normal operation, all residuals are statistically zero (in healthy/nominal case). Figure 5 shows the evolution of the following five residues R_0, R_1, R_2, R_3 and R_4 over a period of 500 s corresponding to the following five redundancy equations.

$$\begin{cases} R_0 = \hat{I}_G - \hat{I}_{stg1} - \hat{I}_{stg2} - \hat{I}_{stg3} - \hat{I}_{stg4} \\ R_1 = \hat{I}_{stg1} - a_{i1}\hat{V} - b_{i1} \\ R_2 = \hat{I}_{stg2} - a_{i2}\hat{V} - b_{i2} \\ R_3 = \hat{I}_{stg3} - a_{i3}\hat{V} - b_{i3} \\ R_4 = \hat{I}_{stg4} - a_{i4}\hat{V} - b_{i4} \end{cases}$$

These observations reflect well the tested faults of our sensors. The above residual relationships contain only measured or known quantities and ignore unknown variables in the previous equations: (8), (9) and (10). The various faults arranged in an occurrence matrix are summarized in Table 2.

The generation of fault signatures based on an interpretation of the content of each column of Table 2 in relation to the state of each residual. Each column being very distinct from the others, gives a strong localization in the signature of each fault. The proposed approach does not allow the simultaneous detection of several faults and several occurrences. This would require increasing the number of residuals, so that the signature of a fault leaves no room for possible confusion.

We randomly choose to present a fault on a current sensor of string 3. This then affects the residuals R3 and R0 as illustrated below. In fact, its value has been subtracted from 10A. Other residuals are not affected.

4.3 Analysis Through DCS Test

Many statistical tests exist to analyze a given residual [13, 14]. Among these tests, our choice fell on the DCS because of its great capacity to detect any change in a signal, in particular when the change affects the mean and the standard deviation [13] of the two sliding windows. To apply the DCS we need the parameters of each of the two sliding

Table 2. Occurrence matrix of PV generator.

	I	V	I_{stg1}	I_{stg2}	I_{stg3}	I_{stg4}
R_0	1	0	1	1	1	1
R_1	0	1	1	0	0	0
R_2	0	1	0	1	0	0
R_3	0	1	0	0	1	0
R_4	0	1	0	0	0	1

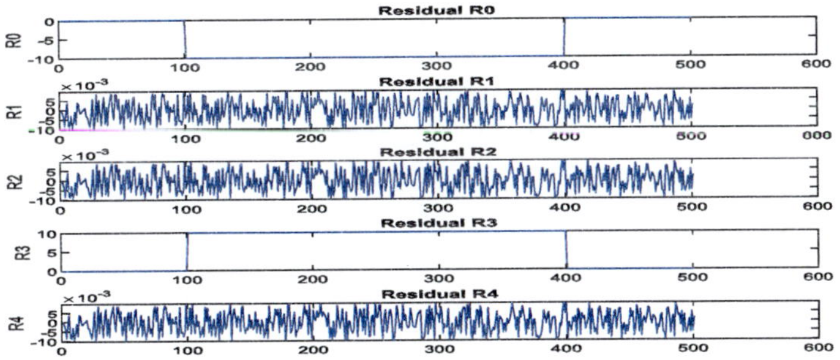

Fig. 5. Residual signals with a fault in the current sensor in the string 3.

windows that we impose throughout our signal.But the importance of this DCS comes more from the fact that its sign changes after the point of change.

We can define the DCS as the sum of the likelihood ratio between two probability density functions f_a^t and f_b^t estimated using the two windows of length W_a^t and W_b^t, i.e. before and after each instant t.

At time j, we define the DCS by relation (12) as the sum of the logarithms of the likelihood ratios from the start of the signal until time j [2]:

$$DCS(t) = \sum\nolimits_{j=t_{p-1}}^{t} log \frac{f_{\theta_a^j}(X_j)}{f_{\theta_b^j}(X_j)} = \sum\nolimits_{j=t_{p-1}}^{t} s_j, t \geq t_{p-1} \tag{12}$$

With s_j the logarithm of the likelihood ratio which is

$$s_j = \frac{1}{2}\left[log \frac{\left(\gamma_a^j\right)^2}{\left(\gamma_b^j\right)^2} + \frac{\left(\varepsilon_a^j\right)^2}{\left(\gamma_a^j\right)^2} - \frac{\left(\varepsilon_b^j\right)^2}{\left(\gamma_b^j\right)^2} \right] \tag{13}$$

$$f_{\theta_a^j}\left(X/X_{j-q}^j\right) = \frac{1}{\sqrt{2\pi(\gamma_a^j)}} e^{-\frac{\left(\varepsilon_a^j\right)^2}{2(\gamma_a^2)^j}} \tag{14}$$

with$\theta_a^j = (\gamma_a^2, a_1^a, \ldots, a_p^a)^j$, the estimate of the first window; γ_a^2, the variance of scalar random variable θ_a and ε_a the mean

$$f_{\theta_b^j}\left(X/X_{j-q}^j\right) = \frac{1}{\sqrt{2\pi(\gamma_b^j)}} e^{-\frac{(\varepsilon_b^j)^2}{2(\gamma_b^2)^j}} \tag{15}$$

with $\theta_b^j = (\gamma_b^2, a_1^b, \ldots, a_p^b)^j$, the estimate of the second window; γ_b^2, the variance of scalar random variable θ_b and ε_b the mean.

This statistical test makes it possible to generate the detection function of the relation (16) which is compared with a decision threshold. The detection function used to estimate the instant of change is given by

$$g(t) = \max_{t_{p-1}<j<t}\left[DCS(j)\right] - DCS(t) \tag{16}$$

The stop instant can therefore be defined as

$$t_p = \inf\{n > t_{p-1} : g(n) > h\} \tag{17}$$

with h the detection threshold to be defined.

The threshold value, as well as the width of the sliding windows of the DCS test, must be fine-tuned to obtain a satisfactory fault.

5 Results and Discussions

The experimental PVG built around 16 PV modules organized in four (4) strings of four (4) modules in series in each string. We incorporated one current sensor per string and one end current sensor to measure the overall current of the PVG. Finally, a single voltage sensor is used to measure the voltage produced by the PVG. The simulation of PVG according to the synoptic of Fig. 3, allows the measurement of the currents I_G and of the voltage V_G having to supply the generation of the residuals.

We placed a sensor fault in string 3, by increasing the measurement value by 10A. This action affects residual R3, but also residual R0 as shown in Fig. 6 and 9. The other residues are not affected. The application of the DCS algorithm, the generation of the detection function by the curve $g(t)$ which will be compared at all times to an imperative threshold, fixed in [7] for interested readers. Also combined with a choice of a window size between 15 and 25, it is possible to have a good compromise between a low rate of false alarms and a high rate of good failure detection.

By putting together all the detection functions as shown in Fig. 11, to be compared to a strongly localizing occurrence matrix, we can deduce the location of the sensor fault of string 3. The diagnostic approach based on the method of parity space, detects and isolates well the faults in our study system that is the PVG.

In each of the Figs. 6, 7, 8, 9, 10, and 11 below, we have grouped, always in the same order:

- the action of the bias on the first curve, representing the residual of order i
- the *DCS(t)* function as a function of time on the second curve
- the detection function *g(t)* on the third curve
- the result of the test di, according to the signature of the fault linked to the residual of order i.

This diagnosis method based on the parity space is quite effective in detecting sensor faults, but does not make it possible to locate system faults regardless of their amplitude, which is sometimes higher than traditional noise levels. It is then obvious that we must consider other approaches to identify and isolate system type faults. Nevertheless, the highlighting of sensor faults initially makes it possible to validate a diagnostic method long used in an industrial environment, but applied to a maximum operating point. This means that the operation of the PVG is constantly subject to special cases, because breakdowns occur at any time. The elimination of any fault hypothesis, therefore a certain failure of the instrumentation sensors necessary for the validation of the data used in this method, ensures the reliability and operating safety of the system. This conclusion on the depth of the diagnosis of sensor faults, then confirms the search for another diagnostic track on system faults through artificial neural networks (ANN) [6] as an extension of our work.

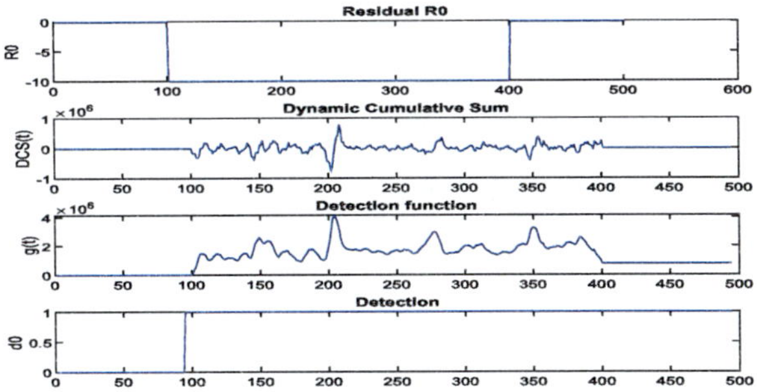

Fig. 6. Detection on R0 residual.

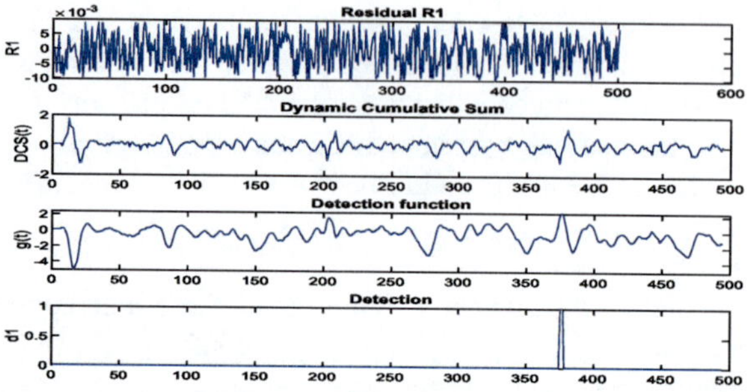

Fig. 7. Detection on R1 residual.

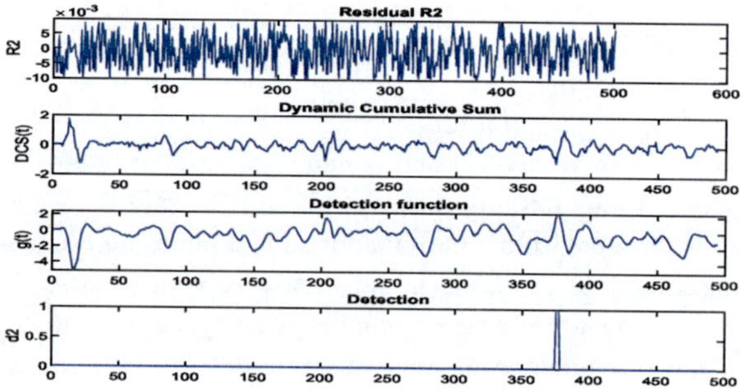

Fig. 8. Detection on R2 residual.

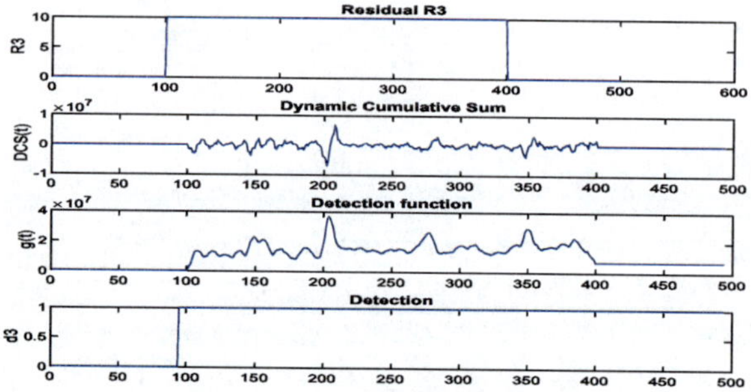

Fig. 9. Detection on R3 residual.

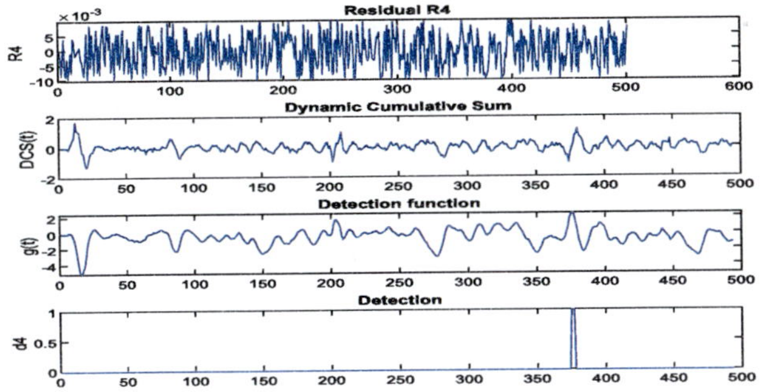

Fig. 10. Detection on R4 residual.

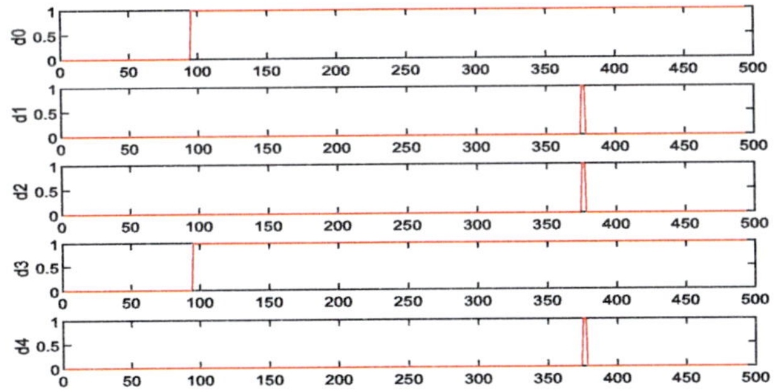

Fig. 11. Signature of faults on the five residuals.

6 Conclusions

In this article, we proposed a diagnostic method based on the parity space method applied to the MPPT of a PVG. The PVG model used here as a complex industrial system perfectly imitates the avalanche phenomenon of a photovoltaic cell. After generating the residuals, we applied DCS for detection, followed by sensor fault isolation, paving the way for accurate predictive maintenance.

Prospects are quite good for the detection, identification and isolation of the sensor faults mentioned above, and for eliminating an instrumentation related fault if necessary. We must improve, among other things, the management of the width of the sliding windows, which has a significant influence on the sensitivity of the detection function. The same goes for the choice of the detection threshold to obtain better performance in the detection of faults. An analysis using the Receiver Operating Characteristics (ROC) curve method is required.

Simultaneous sensor faults can be detected and isolated. An increase in the number of residues improves the fineness of fault isolation. Such an improvement requires a significant increase in the number of sensors and residuals, which are further related to a better knowledge of the system. The simulation results are encouraging and prove the good applicability of the diagnostic method based on a model oriented to the use of the parity space method combined with the DCS around the MPPT of a PVG. The effectiveness of the proposed diagnostic method was verified in detecting damage to sensors on PVG, the nonlinear model of which makes it difficult to apply the method based on observers. It would be desirable to also consider other diagnostic methods but this time for system type faults, in order to be complete on the issue.

References

1. Gao, Z., Cecati, C., Ding, S.X.: A survey of fault diagnosis and fault-tolerant techniques. Part I. Fault diagnosis with model-based and signal-based approaches. IEEE Trans. Indust. Electron. **62**(6), 3757–3767 (2015)
2. El Falou, W., Duchêne, J., Khalil, M.A.: AR-based method for change detection using dynamic cumulative sum. In: Proceedings of the 7th IEEE International Conference on Electronics, Circuits and Systems ICECS, vol. 1, pp. 157–160. Jounieh, Lebanon (2000)
3. Ge, Z.: Review on data-driven modeling and monitoring for plant-wide industrial processes. Chemom. Intell. Lab. Syst. **171**, 16–25 (2017)
4. Dhimish, M.: Fault Detection and Performance Analysis of Photovoltaic Installations. University of Huddershfield Thesis (2018)
5. Tariq, M.F., Khan, A.Q., Abid, M.: Data-driven robust fault detection and isolation of three-phase induction motor. IEEE Trans. Ind. Electron. **66**(6), 4707–4715 (2019)
6. Mansouri, M., Trabelsi, M., Nounou, H., Nounou, M.: Deep learning-based fault diagnosis of photovoltaic systems: a comprehensive review and enhancement prospects. IEEE Access **9** (2021). https://doi.org/10.1109/ACCESS.2021.3110947
7. Abbasi, M.A., et al..: Parity-based robust data-driven fault detection for nonlinear systems using just-in-time learning approach. Trans. Inst. Measure. Control (2021) https://doi.org/10.1177/0142331219894807
8. Chafouk, H., Hoblos, G., Langlois, N., Le Gonidec, S., Ragot, J.: Soft computing algorithm to data validation in aerospace system using parity space approach. J. Aerosp. Eng. **3**(3), 165–171 (2007)
9. Compaoré, O.W., Hoblos, G., Koalaga, Z.: Analysis of the impact of faults in a photovoltaic generator. In: International Conference on Innovation and Intelligence for Informatics, Computing, and Technologies (3ICT), pp. 68–73 (2021). https://doi.org/10.1109/3ICT53449.2021.9581575
10. Mellit, A., Tina, G.M., Kalogirou, S.A.: Fault detection and diagnosis methods for photovoltaic systems: a review. Renew. Sustain. Energy Rev. **91**, 1–17 (2018)
11. Bishop, J.: Computer simulation of the effects of electrical mismatches in photovoltaic cell interconnection circuits. Solar Cells **25**, 73–89 (1988)
12. Chen, Z., Zhang, K., Ding, S.X.: Improved canonical correlation analysis-based fault detection methods for industrial processes. J. Process Control **41**, 26–34 (2016)
13. Basseville, M., Nikiforov, I.: Detection of Abrupt Changes: Theory and Application. Prentice Hall Inc. [ISBN-10 0131267809, ISBN-13:978-0131267800], p. 457 (1993)
14. Zhirabok, A.N., Zuev, A.V., Shumskii, A.E.: Diagnosis of linear systems based on sliding mode observers. J. Comput. Syst. Sci. Int. **58**(6), 898–914 (2019). https://doi.org/10.1134/S1064230719040166

Systems Modeling

A Set-Based Uncertainty Quantification of Evolving Fuzzy Models for Data-Driven Prognostics

Khoury Boutrous[1], Iury Bessa[2(✉)], Fatiha Nejjari[1(✉)], and Vicenç Puig[1(✉)]

[1] Advanced Control Systems, Universitat Politècnica de Catalunya (UPC), Rambla Sant Nebridi 22, 08222 Terrassa, Spain
{boutrous.khoury,fatiha.nejjari,vicenc.puig}@upc.edu
[2] Department of Electricity, Federal University of Amazonas, Manaus, Brazil
iurybessa@ufam.edu.br

Abstract. Recent years have seen a great deal of innovation in the field of systems prognostics and health management. However, even with these advancements, some pertinent issues related with uncertainty in remaining useful life predictions are still open for investigation. One such area of interest is on how to account for the distribution of these predictions such that all uncertainty sources are duly captured and represented. Practically, these uncertainty quantification procedures must be computationally feasible for real-life deployment and reflect real-life situations devoid of strong assumptions. This article thus, proposes a data-based prognostics technique that uses a set-based quantification of uncertainty based on the set-membership paradigm, the interval predictor approach. The methodology is applied in the framework of the Evolving Ellipsoidal Fuzzy Information Granule which has recently proven its potency in prognostics applications. As a case study, the method is tested on the prognostics of insulated bipolar transistors utilising an accelerated aging IGBT dataset from the NASA Ames Research Center.

Keywords: Uncertainty quantification · Data driven prognostics · Set-based propagation · Interval arithmetic · Evolving ellipsoidal fuzzy information granules

1 Introduction

Even though it is common for some articles on prognostics to only yield deterministic remaining useful life (RUL) values, it is emphatically clear that prognostics in engineering is an uncertain endeavour since it involves a prediction of a future event based on current conditions and assumed future measurements. Thus, it is only prudent to present these point-wise estimations with accompanying confidence intervals that afford for a good interpretation of a reliable RUL prediction in real life situations. It could not have been reiterated more by [15] that "it is not even meaningful to make such predictions without

Z. Kowalczuk (Ed.): DPS 2022, LNNS 545, pp. 293–304, 2023.
https://doi.org/10.1007/978-3-031-16159-9_24

computing the uncertainty associated with RUL". Recently though, there have been an interest in the management of uncertainty in prognostics, with different proposed methods of identifying uncertainties, representing them and using resource efficient propagation methods. It is important to highlight early works such as [15,19] that sought to review the subject, emphasising the importance of uncertainty description in prognostics and equally from [1] that goes ahead to propose the inclusion of uncertainty description in some generic prognostic metrics such as the prediction horizon (PH) and $\alpha - \lambda$ performance metrics, offering a method of comparing different quantification methods. Uncertainty in most articles in the literature seek to formalise the types of uncertainties as the epistemic uncertainty, which accounts for uncertainty in modelling, and aleatoric uncertainty which involves exogenous uncertainty sources such as sensor noise and environmental disturbances. But [15] contend this notion for application to condition-based monitoring, rather prescribing for an uncertainty classification involving: (i) Present uncertainty, that is uncertainty pertaining to the estimation of states before prognostics are undertaken, (ii) Future uncertainty, that describes unaccounted conditions during prognostication, (iii) Uncertainty modelling that represents the discrepancy between measured response and the true response and (iv) Prediction method uncertainty, which involves the uncertainty that arises even with the final quantification process of the RUL's uncertainty. In the authors' point of view the second classification offers a more comprehensive description. It is therefore critical to systematically capture all these uncertainty sources. However, some authors [16,21] consider the use of sensitivity analysis to take into account only the dominantly influential input uncertainty sources on the output in the quest to save computational cost. This may however result in under-representation in some cases when propagation times are long, such that the least uncertainty has a major cumulative effect. Essentially, uncertainty management involves a transformation of input PDF to an eventual RUL output PDF [2] such that in the literature, uncertainty quantification more often assumes a Gaussian distribution. However, this assertion is far from the truth as even with a simple prognostics model, the eventual probability distribution of RUL does not yield Gaussian distributions. In this paper, a set-based approach is proposed for uncertainty quantification and subsequent set propagation for prognostics. As a matured field of study, set-based methodology offers ease of quantifying all uncertainty sources which is then propagated with minimal computational effort. The uncertainty is quantified aided by the interval predictor estimation methodology and represented by boxes. Manipulations are undertaken with relevant interval analysis (IA) properties to arrive at an appropriate uncertainty representation. This is applied to an Evolving Ellipsoidal Fuzzy Information Granules (EEFIG) which has proven to be effective in prognostics applications such as in [5] and [14].

2 Evolving Ellipsoidal Fuzzy Information Granules

2.1 Description

The EEFIG and its evolving granular learning algorithm was first introduced by [23]. The learning algorithm is an online data processing that employs evolving fuzzy information granules based on the parametric principle of justifiable granularity [8].

An EEFIG is a collection of N granules $\mathbb{G}_k = \{\mathcal{G}_k^1, \dots, \mathcal{G}_k^N\}$, where each granule is a fuzzy set $\mathcal{G}_k^i = (\mathbb{R}^{n_z}, g_k^i)$, where $g_k^i : \mathbb{R}^{n_z} \to [0,1]$ is the membership function of the EEFIG $\mathcal{G}_k^i$. The membership function ω_k^i is parameterized by the granular prototype $\mathcal{P}_k^i$ of the i-th granule at the time instant k, which is also a numerical evidence basis for the granulation process. The granule prototype is defined as follows:

$$\mathcal{P}_k^i = \left(\underline{\mu}_k^i, \mu_k^i, \overline{\mu}_k^i, \Sigma_k^i\right), \tag{1}$$

where $\underline{\mu}_k^i$, μ_k^i and $\overline{\mu}_k^i$ are the lower, mean and upper bound vectors of the i-th EEFIG at time k and Σ_k^i is the inverse of its covariance matrix. Given the granule prototype $\mathcal{P}_k^i$, the membership function of an EEFIG is parameterized as

$$\omega_k^i(z_k) = \exp\left\{-\left[(z_k - \mu_k^i)^\top \left(\Delta_k^i\right)^{-1}(z_k - \mu_k^i)\right]^{1/2}\right\}, \tag{2}$$

where, for $p \in \mathbb{N}_{\leq n_z}$,

$$\Delta_k^i = \operatorname{diag}\left\{\left(\frac{\overline{\mu}_{k,1}^i - \underline{\mu}_{k,1}^i}{2}\right)^2, \dots, \left(\frac{\overline{\mu}_{k,p}^i - \underline{\mu}_{k,p}^i}{2}\right)^2\right\},$$

being $\overline{\mu}_k^i$, and $\underline{\mu}_k^i$ the semi-axes of the i-th EEFIG prototype such that $\underline{\mu}_k^i < \mu_k^i < \overline{\mu}_k^i$. The normalized membership functions g_k^i at the k-th time instant for i-th granule is

$$g_k^i(z_k) = \frac{\omega_k^i(z_k)}{\sum_{i=1}^N \omega_k^i(z_k)}. \tag{3}$$

Moreover, the distance of a given data sample $z_k \in \mathbb{R}^{n_z}$ to the i-th EEFIG is given by the square of Mahalanobis distance:

$$d(z_k, \mu_k^i) = (z_k - \mu_k^i)^\top \Sigma_k^i (z_k - \mu_k^i). \tag{4}$$

The granulation process is the updating of the EEFIG model-based on the data stream. The updates are performed aiming at improving the so-called granular performance index with respect to a data sample. The performance index of the i-th granule with respect the sample z_k, denoted $\bar{\mathcal{Q}}_k^i$, is defined as

$$\bar{\mathcal{Q}}_k^i(z_k) = d(z_k, \mu_k^i)|\mathcal{G}_k^i| \tag{5}$$

where $|\cdot|$ is the fuzzy cardinality operator of the i-th EEFIG, whose update is performed as follows

$$|\mathcal{G}_k^i| = |\mathcal{G}_{k-1}^i| + g_k^i(z_k) - \frac{\partial g_k^i(z_k)}{\partial \mathcal{P}_k^i}, \tag{6}$$

where the term $\frac{\partial g_k^i(z_k)}{\partial \mathcal{P}_k^i}$ is computed as described in [20]. The total EEFIG performance index is the sum of the data sample contribution index of each granule:

$$\mathcal{Q}_k^i = \frac{1}{k}\sum_{j=1}^{k} \bar{\mathcal{Q}}_j^i(z_j). \tag{7}$$

To decide whether a granule must be updated or not, the concept of data sample admissibility is used. A data sample z_k is said to be admitted by a given granule prototype $\mathcal{P}_k^i$ if it is used to update the granule prototype parameters. In this sense, two criteria are used to evaluate the data sample admissibility:

$$d(z_k, \mu_k^i) < \nu, \tag{8}$$

$$\mathcal{Q}_k^i > \mathcal{Q}_{k-1}^i, \tag{9}$$

where $\nu = \left(\chi^2\right)^{-1}(\gamma, n)$ is a threshold parameterized by the inverse of chi-squared statistic with $n+m$ degrees of freedom, leading EEFIG prototype to cover around $100\gamma\%$ of the stream sample. A data sample z_k which does not meet the first condition (8) for some granule is denominated an anomaly. In parallel, as the data samples are available and evaluated, a structure named tracker whose objective is to follow the data stream dynamics to indicate change points, is established.

The tracker is parameterized by a mean vector μ_k^{tr} and an inverse covariance matrix Σ_k^{tr}, which are recursively updated. A new granule is created if the following conditions hold:

1. The tracker is c-separated from all the existing granule prototypes. The c-separation condition is expressed as follows

$$\|\mu_k^{\text{tr}} - \mu_k^i\| \geq c\sqrt{n_z \max(\bar{\xi}(\Sigma_k^{\text{tr}}), \bar{\xi}(\Sigma_k^i))}, \tag{10}$$

   for all $\mathcal{G}_k^i \in \mathbb{G}_k$, where $\bar{\xi}\left(\Sigma_k^{\text{tr}}\right)$ is the largest eigenvalue of Σ_k and, $c \in [0, \infty)$ specifies the separation level. Here, c is assumed as 2.
2. The number of consecutive anomalies is $n_a > \zeta$ where ζ is a hyper-parameter defined by the user to control the minimum amount of anomalies which may enable the rule creation.

2.2 EEFIG-Based Degradation Modelling and RUL Estimation

Based on the EEFIG model described in the previous section, the following Takagi-Sugeno fuzzy model is proposed for the degradation modelling

$$\begin{aligned} &\textbf{Rule } i: \text{ IF } z_k \text{ is } \mathcal{G}_k^i \\ &\qquad \text{THEN } y_k^i = {\boldsymbol{\theta}_k^i}^\top [y_{k-1}, y_{k-2}, \dots, y_{k-L}]^\top, \end{aligned} \tag{11}$$

for $i \in \mathbb{N}_{\leq C_k}$, where $y_k \in \mathbb{R}$ is the health index, $z_k \in \mathbb{R}^{n_z}$ is the vector of premise variables, $\boldsymbol{\theta}_k^i \in \mathbb{R}^L$ are the coefficients of the i-th fuzzy rule at instant k, $L \in \mathbb{N}$ is the number of regressors in the autoregressive consequent, and $C_k \in \mathbb{N}$ is the number of rules at instant k. Using the center-of-gravity defuzzification for (11), the health index y_k is

$$y_k = \sum_{i=1}^{C_k} g_k^i(z_k) {\boldsymbol{\theta}_k^i}^\top [y_k \; y_{k-1} \; \dots \; y_{k-L+1}]^\top \tag{12}$$

$$\boldsymbol{\Theta}_k h_k(\boldsymbol{y}_k), \tag{13}$$

where

$$\boldsymbol{\Theta}_k = \left[{\boldsymbol{\theta}_k^1}^\top \; \dots \; {\boldsymbol{\theta}_k^{C_k}}^\top \right],$$

$$\boldsymbol{y}_j = \begin{bmatrix} y_j \\ \vdots \\ y_{j-L+1} \end{bmatrix}, \; h_k(\boldsymbol{y}_l) = \begin{bmatrix} g_k^1(z_k)\boldsymbol{y}_j \\ \vdots \\ g_k^{C_k}(z_k)\boldsymbol{y}_j \end{bmatrix}.$$

As described in [23], the consequent parameters $\boldsymbol{\Theta}_k$ are estimated based on recursive least squares methods. In particular, here we use the Sliding-windowed Fuzzily Weighted Recursive Least Squares (SFWRLS) where the weights are the membership degrees and the data window contains the last φ samples:

$$H_k = [h_k(\boldsymbol{y}_{k-1}) \; \dots \; h_k(\boldsymbol{y}_{k-\varphi})] \tag{14}$$

$$X_k = [y_k \; \dots \; y_{k-\varphi+1}] \tag{15}$$

Therefore, the recursive equation for the (SFWRLS) estimator are provided as follows

$$\Upsilon_k = P_k H_k \left(\eta I_\varphi + H_k^\top P_k H_k\right)^{-1} \tag{16}$$

$$P_{k+1} = \eta^{-1}\left(P_k - \Upsilon_k H_k^\top P_k\right) \tag{17}$$

$$\boldsymbol{\Theta}_{k+1} = \boldsymbol{\Theta}_k + (X_k - \boldsymbol{\Theta}_k H_k)^\top \Upsilon_k^\top \tag{18}$$

where $P_k \in \mathbb{R}^{LC_k \times LC_k}$ is an estimate of the inverted regularised data autocorrelation matrix, $\Upsilon_k \in \mathbb{R}^{n_x}$ is the (SFWRLS) gain vector, and $\eta \in (0, 1]$ is the forgetting factor.

Given the estimate of the parameters of (11), the one-step ahead prediction of the degradation at instant k is computed as follows

$$\hat{y}_{k+1|k} = \sum_{i=1}^{C_k} g_k^i(z_k) {\boldsymbol{\theta}_k^i}^\top [y_k \; y_{k-1} \; \dots \; y_{k-L+1}]^\top \tag{19}$$

For any $N \in \mathbb{N}$, define

$$\hat{\boldsymbol{y}}_{k+N|k} = \begin{cases} [y_k, y_{k-1}, \dots, y_u]^\top, & \text{if } N = 1, \\ [\hat{y}_w, \dots, \hat{y}_{k+1}, y_k, \dots, y_u]^\top, & \text{if } 1 < N < L, \\ [\hat{y}_w, \dots, \hat{y}_u]^\top, & \text{if } N \geq L, \end{cases} \tag{20}$$

where $u = k + N - L$ and $w = k + N - 1$. The N-step ahead health index prediction $\hat{y}_{k+N|k}$ is computed as follows:

$$\hat{y}_{k+N|k} = \boldsymbol{A}_k \hat{\boldsymbol{y}}_{k+N|k}, \tag{21}$$

where $\boldsymbol{A}_k = \sum_{i=1}^{C_k} \omega_k^i (z_k) \boldsymbol{\theta}_k^{i\top}$.

Based on the long term prediction described in (21), the RUL can be estimated by predicting the future health state of the system given the current and past system's condition, which are provided by $\hat{\boldsymbol{y}}_{k+N|k}$ and z_k. Indeed, the RUL can be defined as the amount of time until the system's health index reaches a predefined threshold, that is:

$$\hat{\text{RUL}}_k = \inf \{N \in \mathbb{Z}_{\geq 0} : \hat{y}_{k+N|k} \leq \eta\}, \tag{22}$$

where $\hat{\text{RUL}}_k \in \mathbb{Z}_{\geq 0}$ denotes the RUL estimate computed at instant k given the observations of degradation state until k, and η is the end of life threshold, which must be defined based on historic data.

3 Interval Set-Based Uncertainty Description

For quantifying and propagating the uncertainty sets, the following conditions are assumed:

Assumption 1. *The unknown uncertainties on the process and measurement can be represented with an appropriate convex set.*

Assumption 2. *From Assumption 1, since the one-step ahead degradation fuzzy model* (19) *is linear, then a multi-step propagation with a convex uncertainty initial set ensures a conservation of convexity. Thus the final uncertainty set is convex and easily interpretable.*

The initial noise and parametric uncertainties are quantified by finding an optimal bound. This allows to bound all realization of the considered uncertainty sources. For a one-step propagation i.e. $N = 1$, the input as from (20), considers only a certain measurement vector $\hat{\boldsymbol{y}}_{k+N|k} \in \mathbb{R}^{k+N-L}$ with its uncertain component of prediction (21) described as:

$$\Delta\hat{y}_{k+N|k} = \Delta\boldsymbol{A}_k \hat{\boldsymbol{y}}_{k+N|k}, \tag{23}$$

where $\Delta\hat{y}_{k+N|k}$ and $\Delta\boldsymbol{A}_k$ are the uncertain components described as symmetric intervals. For a multi-step forecast to the EOL, the uncertain set is constructed recursively considering a propagation of recursive sum of sets as:

$$\Delta\hat{y}_{k+N|k} = \Delta\boldsymbol{A}_k \hat{\boldsymbol{y}}_{k+N|k} + \Delta\boldsymbol{A}_k \Delta\hat{\boldsymbol{y}}_{k+N|k} + \Delta\boldsymbol{A}_k \Delta\hat{\boldsymbol{y}}_{k+N|k} + \epsilon_k. \tag{24}$$

Where $\Delta\hat{y}_{k+N_{EOL}|k}$ as the uncertainty set at $k + N_{EOL}|k$ of the UUT's EOL and ϵ_k as the bounded noise value during forecasting. For clarity, two noise components are considered, first from sensor noise in the prediction of the states and another ϵ_k for noise in future times of prognostication. The upper and lower bounds of the RUL are therefore given as:

$$\overline{RUL} = \sup\left(x \in \mathbb{R} \cup \{-\infty,\ +\infty\} \mid \Delta\hat{y}_{k+N_{EOL}|k} \in \hat{y}_{k+N_{EOL}|k},\ x \leq \hat{y}_{k+N_{EOL}|k}\right) \tag{25}$$

$$\underline{RUL} = \inf\left(x \in \mathbb{R} \cup \{-\infty,\ +\infty\} \mid \Delta\hat{y}_{k+N_{EOL}|k} \in \hat{y}_{k+N_{EOL}|k},\ \hat{y}_{k+N_{EOL}|k} \leq x\right) \tag{26}$$

Since the parametric uncertainty and the propagation noise (ϵ_k) are propagated in time, the future uncertainty and predicted uncertainty as stated in the introduction are considered in part taken care of in the prognostics process.

4 Case Study

The proposed method is applied to the prognostics of an insulated gate bipolar transistor (IGBT). The IGBT is subjected to rigorous thermal cycles of stress during a run to failure test. This is done till its end of life (EOL), that is when a latch up failure occurs, see [12]. The process involves switching signals steadily between 0 V and 4 V, with temperature controlled between 329 °C and 330 °C outside the rated temperature of the test transistors. The transient data collected when the devices switch are the (i) Collector-emitter turn on Voltage (ii) Gate Voltage and (iii) Collector current. The data set is originally on 4 devices but for our purpose the Collector-emitter turn on voltage (V_{CE}) of 2 devices will be used. The V_{CE} is selected because it has proven to be the best prognostics parameter for IGBT aging tests from evaluations undertaken in the majority of related papers e.g. [26].

For prognostics purposes, features are extracted from the prognostics parameter (i.e.V_{CE}), where the prognostic trend is segmented into batches i according to respective cycles, x_i. Two features are selected as input variables. 1) The energy feature ($\sum_{i=1}^{n} E(x_i)$), for the premise variable of the fuzzy model and 2) a cumulative standard deviation of the trigonometric function (asinh) feature (28) as proposed in [27],

$$X(i) = \sigma\left(\log\left[x_i + (x_i^2 + 1)^{\frac{1}{2}}\right]\right) \tag{27}$$

$$\mathrm{CF}_i(X) = \frac{\sum_{i=1}^{n} X(i)}{|\sum_{i=1}^{n} X(i)|^{\frac{1}{2}}} \tag{28}$$

The auto-regressive consequent of the fuzzy model is selected based on a suitability metric tested on various temporal features such as the cumulative standard deviation of (atanh), smoothed root mean square $\left(\sqrt{\frac{1}{n}\sum_{i=1}^{n} x_i^2}\right)$ etc.

$$\text{Suitability} = \begin{bmatrix} \text{Monotonicity} \\ \text{Trandability} \\ \text{Prognosticability} \end{bmatrix}^{\top} \begin{bmatrix} 1 \\ 0.976 \\ 1 \end{bmatrix} \tag{29}$$

The inverse of the curve is used since it performed well under test and naturally represented degradation for the algorithm better.

The bound on the noise value ϵ_k is taken as the maximum variation of each cycle of stress without smoothing. Whilst the uncertainty $\Delta y_0 \subseteq \sigma(k)$ is taken as a varying error between predicted and original feature at propagation stage.

4.1 Results and Discussion

The prognostic model is trained with data from IGBT1 and two test scenarios considered. First, the model is used for RUL prediction of the IGBT1 itsef and a more realistic prediction on a different device, the IGBT2. Figure 1 depicts the real and the estimated health index, whose error at each prognostic time instant is quantified with the process noise in a hyperbox for propagation, performed for each test scenarios. The final uncertainty set at each prognostic time differs in width which depends on the initial uncertainty set and the duration to the EOL. As evident from Fig. 2, the width of the uncertainty set generally decreases

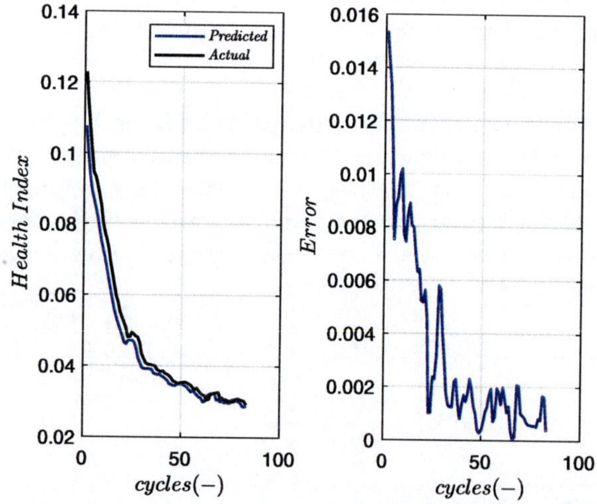

Fig. 1. Predicted and real health index with error at each cycle.

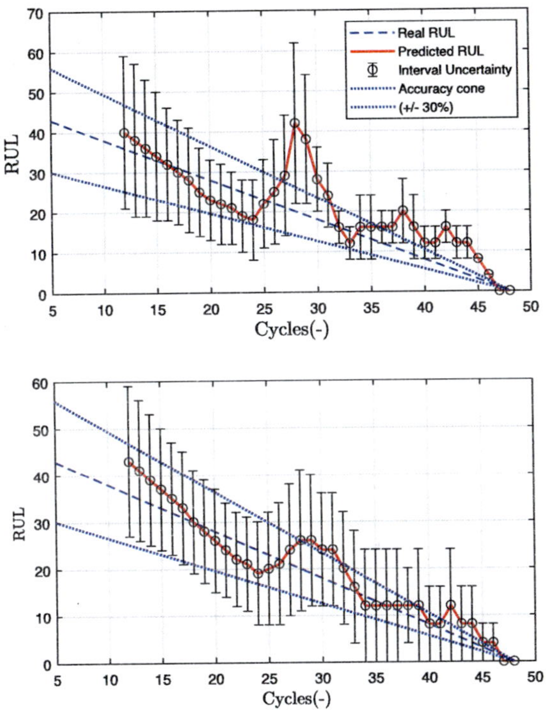

Fig. 2. RUL predictions with uncertainty set description for (top) IGBT1 and (bottom) IGBT2.

as more data are available for a better prediction by the algorithm. That is because the model becomes better as more data is available for training and thus less uncertainties in the prediction stage. Considering that unknown future conditions constitutes uncertainties, the farther away from the EOL, the bigger the size of the interval sets. As expected, the magnitude of prediction error is greater in RUL prediction of the IGBT2, therefore the uncertainty set is wider than the self test case as shown in Fig. 3. The convergence of the sets from both scenarios shows a desirable property of the algorithm.

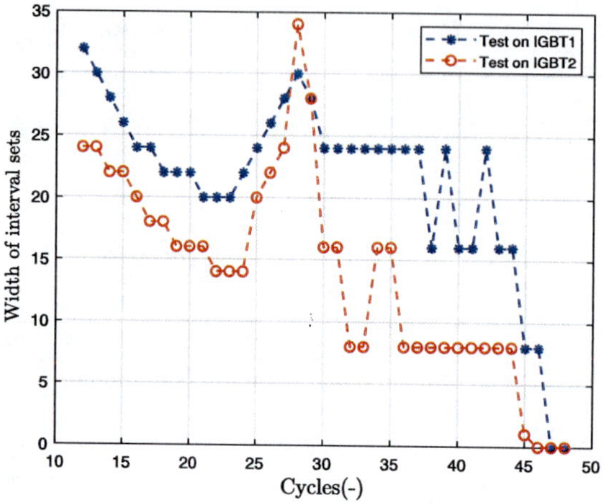

Fig. 3. Width of uncertainty set at each RUL prediction cycle.

5 Conclusions

In this paper, the inevitable presence of uncertainty in prognostics is tackled through set-based quantification and propagation, enabled by set-membership theory. The methodology proves its merits as easy to employ with reasonable uncertainty description. For future research, more complicated geometric options such as zonotopes will be used for less conservative results. It would also be interesting to have a comparative analysis with different uncertainty quantification methods using the metrics as proposed in [1]. The methodology will also be extended to other data-driven techniques such as deep learning.

Acknowledgements. This work has been co-financed by the Spanish State Research Agency (AEI) and the European Regional Development Fund (ERFD) through the project SaCoAV (ref. MINECO PID2020-114244RB-I00), by the European Regional Development Fund of the European Union in the framework of the ERDF Operational Program of Catalonia 2014-2020 (ref. 001-P-001643 Looming Factory), by the DGR of Generalitat de Catalunya (SAC group ref. 2017/SGR/482), by the Brazilian agencies CNPq, FAPEMIG, FAPEAM, and by the PROPG-CAPES/FAPEAM Scholarship Program.

References

1. Saxena, A., Celaya, J., Saha, B., Saha, S., Goebel, K.: Evaluating prognostics performance for algorithms incorporating uncertainty estimates. In: IEEE Aerospace Conference 2010, pp. 1–11 (2010). https://doi.org/10.1109/AERO.2010.5446828
2. Orchard, M., Kacprzynski, G., Goebel, K., Saha, B., Vachtsevanos, G.: Advances in uncertainty representation and management for particle filtering applied to prognostics. In: International Conference on Prognostics and Health Management 2008, pp. 1–6 (2008). https://doi.org/10.1109/PHM.2008.4711433

3. Robinson, E., Marzat, J., Raissi, T.: Model-based prognosis using an explicit degradation model and Inverse FORM for uncertainty propagation. IFAC-PapersOnLine **50**, 14242–14247 (2017). https://doi.org/10.1016/j.ifacol.2017.08.1815
4. Hickey, T., Ju, Q., Emden, M.: Interval arithmetic: from principles to implementation. J. ACM **48**, 1038–1068 (2001). https://doi.org/10.1145/502102.502106
5. Camargos, M., Bessa, I., D'Angelo, M.F.S.V., Cosme, L.B., Palhares, R.M.: Data-driven prognostics of rolling element bearings using a novel error based evolving Takagi-Sugeno fuzzy model. Appl. Soft Comput. **96**, 106628 (2020). https://doi.org/10.1016/j.asoc.2020.106628
6. Blesa, J., Puig, V., Saludes, J.: Identification for passive robust fault detection using zonotope-based set-membership approaches. Int. J. Adapt. Control Signal Process. **25**(9), 788–812 (2011)
7. Tamssaouet, F., Nguyen, T.P.K., Medjaher, K., Orchard, M.E.: Uncertainty quantification in system-level prognostics: application to Tennessee Eastman process. In: The Sixth (6th) Edition in the Series of the International Conference on Control, Decision and Information Technologies (2019)
8. Pedrycz, W., Wang, X.: Designing fuzzy sets with the use of the parametric principle of justifiable granularity. IEEE Trans. Fuzzy Syst. **24**(2), 489–496 (2016)
9. Duong, P.L.T., Raghavan, N.: Uncertainty quantification in prognostics: a data driven polynomial chaos approach. In: IEEE International Conference on Prognostics and Health Management (ICPHM) 2017, pp. 135–142 (2017). https://doi.org/10.1109/ICPHM.2017.7998318
10. Althoff, M., Frehse, G., Girard, A.: Set propagation techniques for reachability analysis. Annu. Rev. Control Robot. Auton. Syst. **4** (2021). https://doi.org/10.1146/annurev-control-071420-081941
11. Huang, D., et al.: A hybrid bayesian deep learning model for remaining useful life prognostics and uncertainty quantification. In: IEEE International Conference on Prognostics and Health Management (ICPHM) 2021, pp. 1–8 (2021). https://doi.org/10.1109/ICPHM51084.2021.9486527
12. Sonnenfeld, G., Goebel, K., Celaya, J.R.: An agile accelerated aging, characterization and scenario simulation system for gate controlled power transistors. In: 2008 IEEE AUTOTESTCON, pp. 208–215 (2008). https://doi.org/10.1109/AUTEST.2008.4662613
13. Cadini, F., Sbarufatti, C., Cancelliere, F., Giglio, M.: State-of-life prognosis and diagnosis of lithium-ion batteries by data-driven particle filters. Appl. Energy **235**, 661–672 (2019). https://doi.org/10.1016/j.apenergy.2018.10.095
14. Camargos, M., Bessa, I., Junior, L.A.Q.C., Coutinho, P., Leite, D.F., Palhares, R.M.: Evolving fuzzy system applied to battery charge capacity prediction for fault prognostics. In: Atlantis Studies in Uncertainty Modelling. Atlantis Press (2021). https://doi.org/10.2991/asum.k.210827.010
15. Sankararaman, S., Goebel, K.: Why is the remaining useful life prediction uncertain? In: PHM 2013 - Proceedings of the Annual Conference of the Prognostics and Health Management Society 2013, pp. 337–349 (2013)
16. Gu, J., Barker, D., Pecht, M.: Uncertainty assessment of prognostics of electronics subject to random vibration. In: AAAI Fall Symposium - Technical Report (2007)
17. Sankararaman, S., Goebel, K.: Uncertainty in prognostics: computational methods and practical challenges. In: IEEE Aerospace Conference 2014, pp. 1–9 (2014). https://doi.org/10.1109/AERO.2014.6836342

18. Wu, R., Ma, J.: An improved LSTM neural network with uncertainty to predict remaining useful life. In: CAA Symposium on Fault Detection. Supervision and Safety for Technical Processes (SAFEPROCESS) 2019, pp. 274–279 (2019). https://doi.org/10.1109/SAFEPROCESS45799.2019.9213408
19. Dewey, H.H., DeVries, D.R., Hyde, S.R.: Uncertainty quantification in prognostic health management systems. In: IEEE Aerospace Conference 2019, pp. 1–13 (2019). https://doi.org/10.1109/AERO.2019.8741821
20. Cordovil, L.A.Q., Coutinho, P.H.S., Bessa, I., Peixoto, M.L.C., Palhares, R.M.: Learning event-triggered control based on evolving data-driven fuzzy granular models. Int. J. Robust Nonlinear Control **32**(5), 2805–2827 (2022)
21. Celaya, J., Saxena, A., Goebel, K.: Uncertainty Representation and Interpretation in Model-based Prognostics Algorithms based on Kalman Filter Estimation (2012)
22. Haque, M.S., Choi, S., Baek, J.: Auxiliary particle filtering-based estimation of remaining useful life of IGBT. IEEE Trans. Industr. Electron. **65**(3), 2693–2703 (2018). https://doi.org/10.1109/TIE.2017.2740856
23. Cordovil, L.A.Q., Coutinho, P.H.S., Bessa, I., D'Angelo, M.F.S.V., Palhares, R.M.: Uncertain data modeling based on evolving ellipsoidal fuzzy information granules. IEEE Trans. Fuzzy Syst. **28**(10), 2427–2436 (2020). https://doi.org/10.1109/TFUZZ.2019.2937052
24. Li, G., Yang, L., Lee, C.-G., Wang, X., Rong, M.: A Bayesian deep learning RUL framework integrating epistemic and aleatoric uncertainties. IEEE Trans. Industr. Electron. **68**(9), 8829–8841 (2021). https://doi.org/10.1109/TIE.2020.3009593
25. Cornelius, J., Brockner, B., Hong, S.H., Wang, Y., Pant, K., Ball, J.: Estimating and leveraging uncertainties in deep learning for remaining useful life prediction in mechanical systems. In: IEEE International Conference on Prognostics and Health Management (ICPHM) 2020, pp. 1–8 (2020). https://doi.org/10.1109/ICPHM49022.2020.9187063
26. Ahsan, M., Stoyanov, S., Bailey, C.: Data driven prognostics for predicting remaining useful life of IGBT. In: 2016 39th International Spring Seminar on Electronics Technology (ISSE), pp. 273–278 (2016). https://doi.org/10.1109/ISSE.2016.7563204
27. Javed, K., Gouriveau, R., Zerhouni, N., Nectoux, P.: Enabling health monitoring approach based on vibration data for accurate prognostics. IEEE Trans. Industr. Electron. **62**(1), 647–656 (2015). https://doi.org/10.1109/TIE.2014.2327917

Qualia: About Personal Emotions Representing Temporal Form of Impressions - Implementation Hypothesis and Application Example

Zdzisław Kowalczuk(✉), Michał Czubenko, and Marlena Gruba

Faculty of Electronics, Telecommunications and Informatics, Department of Robotics and Decision Systems, Gdansk University of Technology, 80-233 Gdańsk, Poland
{kova,micczube,marlena.gruba}@pg.edu.pl

Abstract. The aim of this article is to present the new extension of the xEmotion system as a computerized emotional system, part of an Intelligent System of Decision making (ISD) that combines the theories of affective psychology and philosophy of mind. At the same time, the authors try to find a practical impulse or evidence for a general reflection on the treatment of emotions as transitional states, which at some point may lead to the emergence of new emotional qualities ascribed to phenomena (objects or events) perceived by the agent. Since there is a dispute in the literature about the meaning and usefulness of 'private' emotions, the article attempts to present in a transparent and technical manner the problem of hypothetical qualia against the background of selected theories of emotions, taking into account the cultural aspect or division. By treating qualia as an expression of individual emotional diversity and learning capacity, qualia can be assigned a quite technical role in modeling the human mind and in implementing autonomous robots and agents.

Keywords: Cognitive science · Artificial intelligence · Emotions · Qualia · Mary's room

1 Introduction

The concept of quale has accompanied philosophers since 1929, when it was first used in the current sense [21]. It denotes a certain qualitative feature of the perceived object, but abstract rather than a physical property. Importantly, qualia are, however, a purely subjective concept, albeit reproducible in a given experiment. You can refer to an example: looking at a tomato you can tell that its color wavelength is about 700 nm, alternatively, you can characterize it with a chromatographic computer code (*Red Green Blue*: RGB = #ff0000). Whereas for ordinary people the subjective and qualitative impression of a 'red' color is usually linked to previous subjective experience with that color.

Z. Kowalczuk (Ed.): DPS 2022, LNNS 545, pp. 305–316, 2023.
https://doi.org/10.1007/978-3-031-16159-9_25

The concept of qualia was spread thanks to a thought experiment called the room of Mary [11], presented as an argument against the materialistic theories of the mind. In this experiment, the researcher (Mary) had full knowledge of the color ‘red’ (i.e., what is its wavelength and its determination, what is its perception by neurons in the brain, how it was processed, etc.). In her world, however, that color had not really existed so far; and suddenly at some point the perceived world changed so that ‘that’ red color appeared. The main question is: has the researcher added any element to her knowledge, perceiving the phenomenon of red color? In other words, has the observation of the red color created some quale in the investigator’s mind, or has nothing changed?

From the philosophy of mind standpoint, two approaches collide in this experiment: materialistic and epiphenomenal. Materialism postulates the existence of only matter, according to the laws of physics, and the use of these laws, supported by the language of mathematics, to describe all kinds of forms of perceptions of reality. On the other hand, the epiphenomenal concept claims that any phenomenon (object or event), in addition to its descriptive and possible physical form, can also have parallel mental characteristics [5]. Which inevitably leads to the problem of mental representation of the environment. Meanwhile, according to the consistent attitude of materialists, all mental images, perceived impressions and emotions simply determine the biological state of the human mind, and therefore there is no room for undefined ideas or states, such as, for example, unspoken qualia.

In addition to reflecting on the obtained results of established philosophical concepts and trends, it is worth considering the simple question of subjectivity and privacy. Note that an experiment in which the observer is human and his/her senses are the tool of measurement will always be subjective. Considering the aforementioned Mary’s room, it is enough to agree that each person will perceive the same red object differently.

According to Dennett [6], for an impression to be a quale, it must have the following four features:

1. unspoken: cannot be accurately conveyed (for example, everyone might understand red differently).
2. internally non-relational and unchanging, although dependent on one’s own experience (with any objects).
3. private: it is basically impossible to compare the qualia from person to person
4. self-conscious: to experience the qualia, you must also be aware of them.

In fact, human emotions can be described in a very similar way: they are subjective and unreliable, largely private, and (directly) understandable in consciousness. However, they are also characterized by internal variability: as a rule, they depend on the passage of time and current experience. We can therefore agree with the statement that human emotions can be the result of the projection of qualia [2] onto an established cultural domain.

In view of the above, and to some extent, emotions can be viewed as qualia, certain states of bodily sensation, or a perceived agitation with a sign representing a form of pain or pleasure [9]. The following three differentiators represent a variety of but widely accepted concepts of emotion: somatic [12], cognitive [26], and evolutionary [22].

1.1 Qualia in Computational Models

In general, qualia as an in-depth aspect related to computational models of emotions are very rarely taken into account, although recently some researchers have started to recognize them and appreciate their role. There seem to be only three approaches to dealing with qualia. One approach is related to modeling consciousness [13,25], the second focuses on the temporal context [8,24], and the third one concerns modeling emotions [1,10,20]. In this article, we only cover the last aspect. It should be noted, however, that none of the other aspects mentioned above deal with the topic of mathematical modeling of qualia in any emotional context.

Izard [10] mentions Edelman and Tononi [7] who stated that the first appearance of qualia could be related to a conscious emotional sensation in the presence of the observed object. The same point of view is presented by the authors of [20] who say that qualia can be presented as an emotion broken down into arousal and valence. Similarly, after [1], we can quote that "emotion qualia thus refer to the raw feel of emotion; the actual phenomenon of a particular emotion experienced may actually differ depending on the perception of that emotion by each person". As can be seen from the above, the concept of emotional qualia presented in this article is close to several proposed theories.

1.2 The Contribution

This article proposes a simple extension of the xEmotion system, a computerized emotional system that is part of an Intelligent System of Decision-making (ISD) that combines the theories of affective psychology and philosophy of mind. In particular, we consider the issue of modeling private emotional qualia – hereinafter referred to as *equalia.*

The proposed qualia extension will be tested in a simulation experiment using three different scenarios. On the basis of the obtained results, it can be concluded that the xEmotion system can be rationally enriched with a useful, dynamic mechanism of remembering observable phenomena (objects, events) in the context of emotions and equalia (emotional qualia, which are referred to in the singular as *quale*).

2 Model of Human Emotions

Among other factors, the foundations of theoretical knowledge presented above had a strong influence on the shape of the computational model of emotions called xEmotion [3,16,18]. This model is dedicated to decision-making systems designed for autonomous units [4,14,15,18,19]. xEmotion combines the emotional aspects of objects with their models and other individual features of the system in the memory of an agent (or unit).

As in the Izard model, also in our xEmotion system we use the following four emotional types:

- pre-emotions – related to the somatic theory of emotions and caused by simple features (impressions) of objects and based on known patterns (e.g. red color inducing excitement/anger).
- sub-emotions – related to the appraisal theory and appearing only after the recognition of the entire known object, based of its model stored in the agent's memory (along with the assigned sub-emotion).
- emotion – the result of a synergistic interaction of all sub- and pre-emotions that represents the actual emotional experience.
- mood – derivative emotional state/factor (generally long-term).

The above categorization was used in the dynamic structure of the xEmotion system (processing signals from the environment) shown in Fig. 1. It should be noted that the physiological and sensorimotor aspects have been omitted here due to the intended use in robots and virtual agents, and in particular in their autonomous management/control systems.

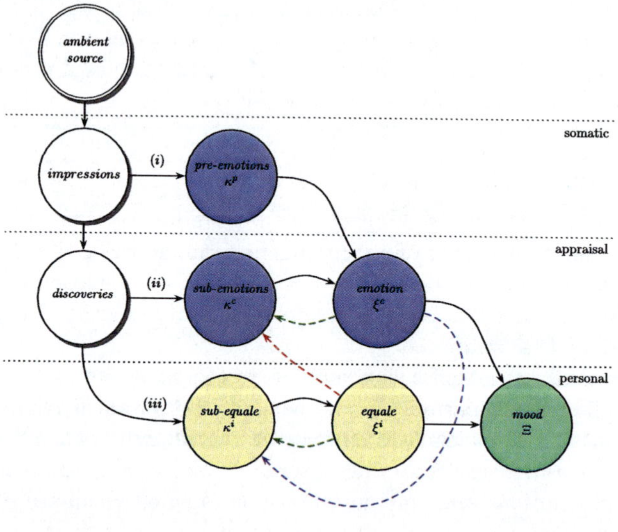

Fig. 1. xEmotion – computational model of emotions for autonomous agents [17,18].

Taking into account the emotional context, two aspects can be distinguished: general or universal – concerning commonly known emotions, and personal or individual – subjective, determined independently by a given agent. General emotions (pre-emotions, sub-emotions and emotions) can be originally mapped on the Plutchik paraboloid [23], and in the case of xEmotion on its modified version called the rainbow of emotions [14,18,19].

On the other hand, personal emotions in the xEmotion system defined as equalia are (similarly to universal emotions built on the basis of sub-emotions) the result of sub-equalia acting on the agent (also directly related to models of known objects). Personal emotions are not assigned to any predetermined (named) state. In the above model of a human emotions, designed for autonomous intelligent systems, equalia (emotions or personal feelings) was used in addition to classical emotions. The implemented equalia represent the agent's personal emotions, which have only an internal temporal interpretation but, with some development or experimentation, can be used to build new emotions.

In general, in a state of very high emotional arousal, the appropriate emotion may be associated with perceived objects (in the form of a sub-emotion or a sub-equale). Thus, on the one hand, the agent can shape its systemic emotional state (quale, including emotion and equale) resulting from the appropriate qualitative characteristics ascribed to the perceived objects. On the other hand, such quale can be appropriately assigned to variant models of perception (as sub-emotion or sub-equale). Note that this solution is completely in line with the concept of emotions as a kind of qualia.

In general, sub-equalia can also be triggered by an internal subsystem that detects that something unexpected is happening. The agent constantly monitors the perceived environment, both in terms of the behavior of objects and the appearance of new objects. An unpleasant sound (named or not) associated with a glass shattered or a feeling of wonder caused by a pleasant sight somewhere in a dumpster are examples of such unexpected phenomena that can generate a new sub-equale.

Like the sub-emotions on the emotion rainbow, the sub-equalia are located on a similar circle of equalia. A single sub-equale is characterized by the intensity (radius) correlated with the weight (importance, strength or probability of occurrence) of a given phenomenon and its sharp emotional color (angle). Certainly, these features must first be associated with the appropriate phenomenon (by means of clustering). Each agent with a different *baggage* of experience will have a different set of sub-equalia which can be considered purely subjective and qualitative sensations.

The xEmotion system also provides further processing of equalia in the emotional context, giving them an emotional interpretation under certain frequency conditions and modifying models of the objects currently perceived and in the agent's attention. In this way, in the next cycle of observation, the object can influence the agent's systemic emotional state, which includes common emotions (characteristic of the nation) and equalia (private emotions). In addition, the appropriate dynamic assessment of the systemic emotion is the basis for changing the agent's mood.

When the systemic equale exceeds a certain intensity threshold (along its radius on the equalia circle), an appropriate sub-emotion is attached to the emotion wheel, which can then be reflected in the sub-emotion of the models of currently perceived objects. In this way, we implement the mechanism of emotional (qualitative) transition from sub-equalia to sub-emotions, by means of which we can strengthen sub-emotions or add new sub-emotions to the models of currently perceived objects.

Accordingly, the equale can be treated as an unconscious impression related to some feature of the perceived object. From a global point of view, this kind of impression usually concerns new (learning) phenomena or situations. As a result, over time or when properly induced, such quale evolves into a specific, already conscious feeling/emotion. A flowchart with a list of triggers that support the creation of emotional traits (sub-emotions and sub-equalia) associated with objects is shown in Fig. 2.

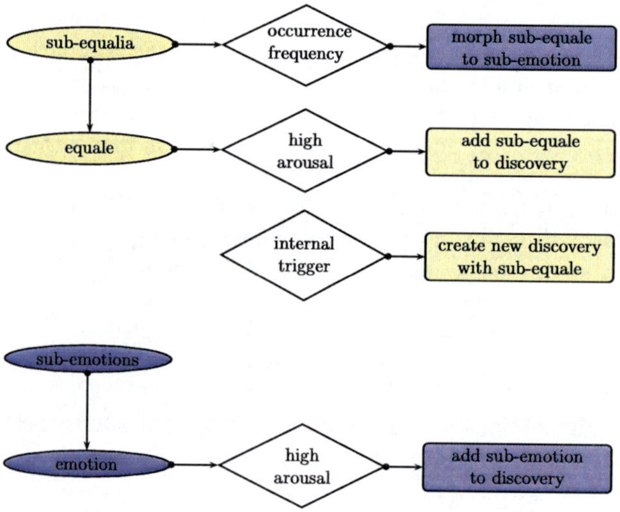

Fig. 2. Emotional components and the appropriate diagram for their creation with the help of triggers.

3 Illustrative Simulation

Due to the great difficulty in obtaining a full data set and carrying out complete verification or emulation, at this stage of development, we present below a simplified simulation study showing the validity of the presented concept of equalia. As a consequence, we will only use specific simulation scenarios to illustrate the entire mechanism of the emergence of sub-emotions and the impact of the perpetrator's perception on his/her emotions. It is equally important that the simulation tests will be carried out in a simple virtual environment designed to simulate human behavior with the help of the ISD system.

So let us now consider the following three scenarios:

1st The agent walks down the city street and notices several objects with an emotional context (i.e. with an assigned sub-emotion). At some point a new emotionally neutral object appears – in our case it is a *song*.

2nd The agent is walking through the forest and hears this song at the same time. When the conditions for creating sub-qualia and sub-emotions are met, the song (as an observation/perception) is expected to receive the corresponding sub-emotion.

3rd The agent walks around the city and encounters similar objects as in the first scenario, but this time the song model already has a sub-emotion and thus affects the emotional state of the agent.

3.1 The Influence of Sub-emotions on the Emotional State of the Agent (1st and 3rd Scenario)

To show a clear difference in the evolution of the agent's emotional state, we first present the results of the first and third scenarios in Fig. 3. Of course, in the first experiment, the emotionally neutral object of the song has no bearing on the computation process of emotions. In turn, in the third scenario, the song object (through its fixed sub-emotion) is already associated with *joy*. Note that the sub-emotion of joy must have arisen earlier (in our case, during the pre-emptive second scenario). As expected, in the third scenario, the subject (agent) reaches a more intense emotional state (experience) that is triggered by the sub-emotion of *joy* (Fig. 3) flowing from the favorite melody. In general, the equalia mechanism thus allows modeling the influence of lived events absorbed as a sub-emotion experience by the subject's emotional system.

3.2 Sub-emotion Creation (2nd Scenario)

A new sub-emotion arises when the intensity of the evolving systemic equale reaches or exceeds a certain threshold τ_{qe}. In the conducted (2nd) experiment, τ_{qe} was set to 70, and the sub-equale threshold τ_q was set to 55. The position (angle and radius) of the systemic equale on the equalia wheel is calculated from the sub-equalia assigned to the objects in the perception scene and in the agent's attention [18]. Moreover, the occurrence of the so-called *emotional arousal* systematically raises the systemic equale whenever the intensity of the classic emotion is greater than a certain threshold τ_e (set up to 40). As a result, the radius ($\overline{r}_{\kappa^i}$) of the systemic equale is increased by an *emotional raise* ($\delta r = 4$) in each calculation step. In the qualia mechanism under discussion, we distinguish the following two regimes as part of the phenomenon of emotional arousal:

- sub-equale creation, when $\tau_q \leq \overline{r}_{\kappa^i} < \tau_{qe}$ (equale being in an active range).
- sub-emotion creation, when $\tau_{qe} \leq \overline{r}_{\kappa^i}$ (equale being in a triggered state).

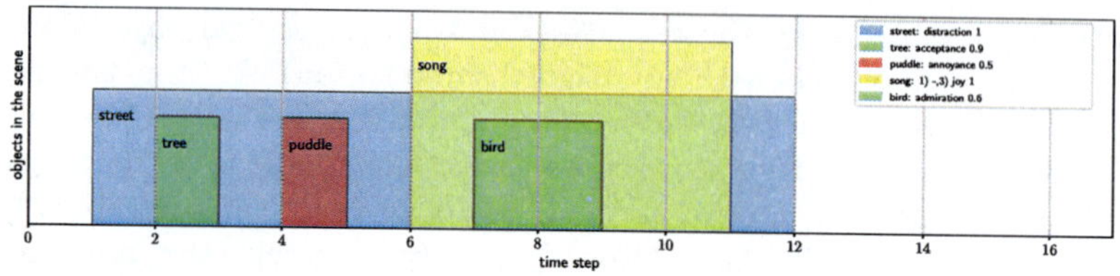

(a) Objects appearing on the scene of perception in the scenarios under consideration along with their sub-emotions and membership values presented in the legend.

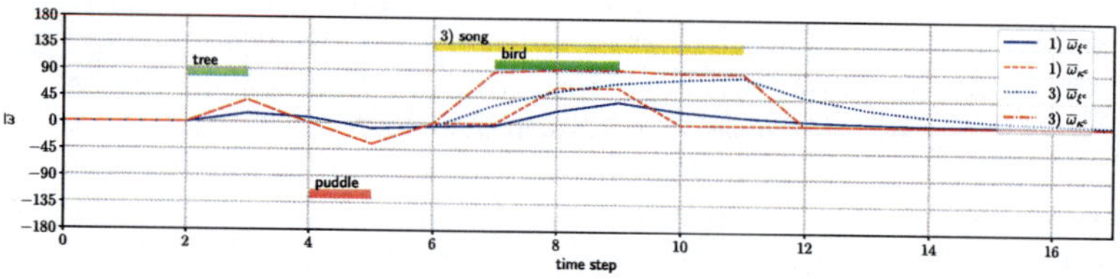

(b) Evolution of the angle of the sub-emotion center ($\overline{\omega}_{\kappa^c}$) and classical emotion ($\overline{\omega}_{\xi^c}$) during the first (1) and third (3) scenarios; the angle of sub-emotions of each object is marked with colored stripes.

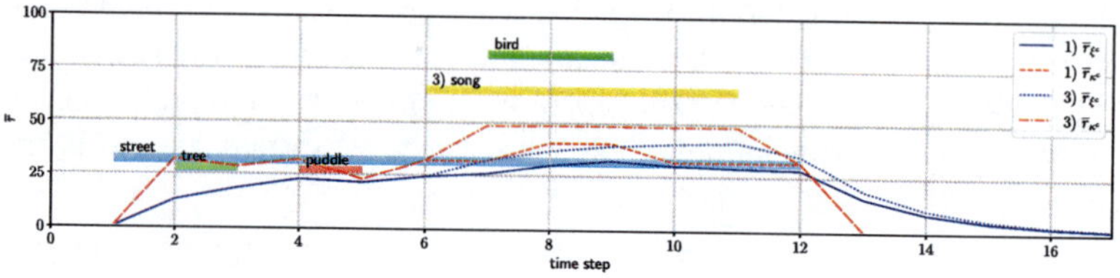

(c) Evolution of the radius of the sub-emotions center ($\overline{r}_{\kappa^c}$) and classical emotion ($\overline{r}_{\xi^c}$) during the first (1) and third (3) scenarios; the radius of sub-emotion of the objects are marked with colored stripes.

Fig. 3. Emotionally exciting objects and the evolution of classical emotions as a result of the perception process in the two scenarios (1st and 3rd).

The simulated results of the process of creating sub-emotions are presented in Fig. 4. The timeframe for the occurrence of each perceived object is presented in subfigure 4a. Note that so far (in this experiment) the object of the song neither attracts the agent's attention nor influences its classical emotions.

The current course of the levels of significance (attention) assigned to selected objects in the perception scene is presented in subfigure 4b. This indicator is normalized, i.e. at any moment the sum of the significance levels of all objects is equal to 1. The relative value of the significance level depends on several distinguishing features (whether the subject is new to this scene, is it moving, does it have bright color, etc.). The significance level is taken into account when calculating the intensity of the systemic equale. In practice, the radius of individual sub-equalia is weighted by the attention factor $\hat{\varrho}_l^{att} = 1 + \varrho_l^{sgfn}$, where ϱ_l^{sgfn} is l-th object's significance level.

Subfigure 4c shows the variability of the angles of the center of emotions, the classic emotion and the system equation. When in the agent's perception there is no object with a defined sub-equale, the angle/color of the system equale $\overline{\omega}_{\kappa^i}$ is zero. In this scenario, only the object of the setting sun (apart from its sub-emotion of ecstasy) contains the sub-equale of serenity (radius 38, angle 130°). Subfigure 4d presents the intensity of the sub-emotional center, the classical emotion, and the systemic equale. In this experiment, before the tenth time step, both activity conditions ($\tau_q \leq \overline{r}_{\kappa^i} < \tau_{qe}$ and $\tau_e \leq \overline{r}_{\xi^c}$) were not met, so no sub-equale was created ($\overline{r}_{\xi^c}$ is the radius/intensity of classical emotion and τ_e is its threshold of excitation). It was only after the 15th time step that such a creation became possible. In step 16, the above conditions were met (because $\overline{r}_{\xi^c} > \tau_e$ and $\overline{r}_{\kappa^i} > \tau_{qe}$), therefore (via sub-equale) an appropriate sub-emotion has been created for the three perceptions in the agent's current attention: the current systemic quale has been converted to (new) sub-emotion and added to the song-object model (with average intensity/radius $r_{\kappa_0^c} = 65$), while the *sunset* and *rainbow* sub-emotions have only been slightly modified by semi-translation operation [19] (see Table 1). The models of the remaining observations (*rain* and *forest*) have not changed because this mechanism of emotions applies only to objects that have appeared recently (in the receding 10-step horizon).

Table 1. The sub-emotion result of the 2nd scenario.

Object	Sub-emotion					
	Before			After		
	Radius	Angle	Linguistic value	Radius	Angle	Linguistic value
Rainbow	90	90	Admiration	89	91.6	Admiration
Sunset	95	135	Ecstasy	93.8	134.8	Ecstasy
Song	–	–	–	65	130	Joy

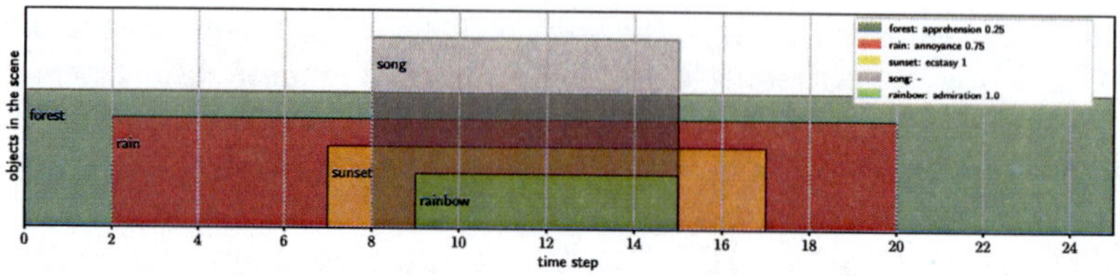

(a) Perceived objects with their sub-emotions and membership values presented in the legend.

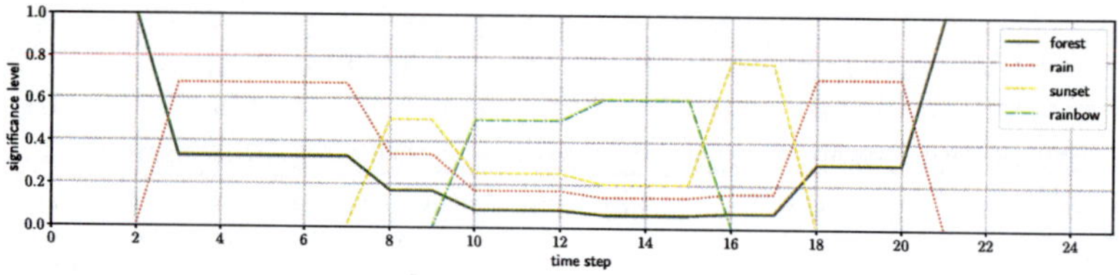

(b) The significance level assigned by the agent's attention to the objects present in the scene.

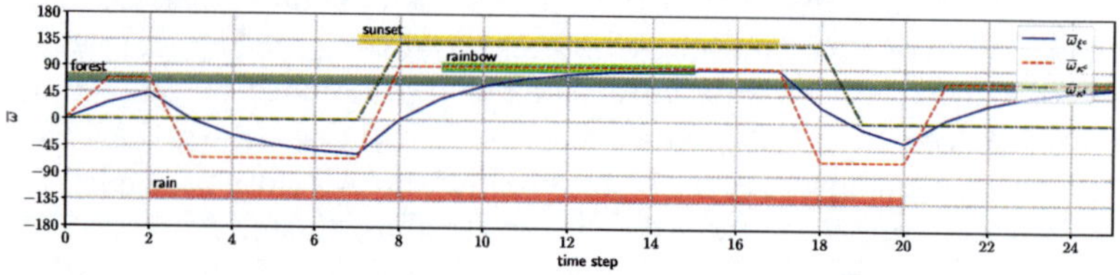

(c) Evolution of the angle of the sub-emotions center ($\overline{\omega}_{\kappa^c}$), classic emotion ($\overline{\omega}_{\xi^c}$), and systemic equale ($\overline{\omega}_{\kappa^i}$) in time, along with colored stripes of the sub-emotion angles of individual objects.

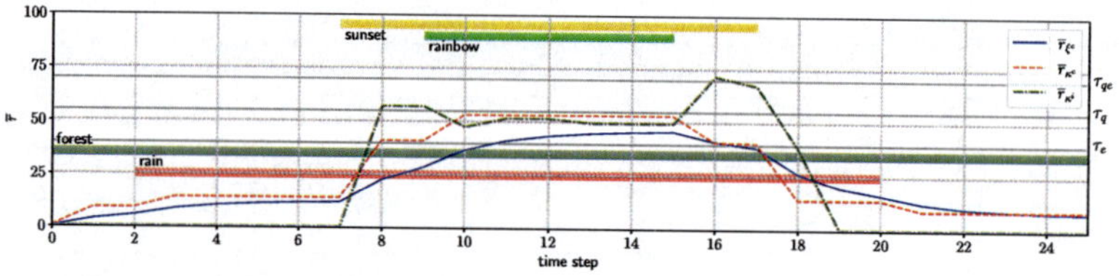

(d) Evolution of the radius of the sub-emotions center ($\overline{r}_{\kappa^c}$), classic emotion ($\overline{r}_{\xi^c}$), and systemic equale ($\overline{r}_{\kappa^i}$) in time, along with colored stripes of objects' sub-emotions and the system thresholds for the emotional raise (τ_e), sub-quale creation (τ_q) and sub-emotion creation (τ_{qe}).

Fig. 4. Perceived objects and the resulting evolution of the classical and private emotions in the second (2nd) scenario, leading to the emergence of sub-emotions.

4 Summary

The presented simulations prove the rational operation of xEmotion mechanisms in the context of the emotional qualia of the objects in the agent's attention. The scenarios used were selected in such a way as to clearly show the influence of sub-emotions on the agent's classical emotion and to illustrate the mechanisms of formation of sub-emotions and the way in which the level of significance (attention) influences the perception of objects within the agent's reach (at the observation stage).

In practice, the presented experiments have shown that the agent can achieve different sensations that are triggered by sub-emotions derived from perceived observations. The mechanism of equalia is also shown, which models the primary impact of experienced events, which are then absorbed by the emotional system of the agent in the form of an appropriately formalized (sub-)emotional experience.

On the other hand, the proposed emotional system fulfills the assumptions of many universally recognized theories of emotions and the philosophical aspects of qualia. In our approach, we treat emotions (called culturally-general and personal-private; or classical and individual) as causative states of the agent defined by emotional qualia related to perceived phenomena (events or objects) called sub-emotions and sub-equalia. The developed xEmotion system (used within the ISD decision-making system) rationally creates a new emotional experience (shaped by a set of impressions and observations) based on the prior emotional (population and personal) experiences of the agent. Importantly, the personal emotional states (abstract features of object models) experienced by the subject take on the appropriate transitional form as impressions or qualia.

References

1. Bann, E.Y., Bryson, J.J.: The conceptualisation of emotion qualia: semantic clustering of emotional tweets. In: Computational Models of Cognitive Processes: Proceedings of the 13th Neural Computation and Psychology Workshop, pp. 249–263. World Scientific (2014)
2. Bermond, B.: The emotional feeling as a combination of two qualia: a neurophilosophical-based emotion theory. Cogn. Emot. **22**(5), 897–930 (2008)
3. Czubenko, M., Kowalczuk, Z.: Elementy psychologii w kontekście autonomii robotów (in English: Elements of Psychology in the Context of Robots' Autonomy), vol. PNT 6. PWNT – Pomorskie Wydawnictwo Naukowo-Techniczne, Gdańsk, Poland (2019)
4. Czubenko, M., Kowalczuk, Z., Ordys, A.: Autonomous driver based on an intelligent system of decision-making. Cogn. Comput. **7**(5), 569–581 (2015). https://doi.org/10.1007/s12559-015-9320-5
5. Damasio, A.: Descartes' Error: Emotion, Reason, and the Human Brain. Gosset/Putnam, New York (1994)
6. Dennett, D.C.: Quining qualia. In: Philosophy of Mind: Classical and Contemporary Readings, pp. 226–246 (2002)

7. Edelman, G.M., Tononi, G.: A Universe of Consciousness: How Matter Becomes Imagination. Basic Books, New York (2008)
8. Farr, M.: Explaining temporal qualia. Eur. J. Philos. Sci. **10**(1), 1–24 (2020). https://doi.org/10.1007/s13194-019-0264-6
9. Frijda, N.: Emotion experience. Cogn. Emot. **19**(4), 473–497 (2005)
10. Izard, C.E.: Emotion theory and research: highlights, unanswered questions, and emerging issues. Annu. Rev. Psychol. **60**, 1–25 (2009)
11. Jackson, F.: What Mary didn't know. J. Philos. **83**(5), 291 (1986)
12. James, W.: What is an emotion? Mind **9**, 188–205 (1884)
13. Kleiner, J.: Mathematical models of consciousness. Entropy **22**(6), 609 (2020)
14. Kowalczuk, Z., Czubenko, M.: Emotions embodied in the SVC of an autonomous driver system. IFAC PapersOnLine **50**(1), 3744–3749 (2017)
15. Kowalczuk, Z., Czubenko, M.: Intelligent decision-making system for autonomous robots. Int. J. Appl. Math. Comput. Sci. **21**(4), 621–635 (2011)
16. Kowalczuk, Z., Czubenko, M.: Computational approaches to modeling artificial emotion-an overview of the proposed solutions. Front. Robot. AI **3**(21), 1–12 (2016)
17. Kowalczuk, Z., Czubenko, M.: An intelligent decision-making system for autonomous units based on the mind model. In: 23rd International Conference on Methods & Models in Automation & Robotics (MMAR), Międzyzdroje, Poland, pp. 1–6. IEEE (2018)
18. Kowalczuk, Z., Czubenko, M.: Inteligentny system decyzyjny jako maszynowa realizacja procesów poznawczych i motywacyjnych (in English: Intelligent System of Decision Making as a Machine Realization of Cognitive and Motivational Processes), vol. PNT 7. PWNT – Pomorskie Wydawnictwo Naukowo-Techniczne, Gdańsk, Poland (2021)
19. Kowalczuk, Z., Czubenko, M., Merta, T.: Interpretation and modeling of emotions in the management of autonomous robots using a control paradigm based on a scheduling variable. Eng. Appl. Artif. Intell. **91**, 103562 (2020)
20. Kron, A., Goldstein, A., Lee, D.H.J., Gardhouse, K., Anderson, A.K.: How are you feeling? Revisiting the quantification of emotional qualia. Psychol. Sci. **24**(8), 1503–1511 (2013)
21. Lewis, C.: Mind and the World-Order: Outline of a Theory of Knowledge. Charles Scribner's Sons, New York (1929)
22. Plutchik, R.: A general psychoevolutionary theory of emotion. In: Plutchik, R., Kellerman, H. (eds.) Emotion: Theory, Research, and Experience, vol. 1, pp. 3–33. Academic, New York (1980)
23. Plutchik, R.: The nature of emotions. Am. Sci. **89**, 344 (2001)
24. Robbins, S.: Form, qualia and time: the hard problem reformed. Mind and Matter **11**(2), 153–181 (2013)
25. Rodger, J.A.: QuantumIS: a qualia consciousness awareness and information theory quale approach to reducing strategic decision-making entropy. Entropy **21**(2), 125 (2019)
26. Russell, J.A.: Core affect and the psychological construction of emotion. Psychol. Rev. **110**(1), 145–172 (2003)

Resistant to Correlated Noise and Outliers Discrete Identification of Continuous Non-linear Non-stationary Dynamic Objects

Janusz Kozłowski(✉) and Zdzisław Kowalczuk

Gdansk University of Technology, Gdańsk, Poland
jk23@eti.pg.gda.pl

Abstract. In the article, specific methods of parameter estimation were used to identify the coefficients of continuous models represented by linear and non-linear ordinary differential equations. The necessary discrete-time approximation of the base models is achieved by appropriately tuned linear FIR "integrating filters". The resulting discrete descriptions, which retain the original continuous parameterization, can then be identified using the classical least squares procedure. Since in the presence of correlated noise, the obtained parameter estimates are biased by an unavoidable asymptotic systematic error (bias), the instrumental variable method is used here to significantly improve the consistency of estimates. The finally applied algorithm based on the criterion of the lowest sum of absolute values is used to identify linear and non-linear models in the presence of sporadic measurement errors. In conclusion, the effectiveness of the proposed solutions is demonstrated using numerical simulations.

Keywords: Non-linear continuous-time models · System identification · Least squares · Least absolute values · Instrumental variable

1 Introduction

In many supervisory and security systems dedicated identification methods (or change detection algorithms) are utilized to obtain relevant diagnostic information concerning the evolution of a monitored process. The dynamics of such a process can conveniently be modeled using simple input-output descriptions or more complex state-space representations. On the one hand, the use of easy-to-use discrete-time representations of the observed processes in the form of difference equations clearly simplifies the computer implementation of appropriate identification systems. On the other hand, these representations are not physically motivated, and the values of their parameters (quantities without units) strongly depend on the sampling frequency used [3, 9].

In the case of continuous-time descriptions, in turn, modeling coefficients can be assigned specific physical units. Moreover, the original parameterization of the model is not affected by the underlying data processing conditions (e.g. selection of the sampling frequency) and the initial verification of the estimation results is found very easy [4, 10] based on intuitive values of interpretable physical parameters.

Z. Kowalczuk (Ed.): DPS 2022, LNNS 545, pp. 317–327, 2023.
https://doi.org/10.1007/978-3-031-16159-9_26

In practical industrial automation systems, the processing of measurement data is usually influenced by noise and different system disturbances. The appearance of such parasitic phenomena often results in a significant deterioration of the quality of industrial processes. In general, high frequency additive noise (e.g., due to quantization by AD converters) can be effectively eliminated by appropriately tuned low pass anti-aliasing pre-filters. In turn, in the case of systematic errors, proper calibration of measuring instruments and the use of dedicated compensation techniques usually solves the problem. Importantly, the outcomes of diagnostic procedures that rely on classical least-squares procedures can turn out to be highly skewed when measurements are contaminated with sporadic errors called outliers. A radical improvement in the estimation quality can be readily obtained by using in practice the identification method resistant to occasional errors in the sense of the least sum of absolute values.

We discuss this issue in the subsequent sections as follows. Discrete-time mechanizations of linear and non-linear differential equation models are presented in Sect. 2. Different parameter estimation procedures are considered in Sect. 3 along with discussions on their asymptotic properties and robustness to outliers in data. As a consequence, Sect. 4 comments on the results of several numerical tests that demonstrate the practical possibilities of the estimation procedures described. Finally, the summary and perspective directions for further research in the field of resistant identification are outlined in Sect. 5.

2 Continuous-Time Modeling

Let the dynamics of the supervised system (the object of parameter estimation) be expressed by an ordinary differential equation

$$y^{(n)} + a_{n-1}y^{(n-1)} + \ldots + a_0 y = b_0 u \tag{1}$$

where n denotes the system order, while $u = u(t)$ and $y = y(t)$ stand for the input and output signals, respectively. The description (1) can be subject to any initial conditions, while the parameterization coefficients $\{a_{n-1}, \ldots, a_0, b_0\}$ are assumed to be unknown. Since the practical identification of continuous-time models must rely on the processing of sampled data, the appropriate discrete-time representation of (1) must be established.

2.1 Discrete-Time Approximation of Differential Equations

There are many practical approaches to the numerical mechanization of differential equations, such as the solution using the *delta* operator for direct approximation of derivatives, the classic simulation method based on multiple integration of both sides of the differential equation, and the technique using properly tuned low-pass *matching* (or *state variable*) filters [9, 10].

Unfortunately, due to the high-pass nature of the delta operator, additive noise is amplified, which can substantially distort the evaluated continuous quantities (derivatives). In the case of pure integration, in turn, the regression data represented by multiple integrals of the input and output signals will certainly go to infinity, while the integrated system free response obviously affects the accuracy of the modeling and cannot simply be ignored. The above enumerated problems can be effectively eliminated by using low-pass matching filters used in the discrete-time approximation synthesis for the model (1).

Among other things, a finite-horizon integrating filter of the FIR type deserves attention [9]. The resulting operator, referred to as the linear integral filter (LIF), takes the form of a multiple integral of the signal (or its i-th derivative) in a finite time horizon $[t-\tau, t]$

$$J_i^n x(t) = \int_{t-\tau}^{t} \int_{t_1-\tau}^{t_1} \cdots \int_{t_{n-1}-\tau}^{t_{n-1}} x^{(i)}(t_n)\, dt_n \cdots dt_2\, dt_1 \tag{2}$$

Applying (2) to both sides of the Eq. (1), we obtain the following "integral" equation with the original parameterization of the system (1):

$$J_n^n y + a_{n-1} J_{n-1}^n y + \ldots + a_0 J_0^n y = b_0 J_0^n u \tag{3}$$

It is important for modeling consistency that with (2) the effect of the "integrated" free response of the system becomes irrelevant after a finite time ($n\tau$). The integrals used above can easily be calculated based on sampled data. With the aid of the bilinear (Tustin's) operator the discrete-time mechanization of (2) can be shown as

$$J_i^n x(t) \underset{t=kT}{\approx} I_i^n x(k) \tag{4}$$

$$I_i^n = (\tfrac{T}{2})^{n-i} (1+q^{-1})^{n-i} (1-q^{-1})^i (1+q^{-1}+\ldots+q^{-L+1})^n \tag{5}$$

where T stands for the sampling time, q^{-1} symbolizes the unit delay operator, L is the length of the integration horizon ($\tau = L\,T$), and for brevity, the sampling moment $t = kT$ is represented by the index k. As a result, the discrete-time counterpart of the original continuous-time model takes a convenient form of regression.

$$I_n^n y(k) = \chi(k) = \boldsymbol{\varphi}^{\mathrm{T}}(k)\,\boldsymbol{\theta} + e(k) \tag{6}$$

$$\boldsymbol{\varphi}(k) = [-I_{n-1}^n y \ \ \ldots \ \ -I_0^n y \ \ I_0^n u]^{\mathrm{T}} \tag{7}$$

$$\boldsymbol{\theta} = [a_{n-1} \ \ \ldots \ \ a_0 \ \ b_0]^{\mathrm{T}} \tag{8}$$

The above model with the equation or residual error $e(k)$ being a component of the stochastic process, includes both system disturbances and modeling inaccuracies. It is worth noting that the model (6)–(8) retains original continuous parameterization, while the "integral" regressors (7) are numerically well conditioned (bounded). According to the rule proposed in [9], the horizon L (or $\tau = L\,T$) should be selected so that the frequency bandwidth of the filter (2) matches as closely as possible the frequency band of the identified system (1). The LIF operator with too narrow a frequency band (large τ and L horizons) simply falsifies the system dynamics (3). On the other hand, when the bandwidth is too large (short τ and L), the broadband noise strongly affects the accuracy of the estimation (due to the asymptotic bias).

2.2 Non-linear Continuous-Time Models

Consider now a continuous differential equation model

$$y^{(n)} + a_{n-1}y^{(n-1)} + \ldots + a_1 y^{(1)} + f(y) = b_0 u \tag{9}$$

with a non-linear term $f(y)$ representing an injection function, such that $f(0) = 0$. Therefore a common series can be used for a practical approximation of this term

$$f(y) = c_1 y + c_2 y^2 + \ldots + c_r y^r \tag{10}$$

Now, using discrete-time FIR integrating filters (5), the counterpart regression representation of the non-linear system (9) can be represented as

$$I_n^n y(k) = \chi(k) = \boldsymbol{\varphi}^{\mathrm{T}}(k)\,\boldsymbol{\theta} + e(k) \tag{11}$$

$$\boldsymbol{\varphi}(k) = [-I_{n-1}^n y \ \ \ldots \ \ -I_1^n y \ \ -I_0^n y \ \ \ldots \ \ -I_0^n y^r \ \ I_0^n u]^{\mathrm{T}} \tag{12}$$

$$\boldsymbol{\theta} = [a_{n-1} \ \ \ldots \ \ a_1 \ \ c_1 \ \ \ldots \ \ c_r \ \ b_0]^{\mathrm{T}} \tag{13}$$

After obtaining the regression representation of non-linear continuous-time systems, identification procedures can be used that will allow us to obtain appropriate estimates of unknown but physically motivated coefficients.

3 Estimation Procedures

As shown, the dynamics of a supervised plant system can be expressed using the regression model (6)–(8) or (11)–(13). Now the underlying parameters can be efficiently estimated with the help of appropriate identification procedures. In the following, three identification methods are considered: the least squares algorithm, the instrumental variable method, and the least absolute values procedure.

3.1 Least-Squares Method

The classical weighted least squares (LS) estimation scheme results from the minimization of the following quadratic index [8]

$$V_{\mathrm{LS}}(\boldsymbol{\theta}) = \sum_{\ell=1}^{k} \lambda^{k-\ell} e^2(\ell) = \sum_{\ell=1}^{k} \lambda^{k-\ell} [\chi(\ell) - \boldsymbol{\varphi}^{\mathrm{T}}(\ell)\,\boldsymbol{\theta}]^2 \tag{14}$$

where the weighing factor λ limited to unity ($0 << \lambda \leq 1$) controls the rate of exponential forgetting in the case of tracking the variable parameters of a non-stationary system.

The above LS criterion can analytically be minimized by nulling its gradient

$$\nabla_{\boldsymbol{\theta}} V_{\mathrm{LS}}(\boldsymbol{\theta}) = (-2) \sum_{\ell=1}^{k} \lambda^{k-\ell} \boldsymbol{\varphi}(\ell)\,[\chi(\ell) - \boldsymbol{\varphi}^{\mathrm{T}}(\ell)\,\boldsymbol{\theta}] = \mathbf{0} \tag{15}$$

As a consequence, the weighted LS estimator takes the following algebraic form

$$\hat{\boldsymbol{\theta}}(k) = \left[\sum_{\ell=1}^{k} \lambda^{k-\ell}\, \boldsymbol{\varphi}(\ell)\, \boldsymbol{\varphi}^{\mathrm{T}}(\ell)\right]^{-1} \left[\sum_{\ell=1}^{k} \lambda^{k-\ell}\, \boldsymbol{\varphi}(\ell)\, \chi(\ell)\right] \tag{16}$$

It is easy to verify that the principal LS procedure using $\lambda = 1$ generates consistent estimates, when the regression data $\boldsymbol{\varphi}(k)$ and the residual error $e(k)$ are uncorrelated: $\mathrm{E}\{\boldsymbol{\varphi}(k)\, e(k)\} = \mathbf{0}$. By definition, this condition is satisfied when $e(k)$ takes form of zero-mean white noise (a sequence of independent random variables).

Most often, however, the residual error turns out to be a correlated process, which makes the LS estimates asymptotically biased. The consistency of the estimation can be substantially improved by using the technique of instrumental variables (IV).

3.2 Instrumental Variable Method

The idea behind the IV method stems from the analysis of the rearrangement (15) that leads to minimization of the LS index (14). Since consistency problems are attributed to the non-zero correlation ($\mathrm{E}\{\boldsymbol{\varphi}(k)\, e(k)\} \neq \mathbf{0}$), one can replace the original regressors $\boldsymbol{\varphi}(k)$ in (15) with a suitably selected instrument $\boldsymbol{\xi}(k)$, provided $\boldsymbol{\xi}(k)$ and $e(k)$ are uncorrelated: $\mathrm{E}\{\boldsymbol{\xi}(k)\, e(k)\} = \mathbf{0}$. The algebraic IV algorithm can be written down as

$$\hat{\boldsymbol{\theta}}(k) = \left[\sum_{\ell=1}^{k} \lambda^{k-\ell}\, \boldsymbol{\xi}(\ell)\, \boldsymbol{\varphi}^{\mathrm{T}}(\ell)\right]^{-1} \left[\sum_{\ell=1}^{k} \lambda^{k-\ell}\, \boldsymbol{\xi}(\ell)\, \chi(\ell)\right] \tag{17}$$

In order to overcome the problem of matrix inversion present in (16) and (17), the matrix inversion lemma [8] can be applied. The resulting recursive form of both algorithms, including the estimation of the prediction error, the update of the covariance matrix and the final correction of the estimation vector, can be shown as

$$\varepsilon(k) = \chi(k) - \boldsymbol{\varphi}^{\mathrm{T}}(k)\, \hat{\boldsymbol{\theta}}(k-1) \tag{18}$$

$$\mathbf{P}(k) = \frac{1}{\lambda}\left[\mathbf{P}(k-1) - \frac{\mathbf{P}(k-1)\, \boldsymbol{\psi}(k)\, \boldsymbol{\varphi}^{\mathrm{T}}(k)\, \mathbf{P}(k-1)}{\lambda + \boldsymbol{\varphi}^{\mathrm{T}}(k)\, \mathbf{P}(k-1)\, \boldsymbol{\psi}(k)}\right] \tag{19}$$

$$\hat{\boldsymbol{\theta}}(k) = \hat{\boldsymbol{\theta}}(k-1) + \mathbf{P}(k)\, \boldsymbol{\psi}(k)\, \varepsilon(k) \tag{20}$$

with $\boldsymbol{\psi}(k) = \boldsymbol{\varphi}(k)$ and $\boldsymbol{\psi}(k) = \boldsymbol{\xi}(k)$ for the LS and IV estimation routines, respectively. Typically the initial diagonal covariance matrix is set as $\mathbf{P}(0) = \mathrm{diag}\,[10^5 \ldots 10^5]$ at the start-up of these estimators.

According to its purpose, the instrumental variable $\boldsymbol{\xi}(k)$ should match the original regressors $\boldsymbol{\varphi}(k)$ as much as possible, while keeping $\boldsymbol{\xi}(k)$ and $e(k)$ uncorrelated: $\mathrm{E}\{\boldsymbol{\xi}(k)\, e(k)\} = \mathbf{0}$. Among the various implementations of such variables, it is preferable [8, 11] to use a method based on simple deterministic filtering. In this approach, the desired noise-free output of the linear system (1) can be obtained from

$$\hat{y}(k) = \frac{\hat{b}_0}{\rho^n + \hat{a}_{n-1}\rho^{n-1} + \ldots + \hat{a}_1\rho + \hat{a}_0} u(k) \tag{21}$$

with the differentiation operator $\rho = {}^{d}/_{dt}$ implemented, for instance, using the well-known Tustin formula: $\rho = [2\,(1 - q^{-1})]/[T\,(1 + q^{-1})]$. Note that the proposed bilinear transformation maps the left-half s-plane onto the unit disc in the z-plane, so the resulting discrete-time filter (21) remains stable.

It should be realized that in the case of the identification of a non-linear model (9), the processing based on the filter (21) used to generate the instrumental variable cannot be directly applied. Therefore, we propose an innovative method IV consisting in using the well-known Taylor series scheme for the numerical approximation of non-linear systems (1). This method can also be used for the numerical evaluation of the noise-free output of a dynamical system as follows:

$$\hat{y}^{(n)}(k-1) = -\hat{a}_{n-1}\hat{y}^{(n-1)}(k-1) + \ldots - \hat{a}_1\hat{y}^{(1)}(k-1) + \hat{c}_1\hat{y}(k-1) + \ldots + \hat{c}_r\hat{y}^r(k-1) + \hat{b}_0 u(k-1)$$

$$\hat{y}^{(n-1)}(k) = \hat{y}^{(n-1)}(k-1) + T\,\hat{y}^{(n)}(k-1)$$

$$\hat{y}^{(n-2)}(k) = \hat{y}^{(n-2)}(k-1) + T\,\hat{y}^{(n-1)}(k-1) + \frac{T^2}{2!}\hat{y}^{(n)}(k-1) \tag{22}$$

$$\hat{y}(k) = \hat{y}(k-1) + T\,\hat{y}^{(1)}(k-1) + \ldots + \frac{T^n}{n!}\hat{y}^{(n)}(k-1)$$

where $\hat{a}_i$, $\hat{c}_i$ and $\hat{b}_0$ are the estimates of the parameters of the model (11)–(13).

Now, the instrumental variable $\boldsymbol{\xi}(k)$ can be determined in a similar fashion as $\boldsymbol{\varphi}(k)$, provided the noise-free output, (21) or (22), replaces the signal $y(t)$ contaminated with noise. This gives

$$\boldsymbol{\xi}(k) = [-I^n_{n-1}\hat{y} \ \ \ldots \ \ -I^n_0\hat{y} \ \ I^n_0 u]^{\mathrm{T}} \tag{23}$$

$$\boldsymbol{\xi}(k) = [-I^n_{n-1}\hat{y} \ \ \ldots \ \ -I^n_1\hat{y} \ \ -I^n_0\hat{y} \ \ \ldots \ \ -I^n_0\hat{y}^r \ \ I^n_0 u]^{\mathrm{T}} \tag{24}$$

for linear (9) and non-linear (22) systems, respectively.

The described IV procedure effectively solves the problem of consistent identification in the presence of correlated noise. Unfortunately, both the LS and IV schemes resulting from minimizing the quadratic criteria are completely ineffective for a reliable estimation of parameters in the presence of large measurement errors. Therefore, to overcome the problem of sporadic outliers, consider an algorithm that penalizes (weighted) errors less efficiently – based on their absolute values rather than squared values.

3.3 Least Absolute Values Method

The deliberated weighted least absolute values (LA) algorithm follows from minimization of the following (non-differentiable) loss index [2]

$$V_{\mathrm{LA}}(\boldsymbol{\theta}) = \sum_{\ell=1}^{k} \lambda^{k-\ell}\,|e(\ell)| = \sum_{\ell=1}^{k} \lambda^{k-\ell}\,|\chi(\ell) - \boldsymbol{\varphi}^{\mathrm{T}}(\ell)\,\boldsymbol{\theta}| \tag{25}$$

In such a situation, the estimation process becomes resistant to large, sporadic outliers that may appear in the processed measurement data. As the index (25) cannot analytically be minimized, we suggest an approximate solution. Assuming that the estimates of residual errors $e(k)$ are available (e.g. from a running-in-parallel LS algorithm), the criterion (25) can be rearranged to the following form:

$$V_{\mathrm{LA}}(\boldsymbol{\theta}) \approx \sum_{\ell=1}^{k} \lambda^{k-\ell} \frac{e^2(\ell)}{|\hat{e}(\ell)|} = \sum_{\ell=1}^{k} \lambda^{k-\ell} \frac{[\chi(\ell) - \boldsymbol{\varphi}^{\mathrm{T}}(\ell)\,\boldsymbol{\theta}]^2}{|\hat{e}(\ell)|} \tag{26}$$

Now, with known estimates of the residual, nulling the gradient of V_{LA} leads to the following estimator:

$$\hat{\boldsymbol{\theta}}(k) = \left[\sum_{\ell=1}^{k} \lambda^{k-\ell} \frac{\boldsymbol{\varphi}(\ell)\,\boldsymbol{\varphi}^{\mathrm{T}}(\ell)}{|\hat{e}(\ell)|}\right]^{-1} \left[\sum_{\ell=1}^{k} \lambda^{k-\ell} \frac{\boldsymbol{\varphi}(\ell)\,\chi(\ell)}{|\hat{e}(\ell)|}\right] \tag{27}$$

This result can be further improved by utilizing the idea of successive approximations [5, 6]. Since the residual error $e(k)$ can be easily evaluated from the recent estimate of the parameter vector $\boldsymbol{\theta}$, the complete iterative identification scheme ($p = 0, 1, \ldots$) thus takes the form

$$\hat{e}^{/p/}(\ell) = \chi(\ell) - \boldsymbol{\varphi}^{\mathrm{T}}(k)\,\hat{\boldsymbol{\theta}}^{/p/} \tag{28}$$

$$\hat{\boldsymbol{\theta}}^{/p+1/} = \left[\sum_{\ell=1}^{k} \lambda^{k-\ell} \frac{\boldsymbol{\varphi}(\ell)\,\boldsymbol{\varphi}^{\mathrm{T}}(\ell)}{|\hat{e}^{/p/}(\ell)|}\right]^{-1} \left[\sum_{\ell=1}^{k} \lambda^{k-\ell} \frac{\boldsymbol{\varphi}(\ell)\,\chi(\ell)}{|\hat{e}^{/p/}(\ell)|}\right] \tag{29}$$

where the easiest way is to use an easily computable standard LS estimate to initiate an iteration loop at $p = 0$:

$$\hat{\boldsymbol{\theta}}^{/0/} = \hat{\boldsymbol{\theta}}_{\mathrm{LS}} \tag{30}$$

The presented iterative implementation (28)–(29) ends when the improvement in minimization drops below the assumed threshold ($\Delta_{\min}$)

$$|V_{\mathrm{LA}}(\hat{\boldsymbol{\theta}}^{/p/}) - V_{\mathrm{LA}}(\hat{\boldsymbol{\theta}}^{/p+1/})| < \Delta_{\min} \tag{31}$$

The above condition is mathematically justified, because the key importance here is the fact of the decreasing sequence of successively calculated values of the underlying LA criterion.

Assuming that the above processing is limited to only one iteration and using the already mentioned matrix inversion lemma [4, 5], the recursive version of the iterative LA scheme (including, as before, the estimation of the prediction error, the update of the covariance matrix and the ultimate correction of the estimation vector) can be roughly approximated as

$$\varepsilon(k) = \chi(k) - \boldsymbol{\varphi}^{\mathrm{T}}(k)\,\hat{\boldsymbol{\theta}}(k-1) \tag{32}$$

$$\mathbf{P}(k) = \frac{1}{\lambda}\left[\mathbf{P}(k-1) - \frac{\mathbf{P}(k-1)\,\boldsymbol{\psi}(k)\,\boldsymbol{\varphi}^{\mathrm{T}}(k)\,\mathbf{P}(k-1)}{\lambda|\varepsilon(k)| + \boldsymbol{\varphi}^{\mathrm{T}}(k)\,\mathbf{P}(k-1)\,\boldsymbol{\psi}(k)}\right] \tag{33}$$

$$\hat{\boldsymbol{\theta}}(k) = \hat{\boldsymbol{\theta}}(k-1) + \mathbf{P}(k)\,\boldsymbol{\psi}(k)\,\mathrm{sgn}\,[\,\varepsilon(k)\,] \tag{34}$$

with $\boldsymbol{\psi}(k) = \boldsymbol{\varphi}(k)$ and $\boldsymbol{\psi}(k) = \boldsymbol{\xi}(k)$ used for the LA-estimation routines in the straight and IV versions, respectively. We recommend starting the procedure (32)–(34) with the LS auxiliary results (18)–(20) for covariance matrices and parameter estimates.

The proposed initiation is indeed crude, but due to the weighing mechanism ($\lambda < 1$) contained in the LA estimation algorithm, such forced start-up data will be gradually eliminated from the finite memory of the estimator. Note that the effective number of observations, referred to as the memory length of the weighted estimator, equals $\Gamma = 1/(1{-}\lambda)$.

All three LS, IV and LA estimation algorithms discussed above are verified in the numerical simulations presented below.

4 Numerical Study

Consider the mechanical system shown in Fig. 1. Such a structure usually constitutes the essential part of the wheeled vehicle suspension system [7].

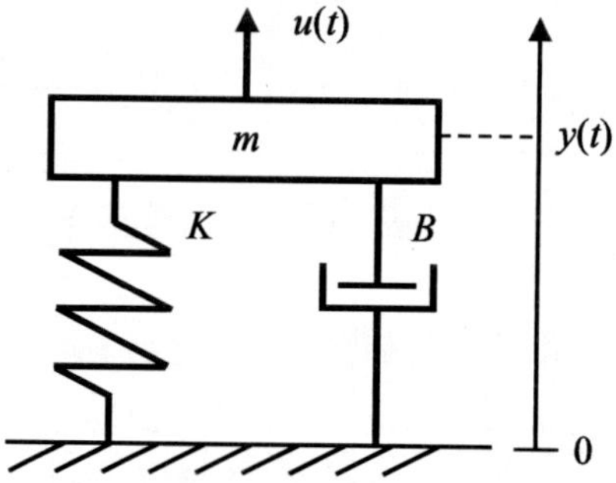

Fig. 1. Suspension dynamics as a non-linear mechanical system.

Given that $u(t)$ represents the external force, the reaction of the damper is proportional to the velocity (with the coefficient B), and the classical spring (K) follows the linear Hooke's rule, the resulting continuous-time model can be described as follows

$$m\ddot{y}(t) = -B\dot{y}(t) - Ky(t) + u(t) \tag{35}$$

Nowadays, manufacturers of automobiles and wheeled vehicles prefer to use the so-called progressive springs as the basis of modern suspension mechanisms, which can be represented as a non-linear model of the following form

$$m\ddot{y}(t) = -B\dot{y}(t) - Ky^3(t) + u(t) \tag{36}$$

Using the idea of integrating filters, the discrete-time counterpart of the non-linear model (36) can be portrayed as (9), where $a_1 = -B/m$, $c_3 = -K/m$, and $b_0 = 1/m$. Therefore, the appropriate regression model of the non-linear dynamics of the mechanical suspension can be presented as (6)–(8).

The instrumental variable procedure which makes our identification process resistant to correlated noise requires an appropriate instrument (24), where the noise-free output for the dynamical system (36) of the order $n = 2$, has the form of (22).

Next, a series of computer simulations of the identification procedures for the non-linear process were performed using the following persistently-exciting input

$$u(t) = \sin 0.7t + \sin 1.1t + \sin 1.4t + \sin 1.7t + \sin 2.1t \tag{37}$$

In the first numerical experiment the non-linear model was identified in presence of an additive correlated noise. The coefficients were assumed as $a_1 = 3.5$, $c_3 = 5$, and $b_0 = 4$, while the residual error $e(k)$ was implemented in the form of a zero-mean correlated process such that the eventual noise-to-signal ratio of the respective standard deviations was N/S = 10%. The discrete-time counterpart of the identified model was obtained using finite-horizon integrating filters with $L = 60$. The sampling time was set to $T = 0.02$ s. Simulation time of the exercised LS and IV recursive estimation schemes (18)–(20) was assumed limited to 200 s (10000 samples).

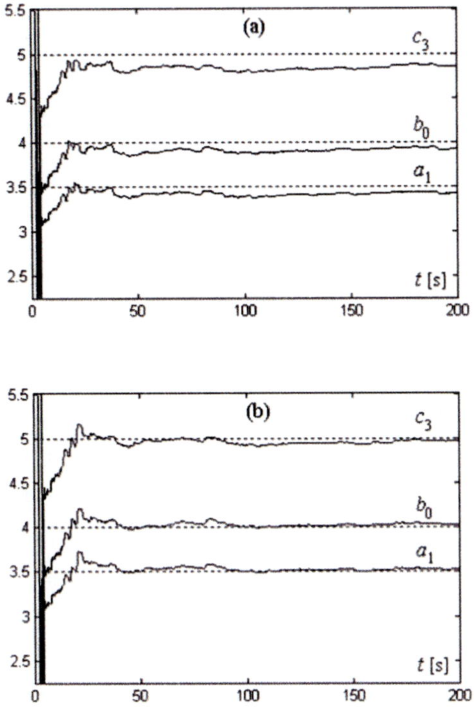

Fig. 2. Parameter estimation of the suspension system using the LS (a) and IV (b) method.

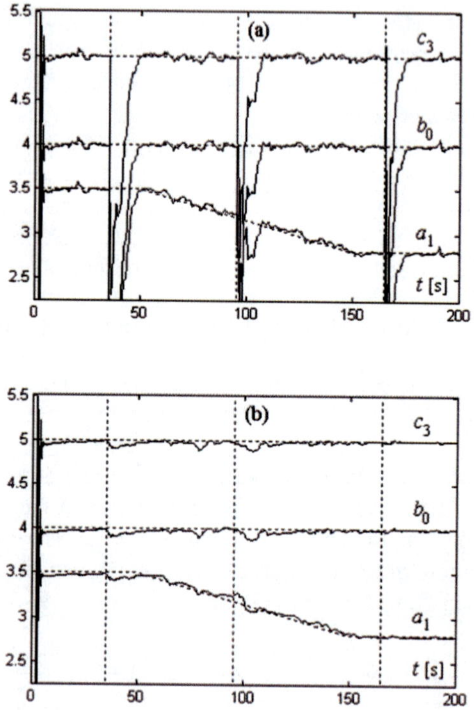

Fig. 3. Tracking variable parameters of the non-linear and non-stationary mechanical suspension system using the LS (a) and LA (b) method.

It is clear from Fig. 2 that the IV scheme using the innovative procedure (22) to generate instrumental variables radically improves the consistency of estimation, while the LS method shows an evident asymptotic bias in estimates.

The same model was then identified using the resistant-to-outliers LA procedure. This time, however, the simulated variable parameter a_1 was changing gradually (3.5 … 2.8) in a restricted time interval [50 s, 150 s], while the other parameters were kept constant ($c_3 = 5, b_0 = 4$). This time the residual error $e(k)$ was represented by a normally distributed zero-mean white-noise sequence with variance selected so that the already mentioned noise-to-signal ratio was N/S = 1%. Measurement faults were represented by missing data ($y(t) = 0$) introduced at $t \in [35$ s, 35.5 s], $t \in [95$ s, 95.5 s], and $t \in [165$ s, 165.5 s]. Again, the integration horizon was set at $L = 60$ and the sampling period was set to $T = 0.02$ s. Yet, the forgetting mechanism with $\lambda = 0.99$ was activated for tracking the time-variant parameter a_1. Simulation time of the recursive LS (18)–(20) and LA (32)–(34) schemes was limited to 200 s (10000 samples).

It is clear from Fig. 3 that the weighted LA procedure allows for reliable tracking of variable parameters regardless of occasional faults. On the other hand, the LS algorithm with pronounced sensitivity to such errors simply fails in this case.

5 Conclusion

The paper considers the problem of parametric estimation in order to obtain identification procedures resistant to correlated noise and occasional outliers in the case of non-stationary and non-linear continuous-time model analysis. The presented numerical simulations show that the applied scheme IV successfully overcomes the problem of correlated noise and therefore can be recommended as a good tool for suppressing the asymptotic bias (systematic error) of the obtained estimates. Moreover, an innovative method of generating instrumental variables adapted to non-linear dynamics was presented, which proved its effectiveness and practical usefulness in experiments. The non-trivial problem of parametric identification resistant to measurement errors is solved by the use of an algorithm that approximates the sum of the absolute values. The weighing mechanism, on the other hand, enables the tracking of the time-varying parameters of non-stationary systems. Since in the presented approach only the static nonlinear term $f(y)$ in the differential equation is considered, an interesting direction for further research seems to be the search for an effective, discrete-time approximation of more complex continuous-time models (i.e. with nonlinear functions corrupting the derivatives).

References

1. Craig, J.J.: Introduction to Robotics: Mechanics and Control. Pearson Education, New Jersey (2014)
2. Janiszowski, K.B.: To estimation in sense of the least sum of absolute errors. In: Proceedings of the 5th International Symposium on Methods and Models in Automation and Robotics, Międzyzdroje, Poland, vol. 2, pp. 583–588 (1998)
3. Kowalczuk, Z., Kozłowski, J.: Continuous-time approaches to identification of continuous-time systems. Automatica **36**(8), 1229–1236 (2000)
4. Kowalczuk, Z., Kozłowski, J.: Non-quadratic quality criteria in parameter estimation of continuous-time models. IET Control Theory Appl. **5**(13), 1494–1508 (2011)
5. Kozłowski, J., Kowalczuk, Z.: Identification of models and signals robust to occasional outliers. In: 12th International Conference on Diagnostics of Processes and Systems, Ustka, Poland (2015)
6. Kozłowski, J., Kowalczuk, Z.: Identification of continuous systems - practical issues of insensitivity to perturbations. In: Proceedings of the 13th International Conference on Diagnostics of Processes and Systems, Sandomierz, Poland (2017)
7. Kozłowski, J., Kowalczuk, Z.: Intelligent monitoring the vertical dynamics of wheeled inspection vehicles. IFAC-PapersOnLine **52**(8), 251–256 (2019)
8. Ljung, L.: System Identification Theory for the User. Prentice-Hall Inc., Englewood Cliffs (1987)
9. Sagara, S., Zhao, Z.Y.: Numerical integration approach to on-line identification of continuous-time systems. Automatica **26**(1), 63–74 (1990)
10. Sagara, S., Yang, Z.J., Wada, K.: Identification of continuous systems using digital low-pass filters. Int. J. Syst. Sci. **22**(7), 1159–1176 (1991)
11. Söderström, T., Stoica, P.: Comparison of some instrumental variable methods – consistency and accuracy aspects. Automatica **17**(1), 101–115 (1981)

Neural Modelling of Dynamic Systems with Time Delays Based on an Adjusted NEAT Algorithm

Krzysztof Laddach and Rafał Łangowski(✉)

Department of Intelligent Control and Decision Support Systems, Digital Technologies Center, Gdańsk University of Technology, G. Narutowicza 11/12, 80-233 Gdańsk, Poland
rafal.langowski@pg.edu.pl

Abstract. A problem related to the development of an algorithm designed to find an architecture of artificial neural network used for black-box modelling of dynamic systems with time delays has been addressed in this paper. The proposed algorithm is based on a well-known NeuroEvolution of Augmenting Topologies (NEAT) algorithm. The NEAT algorithm has been adjusted by allowing additional connections within an artificial neural network and developing original specialised evolutionary operators. This resulted in a compromise between the size of neural network and its accuracy in capturing the response of the mathematical model under which it has been learnt. The research involved an extended validation study based on data generated from a mathematical model of an exemplary system as well as the fast processes occurring in a pressurised water nuclear reactor. The obtaining simulation results demonstrate the high effectiveness of the devised neural (black-box) models of dynamic systems with time delays.

Keywords: Neural modelling · Neural network architecture search · PWR black-box model

1 Introduction

Nowadays, an effective handling of majority industrial plants requires advanced algorithms based on the proper mathematical models of processes that occur in them. These algorithms perform different tasks such as diagnostics, monitoring, estimation, control, etc. It is known that the model makes it possible, e.g., to predict the trajectories of selected process variables, also in unacceptable situations in a real plant, to analyse the behaviour of a given process, or to consider various control strategies, etc. However, the complexity of many processes is significant, and the phenomena occurring in them are sophisticated. Therefore, delivering the system white-box model might become very difficult and even impossible. Thus, developing an alternative model such as a black-box or grey-box may be justified and reasonable [16]. These models are built based on observation of

Z. Kowalczuk (Ed.): DPS 2022, LNNS 545, pp. 328–339, 2023.
https://doi.org/10.1007/978-3-031-16159-9_27

behaviour of a given system. There are different tools to create black-box models. Artificial neural networks (ANNs) are one of the most common and have produced excellent results in many fields of science [1]. The neural (black-box) model is based on ANN, which according to the theorems of Kolmogorov and Cybenko has the ability to represent any mathematical function [4,8]. However, to achieve it, an architecture of ANN (parameters and hyper-parameters) have to be selected for each unique function. This problem is known in the literature, where there are many ways to solve it, although it is still an open issue. One of the common approach to select ANN's architecture is to use neuroevolution [13], i.e. genetic or evolutionary algorithms [5,10,17]. One of the most popular neuroevolutionary algorithms is NeuroEvolution of Augmenting Topologies (NEAT) [18], which has been and still is used and modified to adapt its operation to new tasks [15].

In this paper, the authors' neuroevolution method based on an adaptation of the NEAT algorithm to build a black-box model of a dynamic system with time delays is presented. The NEAT algorithm has been adjusted by allowing additional connections within ANN and developing original specialised evolutionary operators. An algorithm has been developed and verified by simulation that makes it possible to find ANN architecture that represents a compromise between its size and its accuracy in capturing the response of the model under which it has been learnt. The derived algorithm has been marked as dNEAT. To summarise, the main contribution of this work is to develop and verify the dNEAT algorithm that enables the selection of ANN architecture for black-box modelling of dynamic systems with time delays.

As applications, the single-input single-output (SISO) exemplary system and SISO model of the fast processes in a pressurised water reactor (PWR) are taken into account. A PWR is a non-linear, spatial and non-stationary plant whose processes are characterised by multi-scale and complex dynamics and involve delays associated with occurring delayed neutrons. Thus, there are many different mathematical models of PWR that are used depending on the aim, e.g., modelling of physical processes, diagnostics, on-line monitoring, etc. For example, seven modelling methods of processes taking place in a PWR are distinguished in [11]. Starting from the simplest one allows obtaining point-parameters models of low complexity, through a one-dimensional, three-dimensional, multi-point modelling based on fractional-order calculus, up to building models consisting of sub-models and neural models. However, the models built in white-box way 'pay' for their accuracy with a significant degree of complexity. On the other hand, the 'intelligent identification' of ANN-based processes in a PWR, and therefore also the building of ANN-based black-box models, is a topic enjoying considerable activity in the scientific research space [11]. Available in the literature neural reactor models, e.g., [2,6], systems for identifying reactor parameters or states, e.g., [7,14], or neural controllers used in a PWR control structures, e.g., [3] confirm the potential of ANNs in this domain. Also, the fact that the strong demand for models of these plants results from the lack of possibility to perform experiments on a real nuclear reactor. Thus, it seems that models of a

PWR that would allow numerous tests to be carried out in a short time and at a low computational cost would make it possible to design control, diagnostic and safety systems in which various scenarios of events and the controls would be considered. From the research point of view addressed in the paper, a PWR model has been used only to generate learning and validation data. Hence, for this reason, a detailed description of it is not presented.

2 Problem Statement

As it has been mentioned, the main aim of this work is to develop and verify the authors' algorithm - the dNEAT for black-box modelling purposes of dynamic systems with time delays. As a type of ANN, a recurrent network (RNN) has been selected. In RNNs, signals may flow in both directions, i.e. from input to output and vice versa. As a result, RNNs have an internal state that depends on the current input data and the previous network states. A recurrent network has been selected because, similar to considered plants' dynamics, they have an internal feedback. An architecture of exemplary RNN is presented in Fig. 1.

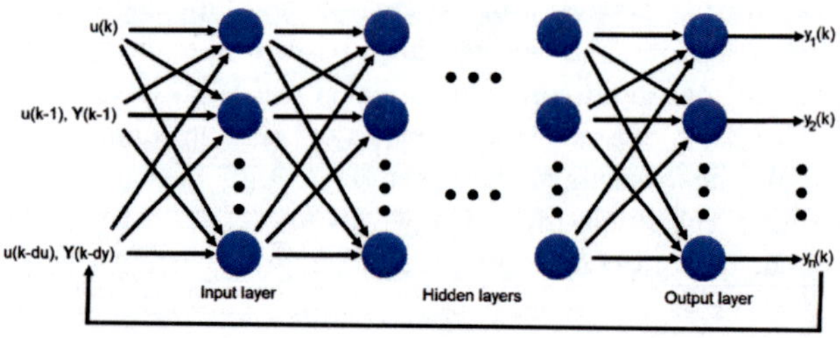

Fig. 1. An architecture of exemplary RNN.

Searching for a network with such architecture as in Fig. 1 is equivalent to looking for a non-linear auto-regressive exogenous discrete model of the process that is based on its time-delayed input and output signals:

$$\begin{aligned} y(k) = f(&y(k-1), y(k-2), y(k-3), ..., y(k-dy), \\ &u(k), u(k-1), u(k-2), u(k-3), ..., u(k-du)), \end{aligned} \tag{1}$$

where: $y(\cdot)$, $u(\cdot)$ are the output and input at the discrete-time instant specified by $(\cdot)$, respectively; $f(\cdot)$ denotes the function specified by $(\cdot)$; du, dy are the maximal levels of input and output (recurrence) delays, respectively; k is the discrete-time instant.

The following parameters and hyper-parameters (architecture of the network) should be determined to create a black-box model using RNN: the number of neurons in assumed single hidden and output layers (so-called hidden and output neurons); the existence of connections between individual neurons; the delay levels of du and dy; the neurons' model, i.e. type and parameters of excitation function and activation (transfer) function; the connections' model; and the values of weights of individual connections. Some of the above have been restricted to predetermined values (hyper-parameters). The excitation function is assumed, classically, as a weighted sum of inputs to a given neuron. In turn, the activation function of neurons in the hidden layer is assumed to a sigmoidal bipolar. This choice has been made because the considered RNNs are intended to represent the operation of continuous plants; thus, the activation function should be also continuous. In turn, the neuron activation function in the output layer is assumed as linear with a directional coefficient equal to one. It is since its task is to linearly transform the sum of the outputs of the previous layer, i.e. the excitation signal. The connection model has been adopted as a unit function without delays. To fully define the architecture of RNN, it is still necessary to select the parameters: the number of hidden neurons, the existence of connections between individual neurons, the values of du and dy, and the weights of individual connections. The proposed dNEAT algorithm is used for these purposes. It is worth adding that the NEAT algorithm (dNEAT as well) provides parallel optimisation of an architecture of the neural network achieved during their evolutionary development [18].

3 dNEAT Algorithm

This section describes the dNEAT algorithm by pointing out the changes made relative to the NEAT algorithm [18]. Hence, it focuses only on the parts that are new concerning the NEAT algorithm. The python code of the NEAT algorithm from [12] has been used and errors in it have been corrected, so that it works according to the original description from [18]. The codes of the dNEAT and mentioned the NEAT are available in [9]. The main algorithms' parameters are:

- $pop_{size} = 100$ - the target number of individuals in population;
- $weight_{init_{mean}} = 0$, $bias_{init_{mean}} = 0$ - the Gaussian distribution mean values for drawing initial values of weights and biases;
- $weight_{init_{stdev}} = 0.5$, $bias_{init_{stdev}} = 0.5$ - the Gaussian distribution standard deviations for drawing initial values of weights and biases;
- $du_{init-max} = 20$; $dy_{init-max} = 20$ - the upper ranges of intervals for drawing values du and dy;
- $bias_{mutate_{power}} = 0.033$ - the zero-centred Gaussian distribution standard deviation for drawing value of bias mutation;
- $bias_{mutate_{rate}} = 0.2$, $bias_{replace_{rate}} = 0.2$ - the probability that mutation will change the bias of a node by adding or assigning a random value;
- $weight_{mutate_{power}} = 0.1$ - the zero-centred Gaussian distribution standard deviation for drawing value of weight mutation;
- $weight_{mutate_{rate}} = 0.6$, $weight_{replace_{rate}} = 0.05$ - the probability that mutation will change the weight by adding or assigning a random value;
- $conn_{add_{prob}} = 0.2$, $conn_{delete_{prob}} = 0.2$ - the probability that mutation will add or delete a connection between existing neurons;

- $enabled_{mutate_{rate}} = 0.2$ - the probability that mutation will change the enabled status of a connection;
- $node_{add_{prob}} = 0.2$, $node_{delete_{prob}} = 0.2$ - the probability that mutation will add a new neuron or delete an existing neuron;
- $compatibility_{threshold} = 2.3$ - the individuals whose genomics distance is less than this threshold are considered to be in the same species;
- $max_{stagnation} = 25$ - the species, in which the best individual has not shown improvement over more than this number of generations will be considered stagnant and removed;
- $species_{elitism} = 3$ - number of species that will be protected from stagnation;
- $elitism = 10$ - the number of most-fit individuals in each species that will be preserved as-is from one generation to the next;
- $survival_{threshold} = 0.25$ - the fraction for each species allowed to reproduce each generation;
- $du_{mutate_{rate}} = 0.2$, $dy_{mutate_{rate}} = 0.2$ - the probability that mutation will change the delay levels;
- $du_{mutate-power} = 2$, $dy_{mutate-power} = 2$ - the specifies the intervals from which the delay mutation values are drawn.

3.1 Initialisation

Each individual is initialised as the RNN consisting of one hidden and one output neuron. The number of output neurons reasons from the number of mapped trajectories. During initialisation, each input neuron is connected to each hidden neuron, and each hidden neuron is connected to each output neuron. The initial delays du and dy are drawn with uniform probability distribution from ranges of $[0, du_{init-max}]$ and $[0, dy_{init-max}]$. The initial values of the weights and biases are drawn using Gaussian distributions with parameters given in Sect. 3.

3.2 Crossover

The crossover operator in respect of neurons and connections between them remained the same as in the NEAT algorithm. In turn, concerning the delays levels of du and dy crossover way yields:

$$d_{child} = round\left(rd_{parent1} + (1-r)d_{parent2}\right), \tag{2}$$

where: d_{child} is the value of du or dy for offspring; $d_{parent1}$, $d_{parent2}$ are the values of du or dy for parents; r is a number drawn using a uniform probability distribution from the range $[0, 1]$.

3.3 Mutation

As for crossover, the mutation concerning neurons and connections between them remained the same as in the NEAT algorithm. The mutation of delays du and dy proceeds as follows. If draw decided that the mutation of the corresponding delay $d(\cdot)$ should occur, the delay change $\delta d(\cdot)$ is drawn with uniform probability distribution from the range $[-d(\cdot)_{mutate-power}, d(\cdot)_{mutate-power}]$. Then $\delta d(\cdot)$ is rounded to an integer value. If $\delta d(\cdot)$ is zero then it is assigned a value of –1 or 1 with equal (50%) probability. The corresponding delay is then changed according to the formula $d(\cdot) = |d(\cdot) + \delta d(\cdot)|$. Hence, if the corresponding delay level du or dy is decreased, then outgoing connections from the given input are removed. On the other hand, if the delay is increased, outgoing connections from the new inputs are added to all hidden neurons in the same way as during initialisation.

3.4 Fitness Function

Since the dNEAT algorithm optimises the structure of individuals directly during its operation, the function determining the fitness of individuals does not need to depend on the RNN architecture. Therefore, the fitness function yields:

$$f_i = -\frac{1000}{N}\sum_{j=1}^{J}(y_{ij} - t_j)^2, \tag{3}$$

where: f_i is the value of fitness function for i-th individual; N is the number of samples during learning phase; y_{ij} are successive j-th samples of the response of the i-th individual; t_j are successive j-th samples of the target trajectory.

It can be noticed that (3) is the mean square error (MSE) of mapping the learning trajectory through the network response. This value is multiplied by a relatively large negative scaling number to magnify the discrepancy in values of the fitness function of individuals with a similar mapping performance. It is a negative number because, by definition, the better individual should achieve a higher fitness function value.

4 Applications

The developed dNEAT algorithm has been simulation-verified on two applications. The first one is the SISO exemplary system, whereas the second is the SISO model of fast processes in a PWR. As a result, using the dNEAT algorithm, neural (black-box) models are created for both plants.

4.1 Application 1

In this application, the black-box model of an exemplary non-linear plant with time delays described by (4) has been created:

$$x(k) = -0.05x(k-1) + 0.02x(k-5) + sin\left(\frac{x(k-10)}{10}\right) + u(k-15). \tag{4}$$

The input ($u(\cdot)$) trajectory during the learning phase, i.e. in the phase of obtaining the model by the NEAT and dNEAT algorithms (red line), and the corresponding target (output - $x(\cdot)$) trajectory (blue line) are presented in Fig. 2a.

In turn, the trajectories used in the verification phase, i.e. the phase in which the responses of the obtained neural model to the trajectories that have been not used in the learning phase are shown in Figs. 2b and 2c. In all cases, the model responses have been normalised by dividing the plant's response by 30. All the figures presented in this paper are available in [9].

4.2 Application 2

The second application is focused on building the SISO neural model of the fast processes in a PWR. The position of the control rods and the thermal power, scaled by dividing by the nominal power of a PWR have been selected as the input and output signals, respectively. The input trajectory during the learning phase (red line) and the corresponding output trajectory (blue line) are presented in Fig. 3a. In turn, the trajectories used in the verification phase are shown in Figs. 3b and 3c. These trajectories are taken from [10].

5 Results

In this section, the simulation results illustrating the performance of the proposed algorithm against the results generated by the NEAT algorithm are presented. First, the learning phase is shown. Next, the verification phase is discussed. The stop condition has taken 2500 generations, and both algorithms have called 10 times. All parameters present in both algorithms have had the same values (see Sect. 3). The average value of the fitness functions of subsequent generations from all NEAT and dNEAT calls in both applications are presented in Fig. 4a. Whereas the average value of fitness functions of the best individuals in subsequent generations from all algorithms is shown in Fig. 4b. The responses of the neural model obtained for the learning data for the best and worst individuals by fitness function values, selected from all obtained the best individuals from each algorithm call for application 1 are shown in Fig. 5a. For simplicity, the best individuals obtained from successive algorithms calls are called 'winners'. The responses of the best and worst individuals from the winners for the verification phase are presented in Figs. 5b and 5c. Analogous results for application 2 are shown in Fig. 6. The MSE values of the best and worst individuals from the winners and the average MSE value of the winners from both phases for both applications are given in Table 1.

Analysing the results obtained, the following conclusions can be drawn. The mean values of the fitness functions of subsequent populations indicate that for the dNEAT algorithm they have been lower than for the NEAT. However, the mean values of the fitness functions of the best individuals show that the dNEAT algorithm provides a better (or comparable) performance in finding the best individuals. Additionally, these individuals are found faster. Therefore, the dNEAT algorithm, despite searching a larger solution space and putting more emphasis on population diversity is able to find individuals better than the NEAT algorithm in the task of creating neural models of dynamic systems with

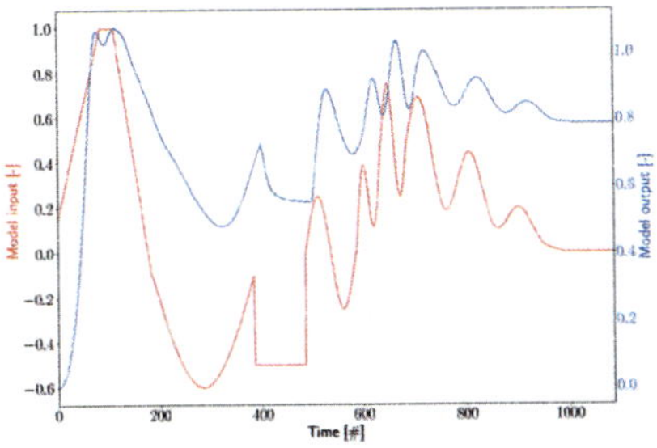

(a) The learning data for application 1.

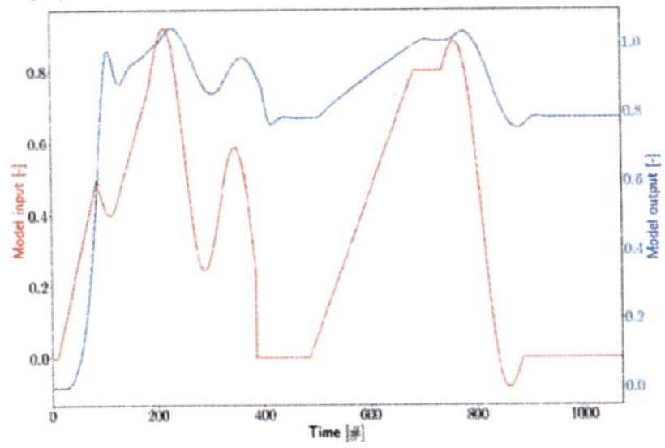

(b) The verification data for application 1 - set 1.

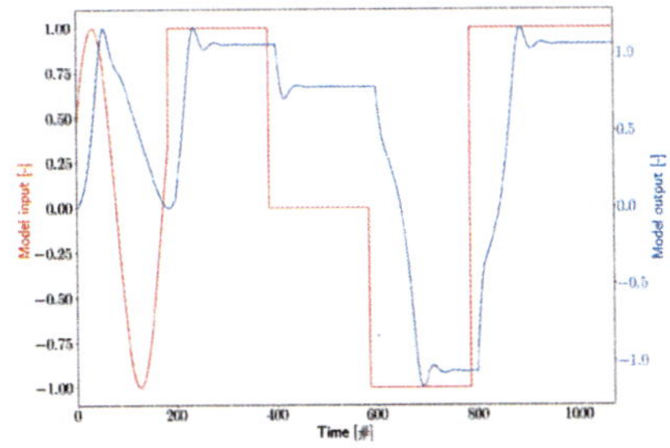

(c) The verification data for application 1 - set 2.

Fig. 2. The learning and verification data for application 1.

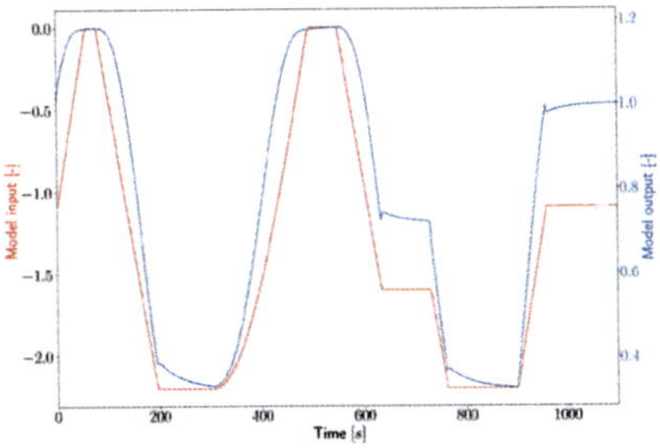

(a) The learning data for application 2.

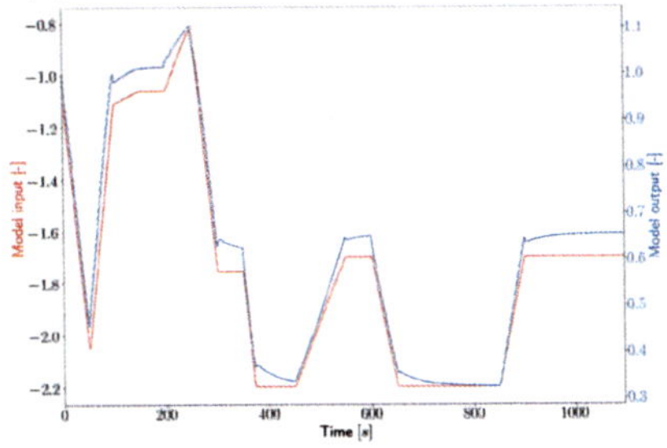

(b) The verification data for application 2 - set 1.

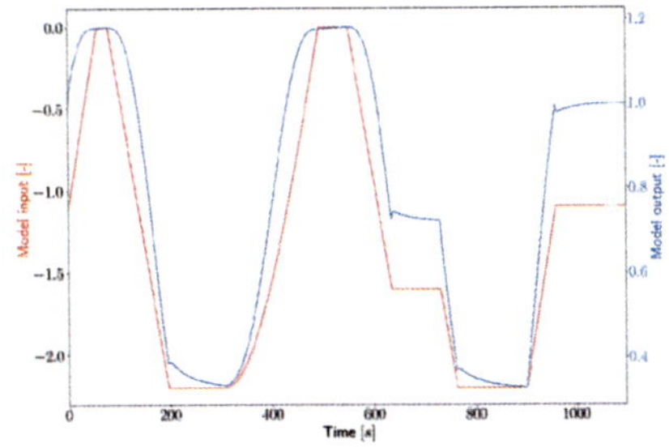

(c) The verification data for application 2 - set 2.

Fig. 3. The learning and verification data for application 2.

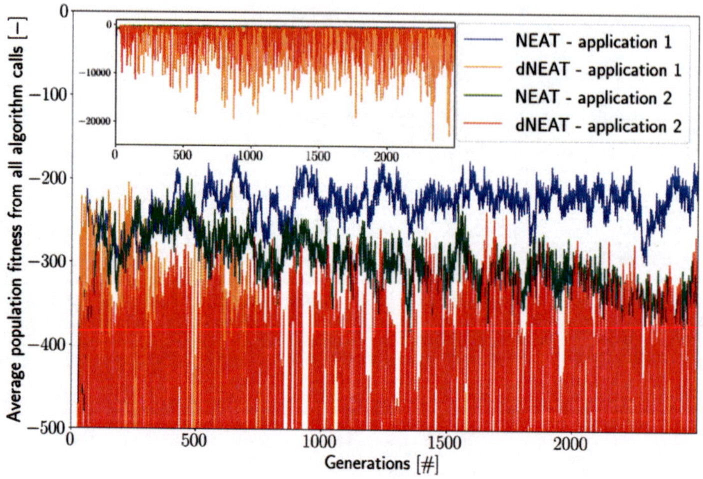

(a) The average value of population fitness function.

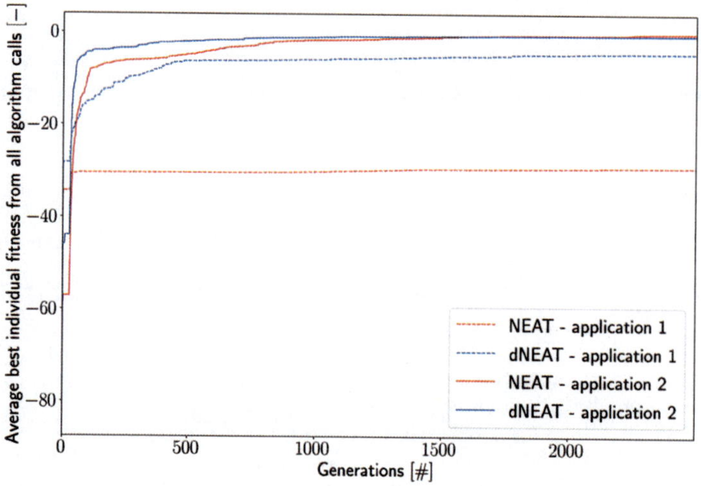

(b) The average value of fitness function of the best individuals.

Fig. 4. The trajectories of average values of the fitness functions.

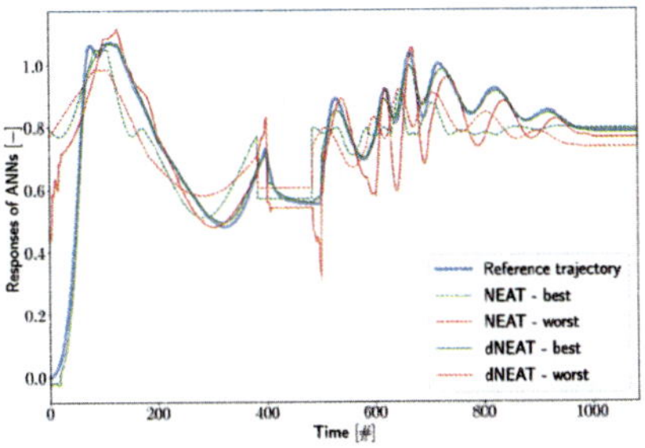

(a) The responses of 'winners' in the learning phase for application 1.

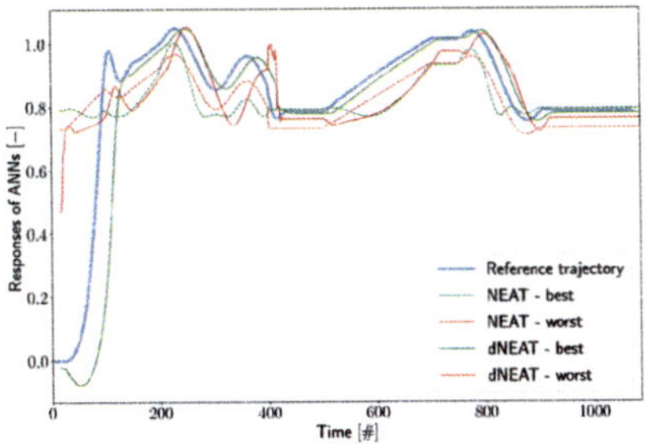

(b) The responses of 'winners' in the verification phase - set 1 for application 1.

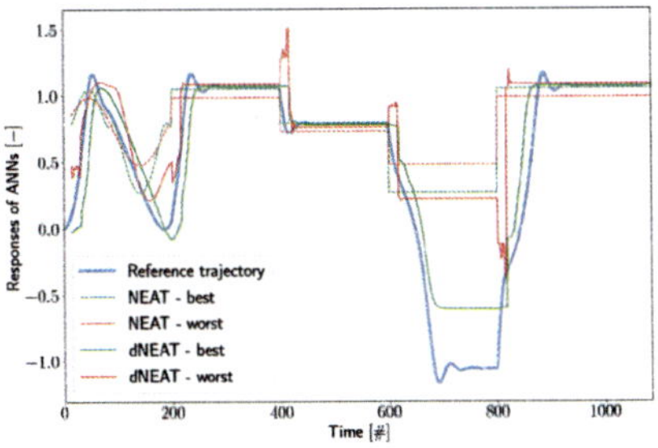

(c) The responses of 'winners' in the verification phase - set 2 for application 1.

Fig. 5. The responses of 'winners' in both phases for application 1.

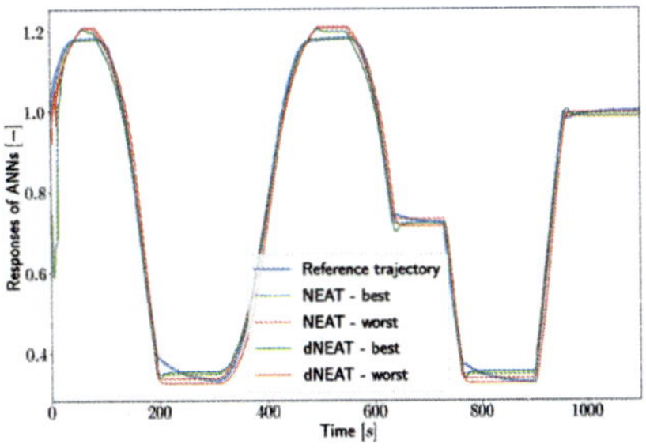

(a) The responses of 'winners' in the learning phase for application 2.

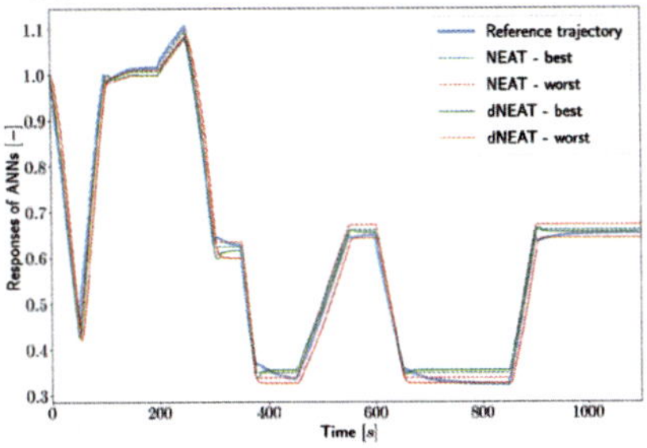

(b) The responses of 'winners' in the verification phase - set 1 for application 2.

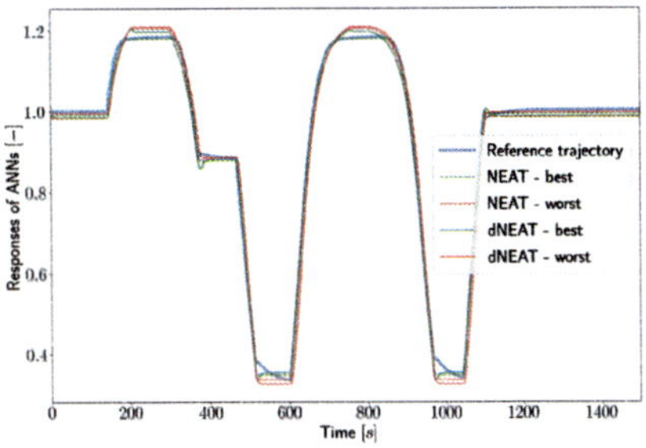

(c) The responses of 'winners' in the verification phase - set 2 for application 2.

Fig. 6. The responses of 'winners' in both phases for application 2.

Table 1. The values of MSE for applications 1 and 2.

Phase	Algorithm	Description	Value $(\cdot 10^{-2})$	
			Application 1	Application 2
Learning	NEAT	The best of the winners	2.85996	0.01335
		The worst of the winners	3.05342	0.0279
		The average value of the winners	2.92178	0.01654
	dNEAT	The best of the winners	0.04096	0.03113
		The worst of the winners	1.7322	0.08256
		The average value of the winners	0.43248	0.05183
Verification - set 1	NEAT	The best of the winners	4.27963	0.02234
		The worst of the winners	4.54949	0.0326
		The average value of the winners	4.45215	0.02278
	dNEAT	The best of the winners	0.47872	0.03478
		The worst of the winners	3.49938	0.12342
		The average value of the winners	0.98783	0.08117
Verification - set 2	NEAT	The best of the winners	41.99314	0.00909
		The worst of the winners	47.21822	0.02924
		The average value of the winners	42.26676	0.014
	dNEAT	The best of the winners	5.58866	0.02675
		The worst of the winners	31.34926	0.09197
		The average value of the winners	16.26694	0.07378

time delays. The individuals found in this algorithm are more diverse through the different levels of input and output delays. Thus, the proposed solution makes it possible to find individuals acting as a black-box model of a dynamic system with time delays more efficiently concerning the NEAT algorithm.

6 Conclusions

In this paper, the problem of developing an algorithm of artificial neural network architecture search for black-box modelling of dynamic systems with time delays has been investigated. In particular, the evolutionary algorithm dNEAT has been devised to build the required models. The specialised evolutionary operators and additional connections within the network ensure the proper operation of the proposed algorithm. It enabled devising the SISO neural model of the exemplary system as well as the fast processes in a PWR. The dNEAT algorithm has been implemented in the computational environment, and obtained simulation results yield satisfying performance of the produced output trajectories. Further work is needed to analyse the impact of the algorithm parameters values on its operation and to attempt to select them automatically, e.g., through machine learning.

Acknowledgements. Financial support of these studies from Gdańsk University of Technology by the DEC-2/2020/IDUB/I.3.3 grant under the Argentum Triggering Research Grants - 'Excellence Initiative - Research University' program is gratefully acknowledged.

References

1. Azar, A.T., Vaidyanathan, S.: Computational Intelligence Applications in Modeling and Control. Springer, Cham (2015). https://doi.org/10.1007/978-3-319-11017-2
2. Boroushaki, M., Ghofrani, M.B., Lucas, C.: A new approach to spatio-temporal calculation of nuclear reactor cores using neural computing. Nucl. Sc. Eng. **155**(1), 119–130 (2017)
3. Coban, R.: Power level control of the TRIGA Mark-II research reactor using the multifeedback layer neural network and the particle swarm optimization. Ann.f Nucl. Energy **69**, 260–266 (2014)
4. Cybenko, G.: Approxmiation by superpositions of a sigmoidal function. Math. Control Signals Syst. **2**, 303–314 (1989)
5. Elsken, T., Metzen, J.H., Hutter, F.: Neural architecture search. In: Hutter, F., Kotthof, L., Vanschoren, J. (eds.) Automated Machine Learning. The Springer Series on Challenges in Machine Learning, pp. 63–77. Springer, Cham (2019)https://doi.org/10.1007/978-3-030-05318-5
6. Khalafi, H., Terman, M.S.: Development of a neural simulator for research reactor dynamics. Progr. Nucl. Energy **51**(1), 135–140 (2009)
7. Kim, H.G., Chang, S.H., Lee, B.H.: Pressurized water reactor core parameter prediction using an artificial neural network. Nucl. Sci. Eng. **113**(1), 70–76 (2017)
8. Kolmogorov, A.N.: On the representation of continuous functions of many variables by superposition of continuous functions of one variable and addition. Doklady Akademii Nauk SSSR **114**(5), 953–956 (1957)
9. Laddach, K.: The developed codes of dNEAT and NEAT algorithms. https://git.pg.edu.pl/p969501/dneat
10. Laddach, K., Łangowski, R., Rutkowski, T.A., Puchalski, B.: An automatic selection of optimal recurrent neural network architecture for processes dynamics modelling purposes. Appl. Soft Comput. **116** (2022)
11. Li, G., Wang, X., Liang, B., Li, X., Zhang, B., Zou, Y.: Modeling and control of nuclear reactor cores for electricity generation: a review of advanced technologies. Renew. Sustain. Energy Rev. **60**, 116–128 (2016)
12. McIntyre, A., Kallada, M., Miguel, C.G., da Silva, C.F.: Neat-python. https://github.com/CodeReclaimers/neat-python. Accessed 16 Feb 2022
13. Miikkulainen, R.: Neuroevolution. In: Sammut, C., Webb, G.I. (eds.) Encyclopedia of Machine Learning. Springer, Boston (2011). https://doi.org/10.1007/978-0-387-30164-8_589
14. Moshkbar-Bakhshayesh, K., Ghofrani, M.B.: Development of an efficient identifier for nuclear power plant transients based on latest advances of error backpropagation learning algorithm. IEEE Transa. Nucl. Sci. **61**(1), 602–610 (2014)
15. Papavasileiou, E., Cornelis, J., Jansen, B.: A systematic literature review of the successors of 'neuroevolution of augmenting topologies'. Evol. Comput. **29**(1), 7–73 (2020)
16. Roffel, B., Betlem, B.: Process Dynamic and Control. Modelling for Control and Prediction. John Wiley & Sons Inc., Chichester (2006)
17. Stanley, K.O., Clune, J., Lehman, J., Miikkulainen, R.: Designing neural networks through neuroevolution. Nat. Mach. Intell. **1**, 24–35 (2019)
18. Stanley, K.O., Miikkulainen, R.: Evolving neural networks through augmenting topologies. Evol. Comput. **10**(2), 99–127 (2002)

A Model-Based Approach for Testing Automotive Embedded Systems – A Preliminary Study

Anna Gnacy–Gajdzik[1,2,3](✉), Marcin Gajdzik[1,2,3], Piotr Przystałka[1], and Kamil Sternal[1,2,3]

[1] Department of Fundamentals of Machinery Design, Silesian University of Technology, 18a Konarskiego Street, 44-100 Gliwice, Poland
anna.gnacy-gajdzik@polsl.pl

[2] Doctoral School, Silesian University of Technology, 2a Akademicka Street, 44-100 Gliwice, Poland

[3] DIP Draexlmaier Engineering Poland Sp. z o. o, Jelenia Góra, Poland

Abstract. The automotive embedded systems are experiencing an increasing transformation to automated cyber–physical systems with an extended reliance on software. Guaranteeing the quality, reliability and safety of both single components and the entire vehicle is a tremendous challenge. They should not only comply with functional and performance requirements, interacting reliably with other real-time components, but also with stringent safety and reliability standards. It is distinguished from comparable software well-known from other devices, like mobile phones. In vehicles there are complex real-time interactions between different mechanical systems: braking, drive, suspension, steering. The automotive system is nothing like any other, characterized by rigorous planning, architecture, development, testing, validation and verification processes. This paper discusses a model-based approach to testing automotive embedded systems, particularly in the context of the occurrence of test result flickering, a phenomenon that affects the reliability, reproducibility and validity of embedded software testing. The aim of the research is to verify simplifications used in Hardware-in-the-Loop simulation during testing. The experiments carried out involved comparing test results from a simplified vehicle simulation and a model-based approach in the context of flickering results. A testing bench was designed, and real experiments were performed. The model-based approach is planned to be developed in the future, as the studies have demonstrated that it can be successfully applied to generate aberrations in the test system and to development of a methodology for test case design.

Keywords: Flickering tests · Model-based testing · Hardware in the Loop · Simulation model · Automotive embedded systems

Z. Kowalczuk (Ed.): DPS 2022, LNNS 545, pp. 340–351, 2023.
https://doi.org/10.1007/978-3-031-16159-9_28

1 Introduction

The automotive industry has recently undergone a tremendous transformation with enormous investment in providing comfort, convenience and safety for both drivers and passengers. At parallel, vehicle producers, responding to ecological trends, implement solutions which reduce the amount of carbon dioxide emitted into the atmosphere. A vehicle has long ceased to be a machine consisting of an engine, steering wheel and wheels – it is no longer just a mechanical device. The automotive sector is facing a progressive evolution from simple embedded control systems to complex and automated cyber–physical systems with an extended reliance on software. High-class vehicles have over 100 million lines of code, distributed on more than 150 Electronic Control Units (ECUs) [14]. Recently, a significant expansion of advanced functions has been observed. It consists in receiving indications from a variety of sensors, camera and radar signals, processing the acquired information by complex algorithms and, because of this processing, controlling the electrical or mechatronic systems.

This has an immense impact on the expanding complexity of the emerging embedded software [2]. All these systems are connected by a data bus together, creating increasingly complex networks of sophisticated electronic and IT solutions. The systems should not only fulfil the functional and performance requirements, reliably interact with other real-time components, but also comply with rigorous safety and reliability requirements [4]. The software of a single ECU is part of a vastly more complex mechatronic architecture responsible for advanced functions at the entire vehicle level. However, even a seemingly basic vehicle function such as steering, opening and closing the vehicle may require interactions between numerous ECUs with embedded software applications, scattered throughout the vehicle and interconnected by different network protocols. Automotive system is unlike any other, characterized by its rigorous planning, architecting, development, testing, validation and verification. The physical task of writing embedded software for automotive applications versus other application areas is not significantly different from other embedded systems, but the key differences are the quality standards which must be followed for any development and test project [7].

Guaranteeing the correct behavior of the electronic devices composing a vehicle is very complex, and its expected behavior must be assured in very harsh environments. In fact, vehicles are exposed to vibration, noise, extreme temperatures and electromagnetic fields that may affect and degrade the electronic components [14]. To test the embedded software as soon as and comprehensively as possible, the verification, and validation of its functionality and quality should start at the very earliest possible phase. According to standard ISO 26262 it is desirable to implement automated regression testing throughout the embedded software development process. In the initial phases of development, pure simulation models (MiL: Model-in-the-Loop, SiL: Software-in-the-Loop) should be used, which, among other things, also simultaneously simulate the hardware of the ECU [5]. The test environment for the Software Qualification Tests (SWE.6) should be as similar as possible to the real one [16]. Vehicle-in-the-Loop simulation is being replaced by Hardware-in-the-Loop (HiL) simulation due to the high hardware costs and problems with triggering errors to check response of embedded software to them (HiL). It tests the actual control unit in interaction with the simulated environment [5]. In the interest of an efficient embedded software testing process, it is required that tests are evaluated automatically, processed systematically and documented on an ongoing basis. Furthermore, all tests and their results must be always traceable and repeatable [5]. Engineering practice shows that a frequent issue that arises during the execution of automated tests in embedded systems is the so-called test flickering. Flickering tests are tests that, without any specified reason, sometimes fail and sometimes succeed, even if we run them on the same compilation server and in the same environment. Whether the failure happens regularly or infrequently, this phenomenon is one of the most worrisome situations for anyone who cares about the results of test automation, as it negatively affects the test reliability [1]. The reasons for the test flickering must be identified. If the cause is in the system under test, it means that there is a defect that is being reported. However, environmental problems are often the source of test flickering. They are usually isolation problems that are caused by automation. Like any other bugs, they should be identified and fixed. This is a time-consuming and resource intensive process. However, it happens that their reasons depend on various elements of the simulation environment and cannot be eliminated. The occurrence of flickering test results affects the quality of the testing process, its reliability, safety and credibility. The automation of the flickering test cause identification process is in the interest of companies involved in the design of embedded systems. It seems reasonable to use augmented intelligence algorithms for this purpose. Therefore, the phenomenon of test flickering has been investigated. Their important step is to check the possibility of controlling the test environment to induce this flickering.

This paper deals with a model-based approach for testing automotive embedded system, especially in the context of the occurrence of flickering test results. The first aim of this work is to verify the simplification used in HiL simulation when testing embedded systems. The research was performed with the electric vehicle model based on the dynamic's equations defining the mechanical phenomenon occurring in the vehicle driving cycle in real environmental conditions. The vehicle speed and forces influencing it were not only modelled, but also the effects observed in the high-voltage battery during vehicle exploitation. It is studied whether modelling an electric vehicle as a static unit (i.e., with fixed parameters such as speed, battery state of charge, battery system voltages and currents, etc.) has an impact on the safety, quality and reliability of the testing process. The second intention was also to prove that the application of a realistic vehicle model enables the flickering phenomenon to be visible in more tests. Furthermore, the possibility of utilizing the model when implementing and learning augmented intelligence algorithms used for identifying aberrations present in the test system, which influence test results, and are not related to the embedded system functions being tested. In addition, the possibility of using the model during the implementation and learning of augmented intelligence algorithms was examined. These algorithms would be used to identify aberrations present in the test system that affect test results but are not related to the embedded system functions under test.

2 Background and Related Works

This section elaborates on the potential application of the vehicle model in the consecutive steps of the embedded software development process and explains the significance and importance of testing embedded software for the safety of vehicle exploitation.

2.1 Modelling Simulations for Embedded Software Development in the Automotive Industry

Engineers working in the automotive industry are observing an increasing use of simulation models in the embedded software development. The reason for the existence of this trend is the shorter delivery time of products to the market, which results in cost reduction [3]. Besides the financial aspects, an undeniable benefit is also the increase in quality, which leads to safer products.

There are two main areas in which simulation models are applicable. The first relates to situations where there is a need to simulate physical phenomena. The indisputable benefit of applying models in this case is the reproducibility of the delivered signals, but also the possibility of fine-tuning them. This is not possible with data that has been collected from real physical objects. This allows embedded controllers to be tested using more scenarios without engaging real hardware.

The second area where modelling simulations are involved is a variation of the first method. Instead of feeding the entire embedded controller with data, part of the source code is used directly in the model. As an example, the functionality responsible for performing the current plausibility can be considered. As shown in Fig. 1, two current signal generators provide input to two blocks that are perform the same function (current plausibility check) but have a different implementation method. This allows us to verify that the algorithm in the embedded code, which is compromised due to limited performance of embedded systems, still fulfil the quality goals.

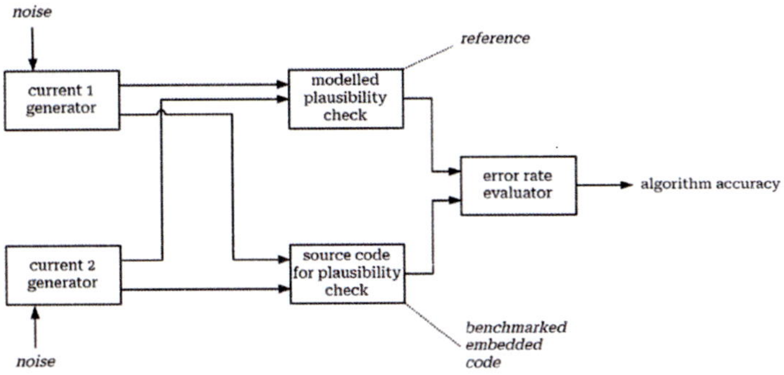

Fig. 1. Accuracy evaluation of current plausibility algorithm

2.2 Embedded Software Testing as an Essential Safety, Quality and Reliability Phase

Erroneous automotive embedded system may cause serious car accident related to human safety. Therefore, it is essential to detect any errors in the embedded software through various test processes [9]. The risks associated with systems errors and random hardware failures are increasing with the tendency to the technological complexity, the software content and the mechatronic implementation. The ISO 26262 standard provides recommendations on how to prevent this risk by ensuring appropriate requirements and supporting processes. According to the aforementioned standard, system safety is achieved through the application of a variety of safety measures, implemented in different technologies (mechanical, hydraulic, pneumatic, electrical, electronic) and at different stages of the development process [16]. Testing, defined as a complex process by ISO 26262 and the ASPICE standard, has become an important part of quality assurance and vehicle operational safety [6]. Embedded software is often evolving to meet customer needs, so it is desirable to automate testing to improve the efficiency of the process [9].

Vehicle manufacturers have to spend lots of time and cost in testifying embedded software which are included in the embedded systems [12]. In automotive systems development a product has to pass tests at several levels of abstraction such as Model-in-the-Loop (MiL), Software-in-the-Loop (SiL) and Hardware-in-the-Loop (HiL) tests [10]. Ensuring quality, reliability and safety of operation of both individual devices and the entire system is a tremendous challenge nowadays. The systems applied must not only comply with the functional and performance requirements imposed on them, reliably cooperate with other components in real time, but also meet stringent safety and reliability requirements [4]. For the qualification tests verifying software safety requirements, real-time software testing in the target environment is required. The test environment recommended by ISO 26262 is the target automotive vehicle. However, due to the cost and difficulty in inducing errors during testing, the commonly used environment is Hardware-In-the-Loop (HiL) simulation [16]. The application of Hardware-In-The-Loop simulation during tests allows to induce errors in the system, and consequently to realize the assumptions of ISO 26262 standard concerning verification and validation of safety functions in the context of reliability and high quality of the embedded software. Thanks to the feedback loop in the HiL simulation, the response of the embedded system to the induced errors can be verified.

When designing test cases in both of these environments, many conditions affecting the test result must be considered. In practice, test case designers focus on the functionality under test, analyzing in the case of a negative test result the reasons for failure and deciding whether to report a defect [6]. Flickering results in the testing process introduce undesirable quality ambiguity into the whole process. The aim of embedded software quality engineers is to eliminate this effect. This is a time-consuming and therefore costly process, so presumptions are made at the very outset to simplify the test environment. However, it is important to have the confidence that their application has no impact on the embedded software functionality under test.

3 Research Methodology

The research presented in this chapter is a part of an industrial PhD project. It will be used to optimize the embedded software testing process at Dreaxelmaier, a company that designs and manufactures embedded high-voltage battery monitoring systems for electric vehicles. For this reason, a specific set of tools and software was chosen for the experiment. The research discussed in this paper are intended to verify whether the simplifications used in the testing process do not negatively affect the testing process quality. Furthermore, they aim to verify the usefulness of the vehicle model in the development of test case design methodology so that augmented intelligence algorithms can be applied to verify the validity of their results.

3.1 Test Setup

The Hardware-In-The-Loop environment showed in Fig. 2, in which the tests were conducted was built using hardware and software from Vector Informatik. The embedded system under test was housed in a Reinhardt needle adapter, allowing connection of any PCB test points with the vTsystem simulator. Configuration of the vTsystem requires a LAN connection to a PC with CANoe simulation software installed. The variables and algorithms implemented in the software provide control of devices connected to the vTsystem. It is possible to generate and send any signals to them and create scripts describing the behavior of simulated system elements [6].

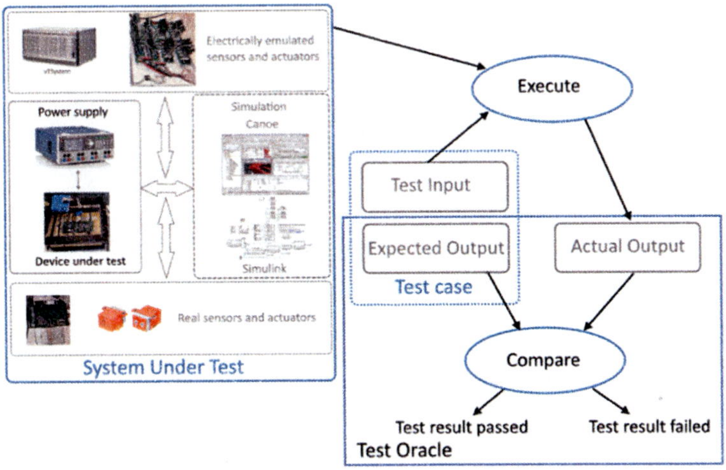

Fig. 2. Block diagram of the test setup

The ability to test the real-time control system under safe conditions is a significant advantage of HiL simulation [13]. The behavior of the embedded system is tested on the target ECU, where the control algorithm is validated in an environment that consists of electronically emulated sensors and actuators. The embedded system used in the tests acts as a unit in the electric vehicle to control the state of charge, efficiency and performance of the high-voltage battery. This controller communicates with the battery's cell controllers, acquiring the information necessary to implement control algorithms. The values of physical parameters (i.e., voltage, current, temperature and isolation resistance) derived from the battery are simulated in the HiL environment. Communication over CAN bus with other ECUs in the vehicle is simultaneously simulated. As a result of the implemented algorithms, through the feedback loop, communication signals are sent to other ECUs in the vehicle over the CAN bus and the appropriate contactors are switched.

3.2 Simulation Model

One of the modern approaches for implementing Hardware-in-the-Loop (HiL) tests is using MATLAB/Simulink in combination with Vector CANoe software. The first one is used for generating simulation data, which is then routed via CANoe to the hardware under test. Figure 3 shows the HiL test composition on the simulation path, excluding testing which is not in the scope of this chapter.

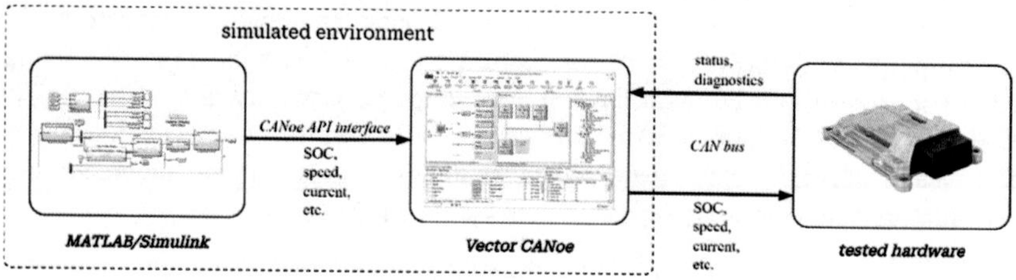

Fig. 3. Hardware-in-the-Loop simulation environment

The first step in building the simulation environment was model selection. It was required to define a set of criteria for the model to ensure that it would fulfil its purpose, which was to deliver dynamic driving data of an electric vehicle with focus on the drivetrain. Criteria used in the model searching:

- simulation of the electric car drivetrain, battery, and body;
- parametrization of battery voltage, capacity, and temperature;
- electric recuperation.

Figure 4 shows the architecture of a theoretical model, with the energy types and flow directions indicated. The black arrows represent energy transition while the vehicle is driving. The opposite situation, braking, is marked in gray. It is worth mentioning that a block called *Motor Drive* performs bidirectional energy conversion, realizing the recuperation requirement. This process is crucial for reflecting the dynamic behavior of physical parameters like battery current and state of charge (SoC). The last type of flow represents energy losses due to various physical properties like the friction inside mechanical component or the self-heating of battery cells caused by current flow and internal resistance.

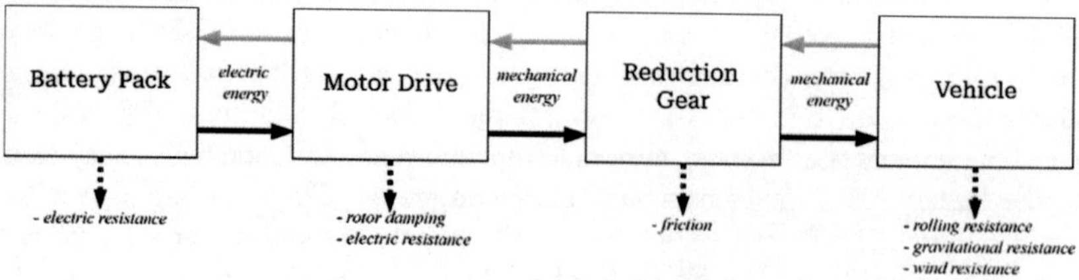

Fig. 4. Architecture of theoretical model

The sentence *mechanical energy* is an abstract concept which should be refined at this point by introducing a the concept of force. To calculate the total vehicle force F_{total} (as shown in Eq. 1) F_{brake} and $F_{resistance}$ must be subtracted from F_{drive} delivered to *Vehicle* block as follows:

$$F_{total} = F_{drive} - F_{brake} - F_{resistance} \tag{1}$$

The component $F_{resistance}$ can be evaluated in such a way:

$$F_{resistance} = -\mu mg - mgSin\theta - \frac{1}{2}\rho C_d A v^2 \tag{2}$$

where μmg is road frictional resistance, $mgSin\theta$ is gravitational resistance if the vehicle is moving up or down a slope, and $\frac{1}{2}\rho C_d A v^2$ is the wind resistance due to air drag. Newton's second law of motion can be applied:

$$ma = F_{drive} - F_{brake} - \mu mg - mgSin\theta - \frac{1}{2}\rho C_d A v^2 \tag{3}$$

and after transformation, the total force can be obtained:

$$F_{total} = (F_{drive} - F_{brake} - \mu mg - mgSin\theta - \frac{1}{2}\rho C_d A v^2) * \frac{1}{ma} \tag{4}$$

with the knowledge of F_{total} value it is possible to calculate the power flow to or from the battery pack depending on the current vehicle acceleration. An open-source model [8] which takes use of the previously mentioned equations and fulfills all the required criteria was used to perform further tests.

3.3 Research Approach

The standard approach to validating embedded software used in the automotive industry is to use HiL simulation, in which external components (electronic, mechatronic, other ECUs in the vehicle) are simulated. To simplify the whole process – from designing the simulation model to verifying the results of negative tests – simulations with constant output values are used. However, it is of course still possible to manipulate these values from the test case script. In the embedded system under test, these parameters are e.g., the current, voltage, temperature of the high-voltage battery, vehicle speed, battery charge level, etc.

The aim of this work is to verify the validity of this approach, whether it does not mask defects present in the embedded software under test. These defects, for example, can be related to the actual communication bus load in the vehicle, emerging time-varying physical volumes. The scientific experiment was organized into two stages. In the first, the standard simulations used in the testing process were applied. In the second stage, the HiL simulation was augmented with a simulation of an electric vehicle and its high-voltage battery prepared in the Simulink application. Time-varying parameters related to the motion of the vehicle and the state of the high-voltage battery, dependent on the vehicle motion cycle were used. The research also aimed to validate the suitability of the dynamic model for further work related to the automation of the verification process of flickering test results with augmented intelligence.

4 Results and Discussion

The first stage of research was carried out with a simplified vehicle simulation, in which output parameters like e.g., motor speed, vehicle speed, battery state of charge, battery current, battery voltage or electrical efficiency were constant during the entire testing process. In the second stage of the research, a vehicle model was implemented considering the laws of dynamics. The aforementioned parameters were time-varying in the model. As an example, Fig. 5 shows some of output parameters obtained by means of the vehicle simulation model described in Sect. 3.2.

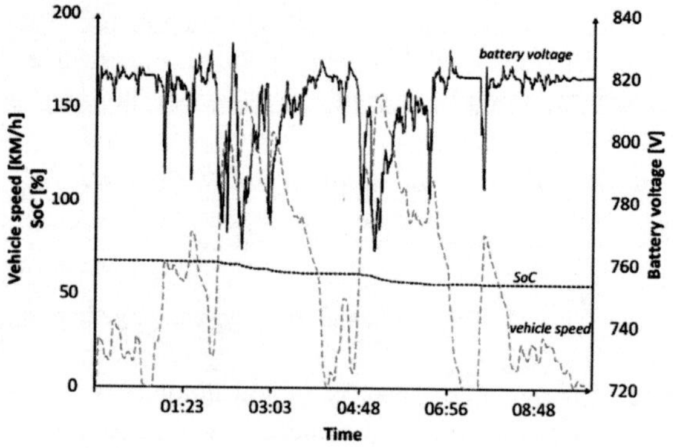

Fig. 5. Example output parameters of the electric vehicle model applied in the research

In this study, 8 test cases were selected from a group of tests with results indicating a tendency to flickering [5]. The details related to the functions tested, the next steps of the test cases as well as the results obtained in them cannot be described in this article due to industrial confidentiality. Also, these details are not relevant to the results of the research described in this paper. The study consisted of performing each of the selected test cases ten times and calculating empirical probability of result flickering (occurrence of a negative result in 10 trials). Each trial was repeated ten times. Examples of particular results for the test case ID = 8 are shown in Table 1.

Average empirical probability of result flickering (occurrence of negative results in 10 trials) was calculated. The average, median and standard deviation for both stages were compared. The results obtained during the research are presented in Table 2.

The comparison of the obtained results reveals that the application of the dynamic vehicle simulation affects the embedded systems test results. It raises the empirical probability of test result flickering. For all test cases, the average empirical probability of test result flickering (occurrence of a negative result in 10 trials) significantly increased when the vehicle model developed in Simulink was used in the simulation. For the tests used in the experiment, 4 flickering result triggers were identified. It was observed that the median of empirical probability of result flickering (occurrence of a negative result in 10 trials) significantly increased when the flickering was induced by the errors associated with the communication on the CAN bus as well as by the undesirable activity on the bus when the ECU was supposed to enter sleep mode. When using dynamic vehicle

Table 1. A summary of data samples for test case ID = 8

Trial	First stage			Second stage		
	Executions	Negative result	Empirical probability	Executions	Negative result	Empirical probability
1	10	6	0,6	10	9	0,9
2	10	0	0	10	9	0,9
3	10	2	0,2	10	9	0,9
4	10	3	0,3	10	10	1,0
5	10	2	0,2	10	9	0,9
6	10	3	0,3	10	10	1,0
7	10	3	0,3	10	9	0,9
8	10	4	0,4	10	9	0,9
9	10	3	0,3	10	9	0,9
10	10	3	0,3	10	9	0,9

Table 2. Comparison of the empirical probability of result flickering (occurrence of negative results in 10 trials) obtained in the first and second stage

Test ID	First stage			Second stage		
	Mean	Median	Standard deviation	Mean	Median	Standard deviation
1	0,08	0,1	0,06	0,14	0,1	0,07
2	0,06	0,05	0,07	0,16	0,1	0,16
3	0,14	0,1	0,08	0,15	0,2	0,07
4	0,36	0,4	0,18	0,42	0,35	0,29
5	0,33	0,35	0,15	0,34	0,4	0,14
6	0,08	0,05	0,12	0,11	0,05	0,13
7	0,07	0,05	0,08	0,08	0,05	0,10
8	0,29	0,3	0,14	0,92	0,9	0,04

simulation, the number of reasons for flickering test results has increased. In addition to the leading cause, there is also a delay in the reading of variable values with the XCP protocol. The reason for this is the increasing CPU load from calculations related to varying values of physical quantities such as current, voltage, temperature from the high-voltage battery. Therefore, the standard deviation of empirical probability of result flickering (occurrence of a negative result in 10 trials) is considerably varying in certain examples.

5 Conclusion

The aim of this work was to verify the simplification used in HiL simulation when testing embedded systems. The research conducted compares test verdicts from a simplified vehicle simulation and a model—based approach in the context of flickering results. The test bench was developed, and real-world experiments were conducted at a company designing and manufacturing embedded systems that supervise high-voltage battery systems in electric vehicles. Statistical analyses were carried out, which revealed that the employment of model-based dynamic vehicle simulation has an impact on the test results of embedded systems. It increases the empirical probability of flickering test results, also the number of reasons for flickering results in a single test increased. This was a consequence of the CPU load of the computations associated with the varying value of the vehicle's physical parameters. It was confirmed that simplification used in HiL simulation could mask defects in embedded software under test.

The described sensitivity of the embedded system to variations in the test environment is a detrimental phenomenon. It is also important to consider it when designing test cases. In the phase of test environment planning, it is reasonable to determine the areas where simplified, static simulation has no impact on the tested functionalities, and consequently on the test results. It may be applied in all those areas.

However, if the testing environment and the changes occurring in it contribute to the occurrence of flickering test results, the use of a dynamic vehicle model should be considered. This will prevent that the influence of the real environment on tested functionalities is neglected in the test results. This requires a significant financial, time and human resource investment.

The research also confirmed the suitability of the dynamic model for further work related to the automation of the verification process of flickering test results with augmented intelligence. The further development of the model-based approach is intended, as the research has proved that it can be applied to the generation of aberrations in the test system. The aim of the further research study is proposing a methodology for the development of test cases to allow the application of augmented intelligence algorithms in the analysis of flickering test results. The model will be utilized in a test cases development methodology for preparing data to implement augmented intelligence algorithms for verification of test results and for identification of cause of flickering results.

Acknowledgment. Research supported by the Polish Ministry of Education and Science Grant No DWD/3/33/2019 and DWD/4/55/2020.This publication is partially supported from the statutory funds of Department of Fundamentals of Machinery Design.

References

1. Axelrod, A.: Complete Guide to Test Automation, PWN, Warszawa (2020)
2. Bień, A., Rzeszutko, J., Szymański, T.: Embedded systems dedicated to measurement and control in the automotive industry. In: The Design Process and the Initial Diagnosis. Pomiary Automatyka Kontrola, pp. 655–659 (2014)

3. Chrisofakis, E., Junghanns, A., Kehrer, Ch., Rink, A.: Simulation-based development of automotive control software with Modelica. In: 8th International Modelica Conference, pp. 1–7 (2011)
4. Gnacy-Gajdzik, A., Przystałka, P.: Ewolucja testowania systemów wbudowanych w motoryzacji wraz z wprowadzeniem nomy ISO 21448:2019 Pojazdy drogowe - Bezpieczeństwo zamierzonej funkcjonalności. Utrzymanie Ruchu **2**, 7–16 (2021)
5. Gnacy-Gajdzik, A., Przystałka, P., Sebzda, W.: Analiza częstości występowania fałszywie negatywnych wyników automatycznych testów układów wbudowanych, In: Metody komputerowe - 2021: Studencka konferencja naukowa, pp. 41–44. Gliwice (2021)
6. Gnacy-Gajdzik, A., Przystałka, P.: Identyfikacja anomalii wywołujących fałszywie negatywne wyniki w procesie testowania oprogramowania wbudowanego, In: Pawliczek, R., Owsiński, R., Łagoda, T. (eds.) Projektowanie, budowa i eksploatacja maszyn, vol. 558, pp. 61–72, Studia i Monografie. Politechnika Opolska (2021)
7. Harris, I.: Embedded software for automotive applications. In: Oshana, R., Kraeling, M. (eds.) Software Engineering for Embedded Systems, Newnes, pp. 767–816 (2013)
8. Ito, I.: Battery Electric Vehicle Model in Simscape. The MathWorks, Inc. (2021). https://github.com/mathworks/Simscape-Battery-Electric-Vehicle-Model. Accessed 20 Feb 2022
9. Kum, D. -h., Son, J., Lee, S. –b., Wilson, I.: Automated testing for automotive embedded systems. In: SICE-ICASE International Joint Conference, pp. 4414–4418 (2006)
10. Krupp, A., Müller, W.: A systematic approach to the test of combined HW/SW systems, 2 design. Automation & Test in Europe Conference & Exhibition **2010**, 323–326 (2010)
11. Mizoguchi, M., Iida, T., Irie, T.: Optimization of automated executions based on integration test configurations of embedded software. In: IEEE International Conference on Software Testing, Verification and Validation Workshops (ICSTW), Porto, pp. 358–363 (2020)
12. Moon, H., Kim, G., Kim, Y., Shin, S., Kim, K., Im, S.: Automation test method for automotive embedded software based on AUTOSAR. In: 2009 Fourth International Conference on Software Engineering Advances, pp. 158–162 (2009)
13. Nikończuk, P., Jaszczak, S.: Stanowisko do testowania ukladów sterowania temperatura w kabinie lakierniczej w trybie Hardware In The Loop. Autobusy: technika, eksploatacja, systemy transportowe **12**, 1244–1247 (2016). https://orcid.org/0000-0002-7809-7653
14. Placho, T., Schmittner, Ch., Bonitz, A., Wana, O.: Management of automotive software updates. Microprocess. Microsyst. **78** (2020)
15. Rebeschieß, S., Liebezeit, T., Bazarsuren, U.: Automatisierter Closed-Loop-Testprozess für Steuergerätefunktionen. ATZ Elektron **2**, 34–41 (2007)
16. Road vehicles, Functional Safety, ISO 26262 (2018)

An Analysis of Observability and Detectability for Different Sets of Measured Outputs – CSTR Case Study

Mateusz Czyżniewski and Rafał Łangowski(✉)

Department of Intelligent Control and Decision Support Systems,
Digital Technologies Center, Gdańsk University of Technology,
G. Narutowicza 11/12, 80-233 Gdańsk, Poland
rafal.langowski@pg.edu.pl

Abstract. A problem of proving the observability/detectability at a given measured output for non-linear biochemical systems has been addressed in this paper. A theory of indistinguishable state trajectories has been used to prove the properties of the observability or detectability of this system. It is related to taking system dynamics into consideration depending on initial conditions and the impact of inputs taking into account a given measured output. An analysis of the observability/detectability has been performed for five cases where the various sets of measured outputs have been considered. The research results obtained are crucial for system state estimation that involves the synthesis of state observer. The proposed method may be successfully applied to the complex biochemical non-linear systems modelled as a continuous stirred tank reactor. Finally, the results obtained demonstrate the high effectiveness of the devised approach.

Keywords: Biochemical system · Indistinguishable sets · Observability

1 Introduction

In general, advanced algorithms performing different tasks, e.g., monitoring, estimation, diagnostics, and control are needed for the effective handling of a majority of modern industrial plants. The proper operation of these algorithms connects with the problem of accessing information about the process variables, e.g., the current state and controlled outputs of a given process. This information is delivered by the measuring devices installed in the plant where the process occurs. In operational practice not all process variables are physically measurable, it is not possible to allocate the necessary number of sensors, or the quality of measurement information provided is not good enough [13]. Thus, the missing information about process variables needs to be recovered by employing, e.g., their estimates. Typically, the estimation is based on the mathematical model

Z. Kowalczuk (Ed.): DPS 2022, LNNS 545, pp. 352–363, 2023.
https://doi.org/10.1007/978-3-031-16159-9_29

of the considered process and measurements of available process variables, and the various state observers are used to reconstruct (estimate) the system state [1,2,5,6,12,17,18]. It is well-known that the possibility of an observer synthesis is strictly related to the observability/detectability properties of the system model. Formally, for a known input-output relationship of the system, the observability addresses the ability to fully and explicitly reconstruct (distinguish) its state in a finite time-horizon, whereas detectability (asymptotic observability) allows only an asymptotic estimation of its state. It should be noted that both properties are more difficult to investigate in non-linear systems [15,16].

The classical approach based on differential geometry tools may be used to prove the observability/detectability of a dynamic system. It is associated with using 'observability maps' that are linked with the Lie differentiation of the measuring output equations along the drift vector field [1,4,8,10]. However, this method is not always convenient. It is due to complicated calculations, especially for the real complex systems such as biochemical systems and the problems with checking the injectivity of the 'observability map' [21]. To overcome these problems, an alternative approach based on a method of indistinguishable state trajectories (indistinguishable dynamics/sets) can be used [1,4,6,8,10]. It is related to the dynamical interpretation of the state behaviour depending on initial conditions and the impact of external inputs considering a given measured output. Technically, it involves the derivation of the 'error system' (differential-algebraic equation - DAE) which represents indistinguishable dynamics [9,15,16,19–21]. The latter methodology with formal mathematical definitions only referenced from [6,8,15–17,20,21] is applied in this paper.

The system under consideration is a part of a wastewater treatment plant (WWTP) which consists of two tanks, i.e. a bioreactor (where the considered biochemical processes appear) and a settler (where treated wastewater is separated from the biomass). The entire system is modelled as a continuous stirred tank reactor (CSTR) with the microbial growth reaction and their mortality with the aggregated substrate and biomass concentrations [3]. This is characterised by taking into account the continuous flow of the biomass which is needed for maintaining appropriate conditions needed to perform biochemical reactions. Therefore, in comparison to, e.g., [16,19–21], the considered system includes essential dynamical relationship between the two mentioned tanks. Hence, the performed research provides the analysis of an interesting case study, where distinctive cross interactions have occurred in the state dynamics. These interactions do not appear in a single bioreactor, thus the observability/detectability analysis must be premised on consideration of more complicated state relations. Moreover, another important issue for the analysis is the choice of the system's measured outputs. Therefore, the various combinations of biomass, substrate and dissolved oxygen concentrations measurements as physically and technologically available in WWTP are considered [3,7]. To summarise, the main aim of this work is to analyse the observability/detectability for a given measured output selection of the biochemical system modelled as the proper CSTR system. This research has been performed with association to further state observer synthesis.

Fundamentally, this research analysis is necessary to determine whether possible to develop a particular state observer dependently on the available measured outputs of the system.

2 Model of CSTR System

The considered biochemical system is highly non-linear and depends on multiple exogenous inputs. Hence, it can be modelled as a multi-input multi-output (MIMO) affine system as follows [10,11]:

$$\begin{cases} \dot{\boldsymbol{x}}(t) & = \boldsymbol{f}(\boldsymbol{x}(t)) + \sum\limits_{i_{\mathrm{p}}=1}^{p} \boldsymbol{b}_{i_{\mathrm{p}}}(\boldsymbol{x}(t))u_{i_{\mathrm{p}}}(t) \\ \boldsymbol{x}(t_0) & = \boldsymbol{x}_0 \\ \boldsymbol{y}(t) & = \boldsymbol{h}(\boldsymbol{x}(t)) \end{cases}, \tag{1}$$

where: $\dot{(\cdot)}$ stands for the derivative with respect to t; $t \in \mathbb{T} \subset \mathbb{R}_{+} \cup \{0\} \subset \mathbb{R}$ is the time instant, $\mathbb{R}_{+}$ denotes the positive part of $\mathbb{R}$; $\forall t \in \mathbb{T}\colon\ \boldsymbol{x}(t) \in \mathbb{X}_{\mathrm{n}} \subset \mathbb{R}^{\mathrm{n}}$ is the vector of state variables, $\mathbb{X}_{\mathrm{n}}$ is a $\mathcal{C}^{\infty}(\cdot)$ manifold of dimension n, $\boldsymbol{x}_0$ signifies the vector of initial conditions; $\forall t \in \mathbb{T}\colon\ \|u_{i_{\mathrm{p}}}(t)\|_{\infty} = \sup\left\{|\ u_{i_{\mathrm{p}}}\ |\colon\ t \in \mathbb{T}\right\} \leq u_{i_{\mathrm{p}}}^{\mathrm{bound}} < \infty, i_{\mathrm{p}} = \overline{1,p}$ denotes the i_{p}th exogenous input; $\forall t \in \mathbb{T}\colon\ \boldsymbol{y}(t) \in \mathbb{Y}_{\mathrm{q}} \subset \mathbb{R}^{\mathrm{q}}$ is the vector of measured outputs; $\forall t \in \mathbb{T}\colon\ \boldsymbol{b}_{i_{\mathrm{p}}}, \boldsymbol{f}\colon\ \mathbb{X}_{\mathrm{n}} \rightarrow \mathrm{T}\mathbb{X}_{\mathrm{n}}$, $\forall t \in \mathbb{T}\colon\ \boldsymbol{h}\colon\ \mathbb{X}_{\mathrm{n}} \rightarrow \mathbb{Y}_{\mathrm{q}}$ are smooth maps representing inputs, internal system dynamics, and measured outputs, respectively; $\mathrm{T}(\cdot)$ is the tangent bundle of vector field.

By invoking mass balance law and based on (1), the model of CSTR system with the microbial growth reaction and their mortality with the aggregated substrate and biomass concentrations yields [3,5,7,20]:

$$\begin{cases} \dot{X}(t) & = \mu(t)X(t) - \beta_{\mathrm{m}}X(t) - (1+r)X(t)D(t) + rX_{\mathrm{r}}(t)D(t) \\ \dot{S}(t) & = -\frac{1}{Y_{\mathrm{s}}}\mu(t)X(t) - m_{\mathrm{s}}X(t) + S_{\mathrm{in}}D(t) - (1+r)S(t)D(t) \\ \dot{DO}(t) & = -\frac{1}{Y_{\mathrm{o}}}\mu(t)X(t) - m_{\mathrm{o}}X(t) + DO_{\mathrm{in}}D(t) \\ & \quad -(1+r)DO(t)D(t) + K_{\mathrm{LA}}(t)(DO_{\mathrm{s}} - DO(t)) \\ \dot{X}_{\mathrm{r}}(t) & = v(1+r)X(t)D(t) - v(w+r)X_{r}(t)D(t) \\ X(t_0) & = X_0,\ S(t_0) = S_0,\ DO(t_0) = DO_0,\ X_{\mathrm{r}}(t_0) = X_{\mathrm{r0}} \end{cases}, \tag{2}$$

where: $\mu(t)$ denotes the reaction kinetics function; $\beta_{\mathrm{m}} \in \mathbb{R}_{+}$ $[h^{-1}]$ is the biomass mortality rate; $m_{\mathrm{s}} \in \mathbb{R}_{+}$ $[h^{-1}]$, $m_{\mathrm{o}} \in \mathbb{R}_{+}$ $[h^{-1}]$ are the maintenance coefficients of the substrate and dissolved oxygen concentrations; $Y_{\mathrm{o}} \in \mathbb{R}_{+}$ $[-]$, $Y_{\mathrm{s}} \in \mathbb{R}_{+}$ $[-]$ denote the yield coefficients of the dissolved oxygen and substrate concentrations; $S_{\mathrm{in}} \in \mathbb{R}_{+}$ $[\mathrm{mg}/L]$, $DO_{\mathrm{in}} \in \mathbb{R}_{+}$ $[\mathrm{mg}/L]$ signify the concentrations of the substrate in inflow to the CSTR and dissolved oxygen; $DO_{\mathrm{s}} \in \mathbb{R}_{+}$ $[\mathrm{mg}/L]$ stands for the dissolved oxygen saturation constant; $K_{\mathrm{LA}}(t)$ denotes the gas-liquid transfer function; $D(t) = Q_{\mathrm{in}}(t)/V_{\mathrm{a}}$ $[h^{-1}]$ is the dilution rate, where $Q_{\text{in}}(t) \in \mathbb{R}_+$ $[m^3/h]$, $V_{\text{a}} \in \mathbb{R}_+$ $[m^3]$ are the inflow rate and bioreactor volume; $r = Q_{\text{r}}(t)/Q_{\text{in}}(t) \in \mathbb{R}_+$ $[-]$ is the constant proportion between the inflow and recycled flow, where $Q_{\text{r}}(t) \in \mathbb{R}_+$ $[m^3/h]$ is the recycled flow rate; $w = Q_{\text{w}}(t)/Q_{\text{in}}(t) \in \mathbb{R}_+$ $[-]$ stands for the constant proportion between the wastewater flow and inflow, where $Q_{\text{w}}(t) \in \mathbb{R}_+$ $[m^3/h]$ is the wastewater flow rate; $v = V_{\text{a}}/V_{\text{s}} \in \mathbb{R}_+$ $[-]$ denotes the constant proportion between the bioreactor's and settler's volumes, where $V_{\text{s}} \in \mathbb{R}_+$ $[m^3]$ is the settler volume; $X(t)$ [mg/L], $S(t)$ [mg/L], $DO(t)$ [mg/L], $X_{\text{r}}(t)$ [mg/L] are the aggregated biomass, aggregated substrate, dissolved oxygen, and aggregated recycled biomass concentrations, respectively. For further considerations the set of all possible system states is defined as:

$$\begin{gathered}\Omega = \{(X(t), S(t), DO(t), X_{\text{r}}(t)) \in \mathbb{R}_+^4 \cup \{\mathbf{0}_{4\times1}\} : \forall t \in \mathbb{T} \\ 0 \le \underline{X}(t) \le X(t) \le \overline{X}, \quad 0 \le \underline{S}(t) \le S(t) \le \overline{S}, \\ 0 \le \underline{DO}(t) \le DO(t) \le \overline{DO}, \quad 0 \le \underline{X}_{\text{r}}(t) \le X_{\text{r}}(t) \le \overline{X}_{\text{r}}\}\end{gathered}, \tag{3}$$

where: $\underline{(\cdot)}$ and $\overline{(\cdot)}$ signify the lower and upper bound of the particular variable. It is worth adding that Ω is an invariant set known with the condition of general theory of biochemical processes dynamics [3,18]: $\Omega \subset \mathbb{X}_{\text{n}} \subset \mathbb{R}^4$. Moreover, according to [17] system (2) is Lyapunov stable and its state trajectories are bounded $\forall t \in \mathbb{T}$.

For imposing presentation related coherency, the particular state variables and external inputs of the non-linear model of the CSTR system are defined as follows:

$$\begin{gathered}\boldsymbol{x}(t) = \left[x_1(t)\ x_2(t)\ x_3(t)\ x_4(t)\right]^{\text{T}} \triangleq \left[X(t)\ S(t)\ DO(t)\ X_{\text{r}}(t)\right]^{\text{T}} \\ \boldsymbol{u}(t) = \left[u_1(t)\ u_2(t)\right]^{\text{T}} \triangleq \left[D(t)\ K_{\text{LA}}(t)\right]^{\text{T}}\end{gathered}. \tag{4}$$

In the considered system, the dilution rate ($0 \le D(t) \le \overline{D} \in \mathbb{R}_+$) and gas-liquid transfer function ($0 \le K_{\text{LA}}(t) \le \overline{K}_{\text{LA}} \in \mathbb{R}_+$) signify the bounded inputs [2,3,14,18]. Moreover, both inputs are the permanent excited positive signals due to the physical properties of the considered system, i.e. [11]:

$$\underline{\alpha}_{\text{u}} \boldsymbol{I}_{2\times2} \le \int_t^{T+t} \boldsymbol{u}(\tau)\boldsymbol{u}^{\text{T}}(\tau)d\tau \le \overline{\alpha}_{\text{u}} \boldsymbol{I}_{2\times2}, \tag{5}$$

where: $\underline{\alpha}_{\text{u}} \in \mathbb{R}_+$, $\overline{\alpha}_{\text{u}} \in \mathbb{R}_+$ are the lower and upper bounds of the permanently excited inputs; $T \in \mathbb{T}$ is the selected time period; $\boldsymbol{I}_{2\times2}$ is the identity matrix.

It is worth adding that according to (5) the $S(t)$ and $DO(t)$ concentrations are always positive (excluding the situation when initial conditions are equal to zero). In turn, the reaction kinetics function ($0 \le \mu(t) \le \overline{\mu} \in \mathbb{R}_+$) is a product of two distinct Monod functions (components) [2,3,5,17,18,20,21]:

$$\mu(t) = \mu_{\max} \frac{S(t)}{K_{\text{s}} + S(t)} \frac{DO(t)}{K_{\text{o}} + DO(t)}, \tag{6}$$

where $\forall t \in \mathbb{T}$: $\mu_{\max} \in \mathbb{R}_+$ $[h^{-1}]$, $K_\mathrm{s} \in \mathbb{R}_+$ $[mg/L]$, $K_\mathrm{o} \in \mathbb{R}_+$ $[mg/L]$ denote the coefficients of maximum specific growth rate, substrate saturation, and dissolved oxygen saturation, respectively.

Thus, the problem of proving observability/detectability of biochemical systems is strictly joined with not only an appropriate selection of measured outputs but also with considering the particular form of the reaction kinetics function. As it has been shown, e.g. in [2,17,20,21], the Monod function entails indistinguishable states of the system for the given initial conditions only in the (biomass) wash-out state ($X(t)$ and $X_\mathrm{r}(t)$ concentrations are zero). In detail, the functional form of Monod kinetics, compared to, e.g. Haldane function, does not affect the verdict of the observability and detectability analysis due to its monotonicity.

3 Analysis of observability and detectability

As mentioned in Sect. 1 the property of observability enables an exact (the system is observable) or asymptotic (the system is detectable) estimation of the system state. In this paper, the observability/detectability analysis is based on a method of indistinguishable state trajectories, and it is given as follows [2,9, 15,16,19–21]. First, the extended system model, which consists of the 'original - $\boldsymbol{x}(t)$' system model and 'copied - $\boldsymbol{z}(t)$' system model is built (see (2)). The extended system model has an analogous form to the general form of affine system (1). Whereas its dynamics is an 8-dimensional manifold composed as the Cartesian product of the original state space and 'error' ($\boldsymbol{\varepsilon}(t) \in \Xi_\mathrm{n}$) space, i.e. $[\boldsymbol{x}^\mathrm{T}(t),\ \boldsymbol{\varepsilon}^\mathrm{T}(t)]^\mathrm{T} \in \Omega \times \Xi_\mathrm{n} = \Psi \subset \mathbb{R}^4_+ \cup \{\mathbf{0}_{4\times 1}\} \times \mathbb{R}^4$. The 'error' is defined as follows:

$$\boldsymbol{\varepsilon}(t) \triangleq \boldsymbol{x}(t) - \boldsymbol{z}(t), \quad \rightarrow \quad \boldsymbol{z}(t) = \boldsymbol{x}(t) - \boldsymbol{\varepsilon}(t). \tag{7}$$

By assumption, the initial conditions of both systems are $\boldsymbol{x}_0 \neq \boldsymbol{z}_0$, and they impose different integral curve trajectories of system state $\boldsymbol{x}(t) \neq \boldsymbol{z}(t)$ $\forall t \in \mathbb{T}$. If the measured outputs and exogenous inputs of both systems are the same in the time interval $\mathbb{T}$, the observability is equivalent to the statement that the only possible state solutions of 'original' and 'copied' state dynamics under these conditions are always equal (the 'error' and its derivatives are zero), which is equivalent to trajectory indistinguishability. In turn, detectability means that mentioned indistinguishable state trajectories mutually tend to one another asymptotically in finite time, which implies asymptotic stability of error dynamics. It should be noticed that, due to measurements, the equivalence of particular elements of the $\boldsymbol{x}(t)$ and $\boldsymbol{z}(t)$ means that corresponding elements of $\boldsymbol{\varepsilon}(t)$ (and also $\dot{\boldsymbol{\varepsilon}}(t)$) are equal to zero $\forall t \in \mathbb{T}$ [15,16,19–21].

In comparison to the classical differential geometry-based method, the several features of applying indistinguishable state trajectories method are worth emphasising [15,16]:

1. Both observability and especially detectability are possible to be prove.
2. The interpretation of the results obtained is relatively simple and based on classical terms of the theory of differential equations.

3. To characterise of particular interactions between system states and exogenous inputs.
4. The results may be directly interpreted as global or local.

The considered dynamics of the 'error' reads:

$$\begin{cases} \dot{\varepsilon}_1(t) &= \mu(t)x_1(t) - \mu_z(t)[x_1(t) - \varepsilon_1(t)] - \beta_m\varepsilon_1(t) \\ & -(1+r)\varepsilon_1(t)u_1(t) + r\varepsilon_4(t)u_1(t) \\ \dot{\varepsilon}_2(t) &= -\frac{1}{Y_s}\left[\mu(t)x_1(t) - \mu_z(t)[x_1(t) - \varepsilon_1(t)]\right] \\ & -m_s\varepsilon_1(t) - (1+r)\varepsilon_2(t)u_1(t) \\ \dot{\varepsilon}_3(t) &= -\frac{1}{Y_o}\left[\mu(t)x_1(t) - \mu_z(t)[x_1(t) - \varepsilon_1(t)]\right] \\ & -m_o\varepsilon_1(t) - (1+r)\varepsilon_3(t)u_1(t) - \varepsilon_3(t)u_2(t) \\ \dot{\varepsilon}_4(t) &= v(1+r)\varepsilon_1(t)u_1(t) - v(w+r)\varepsilon_4(t)u_1(t) \\ \boldsymbol{\varepsilon}(t_0) &= \boldsymbol{\varepsilon}_0 \end{cases}, \tag{8}$$

where $\mu_z(t) \in \mathbb{R}\ [h^{-1}]$ is the reaction kinetics function of 'copied' system, which structure and parametrisation is the same as in the 'original' one, and it is given as:

$$\mu_z(t) = \mu_{\max}\frac{x_2(t) - \varepsilon_2(t)}{K_s + x_2(t) - \varepsilon_2(t)}\frac{x_3(t) - \varepsilon_3(t)}{K_o + x_3(t) - \varepsilon_3(t)}. \tag{9}$$

Further in this section, an analysis of observability/detectability for the considered CSTR model for five various sets of measured outputs are performed.

3.1 Results – CSTR Case Study 1

The set of measured outputs are: $y_1(t) = h_1(\boldsymbol{x}(t)) = x_2(t)$, $y_2(t) = h_2(\boldsymbol{x}(t)) = x_3(t)$. It implies that $\forall t \in \mathbb{T}$ $\varepsilon_2(t) = 0$, $\varepsilon_3(t) = 0$, and $\dot{\varepsilon}_2(t) = 0$, $\dot{\varepsilon}_3(t) = 0$, and $\mu(t) = \mu_z(t)$. Thus, (8) is transformed to the following DAE system:

$$\begin{cases} \dot{\varepsilon}_1(t) &= \mu(t)\varepsilon_1(t) - \beta_m\varepsilon_1(t) - (1+r)\varepsilon_1(t)u_1(t) + r\varepsilon_4(t)u_1(t) \\ 0 &= -\frac{1}{Y_s}\mu(t)\varepsilon_1(t) - m_s\varepsilon_1(t) \\ 0 &= -\frac{1}{Y_o}\mu(t)\varepsilon_1(t) - m_o\varepsilon_1(t) \\ \dot{\varepsilon}_4(t) &= v(1+r)\varepsilon_1(t)u_1(t) - v(w+r)\varepsilon_4(t)u_1(t) \\ \boldsymbol{\varepsilon}(t_0) &= \boldsymbol{\varepsilon}_0 \end{cases}. \tag{10}$$

By taking the second and third equations in (10), $\varepsilon_1(t)$ can be expressed as:

$$(\mu(t) + Y_s m_s)\,\varepsilon_1(t) = 0, \text{ or } (\mu(t) + Y_o m_o)\,\varepsilon_1(t) = 0. \tag{11}$$

Taking into account the boundedness of trajectories of particular variables, $\mu(t)$ is also bounded. When one of state variables $x_2(t)$ or $x_3(t)$ became equal to zero, then the $\mu(t) = 0$. However, this case is omitted due to the physical properties of the permanently excited dynamics (see Sect. 2). Also, this fact does not affect the making the left-hand sides of (11) equal to zero. Hence, to meet equivalence between two sides of (11), $\varepsilon_1(t)$ and $\dot{\varepsilon}_1(t)$ must be equal to zero $\forall t \in \mathbb{T}$. It leads to the conclusion that the first state variable $x_1(t)$ is observable. Therefore, (10) can be transformed to the following DAE system:

$$\begin{cases} 0 & = r\varepsilon_4(t)u_1(t) \\ \dot{\varepsilon}_4(t) & = -v(w+r)\varepsilon_4(t)u_1(t) \\ \boldsymbol{\varepsilon}(t_0) & = \boldsymbol{\varepsilon}_0 \end{cases} . \tag{12}$$

Taking into account that the inputs are the permanent excited positive signals, it can be shown that the solution of the algebraic equation in (12) must always be equal to zero, i.e. due to $\varepsilon_4(t) = 0\ \forall t \in \mathbb{T}$. Hence, the fourth state variable $x_4(t)$ is observable. Thus, the considered model of CSTR is observable for selected measured outputs.

3.2 Results – CSTR Case Study 2

The set of measured outputs are: $y_1(t) = h_1(\boldsymbol{x}(t)) = x_1(t)$, $y_2(t) = h_2(\boldsymbol{x}(t)) = x_4(t)$. It implies that $\forall t \in \mathbb{T}\ \varepsilon_1(t) = 0$, $\varepsilon_4(t) = 0$, and $\dot{\varepsilon}_1(t) = 0$, $\dot{\varepsilon}_4(t) = 0$. Thus, (8) is transformed to the following DAE system:

$$\begin{cases} 0 & = \mu(t)x_1(t) - \mu_z(t)x_1(t) \\ \dot{\varepsilon}_2(t) & = -\frac{1}{Y_s}\Big[\mu(t)x_1(t) - \mu_z(t)x_1(t)\Big] - (1+r)\varepsilon_2(t)u_1(t) \\ \dot{\varepsilon}_3(t) & = -\frac{1}{Y_o}\Big[\mu(t)x_1(t) - \mu_z(t)x_1(t)\Big] - (1+r)\varepsilon_3(t)u_1(t) - \varepsilon_3(t)u_2(t) \\ \boldsymbol{\varepsilon}(t_0) & = \boldsymbol{\varepsilon}_0 \end{cases} . \tag{13}$$

Considering the first equation in (13) this system can be rewritten as:

$$\begin{cases} \dot{\varepsilon}_2(t) & = -(1+r)\varepsilon_2(t)u_1(t) \\ \dot{\varepsilon}_3(t) & = -(1+r)\varepsilon_3(t)u_1(t) - \varepsilon_3(t)u_2(t) \\ 0 & = \mu(t)x_1(t) - \mu_z(t)x_1(t) \\ \boldsymbol{\varepsilon}(t_0) & = \boldsymbol{\varepsilon}_0 \end{cases} . \tag{14}$$

Because of the left-hand side of the algebraic equation in (14) must be equal to zero $\forall t \in \mathbb{T}$, the considerations about observability/detectability of $x_2(t)$ and $x_3(t)$ are divided into two situations. The first is associated with the fact that $x_1(t)$ is zero (either it has become asymptotically or by initial conditions). Whereas the other one is linked with the fact that the kinetics functions $\mu(t)$ and $\mu_z(t)$ are equal to each other $\forall t \in \mathbb{T}$ and $x_1(t)$ is always positive. The first situation is connected with the wash-out state, which is highly undesired for the physical requirements of the entire process. However, even knowing that, it should be analysed.

In the first situation, where $x_1(t)$ becomes zero, the behaviour of kinetics functions is dependent only on inputs and initial conditions of $x_2(t)$ and $x_3(t)$. Taking into account that the inputs are the permanent excited positive signals, the dynamics of $\varepsilon_2(t)$ and $\varepsilon_3(t)$ from (14) will asymptotically tend to zero. Moreover, due to fact that $\mu_z(t)$ and $\mu(t)$ are Monod terms, when $\varepsilon_2(t) \to 0$ and $\varepsilon_3(t) \to 0$ then $\mu_z(t) \to \mu(t)$ asymptotically. Additionally, due to the boundedness of the state trajectories, the kinetics functions do not tend to infinity. Therefore, in the first situation $x_2(t)$ and $x_3(t)$ are detectable variables. In turn, in the second situation, when $x_1(t)$ is always positive the same result is obtained; however, its understanding is slightly different. To make the right-hand side of the third equation in (14) equals zero, both $\varepsilon_2(t)$ and $\varepsilon_3(t)$ and of course its derivatives must tend to zero. Hence, in the second situation, $x_2(t)$ and $x_3(t)$ are detectable. Thus, the considered model of CSTR is detectable for selected measured outputs.

3.3 Results – CSTR Case Study 3

The set of measured outputs are: $y_1(t) = h_1(\boldsymbol{x}(t)) = x_2(t)$, $y_2(t) = h_2(\boldsymbol{x}(t)) = x_4(t)$. It implies that $\forall t \in \mathbb{T}\ \varepsilon_2(t) = 0$, $\varepsilon_4(t) = 0$, and $\dot{\varepsilon}_2(t) = 0$, $\dot{\varepsilon}_4(t) = 0$. Thus, (8) is transformed to the following DAE system:

$$\begin{cases} \dot{\varepsilon}_1(t) &= \mu(t)x_1(t) - \mu_z(t)[x_1(t) - \varepsilon_1(t)] - \beta_m\varepsilon_1(t) - (1+r)\varepsilon_1(t)u_1(t) \\ 0 &= -\dfrac{1}{Y_s}\Big[\mu(t)x_1(t) - \mu_z(t)[x_1(t) - \varepsilon_1(t)]\Big] - m_s\varepsilon_1(t) \\ \dot{\varepsilon}_3(t) &= -\dfrac{1}{Y_o}\Big[\mu(t)x_1(t) - \mu_z(t)[x_1(t) - \varepsilon_1(t)]\Big] \\ & -m_o\varepsilon_1(t) - (1+r)\varepsilon_3(t)u_1(t) - \varepsilon_3(t)u_2(t) \\ 0 &= v(1+r)\varepsilon_1(t)u_1(t) \\ \varepsilon(t_0) &= \varepsilon_0 \end{cases} . \tag{15}$$

By invoking to the last equation in (15), if $\forall t \in \mathbb{T}\ u_1(t) \neq 0$ then $\forall t \in \mathbb{T}\ \varepsilon_1(t) = 0$ and $\dot{\varepsilon}_1(t) = 0$, hence $x_1(t)$ is observable. It leads to the following transformation of (15):

$$\begin{cases} 0 &= \mu(t)x_1(t) - \mu_z(t)x_1(t) \\ 0 &= -\dfrac{1}{Y_s}\Big[\mu(t)x_1(t) - \mu_z(t)x_1(t)\Big] \\ \dot{\varepsilon}_3(t) &= -\dfrac{1}{Y_o}\Big[\mu(t)x_1(t) - \mu_z(t)x_1(t)\Big] - (1+r)\varepsilon_3(t)u_1(t) - \varepsilon_3(t)u_2(t) \\ \varepsilon(t_0) &= \varepsilon_0 \end{cases} . \tag{16}$$

Noting that the first two equations in (16) are essentially the same mathematically, thus (16) can be rewritten as:

$$\begin{cases} \dot{\varepsilon}_3(t) &= -(1+r)\varepsilon_3(t)u_1(t) - \varepsilon_3(t)u_2(t) \\ 0 &= \mu(t)x_1(t) - \mu_z(t)x_1(t) \\ \varepsilon(t_0) &= \varepsilon_0 \end{cases} . \tag{17}$$

It should be noticed that the right-hand side of the algebraic equation in (17) is the same as in (14). Therefore, the additional considerations are omitted and the conclusions are drawn. When $\varepsilon_2(t)$ and its derivative are equal to zero $\forall t \in \mathbb{T}$, $x_1(t)$ and $x_3(t)$ are observable. Since the kinetics functions are equal to each other the following understanding is given. If only one 'error' variable appears in 'copied' kinetics function, the equivalence imposes that $\varepsilon_3(t)$ must be equal to zero $\forall t \in \mathbb{T}$:

$$0 = (\mu(t) - \mu_z(t))\, x_1(t) \equiv \frac{x_3(t)}{K_o + x_3(t)} = \frac{x_3(t) - \varepsilon_3(t)}{K_o + x_3(t) - \varepsilon_3(t)}. \tag{18}$$

Hence, when biomass is not washed from the bioreactor, all of the states are observable. Thus, the considered model of CSTR is observable for selected measured outputs. In turn, in the wash-out state, due to permanent excitation of the inputs and the boundedness of both kinetics functions, $x_1(t)$ is observable but $x_3(t)$ is only detectable.

3.4 Results – CSTR Case Study 4

The set of measured outputs are: $y_1(t) = h_1(\boldsymbol{x}(t)) = x_3(t)$, $y_2(t) = h_2(\boldsymbol{x}(t)) = x_4(t)$. It implies that $\forall t \in \mathbb{T}$ $\varepsilon_3(t) = 0$, $\varepsilon_4(t) = 0$, and $\dot{\varepsilon}_3(t) = 0$, $\dot{\varepsilon}_4(t) = 0$. Thus, (8) is transformed to the following DAE system:

$$\begin{cases} \dot{\varepsilon}_1(t) &= \mu(t)x_1(t) - \mu_z(t)[x_1(t) - \varepsilon_1(t)] - \beta_m\varepsilon_1(t) - (1+r)\varepsilon_1(t)u_1(t) \\ \dot{\varepsilon}_2(t) &= -\frac{1}{Y_s}\Big[\mu(t)x_1(t) - \mu_z(t)[x_1(t) - \varepsilon_1(t)]\Big] - m_s\varepsilon_1(t) - (1+r)\varepsilon_2(t)u_1(t) \\ 0 &= -\frac{1}{Y_o}\Big[\mu(t)x_1(t) - \mu_z(t)[x_1(t) - \varepsilon_1(t)]\Big] - m_o\varepsilon_1(t) \\ 0 &= v(1+r)\varepsilon_1(t)u_1(t) \\ \boldsymbol{\varepsilon}(t_0) &= \boldsymbol{\varepsilon}_0 \end{cases} . \tag{19}$$

By invoking to the last equation in (19), if $\forall t \in \mathbb{T}$ $u_1(t) \neq 0$ then $\forall t \in \mathbb{T}$ $\varepsilon_1(t) = 0$ and $\dot{\varepsilon}_1(t) = 0$, hence $x_1(t)$ is observable. It leads to the following transformation of (19):

$$\begin{cases} 0 &= \mu(t)x_1(t) - \mu_z(t)x_1(t) \\ \dot{\varepsilon}_2(t) &= -\frac{1}{Y_s}\Big[\mu(t)x_1(t) - \mu_z(t)x_1(t)\Big] - (1+r)\varepsilon_2(t)u_1(t) \\ 0 &= -\frac{1}{Y_o}\Big[\mu(t)x_1(t) - \mu_z(t)x_1(t)\Big] \\ \boldsymbol{\varepsilon}(t_0) &= \boldsymbol{\varepsilon}_0 \end{cases} . \tag{20}$$

Noting that the first and third equation in (20) are essentially the same mathematically, thus (20) can be rewritten as:

$$\begin{cases} \dot{\varepsilon}_2(t) &= -(1+r)\varepsilon_2(t)u_1(t) \\ 0 &= \mu(t)x_1(t) - \mu_z(t)x_1(t) \\ \boldsymbol{\varepsilon}(t_0) &= \boldsymbol{\varepsilon}_0 \end{cases} . \tag{21}$$

It should be noticed that the right-hand side of the algebraic equation in (21) is the same as in (17). Therefore, the additional considerations are omitted and the conclusions are drawn. When $\varepsilon_3(t)$ and its derivative are equal to zero $\forall t \in \mathbb{T}$, $x_1(t)$ and $x_2(t)$ are observable. The considerations concerning the kinetics functions are analogous to those presented in Sect. 3.3. Hence, when biomass is not washed from the bioreactor, the considered model of CSTR is observable for selected measured outputs. In turn, in the wash-out state, due to permanent excitation of the inputs and the boundedness of both kinetics functions, $x_1(t)$ is observable but $x_2(t)$ is only detectable.

3.5 Results – CSTR Case Study 5

The set of measured outputs are: $y_1(t) = h_1(\boldsymbol{x}(t)) = x_1(t)$. It implies that $\forall t \in \mathbb{T}$ $\varepsilon_1(t) = 0$, and $\dot{\varepsilon}_1(t) = 0$. Thus, (8) is transformed to the following DAE system:

$$\begin{cases} 0 &= \mu(t)x_1(t) - \mu_z(t)x_1(t) + r\varepsilon_4(t)u_1(t) \\ \dot{\varepsilon}_2(t) &= -\frac{1}{Y_s}\Big[\mu(t)x_1(t) - \mu_z(t)x_1(t)\Big] - (1+r)\varepsilon_2(t)u_1(t) \\ \dot{\varepsilon}_3(t) &= -\frac{1}{Y_o}\Big[\mu(t)x_1(t) - \mu_z(t)x_1(t)\Big] - (1+r)\varepsilon_3(t)u_1(t) - \varepsilon_3(t)u_2(t) \\ \dot{\varepsilon}_4(t) &= -v(w+r)\varepsilon_4(t)u_1(t) \\ \varepsilon(t_0) &= \varepsilon_0 \end{cases} \cdot \tag{22}$$

Combining the first equation with the second and third, (22) is rewritten as:

$$\begin{cases} 0 &= \mu(t)x_1(t) - \mu_z(t)x_1(t) + r\varepsilon_4(t)u_1(t) \\ \dot{\varepsilon}_2(t) &= \frac{1}{Y_s}r\varepsilon_4(t)u_1(t) - (1+r)\varepsilon_2(t)u_1(t) \\ \dot{\varepsilon}_3(t) &= \frac{1}{Y_o}r\varepsilon_4(t)u_1(t) - (1+r)\varepsilon_3(t)u_1(t) - \varepsilon_3(t)u_2(t) \\ \dot{\varepsilon}_4(t) &= -v(w+r)\varepsilon_4(t)u_1(t) \\ \varepsilon(t_0) &= \varepsilon_0 \end{cases} \cdot \tag{23}$$

Taking into account that the inputs are the permanent excited positive signals, the dynamics expressed by the last equation in (23) is asymptotically stable, therefore $\varepsilon_4(t) \to 0$ and $x_4(t)$ is detectable. Hence, (23) can be rewritten as:

$$\begin{cases} 0 &= \mu(t)x_1(t) - \mu_z(t)x_1(t) + r\gamma(t)u_1(t) \\ \dot{\varepsilon}_2(t) &= \frac{1}{Y_s}r\gamma(t)u_1(t) - (1+r)\varepsilon_2(t)u_1(t) \\ \dot{\varepsilon}_3(t) &= \frac{1}{Y_o}r\gamma(t)u_1(t) - (1+r)\varepsilon_3(t)u_1(t) - \varepsilon_3(t)u_2(t) \\ \varepsilon(t_0) &= \varepsilon_0 \end{cases}, \tag{24}$$

where the solution of the third differential equation is as follows:

$$\varepsilon_4(t) \equiv \gamma(t) = \varepsilon_4(t_0)\exp\left[-v(w+r)\int_{t_0}^{t} u_1(\tau)d\tau\right] \to 0 \text{ for } t \to \infty.$$

According to permanent excitation of the inputs and since $x_1(t)$ is always positive, the dynamics from (24) will asymptotically tend to zero. Moreover, due to fact, that both kinetics functions $\mu_z(t)$ and $\mu(t)$ are Monod terms, when $\varepsilon_2(t) \rightarrow 0$ and $\varepsilon_3(t) \rightarrow 0$ then $\mu_z(t) \rightarrow \mu(t)$ asymptotically. Hence, all of the unmeasured states are detectable. Thus, the considered model of CSTR is detectable for selected measured outputs. The fact that the right-hand sides of differential equations in (24) would not resemble to be negative for some time and what would affect the detectability should be explained. Indeed, the essence of detectability property justification could be explained analogously as it has been performed in [21], where the method of constants variation has been applied. By calculating another integrals, it is possible to show that these two differential equations are asymptotically convergent to zero point.

For the situation where $x_1(t)$ is equal to zero, the first equation in (23) must be equal to zero, and the only $r\gamma(t)u_1(t)$ term appears. Therefore, due to positiveness of $u_1(t)$, $\gamma(t)$ must be equal to zero $\forall t \in \mathbb{T}$. Hence, $x_4(t)$ is observable. By taking into account conditions mentioned above, $x_2(t)$ and $x_3(t)$ are detectable due to asymptotically stable dynamics.

4 Conclusions

In this paper, the problem of the observability/detectability analysis at the given measured outputs for non-linear biochemical systems has been investigated. In particular, a method of indistinguishable state trajectories (indistinguishable dynamics/sets) has been used to prove these properties for a continuous stirred tank reactor. During the research, five different cases covering a wide range of possible combinations of the measured outputs have been considered. The delivered comprehensive analysis shows, how the given structure of the system mathematical model and the selection of particular measurements affect the observability/detectability properties. In detail, the non-trivial interactions between system state variables and exogenous inputs have been discussed. The results obtained are necessary to determine whether possible to develop a particular state observer dependently on the available measured outputs of the considered system. Thus they may be used for the state observer synthesis and diagnostic systems development for biochemical systems.

Acknowledgements. Financial support of these studies from Gdańsk University of Technology by the DEC-2/2020/IDUB/I.3.3 grant under the Argentum Triggering Research Grants - 'Excellence Initiative - Research University' program is gratefully acknowledged.

References

1. Besancon, G.: Nonlinear Observers and Applications. Springer-Verlag, Berlin (2007). https://doi.org/10.1007/978-3-540-73503-8

2. Czyżniewski, M., Łangowski, R.: A robust sliding mode observer for non-linear uncertain biochemical systems. ISA Trans. **123**, 25–45 (2022). https://doi.org/10.1016/j.isatra.2021.05.040
3. Dochain, D., Vanrolleghem, P.: Dynamical modelling and estimation in wastewater treatment processes. IWA Publishing, London, UK (2001)
4. Gauthier, J.P., Kupka, I.A.K.: Observability and observers for nonlinear systems. SIAM J. Control Optim. **32**, 975–994 (1994)
5. Hadj-Sadok, M.Z., Gouzé, J.L.: Estimation of uncertain models of activated sludge processes with interval observers. J. Process Control **11**, 299–310 (2005)
6. Hammouri, H., Tmar, Z.: Unknown input observer for state affine systems: a necessary and sufficient condition. Automatica **46**, 271–278 (2010)
7. Henze, M., Gujer, W., Mino, T., Van Loosdrecht, M.: Activated sludge models ASM1, ASM2, ASM2d and ASM3. IWA Publishing, London (2000)
8. Hermann, R., Krener, A.: Nonlinear controllability and observability. IEEE Trans. Autom. Control **22**, 728–740 (1977)
9. Ibarra-Rojas, S., Moreno, J.A., Espinosa-Pérez, G.: Global observability analysis of sensorless induction motors. Automatica **40**, 1079–1085 (2004)
10. Isidori, A.: Nonlinear Control Systems. Springer-Verlag, London (1995). https://doi.org/10.1007/978-3-540-29678-2_4021
11. Khalil, H.: Nonlinear Systems, 3rd edn. Prentice-Hall Inc., Upper Saddle River (2002)
12. Łangowski, R., Brdys, M.A.: An interval estimator for chlorine monitoring in drinking water distribution systems under uncertain system dynamics, inputs and chlorine concentration measurement errors. Int. J. Appl. Math. Comput. Sc. **27**(2), 309–322 (2017)
13. Łangowski, R., Brdys, M.A.: An optimised placement of the hard quality sensors for a robust monitoring of the chlorine concentration in drinking water distribution systems. J. Process Control **68**, 52–63 (2018)
14. Lindberg, C.F.: Control and estimation strategies applied to the activated sludge process. Ph.D. thesis, Uppsala University (1997)
15. Moreno, J.A., Dochain, D.: Global observability and detectability analysis of uncertain reaction systems and observer design. Int. J. Control **81**, 1062–1070 (2008)
16. Moreno, J.A., Rocha-Cózatl, E., Wouwer, A.V.: A dynamical interpretation of strong observability and detectability concepts for nonlinear systems with unknown inputs: application to biochemical processes. Bioprocess Biosyst. Eng. **37**(1), 37–49 (2013). https://doi.org/10.1007/s00449-013-0915-5
17. Moreno, J.A., Vargas, A.: Nonlinear observer design and observability of an aerobic bioreactor using respirometry. In: 1999 European Control Conference (ECC). IEEE, Karlsruhe, Germany (1999)
18. Rapaport, A., Dochain, D.: Interval observers for biochemical processes with uncertain kinetics and inputs. Math. Biosci. **193**, 235–253 (2005)
19. Saraiva, I., Wande-Wouwer, A., Hantson, A.L., Moreno, J.A.: Observability analysis and software sensor design for an animal cell culture in perfusion mode. IFAC Proc. Vol. **46**, 255–260 (2013)
20. Schaum, A., Moreno, J.A.: Dynamical analysis of global observability properties for a class of biological reactors. IFAC Proc. Vol. **40**(4), 213–218 (2007)
21. Schaum, A., Moreno, J.A., Vargas, A.: Global observability and detectability analysis for a class of nonlinear models of biological processes with bad inputs. In: 2nd International Conference on Electrical and Electronics Engineering. IEEE, Mexico City, Mexico (2005)

Adaptive, Robust and FTC Systems

The 'Sense and Avoid' Aircraft System Based-on a Monocular Camera as the Last Chance to Prevent Accidents

Peter Bauer(✉)

Institute for Computer Science and Control (SZTAKI), Budapest, Hungary
bauer.peter@sztaki.hu

Abstract. A monocular camera-based sense and avoid (S&A) system can possibly be the last chance to avoid nearby intruders or ground obstacles during an unmanned aerial vehicle (UAV) mission if the other systems cannot detect it. This article targets to summarize the main results of decade long S&A research in Institute for Computer Science and Control (SZTAKI) Hungary. After giving motivation and a historical review of SZTAKI S&A research it will first overview the object parameters measurable by a monocular camera. Detectability of different kinds of intruder vehicles and their avoidability will be discussed after. Then it discusses single camera system extension to a multi-camera one. The projection model of real world objects is the next topic as textbook projection models cannot be successfully applied so disc and rectangle projection models are proposed. The first approach of S&A decision is to assume linear, constant velocity own and aerial intruder trajectories, define time-to-closest-point-of-approach (TTCPA) and closest-point-of-approach (CPA) and estimate their values based-on the camera measurements. This is discussed in 2D (horizontal plane) and then in 3D. In the next step the extension of the method to steady and moving obstacles and non-straight, variable speed own flight trajectories is presented targeting to estimate also their size and global position. The first approach is compared to the new one giving application guidelines. A crucial issue is the decision about the type of the own flight trajectory for which a trajectory measure is introduced. After the summary of achieved results the future challenges are discussed.

Keywords: Aircraft sense and avoid · Monocular camera · Aerial and ground obstacles

1 Introduction

Nowadays we are witnessing the rapid development of unmanned aerial vehicle (UAV) technology. However, a significant barrier to the introduction of these

The numerous supporting organizations and grants of this decade long research are presented in the Acknowledgement section.

Z. Kowalczuk (Ed.): DPS 2022, LNNS 545, pp. 367–385, 2023.
https://doi.org/10.1007/978-3-031-16159-9_30

devices to the airspace is missing safety features such as Sense and avoid (S&A) capability. It is vital to integrate civilian and governmental UAVs into the common airspace according to [17]. The most critical question is the case of non-cooperative S&A. In case of small UAVs the size, weight and power consumption of the onboard S&A system should be minimal. Monocular vision based solutions can be cost and weight effective therefore especially good for small UAVs [15,19,23,24,30,31,33]. So S&A research started at Institute for Computer Science and Control (SZTAKI) Hungary focusing on monocular camera solutions in frame of an Office of Naval Research (ONR, USA) grant GR-ONR.

These systems basically measure the bearing and size of intruder aircraft (A/C) image without range information. This leads to the question of bearing-only target motion analysis (see e.g. [21] and [26]) and in certain circumstances to a scale ambiguity making intruder size and absolute position unobservable. One possibility for collision detection is to recover range information from own craft maneuvers by simulating stereo vision with ego motion (see e.g. [20,32]). This was the first concept in SZTAKI research [30]. However, this interrupts the mission of the own craft and forces it to make unnecessary motion which has an adverse effect on intruder image tracking (see [13] about the effective field of view issue), energy consumption of own A/C and observation or measurement missions. So its better to use only the bearing (centroid position) and size information without interrupting the mission.

[15] points out that the relative distance of intruder from own A/C called the closest point of approach (CPA) well characterizes the possibility of collision together with the time to closest point of approach (TTCPA) and there is a chance to estimate these parameters despite the scale ambiguity. In the literature there are several works dealing with TTCPA estimation based-on optic flow or first order motion models such as [25,28] however, it is hard to find references for CPA estimation. That's why the next stage of research focused on TTCPA and CPA estimation assuming straight line, constant velocity flight trajectories [5,8–11]. This part was partially financed by SZTAKI (see GRSZTAKI) and finally real flight evaluation of the system was done.

[29] explicitly requires 360° horizontal field of regard from a sense and avoid camera system so a multi-camera omnidirectional system covering 360° horizontal field of view (FOV) is required. This is underlined by Federal Aviation Administration (FAA) rules (FAR Part 91.113) regarding general aviation (GA) which define that in case of overtaking 'Each aircraft that is being overtaken has the right-of-way and each pilot of an overtaking aircraft shall alter course to the right to pass well clear' (see e.g. [18]). It is well known that it is really hard for a GA pilot to observe a small size UAV so in this case his/her responsibility should be withdrawn and the UAV should provide clearance. Omnidirectional S&A system development can be found in e.g. [22] with fisheye lens where distortion causes larger errors as one approaches the boundary of the lens than with standard lenses. That's why its worth to examine the possibility to create an omnidirectional system from a set of cameras with standard lenses (see [8,9]). This research was partially financed by grant GR-VISION.

Usually S&A is understood as the sensing and avoidance of aerial vehicles, however in case of low level flight with small UAVs the avoidance of ground obstacles - such as transmission towers, tower-cranes, smokestacks or even tall tress - can be also vital to integrate them into the airspace. This means that a small UAV's S&A system should be prepared also to detect and avoid ground obstacles. In the literature, the detection and avoidance of ground obstacles is discussed e.g. in [3,16] and [27]. In this field SZTAKI research gives solutions considering a monocular camera without differentiation of any noisy signal (contrary to [27]) and both for circular and rectangular cross section obstacles: [4,6,12]. This part of research was financed by grants GR-VISION, GR-BOLYAI, GR-BME-TKP and GR-UNKP. Through this period of research the evaluation of detectability and avoidability was also done in [14].

The latest work [7] summarized all previous methods targeting to estimate absolute distance and size of static and moving obstacles. Relative methods targeting TTCPA and CPA are compared to artificial stereo vision (proper own aircraft motion) methods giving application guidelines and a trajectory (own aircraft) measure to decide between the methods. New financing was given by grant GR-ARNL.

The current article as the brief summary of the author's plenary presentation targets to overview all of this research starting with the monocular camera measurable parameters (Sect. 2) then discussing detectability, avoidability and multi camera systems (Sects. 3, 4). In Sect. 5 horizontal disc and rectangle projection models are introduced which are the basis of all solutions. Section 6 summarizes the 2D and 3D methods targeting to estimate relative trajectory parameters. Section 7 presents a method to solve for absolute intruder distance and size, relates it to the relative solution and gives application guidelines for the different methods. Section 8 presents a sample real flight test result and finally Sect. 9 concludes the paper summarizing the work and future challenges.

2 Measurable Image Parameters

First of all the applied coordinate systems in S&A should be defined as shown in Fig. 1 including the N, E North-East (NE) and the X_T, Y_T trajectory, the X, Z A/C body and X_C, Z_C camera systems. The trajectory is aligned with the average linear direction of the own flight in case of nonlinear flight path. Transformation angles between the systems (χ_T, ψ, β_C) are also shown.

Figure 2 shows the images of a UAV in real flight test (collected by SZTAKI). It is obvious that only the bounding box of the image can be determined especially for the distant image (where its hard to recognize the shape of the A/C). The resulting basic parameters of the bounding box of an intruder A/C are shown in Fig. 3. In X_C, Y_C, Z_C camera frame x, y are the positions of intruder image centroid and S_x, S_y are the intruder image sizes (horizontal/vertical). The f focal length of the camera can be calibrated in advance. Knowing only the size of the intruder image and applying pinhole camera projection there is a well known scale ambiguity.

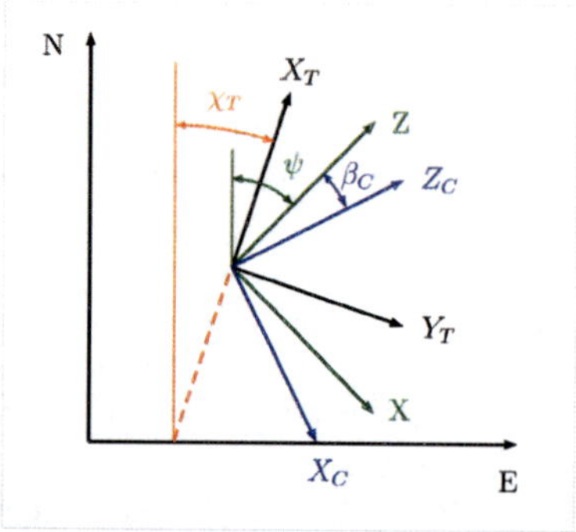

Fig. 1. The applied coordinate systems

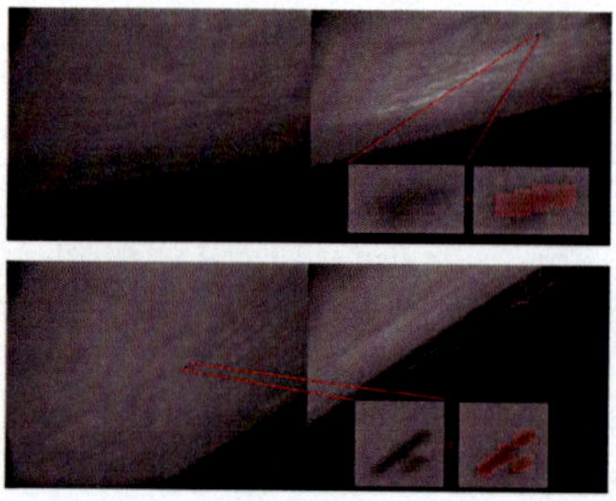

Fig. 2. Object detection examples

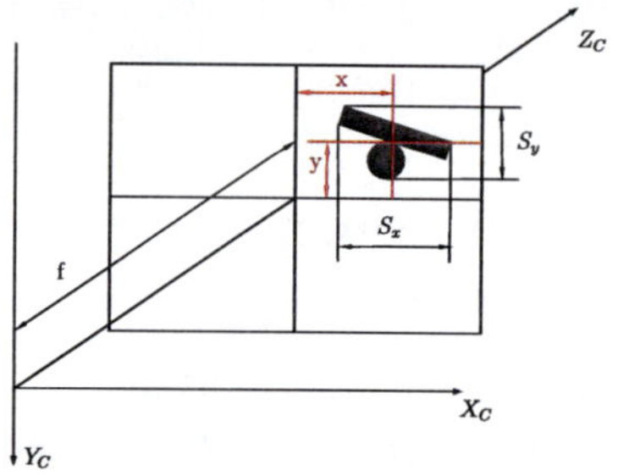

Fig. 3. Considered image parameters

3 Detectability and Avoidability

The critical question of an electro-optical system application in S&A is the detectability of intruders. For detailed evaluation of detectability four different cameras were selected (see Table 1). Only camera resolution and a minimum area on the image sensor to be occupied by the object are considered neither homogeneity of the background nor contrast parameters. Based on camera parameters a maximum distance can be calculated for any object size from which the object is detectable and also trackable. In real flight tests (see e.g. Fig. 2) during tracking of the intruder with our Basler dcA 1280 camera system the minimum occupied sensor size was $8 \cdot 3.75 \cdot 10^{-6} = 30\,\mu\text{m}$. From this occupied size the minimum trackable pixel numbers (MIN size S) can be calculated for the other cameras in Table 1 considering their pixel pitches and assuming that trackability depends on the occupied physical sensor size.

Table 1. Camera data

Type	Basler dcA 1280	Basler acA 1440	Basler acA 2040	Basler acA 2500
Camera resolution	1280 × 960	1440 × 1080	2048 × 1536	2592 × 1944
Pixel pitch [μm]	3.75	3.45	3.45	2.2
Focal length f [px]	997.3	1217.4	1739.1	1818.2
View angle [°]	65.4	61.2	61	71
MIN size S [px]	8	9	9	14

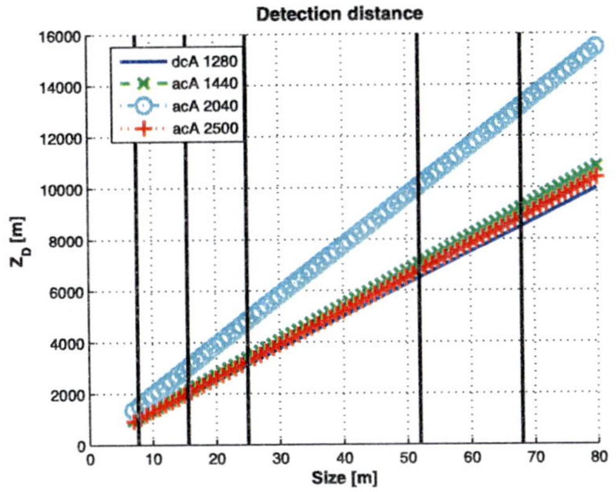

Fig. 4. Detection distances for different intruder sizes with different cameras

From this point the maximum detection distance of different size intruders can be calculated as $Z_D = f\frac{R}{S}$ for the size range $R = 7 \dots 80\,\text{m}$ covering real aircraft sizes. These calculations were done for all four cameras from Table 1 and the results are shown in Fig. 4. For more details see [14].

Avoidability means the capability of our system to avoid the intruder in time, this depends on the detection distance, the relative velocity between the two aircraft and the dynamical capabilities of our own aircraft. We did several in-flight S&A tests between two UAVs from which the dynamic capabilities of our own Sindy aircraft (see Fig. 14) can be estimated. We executed avoidance as a maximum bank turn into the opposite direction for a given time and then a slow turn back to the original trajectory.

Table 2 shows the gradual increase of absolute cross track distance relative to the own trajectory which characterizes the increase of distance also between the two aircraft. Is shown that cross track distance starts to highly increase only after 2 s, below this time the avoidance will not be effective. Considering also the 2–3 s required time between first detection and decision about avoidance at least 5–6 s are required between detection and collision (or near collision) to safely avoid a small aircraft.

Table 2. Gained average cross track distances with Sindy UAV

Time [s]	0	1	2	3	4	5	6
Distance [m]	0	2.3	8.1	22.7	42.54	58.7	74.8

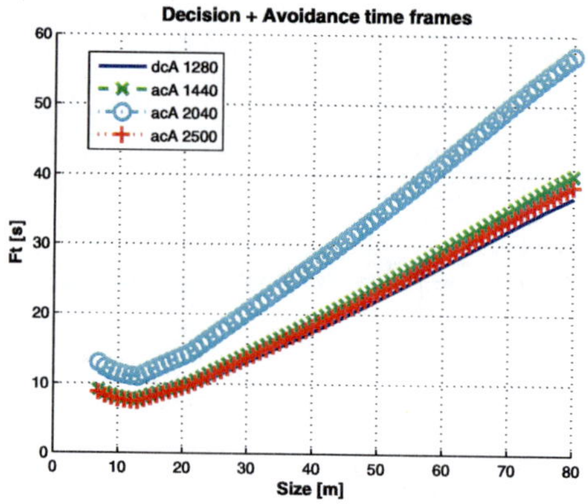

Fig. 5. Decision plus avoidance time frames for different intruder sizes with different cameras

Considering the 20 m/s velocity of the Sindy own aircraft and the maximum cruise velocities of the considered intruders its easy to calculate the maximum relative velocities and from these and the detection distance the minimum time frames available to make the decision and execute avoidance (Ft). These results are shown in Fig. 5 for the different cameras. The figure shows that at least 7 s are available with any camera for any intruder size. This means that if the decision about the possibility of collision requires 3 s than Sindy has at least 4 s for the avoidance and so can gain at least about 40 m distance from the intruder. This can be sufficient to avoid collision.

4 Multi Camera Systems

An example six camera system (each camera with 71° FOV as Basler acA 2500 which means a 11° overlap between them) to cover the required whole 360° around the A/C is shown in Fig. 6 together with camera angles relative to the front body frame. The image of the intruder will usually move from one camera FOV to another during an approach however one needs multiple images to estimate TTCPA and CPA (see later) so transformation of the past information (points) to the new frame should be solved. The representation of one point (O object) is visualized in Fig. 7. It is assumed that the camera frame origins are in one point. In a real system their distances can be a few centimeters so the assumption is approximately true compared to the few tens to hundreds meters distance of an intruder. Considering the figure the common information about O in the horizontal plane is the angle of the horizontal projection of the frame origin - O vector relative to the two frames (β_1 and β_2). If the object moves from frame 1 to 2 considering the camera setting angles (β_C^1 and β_C^2) the transformation of measured β_1 values to β_2 is straightforward $\beta_2 = \beta_1 + \beta_C^1 - \beta_C^2$. For more details see [8] and [9].

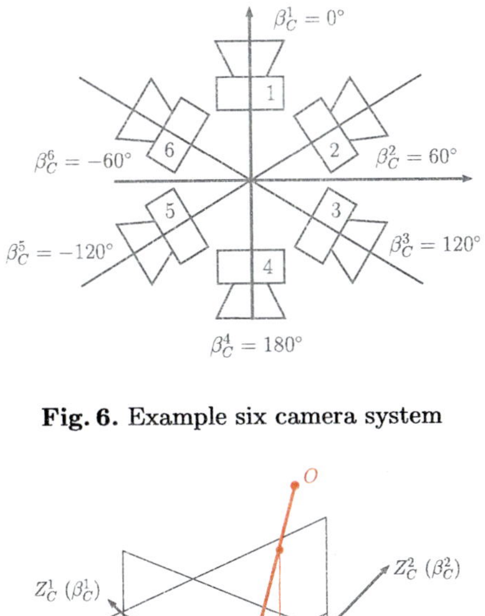

Fig. 6. Example six camera system

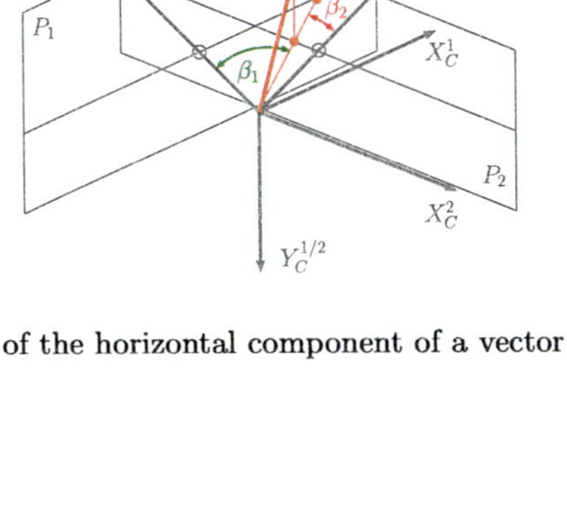

Fig. 7. Representation of the horizontal component of a vector in two camera frames

5 Projection Models

A pinhole camera model is used which relates image parameters (x, y, S_x, S_y) to own A/C camera focal length f, intruder position (X_C, Y_C, Z_C) in camera frame and intruder size $R_{x/y}$ (horizontal/vertical). The basic equations of pinhole camera horizontal projection are:

$$x = f\frac{X_C}{Z_C}, \quad S_x = f\frac{R_x}{Z_C} \tag{1}$$

However, these Eqs. (1) usually published in the literature are derived for a line approaching the image plane parallel to it. With real 3D objects two problems can arise. One is the rotation of the object, the other is the depth information. This was examined in detail in [5]. Considering data about several aircraft from [2] the *length/wingspan* ratio gives a mean value of 0.93 which is not very far from 1. This means that a disc can well approximate the horizontal contour of an aircraft solving both rotation and depth problem.

5.1 Disc Projection Model for Oblique Camera

The disc projection model is derived considering camera orientation not aligned with the A/C body system (see Fig. 6). An oblique camera setup is shown in Fig. 8 with β_C camera angle relative to the body. The disc represents the intruder aircraft (with half size $r = R/2$).

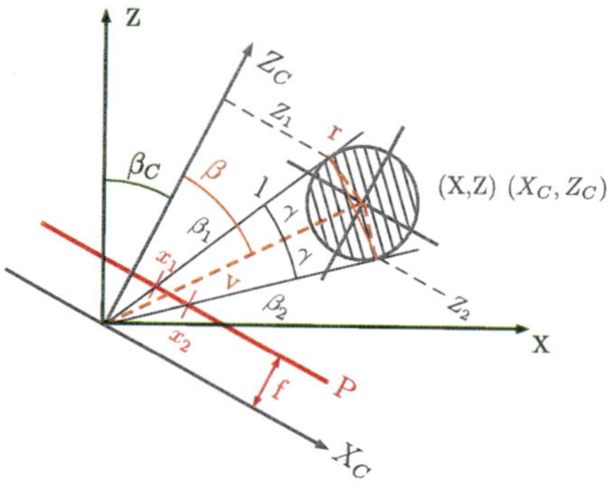

Fig. 8. Oblique camera disc model

Considering the measurable coordinates (S_x, β_1, β_2, x ($S_x = x_2 - x_1$ intruder image size and $x = (x_2 + x_1)/2$ intruder image centroid position)) in the image they can not be directly related to intruder distances X_C, Z_C in the camera frame. However, their transformed values $\bar{S}_x$, $\bar{x}$ has a close connection to X_C, Z_C as shown below.

$$\bar{S}_x = S_x(\cos\beta_1 + \cos\beta_2) = \frac{2fR}{Z_C}$$
$$\bar{x} = x\left(1 - \frac{\bar{S}^2}{16f^2}\right) = f\frac{X_C}{Z_C} \tag{2}$$

Considering the vertical plane there, the intruder is rather an ellipse than a disc so a special transformation to disc and at the end of calculations a reverse transformation should be applied. For details see [10].

5.2 Rectangle Projection Model for Oblique Camera

The disc projection model considers the tangents of the assumed disc in deriving the relations between object size and distance. However, this can lead to false size estimates if the projection of a rectangle is done as shown in Fig. 9. To obtain better results for square objects the projection of the line edges should be considered. However, this gives overly complicated formulas if the edge is not aligned with the X_C axis as discussed in [5]. This means that at first, the camera system should be aligned with the square object and then the line projection formulas can be applied. By applying ego motion transformation of the image parameters from camera to body and then to trajectory system (for details see [12]) we can virtually align the camera with the trajectory system. Assuming flat ground and that the object edges were detected its easy to determine object orientation relative to the trajectory considering the vanishing point along the trajectory ((0, 0) image point in the aligned camera system). This is shown in Fig. 10. The edge of the object is parallel with camera Z_C axis if it points into the (0, 0) point. This gives the possibility to fit a line on the edges (1–2 and 2–3 in the figure) and determine their zero crossings (Δx_{ij} distance) with X_C. From this the required rotation angle ($\Delta\beta_C = atan\,(\Delta x/f)$) to align the camera can be calculated. Finally, $\beta'_C = \beta_C + \Delta\beta_C$ can be used to virtually align the coordinate system. After the virtual camera system alignment, the original pinhole projection formulae (1) can be applied with $R = W$ object size. For details see [4].

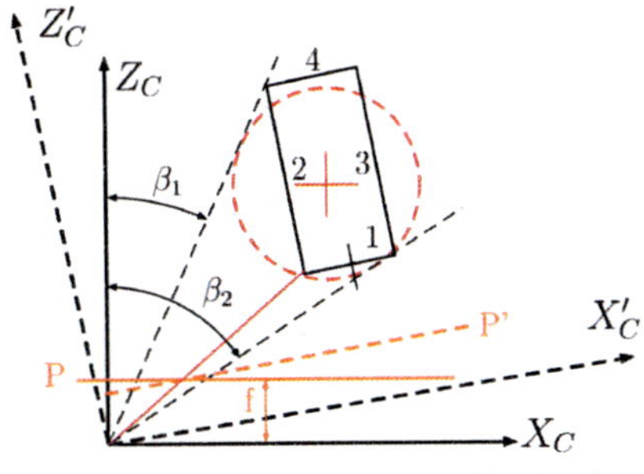

Fig. 9. Projection of disc or rectangle

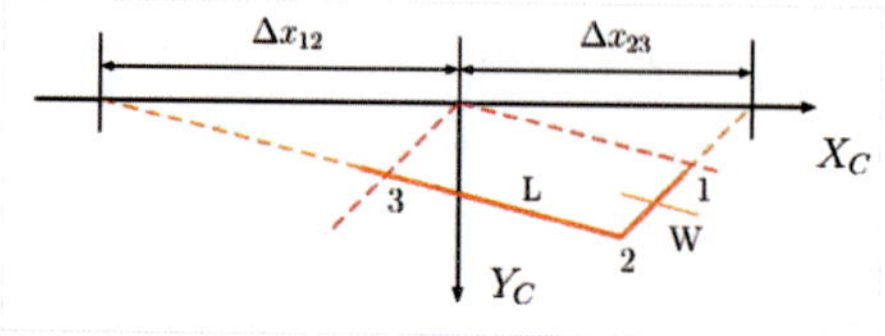

Fig. 10. Camera alignment based-on vanishing point

6 TTCPA and CPA Calculation in 2D and 3D

Considering now the (X, Y, Z) intruder coordinates in the body frame characterized by the X_a, Y_a miss distances (see Fig. 11 for example), the V_x, V_y, V_z relative velocities and t_{CPA} and executing the body to camera frame transformation one gets the X_C, Y_C, Z_C coordinates. Y_a is defined similarly to X_a but in the Y, Z plane.

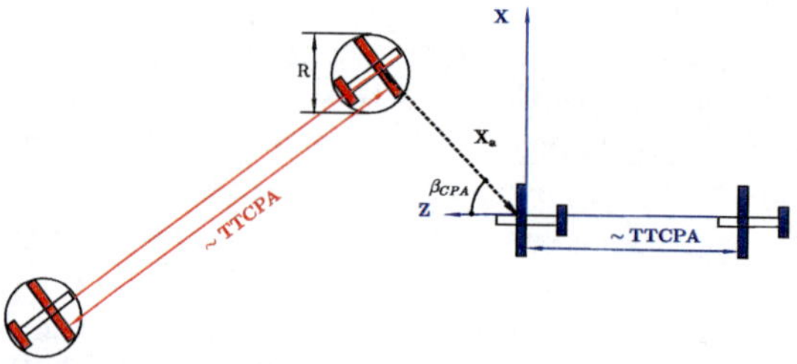

Fig. 11. Define TTCPA, $CPA = X_a/R$ and β_{CPA} (intruder red from left, own aircraft blue from right)

$$
\begin{aligned}
&X = X_a - V_x t_{CPA}, Y = Y_a - V_y t_{CPA}, Z = -V_z t_{CPA} \\
&X_C = X_a \cos\beta_C - (V_x \cos\beta_C - V_z \sin\beta_C) t_{CPA} \\
&Y_C = Y_a - V_y t_{CPA} \\
&Z_C = X_a \sin\beta_C - (V_x \sin\beta_C + V_z \cos\beta_C) t_{CPA}
\end{aligned}
\tag{3}
$$

Substituting now the expressions of X_C and Z_C into the reciprocal and ratio of the expressions for $\bar{x}$ and $\bar{S}_x$ in (2) and considering $CPA_{x/x} = \frac{X_a}{R_x}$ and $t_{CPA} = t_{CPA/x}$ in the horizontal plane one gets:

$$
\begin{aligned}
\frac{1}{\bar{S}_x} &= \frac{CPA_{x/x}}{2}\frac{\sin\beta_C}{f} - \frac{V_x \sin\beta_C + V_z \cos\beta_C}{2fR} t_{CPA/x} \\
\frac{\bar{x}}{\bar{S}_x} &= \frac{CPA_{x/x}}{2}\cos\beta_C - \frac{V_x \cos\beta_C - V_z \sin\beta_C}{2R} t_{CPA/x}
\end{aligned}
\tag{4}
$$

In this system of equations the unknowns are $CPA_{x/x}$ and $t_{CPA/x}$ and the time varying terms are $\bar{x}, \bar{S}_x, t_{CPA/x}$. The other terms such as the camera focal length f, the camera angle β_C, the relative velocities V_x, V_z and the intruder size R_x are all constant. Considering this and $t_{CPA/x} = t_{C/x} - t$ one gets (t is actual time, $t_{C/x}$ is the time when intruder is closest to own aircraft (it is constant)):

$$
\begin{aligned}
\frac{1}{\bar{S}_x} &= \frac{\sin\beta_C}{f}\frac{CPA_{x/x}}{2} - a_1 t_{C/x} + a_1 t = c_1 + a_1 t \\
\frac{\bar{x}}{\bar{S}_x} &= \cos\beta_C \frac{CPA_{x/x}}{2} - a_2 t_{C/x} + a_2 t = c_2 + a_2 t
\end{aligned} \tag{5}
$$

Fitting least squares (LS) optimal linear curves to the expressions on the left side of (5) with independent variable t gives a_1, a_2, c_1, c_2. From the estimated coefficients a system of linear equations results for $CPA_{x/x}/2$ and $t_{C/x}$:

$$
\begin{bmatrix} \frac{\sin\beta_C}{f} & -a_1 \\ \cos\beta_C & -a_2 \end{bmatrix} \begin{bmatrix} \frac{CPA_{x/x}}{2} \\ t_{C/x} \end{bmatrix} = \begin{bmatrix} c_1 \\ c_2 \end{bmatrix} \tag{6}
$$

The $CPA_{x/x}$ solution will give the closest point of approach when the intruder crosses the X axis of the body coordinate frame in Fig. 8 and this is not always the smallest $CPA_{x/x}$ in the horizontal plane as Fig. 11 visualizes. Usually the body system should be virtually rotated by a $-\Delta\beta_{C/x}$ angle to have an orientation which gives $\beta_{CPA/x}$ and the real minimum $CPA_{x/x}$. This gives a different $\beta'_{C/x} = \beta_C + \Delta\beta_{C/x}$ camera angle. Reformulating now (6) with the new angle gives:

$$
\begin{bmatrix} \frac{\sin(\beta_C + \Delta\beta_{C/x})}{f} & -a_1 \\ \cos(\beta_C + \Delta\beta_{C/x}) & -a_2 \end{bmatrix} \begin{bmatrix} \frac{CPA_{x/x}}{2} \\ t_{C/x} \end{bmatrix} = \begin{bmatrix} c_1 \\ c_2 \end{bmatrix} \tag{7}
$$

A detailed derivation in [9] gives $\Delta\beta_{C/x}$ resulting in the smallest $CPA_{x/x}$:

$$
\tan\Delta\beta_{C/x} = \frac{\left(\frac{-a_2\cos\beta_C}{f} - a_1\sin\beta_C\right)}{\left(\frac{-a_2\sin\beta_C}{f} + a_1\cos\beta_C\right)} \tag{8}
$$

By substituting $\Delta\beta_{C/x}$ back to (7) one gets the real minimum $CPA_{x/x}$ solution and the related $t_{C/x}$ and $t_{CPA/x}$ values and finally $\beta_{CPA/x}$ also (for details see [9]):

$$
\beta_{CPA/x} = -\Delta\beta_{C/x} + sign(CPA_{x/x})\frac{\pi}{2} \tag{9}
$$

In summary, it can be stated that TTCPA, global minimum CPA and direction of CPA (β_{CPA}) can all be calculated in the horizontal plane based solely on the image parameters obtained from a monocular camera system.

In the vertical plane the calculations are similar except for that there is no need for virtual rotation. They are detailed in [10].

Considering both the horizontal and the vertical planes finally eight different parameters can be determined: $t_{CPA/x}$ $(t_{C/x}), CPA_{x/x}, CPA'_{y/x}, \beta_{CPA/x}$ and

$t_{CPA/y}$ $(t_{C/y})$, $CPA_{x/y}$, $CPA'_{y/y}$, $\beta_{CPA/y}$ which give two points P_1 and P_2 in the 3D space (see Fig. 12).

6.1 TTCPA and CPA in 3D

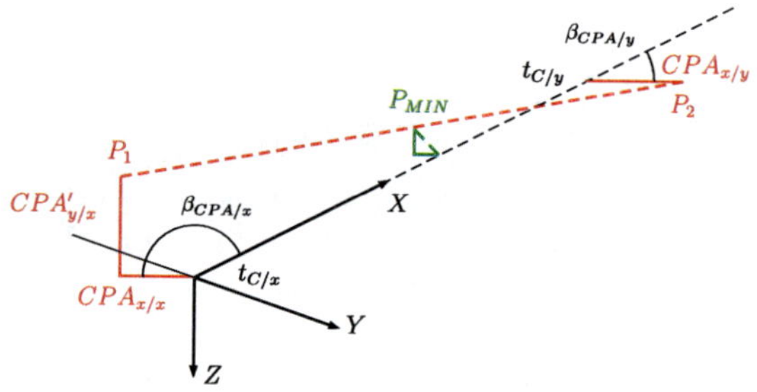

Fig. 12. Special points along the 3D aircraft trajectories

In Fig. 12 $\beta_{CPA/x}$ should be understood in the horizontal (X-Y) plane. The coordinates of points P_1 and P_2 in the body coordinate system (X, Y, Z in Fig. 12) result as:

$$
\begin{aligned}
&P_1 : \\
&X_1 = \cos(\beta_{CPA/x})CPA_{x/x} \\
&Y_1 = \sin(\beta_{CPA/x})CPA_{x/x} \\
&Z_1 = CPA'_{y/x} \\
&P_2 : \\
&X_2 = \cos(\beta_{CPA/y})CPA_{x/y} \\
&Y_2 = \sin(\beta_{CPA/y})CPA_{x/y} \\
&Z_2 = 0
\end{aligned}
\tag{10}
$$

The coordinates of an interpolated point between P_1 and P_2 can be represented as follows (Y and Z expressions are similar):

$$X_M = X_1 + t_M \frac{X_2 - X_1}{t_{C/y} - t_{C/x}} = X_1 + t_M \Delta X \tag{11}$$

From this the t_M value giving minimum 3D distance can be derived. Considering this result the time of minimum 3D distance and the related relative CPA values can be determined as follows:

$$\begin{aligned} t_{CMIN} &= t_{C/x} + t_M \\ t_{CPAMIN} &= t_{CMIN} - t \\ CPA'_{yMIN} &= Z_M \\ CPA_{xMIN} &= \sqrt{X_M^2 + Y_M^2} \\ \tan(\beta_{CPAMIN}) &= \frac{Y_M}{X_M} \end{aligned} \tag{12}$$

Finally, the vertical CPA should be transformed considering the difference between intruder horizontal and vertical size (size ratio) giving CPA_{yMIN} as published in [10].

7 Extension of Method for Absolute Distance and Size and Application Guidelines for the Methods

Formulating the camera measurements-based system of equations in the NE system gives a chance to estimate absolute distance and size of the intruder when possible. Regarding the intruder steady or linear, constant velocity trajectories are considered with Ni_0, Ei_0 starting position and Vi_N, Vi_E velocity components. The own aircraft can fly straight with constant velocity (linear trajectory) or non-straight with any velocity (nonlinear trajectory) having initial position No_0, Eo_0 and $Vo_N(t), Vo_E(t)$ velocity components (here t shows the possible time dependence). For illustration see Fig. 13.

To formulate the disc or rectangle (here the disc is considered) projection model in the camera system the relative distance between own and intruder should be calculated in NE and transformed into the camera frame. The transformation angle is $\beta_\psi = \psi + \beta_C$ according to Fig. 1. The relative distance in NE frame can be calculated as:

$$\begin{aligned} N &= Ni_0 + Vi_N t - No_0 - Vo_N t \\ E &= Ei_0 + Vi_E t - Eo_0 - Vo_E t \end{aligned} \tag{13}$$

Note that in case of time-varying $Vo_N(t), Vo_E(t)$ $Vo_N t$ and $Vo_E t$ should be substituted with $\int_{\tau=0}^{t} Vo_N(\tau)d\tau$ and $\int_{\tau=0}^{t} Vo_E(\tau)d\tau$. In camera frame the coordinates are (s and c are shorthand for sin and cos in the sequel):

$$\begin{aligned} X_C =& - s\beta_\psi(Ni_0 + Vi_N t - No_0 - Vo_N t) \\ &+ c\beta_\psi(Ei_0 + Vi_E t - Eo_0 - Vo_E t) \\ Z_C =& c\beta_\psi(Ni_0 + Vi_N t - No_0 - Vo_N t) \\ &+ s\beta_\psi(Ei_0 + Vi_E t - Eo_0 - Vo_E t) \end{aligned} \tag{14}$$

Considering the special forms derived from (2) $2f/\bar{S}_x = Z_C/R_x, 2\bar{x}/\bar{S}_x = X_C/R_x$ and the current position of the own aircraft ($No(t) = No_0 + Vo_N t, Eo(t) = Eo_0 + Vo_E t$ or with the integrals) and grouping the unknown variables gives a system of equations (SYS) with five unknowns (see (15)).

In case of steady obstacles the Vi_N, Vi_E terms can be removed reducing the system to three unknowns (RSYS).

$$\begin{bmatrix} c\beta_\psi & s\beta_\psi & c\beta_\psi t & s\beta_\psi t & -(c\beta_\psi No(t) + s\beta_\psi Eo(t)) \\ -s\beta_\psi & c\beta_\psi & -s\beta_\psi t & c\beta_\psi t & -(-s\beta_\psi No(t) + c\beta_\psi Eo(t)) \end{bmatrix} \begin{bmatrix} \frac{Ni_0}{R_x} & \frac{Ei_0}{R_x} & \frac{Vi_N}{R_x} & \frac{Vi_E}{R_x} & \frac{1}{R_x} \end{bmatrix}^T = \begin{bmatrix} \frac{2f}{S_x} \\ \frac{2\bar{x}}{S_x} \end{bmatrix} \tag{15}$$

In case of linear own and intruder trajectories ($No(t) = No_0 + Vo_N t, Eo(t) = Eo_0 + Vo_E t$) the coefficient matrix of SYS will be rank deficient (the fifth column is a linear combination of the others) so the system is unsolvable. In this case only a relative solution including CPA and t_{CPA} (referenced as TTCPA solution) can be obtained as shown in Sect. 6.

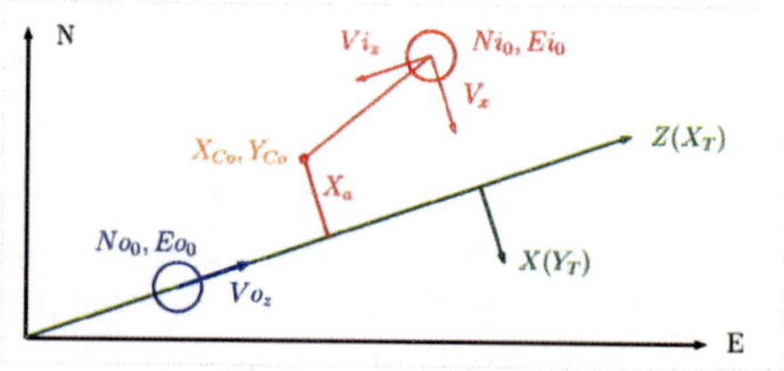

Fig. 13. Parameters in trajectory system

After evaluating the possible formulae and their limitations the S&A scenarios and calculation possibilities can be summarized in Table 3 (for details see [7]).

Table 3. S&A scenarios and calculation possibilities

Own trajectory	Linear		Nonlinear	
Intruder trajectory	Steady	Linear	Steady	Linear
TTCPA calculation	Valid	Valid	Invalid	Invalid
RSYS	Valid	Invalid	Valid	Invalid
SYS	Invalid	Invalid	≈Valid	Valid

Here RSYS means (15) without $Vi_N/R_x, Vi_E/R_x$ (only position and size are unknown). ≈ means approximate validity (system solvable but may be inaccurate). The table shows that to select the proper method one has to decide about the own trajectory (linear or nonlinear) and the state of the intruder (steady or moving on linear path). To decide about the linearity of the own trajectory a measure is introduced. It is calculated from the moving window (N frames) own velocity and position data. The average flight direction can be calculated from the mean velocity components as (note that the N division of the means cancels out):

$$\chi_{AV} = \arctan\left(\frac{\sum_{i=1}^{N} V_{OE}(i)}{\sum_{i=1}^{N} V_{ON}(i)}\right)$$

Calculating also the average position a line can be defined through it pointing in the average direction. Summing up the absolute perpendicular distances of every position from this line gives a measure for the nonlinearity of the trajectory (having zero value if the trajectory is a line). Above a given threshold the own trajectory can be considered nonlinear.

Returning to Table 3 and summing up the results the limitations of the methods are:

- SYS is unsolvable for linear own trajectories and inaccurate for steady intruder position.
- RSYS is solvable for any own trajectory, gives superior results for steady intruder but violation of the steady intruder assumption can not always be detected.
- TTCPA is solvable only for linear own trajectories and both for steady and linear intruders. It basically gives CPA and t_{CPA}. In case of steady intruder absolute coordinates and size can be determined from own velocity but in this case RSYS method is valid for any own trajectory so it is better to be applied.

The resulting application guidelines:

- In case of linear own trajectory apply the TTCPA method. If steady intruder is detected from e.g. image background relative motion switch to the RSYS method.
- In case of nonlinear own trajectory apply the RSYS method and run detection for linear intruder trajectory. If linear intruder trajectory is detected switch to the SYS method.

8 Real Flight Results

Real flight tests were done in 2015–2016 with a large twin engine UAV (Sindy UAV of SZTAKI, Nr. 1 in Fig. 14) with a two camera system applying on-board image processing, decision and maneuvering and a small intruder aircraft (E-flite Ultrastick, Nr. 2) with 1.27 m wingspan which made its detection challenging. Straight, constant velocity, parallel flight trajectories were defined and the TTCPA method was applied in 2D having a safety altitude gap between the A/C. Upon detection of collision possibility a maximum bank turn away from the intruder aircraft was executed returning to the original straight trajectory after a few seconds. In case of close trajectories several successful avoidances were done (one pair of real flight trajectories is shown in Fig. 14) while in case of far trajectories there was no avoidance. Video recordings can be seen on our Youtube channel [1].

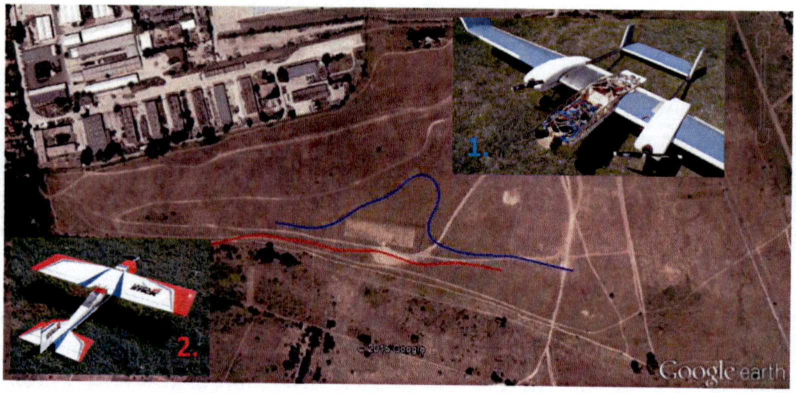

Fig. 14. Real flight test aircraft and trajectories

9 Conclusion

The present paper gives a brief overview about the decade long monocular camera-based sense and avoid research in Institute for Computer Science and Control (SZTAKI) Hungary. All the detailed works of the researchers are referenced. The main discussed topics are detectability and avoidability of intruders, multi camera systems, disc and rectangle projection models, a relative calculation method giving time to closest point of approach and closest point of approach in 2D or 3D and a method giving absolute distance and real intruder size. Properties and limitations of the methods are discussed and application guidelines are given at the end.

Future challenges include real flight test of ground obstacle avoidance, the theory and practice for circular intruder trajectories and swarm sense and avoid with multiple mono camera equipped aerial vehicles.

Acknowledgements. The funding organizations and grants of the author's decade long S&A research are listed here together with the people involved in the research.

GR-ONR. This work was supported by the Office of Naval Research Global, Grant Number N62909-10-1-7081, Dr. Charles Holland program officer (2011–2013).

GR-SZTAKI. This work was supported by the Institute for Computer Science and Control, Hungarian Academy of Sciences (MTA SZTAKI) Grant Number 008/2016 (2016–2017).

GR-VISION. The research leading to these results has received funding from the European Union's Horizon 2020 research and innovation programme under grant agreement No. 690811 and the Japan New Energy and Industrial Technology Development Organization under grant agreement No. 062600 as a part of the EU/Japan joint research project entitled 'Validation of Integrated Safety-enchanced Intelligent flight cONtrol (VISION)' (2016–2019).

GR-BOLYAI. This paper was supported by the Janos Bolyai Research Scholarship of the Hungarian Academy of Sciences (2018–2021).

GR-BME-TKP. Part of the research reported in this paper and carried out at BME has been supported by the NRDI Fund (TKP2020 IES, Grant No. BMEIEMIFM) based on the charter of bolster issued by the NRDI Office under the auspices of the Ministry for Innovation and Technology (2020–2021).

GR-UNKP. Part of the research was SUPPORTED BY THE ÚNKP-20-5 NEW NATIONAL EXCELLENCE PROGRAM OF THE MINISTRY FOR INNOVATION AND TECHNOLOGY FROM THE SOURCE OF THE NATIONAL RESEARCH, DEVELOPMENT AND INNOVATION FUND (2020–2021).

GR-ARNL. The research was supported by the European Union within the framework of the National Laboratory for Autonomous Systems (RRF-2.3.1-21-2022-00002).

Contributors of the project were:

Systems and Control Laboratory:
József Bokor, Bálint Vanek, Tamás Péni, István Gőzse, Péter Ónodi, Máté Kisantal, Márk Melczer, Tamás Baár, Soma Koltai
Computational Optical Sensing and Processing Laboratory:
Ákos Zarándy, Antal Hiba, Borbála Pencz, Pázmán Koppány
Informatics Laboratory:
András Benczúr, Bálint Daróczy, Rita Aleksziev
Pázmány Péter Catholic University:
Tamás Zsedrovits, Tamás Roska†

References

1. Aero GNC: SZTAKI Aerospace Guidance Navigation and Control Research Group YouTube Channel. https://www.youtube.com/channel/UCQMpnOuOMCiodDKQw8_hf5A
2. airliners.net: aircraft-data (2015). http://www.airliners.net/aircraft-data/
3. Badal, S., Ravela, S., Draper, B., Hanson, A.: A practical obstacle detection and avoidance system. In: IEEE WACV, Sarasota, FL (1994)
4. Bauer, P.: Position, size and orientation estimation of ground obstacles in sense and avoid. In: Proceedings of 21st IFAC Symposium on Automatic Control in Aerospace, ACA 2019 (2019)

5. Bauer, P., Hiba, A., Vanek, B., Zarandy, A., Bokor, J.: Monocular image-based time to collision and closest point of approach estimation. In: Proceedings of 24th Mediterranean Conference on Control and Automation (MED 2016), Athens, Greece (2016)
6. Bauer, P.: Improvements and detailed evaluation of ground obstacle position, size and orientation estimation. In: Proceedings of 21st IFAC World Congress (Virtual), pp. 14937–14942 (2020)
7. Bauer, P.: Applicability of different mono camera-based aircraft sense and avoid methods for steady and linear path intruders. In: 2021 29th Mediterranean Conference on Control and Automation (MED), pp. 138–143 (2021). https://doi.org/10.1109/MED51440.2021.9480201
8. Bauer, P., Hiba, A.: Vision only collision detection with omnidirectional multi-camera system. In: Proceedings of the 20th World Congress of the International Federation of Automatic Control, IFAC, Toulouse, France, pp. 15780–15785, July 2017
9. Bauer, P., Hiba, A., Bokor, J.: Monocular image-based intruder direction estimation at closest point of approach. In: Proceedings of the International Conference on Unmanned Aircraft Systems (ICUAS 2017), Miami, FL, USA, pp. 1108–1117. ICUAS Association, June 2017
10. Bauer, P., Hiba, A., Bokor, J., Zarandy, A.: Three dimensional intruder closest point of approach estimation based-on monocular image parameters in aircraft sense and avoid. J. Intell. Robot. Syst. **93**, 261–276 (2019). https://doi.org/10.1007/s10846-018-0816-6
11. Bauer, P., Hiba, A., Zsedrovits, T.: Real flight application of a monocular image-based aircraft collision decision method. Int. J. Aerosp. Eng. **2019**, 16 (2019). https://doi.org/10.1155/2019/7191839
12. Bauer, P., Vanek, B., Bokor, J.: Monocular vision-based aircraft ground obstacle classification. In: Proceedings of European Control Conference 2018 (ECC 2018), pp. 1827–1832 (2018)
13. Bauer, P., et al.: Aircraft trajectory tracking with large sideslip angles for sense and avoid intruder state estimation. In: Proceedings of 22nd Mediterranean Conference on Control & Automation (MED 2014), Palermo, Italy, pp. 1287–1292. IEEE, June 2014
14. Bauer, P., Zarandy, A., Hiba, A., Bokor, J.: Camera-based in-time detection of intruder aircraft. In: Proceedings of IEEE CNNA 2018, Budapest, Hungary, pp. 82–85. VDE Verlag Gmbh (2018)
15. Degen, S.: Reactive image-based collision avoidance system for unmanned aircraft systems. Master's thesis, Australian Research Centre for Aerospace Automation, May 2011
16. Esrafilian, O., Taghirad, H.D.: Autonomous flight and obstacle avoidance of a quadrotor by monocular SLAM. In: 2016 4th International Conference on Robotics and Mechatronics (ICROM), pp. 240–245, October 2016. https://doi.org/10.1109/ICRoM.2016.7886853
17. EU: Roadmap for the integration of civil Remotely-Piloted Aircraft Systems into the European Aviation System. Technical report, European RPAS Steering Group (2013)
18. FAA: Federal Aviation Regulation 14 CFR Part 91. Federal Aviation Administration (FAA) (2016)
19. Forlenza, L.: Vision based strategies for implementing sense and avoid capabilities onboard unmanned aerial systems. Ph.D. thesis, Universitá degli Studi di Napoli Federico II (2012)

20. Frew, E.W.: Observer trajectory generation for target-motion estimation using monocular vision. Ph.D. thesis, Stanford University (2003)
21. Jauffret, C., Pillon, D.: Observability in passive target motion analysis. IEEE Trans. Aerosp. Electron. Syst. **32**(4), 1290–1300 (1996). https://doi.org/10.1109/7.543850
22. Luis, M., Ivan, M., Pascual, C.: Omnidirectional bearing-only see-and-avoid for small aerial robots. In: Proceedings of the 5th International Conference on Automation, Robotics and Applications (2011)
23. Lyu, Y., Pan, Q., Zhao, C., Zhang, Y., Hu, J.: Feature article: vision-based UAV collision avoidance with 2D dynamic safety envelope. IEEE Aerosp. Electron. Syst. Mag. **31**(7), 16–26 (2016). https://doi.org/10.1109/MAES.2016.150155
24. Mejias, L., McFadyen, A., Ford, J.J.: Feature article: sense and avoid technology developments at Queensland University of Technology. IEEE Aerosp. Electron. Syst. Mag. **31**(7), 28–37 (2016). https://doi.org/10.1109/MAES.2016.150157
25. Meyer, F., Bouthemy, P.: Estimation of time-to-collision maps from first order motion models and normal flows. In: Proceedings of 11th IAPR International Conference on Pattern Recognition (1992)
26. Nardone, S.C., Aidala, V.J.: Observability criteria for bearings-only target motion analysis. IEEE Trans. Aerosp. Electron. Syst. **AES-17**(2), 162–166 (1981). https://doi.org/10.1109/TAES.1981.309141
27. Saunders, J., Beard, R., Byrne, J.: Vision-based reactive multiple obstacle avoidance for micro air vehicles. In: 2009 American Control Conference, pp. 5253–5258, June 2009. https://doi.org/10.1109/ACC.2009.5160535
28. Schaub, A., Burschka, D.: Spatio-temporal prediction of collision candidates for static and dynamic objects in monocular image sequences. In: Proceedings of IEEE Intelligent Vehicles Symposium IV (2013)
29. Speijker, L., Verstraeten, J., Kranenburg, C., van der Geest, P.: Scoping improvements to 'See and Avoid' for general aviation (SISA). Technical report, European Aviation Safety Agency (EASA), December 2012
30. Vanek, B., Peni, T., Zarandy, A., Bokor, J., Zsedrovits, T., Roska, T.: Performance characteristics of a complete vision only sense and avoid system. In: Proceedings of AIAA GNC 2012 (Guidance, Navigation and Control Conference), Minneapolis, Minnesota, pp. 1–15. No. AIAA 2012-4703, August 2012
31. Watanabe, Y.: Stochastically optimized monocular vision-based navigation and guidance. Ph.D. thesis, Georgia Institute of Technology, Atlanta, Georgia, USA (2008)
32. Zarandy, A., Nagy, Z., Vanek, B., Zsedrovits, T., Kiss, A., Nemeth, M.: A five-camera vision system for UAV visual attitude calculation and collision warning. In: Chen, M., Leibe, B., Neumann, B. (eds.) ICVS 2013. LNCS, vol. 7963, pp. 11–20. Springer, Heidelberg (2013). https://doi.org/10.1007/978-3-642-39402-7_2
33. Zsedrovits, T., et al.: Onboard visual sense and avoid system for small aircraft. IEEE A&E Syst. Mag. **31**, 18–27 (2016)

Dynamic Positioning Capability Assessment for Ship Design Purposes

Agnieszka Piekło[1,2](✉), Anna Witkowska[1], and Tomasz Zubowicz[1]

[1] Gdańsk University of Technology, ul. G. Narutowicza 11/12, 80-233 Gdańsk, Poland
[2] Maritime Advances Research Centre, Gdańsk, Poland
agnieszka.pieklo@cto.gda.pl

Abstract. The article focuses on solving a problem of optimal thrust distribution over the actuators in a ship Dynamic Positioning, according to DNV-ST-0111 standard, *Level 1*. The classic Quadratic Programming approach is combined with the numerical solusion used to handle the propeller with the rudder constraints in the optimization task and the influence between thrusters and skeg. It is presented as an efficient method of minimizing the power consumption. The resulting tool for performing a Dynamic Positioning capability assessment allows for fast calculations to qualitatively compare different designs. The study has proven that the Quadratic Programming based method gives less optimistic solution in comparison with DNV tool and can be safely applied at an early design stage. Further validation of the tool with the time-domain simulations would contribute to increasing confidence in its application to the daily routine calculations.

Keywords: Optimal thrust allocation · Dynamic positioning · Quadratic programming

1 Introduction

Dynamic positioning (DP) is one of the ship's operational states in which its relative or absolute position and heading are automatically maintained at desired set-points, by using only its own, active thrusters without any mooring lines or other equipment. A DP control system (DPCS) that handles this task can be classified as an advanced, model-based and over-actuated closed-loop stabilizing with the aim of achieving high disturbance rejection capabilities.

A DP capability defines a ship ability to station keeping under given environmental conditions. The assessment of a vessel's capability to keep the position is critical for planning and executing safe and reliable dynamic positioning operations (DP). A leading classification society, Det Norske Veritas (DNV) has developed a method for DP station-keeping capability assessments provided in DNV-ST-0111 standard [5]. The results are documented by DP capability numbers (corresponding to the Beaufort scale) and capability plots (in polar form). The standard defines three different DP capability levels, each requiring a specific

Z. Kowalczuk (Ed.): DPS 2022, LNNS 545, pp. 386–397, 2023.
https://doi.org/10.1007/978-3-031-16159-9_31

assessment method. The *Level 1* considered in this paper is specified for monohull ships. The calculation method at this level shall be based on a static balance of environmental and the vessel's actuator forces, assuming the same specified environmental data for all vessels. The static balance shall determine the thrust distribution among thrusters (both magnitude and direction) called thrust allocation. This task, however, has no unique solution since the DP-capable ships are over-actuated. Considering thrust x and y components as more than a number of equilibrium equations the task of DP capability assessment evolves to an optimal thrust allocation problem.

This paper focuses on an optimal thrust allocation problem based on its approximation to quadratic optimisation problem. The objective of this task is to minimize the ship's propulsion power consumption in DP operations. The effect of thruster failure can also be analyzed by using DP capability plot. The proposed approach meets DNV standards and can be easily adapted to mono - shaped ships for DP capability assesement. The results of DP capability were compared with existing tools offered by DNV.

1.1 Related Works

A general overview of marine control systems and optimal thrust allocation problem is given in [6,8]. The applicable methods division into two categories may be observed in the literature, based on the task formulation and the used method. The first group of methods consists of gradient-based approaches. Among this group the most commonly used are the ones based on the Quadratic Programming (QP) or Lagrangian multipliers [4]. Both approaches assume that the objective function and constraints are smooth and they guarantee reaching a global minimum. In turn, the second group of methods consists of the so called non-gradient (derivative-free) methods. This group is represented by the so-called meta-heuristic algorithms, such as particle swarm optimization, genetic or evolutionary algorithms [11] or direct-search algorithms [2]. In both cases, if certain requirements are met, algorithms only tend to converge global optimum. Also, both groups of algorithms converge relatively slowly.

In addition to the mentioned above, a different allocation strategies without optimization are also presented. Those include deterministic and pseudo-inverse based matrix methods [14,18]. The model predictive control allocation [16] and adaptive control allocation [15,17] are quite complex and time-consuming, which allow to take into account the actuator dynamic and uncertainties in the calculations.

Literature that covers the subject of *optimal thrust allocation* using QP approximation, [4,7,9,10]. Other articles related to QP application in *optimal thrust allocation* are [12,13]. Related convex optimization problem is discussed in [3].

1.2 Motivation and Contribution

DNV standard is a widely used method for DP capability assessment at an early stage of ship design, yet there is no proposition in the literature how to solve the thrust allocation problem with the conditions imposed by it. The method gives the procedure to calculate environmental and thruster forces as well as thruster losses but does not proposes a way of solving the over-actuated system problem. A typical ship design office uses a purchased input-output software with no awareness on the optimization applied. This paper presents a step by step Quadratic Programming application in DP capability assessment with DNV standard method which can be directly applied by a skilled engineer in the ship concept design process.

The quadratic programming algorithms can guarantee that an optimal solution will be found in the finite time (which can be very important for online applications) or that a solution does not exist, whereas no such guarantees can be given for nonlinear optimization techniques. The problem is simple to define in quadratic form by approximation techniques and relatively easy to use. The presented optimal thrust allocation method is build on foundation of the QP approach described in [4]. The azimuth and tunnel thruster constraints as well as objective function are adopted here accordingly. For propeller with the rudder a new approach was proposed to fulfill the DNV standard description.

The presented methodology allows for a complete (all kinds of thrusters, including propeller with the rudder) and fast in-house DP capability assessment for ship concept design.

1.3 Structure of the Paper

The remaining section of this research work is organised in the following manner. In Sect. 2, the DP capability assessment problem has been formulated. Section 3, provides information on the implied method used to propose a solution to the formulated problem. The results obtained with comparison to available external tools are discussed in Sect. 4. Section 5, concludes the paper.

2 Problem Definition

DP capability analysis can assist in determining the maximum environmental forces and moment that DP system can counteract for the given headings (from 0° to 360°). The environmental conditions are statically balanced by thrust forces and moment provided by the propulsion system. Considering a planar movement of the ship, the balance equations yield:

$$\begin{aligned}\sum_{i=1}^{N} T_{\mathrm{x}\,i} &= F_{\mathrm{env\,x}},\\ \sum_{i=1}^{N} T_{\mathrm{y}\,i} &= F_{\mathrm{env\,y}},\\ \sum_{i=1}^{N} \left(-T_{\mathrm{x}\,i}\, y_i + T_{\mathrm{y}\,i}\, x_i\right) &= M_{\mathrm{env\,z}},\end{aligned} \tag{1}$$

where $F_{\text{env x}}$, $F_{\text{env y}}$ and $M_{\text{env z}}$ denote x and y direction net force components and torques resulting from the environmental influences (wind, wave, current), expressed in [N] and [Nm], respectively, also considered as disturbance inputs; $T_{\text{x}\,i}$ and $T_{\text{y}\,i}$ are forces generated by the ith thruster in [N], considered as control inputs; x_i and y_i define position of the ith thruster in the ship centered coordinate frame [5]; N denotes the total number of thrusters.

Since DPCS belongs to a class of an over-actuated system, the equation (1) has no unique solution in terms of thruster generated forces $T_{\text{x}\,i}$ and $T_{\text{y}\,i}$, $\forall i$. Therefore, instead of solving (1), the thrust is to be allocated while optimizing the total power consumption, taking into account balance equation as a one of thrust constraints. To that goal, as indicated in the introductory part of the paper, the method applied is to utilize a constrained QP formulation of the problem under consideration.

For the mentioned purpose, the following set of assumptions is considered:

Assumption 1. *The DP capability is achieved at given operational conditions whenever the balance* (1) *holds.*

The Assumption 1 is a consequence of physical laws and does not introduce artificial limitations to the problem.

Assumption 2. *The environmental forces in* (1) *are assumed to be scenario driven.*

The static balance equation is used to calculate needed thrust for one scenario of waves, wind and current (e.g., by using Beaufort scale) to hold the ship in position. DP capability assessment needs to be studied various scenarios and environmental conditions. The results are given in the form of DP capability polar plot which presents the maximum weather condition that the ship can withstand for each environmental angle assuming the worst case scenario where the wind, current and waves are collinear.

Assumption 3. *The relation between power and thrust can be expressed as a quadratic function with a satisfactory accuracy.*

From [6] it is found that the physical relationship between produced thrust T and consumed power P can be given by non-linear relation $P = |T|^{3/2}$ but it can be effectively approximated by a function of at most second degree.

Assumption 4. *The thrust constraints imposed by its physical limits, the propulsion configuration and type can be approximated by a set of convex polygons.*

Different thruster types will have different thrust region shapes. In the general case, the constraints on the thrust form non-convex regions. By the virtue of Assumptions 3 and 4, the problem is reduced to convex. However, this simplification can lead to precision loss in the power assessment.

3 Methodology

In the following subsection, the QP approach to thrust allocation has been defined, including propeller thrust with rudder angle and thrust loss of spoiled zone of azimuth thruster.

3.1 Decision Variables

Considering the alluded issue problem and (1), a vector of decision variables is defined in the following lines.

$$\boldsymbol{u} \stackrel{\text{def}}{=} \left[T_{\mathrm{x}\,1}, T_{\mathrm{y}\,1}, T_{\mathrm{x}\,2}, T_{\mathrm{y}\,2}, \ldots, T_{\mathrm{x}\,N}, T_{\mathrm{y}\,N}\right]^{T}. \tag{2}$$

The choice of thruster forces is not arbitrary but is subject to constraints resulting from propulsion type and respective mount points in the considered coordinate frame. Hence, the following constraints need to be taken into account.

3.2 Constraints

In the described thruster allocation task, it can be observed that the direct formal description of the problem leads to nonlinear constraints. Invoking Assumption 4, by utilizing linearization mechanisms, with respect to the constraints resulting from the use of azimuth thrusters and propellers with rudder, allows one to preserve the linear nature of the problem constraints, as it has been shown in the following lines.

Azimuth Thruster. The constraints of the azimuth thruster arise from operational restrictions. First, flushing another operating thruster is forbidden (Fig. 1a). Second, the method and location of mounting face limitations on the generated thrust in some directions (Fig. 1a). The first case, introduces the so-called forbidden zones, where the thruster capacity is assumed to be equal to zero (Fig. 1b). The second one, introduces spoiled zones where the capacity of the thruster is reduced to a fraction of its maximum value (Fig. 1b).

An example of handling constraints for azimuth thruster is shown in Fig. 1b. In general, the approach is to divide the thrust constraint region into convex sets, in this case labeled I, II, III, IV, and then into polygons that can be defined by linear equations. From these considerations, the zone boundary and saturation inequality constraints, arise.

Zones Boundary Inequality Constraint. Following Assumption 4, boundary conditions are formulated as linear inequalities [4]:

$$\begin{aligned} T_{\mathrm{x}\,i}\, \sin(\beta_{\mathrm{start}\,n}) - T_{\mathrm{y}\,i}\, \cos(\beta_{\mathrm{start}\,n}) &\le 0, \\ -T_{\mathrm{x}\,i}\, \sin(\beta_{\mathrm{end}\,n}) + T_{\mathrm{y}\,i}\, \cos(\beta_{\mathrm{end}\,n}) &\le 0, \end{aligned} \tag{3}$$

where $\beta_{\mathrm{start}\,n}$ and $\beta_{\mathrm{end}\,n}$ denote angles at which the nth zone starts and ends, respectively.

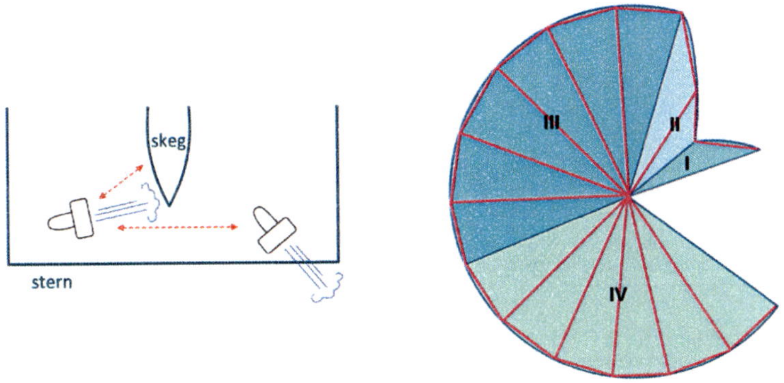

(a) Thruster flushing skeg and another working thruster

(b) Thruster capacity (background) and convex safe zones

Fig. 1. Forbidden, spoiled zones and safe zones of azimuth thruster

Saturation Inequality Constraints. For convenience, two mechanisms of description are provided. First, for the circle-shaped zones (e.g., Fig. 1b, zones III, IV) [4]:

$$T_{\mathrm{x}\,i}\,\cos(\varphi_j) + T_{\mathrm{y}\,i}\,\sin(\varphi_j) \leq r_i, \tag{4}$$

where r_i is the maximum effective thrust of the ith thruster and φ_j refers to each polygon middle angle. An illustration of the resulting polygon is provided in Fig. 2a. Second, for the spoiled zones (e.g., Fig. 1b, zones I, II) [4]:

$$T_{\mathrm{x}\,i}\,(y_{k+1} - y_k) + T_{\mathrm{y}\,i}\,(x_{k+1} - x_k) \leq x_k\,y_{k+1} - x_{k+1}\,y_k, \tag{5}$$

where: x_k and y_k are coordinates of the first point while x_{k+1}, y_{k+1} are coordinates of the subsequent point. An illustration is provided in Fig. 2b.

In practice, the number of polygons, which the zone is divided to, determines the accuracy of the method and shall be chosen individually, however, an error of 1% is recommended [4].

Propeller with Rudder. In the case of a propeller with a rudder, specific coefficients for x and y thrust forces depending on the rudder angle are provided in [5]. In the presented approach, rudder angle (α) is not a decision variable. Therefore, only the relation between x and y direction thrust components, is considered to complement the constraints set.

Remark 1. The consequence of the approach is that α remains a free variable to be handled outside the optimisation task.

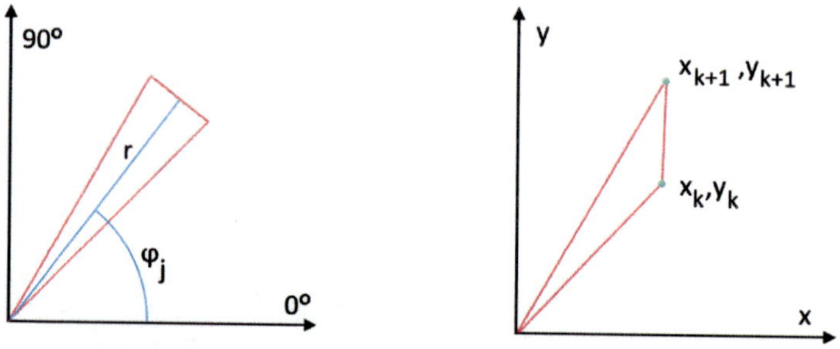

(a) Saturation - circle shaped zone

(b) Saturation - spoiled zone

Fig. 2. Inequality constraints - azimuth thruster

Propeller Equality Constraints. The forces produced by the sum of the propeller and the rudder are satisfied by equality constraints [5]:

$$T_{\mathrm{x}\,i} - \frac{a_{\mathrm{x}\,i}(\alpha)}{a_{\mathrm{y}\,i}(\alpha)} T_{\mathrm{y}\,i} = 0, \tag{6}$$

where $a_{x\,i}(\alpha)$ and $a_{y\,i}(\alpha)$ are coefficients dependant on the rudder angle α.

Propeller Inequality Constraints. The propeller and rudder thrust inequality constraints yield [5]:

$$\begin{aligned} T_{\mathrm{x}\,i} &\leq T_{\max\,i}\, a_{\mathrm{x}\,i}(\alpha), \\ T_{\mathrm{y}\,i} &\leq T_{\max\,i}\, a_{\mathrm{y}\,i}(\alpha). \end{aligned} \tag{7}$$

3.3 Objective Function

By the virtue of Assumption 3, the allocation of thrust is assessed by considering an objective function (P_{total}) which is given as a quadratic form and represents relation between thruster generated forces (control inputs) and total power consumption, which yields:

$$P_{\mathrm{total}}(\boldsymbol{u}) \stackrel{\mathrm{def}}{=} \boldsymbol{u}^T \boldsymbol{W} \boldsymbol{u}, \tag{8}$$

where $\boldsymbol{W} \stackrel{\mathrm{def}}{=} \mathrm{diag}\,(w_1,\, w_1,\, w_2,\, w_2,\, \ldots,\, w_N,\, w_N)$ is a diagonal matrix of weight coefficients w_i, $\forall i$, corresponding to each mounted thruster. Weight coefficients are determined according to [5] with the exception that the maximum thrust loss in case of spoiled zones is also taken into account.

Remark 2. The method used for weight selection is a type of worst-case approach, hence it introduces certain level of conservatism.

3.4 Optimization Task

Taking into account the contribution of the Subsects. 3.1–3.3, the optimal thrust allocation task yields:

$$\begin{aligned} \min_{\boldsymbol{u}} \quad & P_{\text{total}}(\boldsymbol{u}) \\ s.t. \quad & \boldsymbol{A}(\alpha)\,\boldsymbol{u} = \boldsymbol{b}\,, \\ & \boldsymbol{G}\boldsymbol{u} \leq \boldsymbol{h}(\alpha) \end{aligned} \tag{9}$$

where: $\boldsymbol{A}(\alpha)$ and $\boldsymbol{G}$ are the equality and inequality constraints matrices, respectively; $\boldsymbol{b}$ denotes a vector encompassing forces and torques; $\boldsymbol{h}(\alpha)$ represents the thrust saturation and limiting operation angle. The internal structure of vectors and matrices results directly from (3)–(7).

3.5 DP Capability Assessment

Finally, the ship's DP capability is assessed in the following manner. Considering Remark 1, the problem (9) is solved for discrete values of angle of environmental impact, from 0° to 360°, and rudder angle, which varies from ship to ship, to cover whole considered domain of interest. Consequently, a population of results is obtained for each considered environmental angle. In each case, the final result is obtained by applying the minimum operator. The results of DP capability are typically presented in a graphic form, as a polar plot, where each circle in the plot represents a DP number corresponding to specific weather condition [5].

4 Results

The case study is being used to illustrate the use of the proposed approach in assessing the DP capability of a rescue ship which total length can reach 96 m. The ship is equipped with five thrusters layout which has been illustrated in Fig. 3a. The crucial ship's parameters have been included in Table 1. The DP capability has been assessed using the approach described in Sect. 3 by the means of a dedicated tool developed using Python programming language. To improve legibility, first in Subsect. 4.1, an elementary example for two arbitrarily selected angles is presented. Second, in Subsect. 4.2 the DP capability assessment results are discussed. The obtained results have been compared with the results obtained using DNV tool available on-line[1].

4.1 Optimal Thrust Allocation

In Fig. 3 an example of optimal thrust allocation is presented. It results from solving (9), for two distinct angles, namely 30° and 150° indicated by blue arrow, has been presented. For the illustrated test, the environmental conditions were

[1] The free version of the application is limited to analysis of maximum four thrusters and does not share detailed results, just the DP capability plot.

Table 1. Ship main data

Symbol	Value	Unit	Description
L_{pp}	86.6	m	Length between perpendiculars
B	18.8	m	Maximum breadth at waterline
T	5.0	m	Summer load line draft
$A_{\mathrm{F,\,wind}}$	392.0	m^2	Frontal projected wind area
$A_{\mathrm{L,\,wind}}$	1203.0	m^2	Longitudinal projected wind area
$A_{\mathrm{L,\,current}}$	441.0	m^2	Longitudinal projected submerged current area

set to standardized level DP number 8 according to [5]. The mounting points of the thrusters have been indicated with blue numbered dots. The red areas near the rear thrusters indicate the forbidden zones. The result of thrust allocation has been indicated by magenta arrows attached to thrusters. For legibility of the result, the percentage of maximum thrust has been display to clearly characterize the magnitude of the thrust vectors.

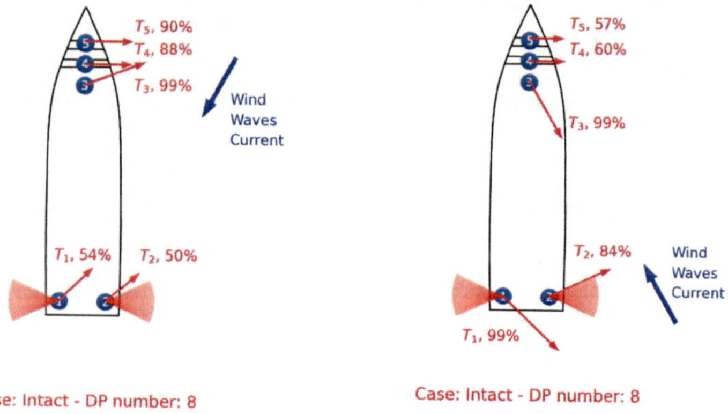

(a) Environmental angle 30° (b) Environmental angle 150°

Fig. 3. Optimal thrust allocation

As a result of the analysis, one is able to clearly identify that the environmental impact from the direction coming from the bow (30°) causes higher engagement of the fore thrusters. In turn, the environmental impact from the direction from the stern (150°) engages stern thrusters more. This directly relates to the ship's geometry, both under and above water level, and exerted environmental forces and torques that depend on it. Therefore, this tendency will vary from ship to ship and is considered as crucial during DP ship design.

4.2 DP Capability Assessment

In Fig. 4 the DP capability assessment results have been presented. Two DP operation modes have been investigated. First, 'intact', where all the thrusters are considered operational. Second, single failure, where one of the thrusters failed. In the later case, two scenarios were explored. One considering failure of thruster 2 and second of thruster 5. In both failure scenarios, the DP capability is significantly reduced. In case of starboard azimuth thruster's failure, the capability the capability plot is no longer symmetric. This is an indication of the effective thrust reduction due to flushing a dead thruster by the port azimuth thruster. When azimuth thruster angle is within a spoiled zone, the efficiency of a generated thrust decreases.

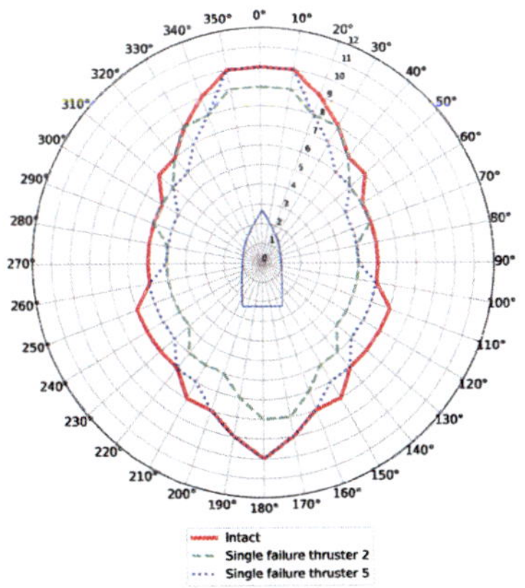

Fig. 4. DP capability plot

The results from the developed tool were compared to the ones obtained using the on-line application by DNV (Fig. 5). It has been found that in case of using azimuth thrusters as main propulsion, the results obtained from both tools are very similar (Fig. 5a). The only difference is at the environmental angles of 160° and 200°, respectively. Since the information on the optimization algorithm used by DNV free web application is not available, it is hypothesized that this may be the consequence of different handling of the spoiled zones due to skeg. In turn, the results, obtained considering propellers with the rudder as main thruster, vary significantly (Fig. 5b). In general, the methodology presented in this work shows a more pessimistic outcome in comparison to the DNV web application. Further investigation of the discrepancies is required to gain better understanding of the matter.

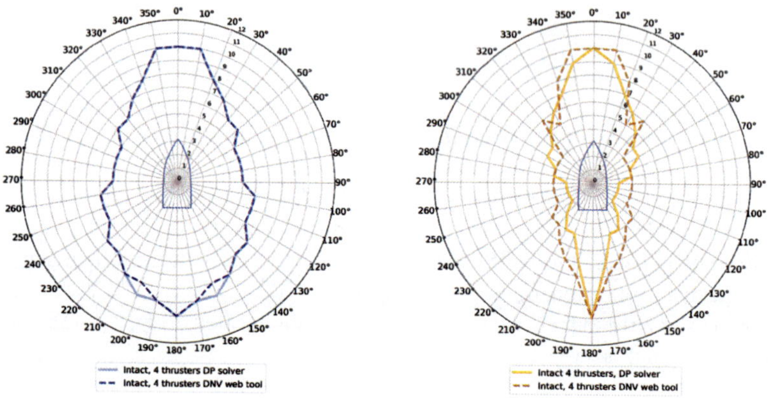

(a) 4 thrusters - two aft propulsors: azimuth thrusters

(b) 4 thrusters - two aft propulsors: propellers with rudders

Fig. 5. Comparison of DP capability plots

5 Conclusions

The recently increasing need for DP assessment tools, both fast rough calculations as well as time-domain simulations for the early stage of the design, was the motivation for the study. The Quadratic Programming method used in *optimal thrust allocation* proved relevant results when applied to DNV rules (DNV-ST-0111 standard). In particular, it was evident while concerning handling the influences between thrusters and skeg. The presented method could be effectively used by design offices around the world for rough initial calculation without a need for investment in expensive software. An important element of future research will be the evaluation of the accuracy of the method based on time - domain simulation and model tests. The verification of simulation is planned on the basis of experiments on dedicated test stand of Maritime Advanced Research Centre, using the physical model of the ship, equipped with DP system.

References

1. Arditti, F., Souza, F., Martins, T., Tannuri, E.: Thrust allocation algorithm with efficiency function dependent on the azimuth angle of the actuators. Ocean Eng. **105**, 206–216 (2015)
2. Baeyens, E., Herreros, A., Perán, J.R.: A direct search algorithm for global optimization. Algorithms **9**(2), 40 (2016)
3. Boyd, S., Vanderberghe, L.: Convex Optimisation, 7th edn. Cambridge University Press, New York (2009)
4. de Wit, C.: Optimal thrust allocation methods for dynamic positioning of ships. M.SC. Thesis. Delft University of Technology (2009)

5. DNV. : DNV-ST-0111, Assessment of station keeping capability of dynamic positioning vessels. In: DNV AS, December 2021
6. Fossen, T.: Handbook of Marine Craft Hydrodynamics and Motion Control, 1st edn. John Wiley & Sons Ltd., New York (2011)
7. Ruth, E.: Propulsion control and thrust allocation on marine vessels. Doctoral thesis. Norwegian University of Science and Technology (2008)
8. Sørensen, A.: Marine Control Systems. Propulsion and Motion Control of Ships and Ocean Structures. Lecture Notes. Department of Marine Technology. Norwegian University of Science and Technology (2013)
9. Valčić, M.: Optimization of thruster allocation for dynamically positioned marine vessels. Doctoral Thesis. University of Rijeka (2020)
10. Wang, L., Yang, J., Xu, S.: Marine control systems. Propulsion and motion control of ships and ocean structures. China Ocean Eng. **32**(1), 90–98 (2018). https://doi.org/10.1007/s13344-018-0010-4
11. Yang, X.S.: 1 - optimization and metaheuristic algorithms in engineering. In: Metaheuristics in Water, Geotechnical and Transport Engineering, pp. 1-23. Elsevier (2013)
12. Zalewski, P.: Constraints in allocation of thrusters in a DP simulator. Sci. J. Mar.Univ. Szczecin **52**(124), 45–50 (2017)
13. Zalewski, P.: Convex optimization of thrust allocation in a dynamic positioning simulation system. Sci. J. Mar.Univ. Szczecin **48**(120), 58–62 (2016)
14. Mauro, F., Nabergoj, R.: Advantages and disadvantages of thruster allocation procedures in preliminary dynamic positioning predictions. Ocean Eng. **123**, 96–102 (2016)
15. Witkowska, A., Smierzchalski, R.: Adaptive backstepping tracking control for an over-actuated DP marine vessel with inertia uncertainties. Int. J. Appl. Math. Comput. Sci. **28**(4), 679–693 (2018)
16. Luo, Y., Serrani, A., Yurkovich, S., Doman, D. B., Oppenheimer, M. W.: Model predictive dynamic control allocation with actuator dynamics. In: IEEE Proceedings of the 2004 American control conference, pp. 1695–1700, June 2004
17. Tjønnås, J., Johansen, T.A.: Adaptive control allocation. Automatica **44**, 2754–2766 (2008)
18. Harkegard, O.: Dynamic control allocation using constrained quadratic programming. J. Guid. Contr. Dynam. **27**(6), 1028–1034 (2004)

Degradation Tolerant Optimal Control Design for Linear Discrete-Times Systems

Soha Kanso, Mayank Shekhar Jha, and Didier Theilliol(✉)

Centre de Recherche en Automatique de Nancy (CRAN), UMR 7039, CNRS, Université de Lorraine, 54506 Vandoeuvre-lès-Nancy Cedex, France
{soha.kanso,mayank-shekhar.jha,didier.theilliol}@univ-lorraine.fr
http://www.cran.univ-lorraine.fr/

Abstract. This paper develops a degradation tolerant approach based on optimal control. Compared to classical control design, the aim of this work is to decelerate the rate of evolution of degradation by minimizing a quadratic cost function of control input, tracking error and rate of degradation. A linear quadratic regulator (LQR) and tracker (LQT) are expanded, in a finite and infinite-horizon, for a discrete-time linear system in presence of degradation of its components. The performance of the systems is affected by the degradation in an affine manner. First, a LQR is developed for finite and infinite horizon, then the structure of a LQT is designed for finite-horizon. By tuning the weighting matrices, the performance of the system in closed loop can be modified so that the system can achieve its mission before the occurrence of component failure and the remaining useful life can be extended. Electro-Mechanical Actuator (EMA) system is widely used in modern automobiles, transportation and industrial processes and is adopted to verify the effectiveness of the proposed control scheme.

Keywords: Linear quadratic tracker · Fault-tolerant control · Health management

1 Introduction

The evolution of complex and autonomous systems, such as modern aircraft, unmanned aerial vehicles, and automated industrial processes requires the development and the implementation of new control technologies that maintain system stability and that address incipient failure. Traditional control system design focuses only on performance and stability without taking into account the effects of aging, fatigue, and damage to the concerned components. For this purpose, the implementation of Prognostics and Health Management (PHM) systems [12] turns out to be an important part of maintenance activities and the prognostics process becomes one of the main action levers in the search for global performance.

Z. Kowalczuk (Ed.): DPS 2022, LNNS 545, pp. 398–409, 2023.
https://doi.org/10.1007/978-3-031-16159-9_32

Recent approaches have applied modern control techniques such as adaptive or robust control to address situations where the degree of failure may be unknown. In [2], an adaptive failure control is developed for incipient failure modes, which do not affect the stability of the system, but lead to a catastrophic failure. For various industrial and mission critical systems that operate in closed loop, Fault Tolerant control design has been developed [1,10] in order to compensate under passive or active approaches fault occurrence but more recently new methods are sought such that useful life of critical systems can be extended. In this context, health-aware control has recently emerged as one of the domains where control synthesis is sought, based upon the state of health (SOH) and/or Remaining Useful Life (RUL) prognostics of critical components [8]. RUL prognostics are obtained using degradation models which estimate their parameters online and adapt the degradation dynamics based upon individual assets. Few notable works have proposed methods to design control laws that seek to extend the RUL of component/system as [13,14] and [5]. Also, [4] proposed Reliability or RUL based constraints within MPC. In such a framework, a model predictive control (MPC) framework is adopted to produce a controller that adjusts its reference points and ensures robustness to particular failures, thus reducing its impact on the system. In the same framework, [3] presents a new methodology for failure-tolerant design of electromechanical actuator (EMA), it takes advantage of online estimates of the remaining useful life (RUL) of a defaulted component and reconfigures the existing MPC control authority. In [11], a comprehensive framework for monitoring the RUL of the system is developed by using the post-prognostic information. The RUL controller uses a cost function that adjusts the performance requirements and the desired RUL values.

This paper examines a degradation tolerant approach based on optimal control [9,15] by building a quadratic cost function of control input, tracking error and rate of degradation. First, it is developed for a general discrete-time linear system affected by a linear degradation in an affine manner, then implemented for a specific application. The application example is an EMA, which is widely used in aerospace and industrial processes in place of hydraulic actuators, where a system failure would be potentially dangerous and extremely costly. The usage of EMA brings some specific problems associated to thermal balance, reflected inertia, parasitic motion due failure response [7]. Another important damage mechanism for EMA is the degradation of friction coefficient due to usury. Sliding and rolling friction cause material wear, which enhances the coefficient of friction [6].

This paper is arranged as follows. Section 2 introduces the problem statement and the issue addressed. Section 3 presents the proposed reconfiguration approach in finite and infinite horizon. Section 4 examines the feasibility of the proposed approach using an example of an EMA. Finally, the conclusion summarizes the significant advances and includes plans for future work.

2 Problem Formulation

The degradation of components of an active system affects directly the remaining useful life of the system and consequently its usability and productivity. The state of degradation or deterioration considered as a health indicator are affected by the action of the controller. Hence the importance of development of an optimal approach for performing a control action that takes into account the stability and the performance requirements of the system and also the SOH.

This paper focuses on linear MIMO discrete-time systems represented by the state transition, control and observation matrices, $A_1 \in \mathbb{R}^{n\times n}$, $A_2 \in \mathbb{R}^{n\times l}$, $B_1 \in \mathbb{R}^{n\times m}$ and $C_1 \in \mathbb{R}^{p\times n}$.

$$x_{k+1} = A_1 x_k + A_2 d_k + B_1 u_k \tag{1}$$

$$y_k = C_1 x_k \tag{2}$$

where $u \in \mathbb{R}^m$, $x \in \mathbb{R}^n$, $d \in \mathbb{R}^l$ and $y \in \mathbb{R}^p$ correspond respectively to the input, state of the system, state of degradation and measurement vectors. The system is supposed to be affected by the degradation in an affine manner and the degradation evolution is described by the following state-space representation:

$$d_{k+1} = A_3 x_k + A_4 d_k \tag{3}$$

with $A_3 \in \mathbb{R}^{l\times n}$ and $A_4 \in \mathbb{R}^{l\times l}$. In most cases, the evolution of the degradation is monotonic and irreversible, moreover, it is unknown. In this work, the current state of degradation is assumed to be dependent on the previous state of degradation and also the previous state of the system. The evolution of the degradation is controlled implicitly by the input since it is sensitive to the state of the system.

In order to maintain the performance of the system while minimizing the energy and the speed of evolution of degradation, a quadratic utility function is defined by:

$$\mathscr{U}_k = x_k^T Q x_k + u_k^T R u_k + \Delta d_k^T Q_1 \Delta d_k \tag{4}$$

where Δd_k is the rate of evolution of degradation described by (5).

$$\Delta d_k = d_{k+1} - d_k = (A_4 - I)d_k + A_3 x_k \tag{5}$$

The utility function (4) is used to develop the performance index of a linear quadratic regulator problem, which gives the following quadratic cost function:

$$J_0 = \frac{1}{2}(x_N^T \bar{S}_N x_N + \Delta d_N^T \bar{P}_N \Delta d_N) + \frac{1}{2}\sum_{k=0}^{N-1} x_k^T Q x_k + u_k^T R u_k + \Delta d_k^T Q_1 \Delta d_k \tag{6}$$

Q, Q_1, R, $\bar{S}_N$ and $\bar{P}_N$ are symmetric positive definite cost-weighting matrices and $|R| \neq 0$. The initial plant and degradation state are given as x_0 and d_0 respectively.

3 Optimal Reconfiguration Control

In this bellowing section, an optimal control based approach is developed that allows the synthesis of a state feedback control law using the minimization of a quadratic criterion involving the state, the control and the rate of evolution of degradation. The problem posed is to bring the state of equilibrium (zero) starting from a non-zero initial condition. This problem is equivalent to bringing the state to any reference track. Hence, in Sect. 3.1, the solution of the problem is developed for a LQR and then in Sect. 3.2 the solution is extended to a LQT problem.

3.1 Linear Quadratic Regulator

The solution for the LQR is reached by determining the control sequence u_0, u_1, ..., u_{N-1} that minimizes J_0 in (6). Using (5) to eliminate Δd in (6) gives:

$$\begin{aligned} J_0 = \frac{1}{2}[x_N^T(\bar{S}_N + A_3^T P_N A_3)x_N + d_N^T(A_4 - I)^T \bar{P}_N(A_4 - I)d_N \\ + d_N^T(A_4 - I)^T \bar{P}_N A_3 x_N + x_N^T A_3^T \bar{P}_N(A_4 - I)d_N] \\ + \frac{1}{2}\sum_{k=0}^{N-1}[x_k^T(Q + A_3^T Q_1 A_3)x_k + u_k^T R u_k + x_k^T A_3^T Q_1(A_4 - I)d_k \\ + d_k^T(A_4 - I)^T Q_1 A_3 x_k + d_k^T(A_4 - I)^T Q_1(A_4 - I)d_k] \end{aligned} \tag{7}$$

To solve the problem of LQR, we have first to begin with the Hamiltonian in order to derive the necessary conditions. The Hamiltonian function is defined by the following equation:

$$\begin{aligned} H_k = \tfrac{1}{2}[x_k^T(Q + A_3^T Q_1 A_3)x_k + u_k^T R u_k + x_k^T A_3^T Q_1(A_4 - I)d_k + d_k^T(A_4 - I)^T Q_1 A_3 x_k \\ + d_k^T(A_4 - I)^T Q_1(A_4 - I)d_k] + \lambda_{k+1}[A_1 x_k + A_2 d_k + B_1 u_k] \end{aligned} \tag{8}$$

where the costate of the system $\lambda_k \in \mathbb{R}^n$ and it's given by:

$$\lambda_k = \frac{\partial H_k}{\partial x_k} = (Q + A_3^T Q_1 A_3)x_k + A_1^T \lambda_{k+1} + A_3^T Q_1(A_4 - I)d_k \tag{9}$$

Solving the stationarity condition $\frac{\partial H_k}{\partial u_k} = 0$, yields to:

$$u_k = -R^{-1} B_1^T \lambda_{k+1} \tag{10}$$

If the optimal λ_k can be found, (10) can therefore be used to find the optimal control. Moreover, the boundary condition is given by:

$$\lambda_N = \frac{\partial \Phi_N}{\partial x_N} = (\bar{S}_N + A_3^T \bar{P}_N A_3)x_N + A_3^T \bar{P}_N(A_4 - I)d_N \tag{11}$$

with

$$\begin{aligned} \Phi_N = \frac{1}{2}[x_N^T(\bar{S}_N + A_3^T \bar{P}_N A_3)x_N + d_N^T(A_4 - I)^T \bar{P}_N(A_4 - I)d_N \\ + d_N^T(A_4 - I)^T \bar{P}_N A_3 x_N + x_N^T A_3^T \bar{P}_N(A_4 - I)d_N] \end{aligned} \tag{12}$$

Thus, assuming that a linear relation like (11) holds for all times $k \leq N$:

$$\lambda_k = S_k x_k + P_k d_k \tag{13}$$

for a sequence of $n \times n$ matrices S_k and $n \times l$ vectors P_k. If a consistent formula for these postulated S_k and P_k can be found, then (13) is a valid assumption. To do this, use (10) and (13) in (1) to get:

$$\begin{aligned} x_{k+1} = (I + B_1 R^{-1} B1^T S_{k+1})^{-1}[(A_1 - B_1 R^{-1} B_1^T P_{k+1} A_3)x_k \\ + (A_2 - B_1 R^{-1} B_1^T P_{k+1} A_4)d_k] \end{aligned} \tag{14}$$

Using (13) and (14) in the costate equation (9) gives:

$$\begin{aligned} S_k x_k + P_k d_k = (Q + A_3^T Q_1 A_3)x_k \\ + A_1^T S_{k+1}(I + B_1 R^{-1} B_1^T S_{k+1})^{-1}(A_1 - B_1 R^{-1} B_1^T P_{k+1} A_3)x_k \\ + A_1^T S_{k+1}(I + B_1 R^{-1} B_1^T S_{k+1})^{-1}(A_2 - B_1 R^{-1} B_1^T P_{k+1} A_4)d_k \\ + A_1^T P_{k+1} A_4 d_k + A_1^T P_{k+1} A_3 x_k + A_3^T Q_1 (A_4 - I)d_k \end{aligned} \tag{15}$$

This equation must hold for all state sequences x_k and d_k given any x_0 and d_0, therefore we can write:

$$S_k = Q + A_3^T Q_1 A_3 + A_1^T S_{k+1}(I + B_1 R^{-1} B_1^T S_{k+1})^{-1}(A_1 - B_1 R^{-1} B_1^T P_{k+1} A_3) + A_1^T P_{k+1} A_3 \tag{16}$$

and

$$P_k = A_1^T S_{k+1}(I + B_1 R^{-1} B_1^T S_{k+1})^{-1}(A_2 - B_1 R^{-1} B_1^T P_{k+1} A_4) + A_1^T P_{k+1} A_4 + A_3^T Q_1 (A_4 - I) \tag{17}$$

By comparing (11) and (13), the boundary conditions for these recursions are $S_N = \bar{S}_N + A_3^T \bar{P}_N A_3$ and $P_N = A_3^T \bar{P}_N (A_4 - I)$. Since the auxiliary sequences S_k and P_k can now be computed, assumption (13) was a valid one, and the optimal control is:

$$u_k = -R^{-1} B_1^T (S_{k+1} x_{k+1} + P_{k+1} d_{k+1}) \tag{18}$$

Substituting (1) and (3) in (18) yields to:

$$\begin{aligned} u_k = -(R + B_1^T S_{k+1} B_1)^{-1} B^T (S_{k+1} A_1 + P_{k+1} A_3)x_k \\ -(R + B_1^T S_{k+1} B_1)^{-1} B^T (S_{k+1} A_2 + P_{k+1} A_4)d_k \end{aligned} \tag{19}$$

Defining the gains sequences:

$$K_k^x = (R + B_1^T S_{k+1} B_1)^{-1} B^T (S_{k+1} A_1 + P_{k+1} A_3) \tag{20}$$

$$K_k^d = (R + B_1^T S_{k+1} B_1)^{-1} B^T (S_{k+1} A_2 + P_{k+1} A_4) \tag{21}$$

The control takes the form:

$$u_k = -K_k^x x_k - K_k^d d_k \tag{22}$$

Equations (16) and (17) are solved backwards in time starting from time N. As $k \to -\infty$, the sequences S_k and P_k converge to a steady-state matrices S_∞ and P_∞, then the corresponding steady-state gains are:

$$K_\infty^x = (R + B_1^T S_\infty B_1)^{-1} B^T (S_\infty A_1 + P_\infty A_3) \tag{23}$$

$$K_\infty^d = (R + B_1^T S_\infty B_1)^{-1} B^T (S_\infty A_2 + P_\infty A_4) \tag{24}$$

Note: The augmented system is formed by the state of the system and the degradation, and $A \in \mathbb{R}^{(n+l)\times(n+l)}$, $B \in \mathbb{R}^{(n+l)\times m}$ and $C \in \mathbb{R}^{p\times(n+l)}$ are respectively the state transition, the control and the observation matrices of the augmented system. If (A, B) is stabilizable and (A, C) is detectable, S_k and P_k converge to unique steady-state matrices S_∞ and P_∞.

3.2 Linear Quadratic Tracker

This section synthetizes an optimal control law that forces the system to track a desired reference trajectory r_k over a specified time interval $[0, N]$. The cost function must involve the tracking error, the input, and Δd to force the state to reach the reference and to decelerate the speed of evolution of degradation:

$$\begin{aligned} J_0 &= \frac{1}{2}[(C_1 x_N - r_N)^T \bar{S}_N (C_1 x_N - r_N) + \Delta d_N^T \bar{P}_N \Delta d_N] \\ &+ \frac{1}{2}\sum_{k=0}^{N-1} (C_1 x_k - r_k)^T Q (C_1 x_k - r_k) + u_k^T R u_k + \Delta d_k^T Q_1 \Delta d_k \end{aligned} \tag{25}$$

In this case, the Hamiltonian function becomes:

$$\begin{aligned} H_k &= \frac{1}{2}[x_k^T (C_1^T Q C_1 + A_3^T Q_1 A_3) x_k + u_k^T R u_k + d_k^T (A4 - I)^T Q_1 (A_4 - I) d_k \\ &+ r_k^T Q r_k - x_k^T C_1^T Q r_k - r_k^T Q C_1 x_k + d_k^T (A_4 - I)^T Q_1 A_3 x_k \\ &+ x_k^T A_3^T Q_1 (A_4 - I) d_k] + \lambda_{k+1}[A_1 x_k + A_2 d_k + B_1 u_k] \end{aligned} \tag{26}$$

The costate equation (9) and the boundary condition (11) are replaced by:

$$\lambda_k = \frac{\partial H_k}{\partial x_k} = (C^T Q C + A_3^T Q_1 A_3) x_k + A_1^T \lambda_{k+1} + A_3^T Q_1 (A_4 - I) d_k - C_1^T Q r_k \tag{27}$$

$$\lambda_N = \frac{\partial \Phi_N}{\partial x_N} = (C_1^T \bar{S}_N C_1 + A_3^T \bar{P}_N A_3) x_N + A_3^T \bar{P}_N (A_4 - I) d_N - C_1^T \bar{S}_N r_N \tag{28}$$

with

$$\begin{aligned} \Phi_N &= \frac{1}{2}[x_N^T (C_1^T \bar{S}_N C_1 + A_3^T \bar{P}_N A_3) x_N + d_N^T (A4 - I)^T \bar{P}_N (A_4 - I) d_N \\ &+ r_N^T \bar{S}_N r_N x_N^T C_1^T \bar{S}_N r_N - r_N^T \bar{S}_N C_1 x_N + d_N^T (A_4 - I)^T \bar{P}_N A_3 x_N \\ &+ x_N^T A_3^T \bar{P}_N (A_4 - I) d_N] \end{aligned} \tag{29}$$

From the looks of (28), it seems reasonable to assume that for all $k \leq N$, the costate equation can be written as:

$$\lambda_k = S_k x_k + P_k d_k - v_k \tag{30}$$

Using (30) in the state equation (1) to get:

$$\begin{aligned} x_{k+1} = (I + B_1 R^{-1} B1^T S_{k+1})^{-1} [(A_1 - B_1 R^{-1} B_1^T P_{k+1} A_3) x_k \\ + (A_2 - B_1 R^{-1} B_1^T P_{k+1} A_4) d_k + B_1 R^{-1} B_1^T v_{k+1}] \end{aligned} \tag{31}$$

Using (31) and (30) in the costate equation (27) gives

$$\begin{aligned} S_k x_k + P_k d_k - v_k = [C_1^T Q C_1 + A_3^T Q_1 A_3] x_k \\ + A_1^T S_{k+1} [I + B_1 R^{-1} B_1^T S_{k+1}]^{-1} [A_1 - B_1 R^{-1} B_1^T P_{k+1} A_3] x_k \\ + A_1^T S_{k+1} [I + B_1 R^{-1} B_1^T S_{k+1}]^{-1} [A_2 - B_1 R^{-1} B_1^T P_{k+1} A_4] d_k \\ + A_1^T S_{k+1} [I + B_1 R^{-1} B_1^T S_{k+1}]^{-1} B_1 R^{-1} B_1^T v_{k+1} \\ + A_1^T P_{k+1} A_4 d_k + A_1^T P_{k+1} A_3 x_k \\ + A_3^T Q_1 (A_4 - I) d_k - A_1 v_{k+1} - C_1^T Q r_k \end{aligned} \tag{32}$$

This equation must hold for all state sequences x_k and d_k given any x_0 and d_0, therefore allows to write:

$$S_k = C^T Q C + A_3^T Q_1 A_3 + A_1^T S_{k+1} (I + B_1 R^{-1} B_1^T S_{k+1})^{-1} (A_1 - B_1 R^{-1} B_1^T P_{k+1} A_3) + A_1^T P_{k+1} A_3 \tag{33}$$

$$P_k = A_1^T S_{k+1} (I + B_1 R^{-1} B_1^T S_{k+1})^{-1} (A_2 - B_1 R^{-1} B_1^T P_{k+1} A_4) + A_1^T P_{k+1} A_4 + A_3^T Q_1 (A_4 - I) \tag{34}$$

$$v_k = A_1^T v_{k+1} - A_1^T S_{k+1} (I + B_1 R^{-1} B_1^T S_{k+1})^{-1} B_1 R^{-1} B_1^T v_{k+1} + C^T Q r_k \tag{35}$$

By comparing (28) and (30), the boundary conditions for these recursions are $S_N = C_1^T \bar{S}_N C_1 + A_3^T \bar{P}_N A_3$, $P_N = A_3^T \bar{P}_N (A_4 - I)$ and $v_N = -C^T \bar{S}_N r_N$. Since the sequences S_k, P_k and v_k can be computed, assumption (30) is valid, and the optimal control is:

$$u_k = -K_k^x x_k - K_k^d d_k + K_k^v v_{k+1} \tag{36}$$

with

$$K_k^x = (R + B_1^T S_{k+1} B_1)^{-1} B^T (S_{k+1} A_1 + P_{k+1} A_3) \tag{37}$$

$$K_k^d = (R + B_1^T S_{k+1} B_1)^{-1} B^T (S_{k+1} A_2 + P_{k+1} A_4) \tag{38}$$

$$K_k^v = (R + B_1^T S_{k+1} B_1)^{-1} B^T \tag{39}$$

If (A, B) is stabilizable and (A, C) is detectable, the tracker gains K_k^x, K_k^d and K_k^v reach steady-state values K_∞^x, K_∞^d and K_∞^v as $N \rightarrow \infty$.

A disadvantage of this formulation is the need of computing v_k using backward recursion (35) (Fig. 1). To solve this problem, an infinite-horizon LQT in a causal manner was proposed in [6]. In the following, to verify the effectiveness of the developed control schemes, a finite horizon LQT is implemented on an EMA linear model.

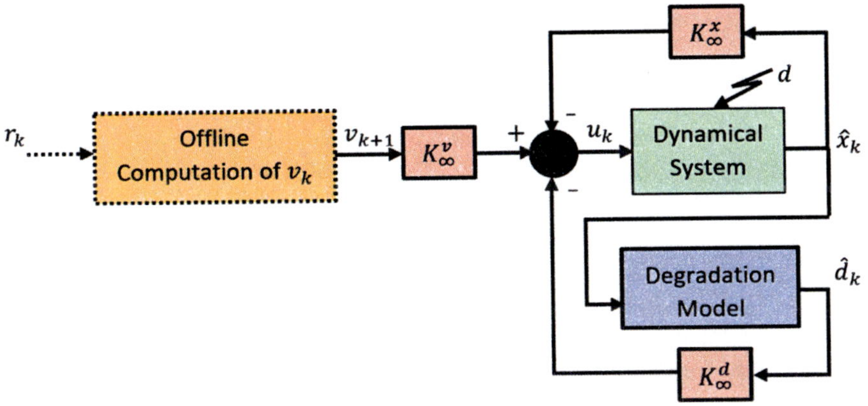

Fig. 1. LQT design

4 EMA Application Example

4.1 Actuator Model

A 5th order state-space model [4] was developed by using (1) and (2) to represent the actuator dynamics. The state vector $x = [i_m\ \theta_m\ \omega_m\ \theta_l\ \omega_l\ ]^T$ is described by the motor current, motor position, motor speed, load position and load speed, the control input $u = [\theta_{ref}]$ is constitute of the reference position (the external load disturbance is supposed to be negligible), and the output vector $y = [i_m\ \theta_l]^T$ is defined by the motor current and the load position. The coefficients of the EMA model are given in Table 1. The transition and the control matrices are provided respectively in (40), (41).

$$A_1 = \begin{bmatrix} -\frac{R_{tt}}{L_{tt}} & -\frac{K_{p1}K_{p2}}{L_{tt}} & -\frac{K_{p1}+K_e}{L_{tt}} & 0 & 0 \\ 0 & 0 & 1 & 0 & 0 \\ \frac{K_t}{J_m} & -\frac{K_{cs}}{J_m N_{cm}^2} & -\frac{b_m}{J_m} & \frac{K_{cs}N_{cl}}{J_m N_{cm}} & 0 \\ 0 & 0 & 0 & 0 & 1 \\ 0 & \frac{K_{cs}N_{cl}}{J_l N_{cm}} & 0 & -\frac{K_l+K_{cs}N_{cl}^2}{J_l} & -\frac{b_l}{J_l} \end{bmatrix} \quad (40)$$

$$B_1 = \begin{bmatrix} \frac{K_{p1}K_{p2}N_{cm}}{L_{tt}N_{cl}} & 0 & 0 & 0 & 0 \end{bmatrix}^T \quad (41)$$

All the state of the system and the degradation are supposed to be measurable, but we only want to track the load position, thus:

$$C_1 = \begin{bmatrix} 0 & 0 & 0 & 1 & 0 \end{bmatrix} \quad (42)$$

4.2 Model of Degradation

The winding temperature is proportional to the power loss in the copper windings [2–4], thus proportional to the absolute value of motor current. The relation between the motor current and the winding-to-ambient temperature can be described by a first order thermo-electrical model with $R_0 = 2.4 \times 10^{-1} A.K/W.s$ and $T_0 = 10^{-3} s$:

$$\dot{d}(t) = R_0|i_m(t)|+T_0 d(t) \tag{43}$$

Temperature and time affect the electrical endurance qualities of insulation. Moreover, an increase of temperature leads to winding insulation breakdown, which is a primary failure mechanism for the EMA's motors. In the following, only the rate of evolution of temperature is intended to be decelerated.

Remark: The augmented system formed by the EMA system and the state of the winding temperature is stabilizable and detectable. The continuous augmented system is discretized using the Zero-Order Hold method (ZOH) for $T_s = 0.01s$.

4.3 Results and Simulation

The weight matrices Q and R are chosen in such a way to get a null steady state error. When increasing the value of matrix R, a larger penalty is applied to the aggressiveness of the control action, and the control gains are decreasing. Choosing a large value of R means trying to stabilize the system with less energy and by choosing a large value of Q, the error between the load position and the reference has vanished. The main purpose of this work is to decelerate the rate of evolution of winding temperature, this can be achieved by increasing the value of Q_1. In the following, Q and R are fixed and have the values:

$$R = I[1]; \qquad Q = 40.10^2.I[5]$$

and

$$\bar{S}_N = Q; \qquad \bar{P}_N = Q_1$$

For different value of Q_1 and for a finite horizon $N = 30000$, the results are represented in Fig. 2. It can be seen that by increasing the value of Q_1 the steady state error increases and the maximal temperature value decreases.

Moreover, for a constant reference track and for $N = 300000$, Fig. 3 shows that for $Q_1 = 10^9$ the settling time is greater than for $Q_1 = 10^2$ and $Q_1 = 10^8$, and the steady state error is bigger, than it starts to decrease with time. To reduce more the rate of evolution of degradation, the value of Q_1 can be increased but this will increase the rising and the response time.

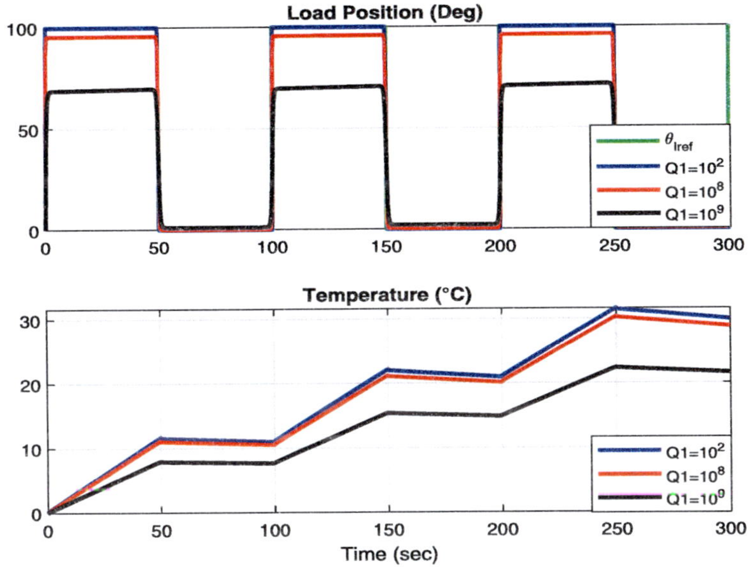

Fig. 2. Evolution of load position and winding temperature for different Q_1

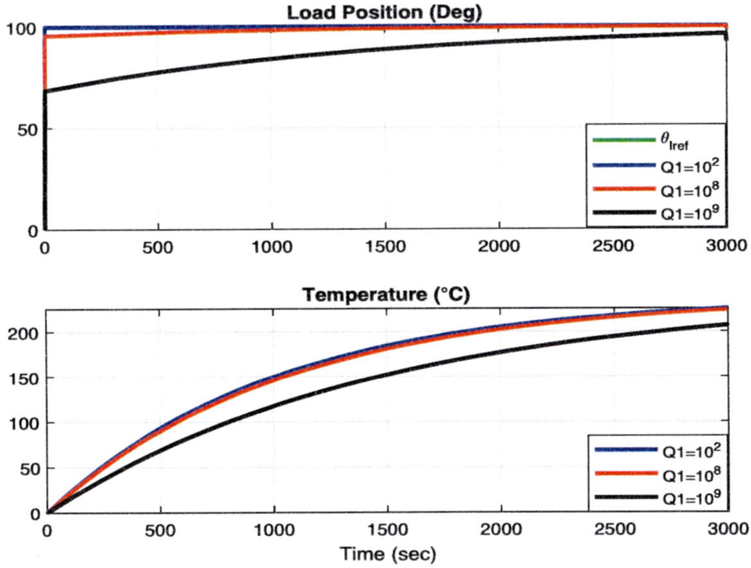

Fig. 3. Evolution of load position and winding temperature for a constant reference track

Table 1. List of used symbols and constants [4].

Sym	Description	Units	Value
b_l	Load damping	in·lbf/rad/s	2.5×10^{-1}
b_m	Motor damping	in·lbf/rad/s	1×10^{-4}
K_{cs}	Coupling stiffness	rad/rad	1×10^{5}
K_e	Back-emf coef.	V/rad/s	1.1×10^{-1}
K_l	Load stiffness	in·lbf/rad/s	2×10^{-3}
K_{p1}	Motor speed gain	V/rad/s	1
K_{p2}	Motor position gain	s^{-1}	1
K_t	Motor torque coef.	in·lbf/A	1.01
J_l	Load inertia	in·lbf·^{2}	2×10^{-3}
J_m	Motor inertia	in·lbf·^{2}	2.1×10^{-3}
L_{tt}	Turn-to-turn induct.	H	3×10^{-4}
R_{tt}	Turn-to-turn resist	Ω	1.6×10^{-1}
N_{cl}	Load coupling	-	1
N_{cm}	Motor coupling	-	1

5 Conclusion and Future Work

Reconfigurable control strategies have gained attention in the control community in recent years to improve the survivability of critical systems under failure conditions. This paper proposed a degradation tolerant control design, a finite and an infinite horizon optimization approach was developed. This approach was examined for a linear electromechanical actuator system and simulation results show its feasibility. By tuning the matrix Q_1, the rate of evolution of winding temperature can be decreased thus prevents winding insulation breakdown.

Future work will focus on how to implement LQG control to estimate non-measurable states. Furthermore, the integration of remaining useful life in the cost function to extend its value.

References

1. Blanke, M., Kinnaert, M., Lunze, J., Staroswiecki, M., Schröder, J.: Diagnosis and Fault-Tolerant Control, vol. 2. Springer, Heidelberg (2006). https://doi.org/10.1007/978-3-540-35653-0
2. Bole, B.M., Brown, D.W., Pei, H.L., Goebel, K., Tang, L., Vachtsevanos, G.: Fault adaptive control of overactuated systems using prognostic estimation (2010)
3. Brown, D.W., et al.: Prognostics enhanced reconfigurable control of electro-mechanical actuators (2021)
4. Brown, D., Bole, B., Vachtsevanos, G.: A prognostics enhanced reconfigurable control architecture, pp. 1061–1066. IEEE (2010). https://doi.org/10.1109/MED.2010.5547651

5. Chamseddine, A., Theilliol, D., Sadeghzadeh, I., Zhang, Y., Weber, P.: Optimal reliability design for over-actuated systems based on the MIT rule: application to an octocopter helicopter testbed. Reliab. Eng. Syst. Saf. **132**, 196–206 (2014)
6. Fu, J., Hazyuk, I., Maré, J.C.: Preliminary design rules for electromechanical actuation systems -effects of saturation and compliances. In: 5th European Air and Space Conference, pp. 1–14 (2015)
7. Fu, J., Maré, J.C., Fu, Y.: Modelling and simulation of flight control electromechanical actuators with special focus on model architecting, multidisciplinary effects and power flows. Chin. J. Aeronaut. **30**, 47–65 (2017). https://doi.org/10.1016/j.cja.2016.07.006
8. Jha, M.S., Weber, P., Theilliol, D., Ponsart, J.C., Maquin, D.: A reinforcement learning approach to health aware control strategy. In: 2019 27th Mediterranean Conference on Control and Automation (MED), pp. 171–176. IEEE (2019)
9. Lewis, F.L., Vrabie, D.L., Syrmos, V.L.: Optimal Control (2012)
10. Noura, H., Theilliol, D., Ponsart, J.C., Chamseddine, A.: Fault-Tolerant Control Systems: Design and Practical Applications. Springer, London (2009). https://doi.org/10.1007/978-1-84882-653-3
11. Obando, D.J.R.: From deterioration modeling to remaining useful life control: a comprehensive framework for post-prognosis decision-making applied to friction drive system (2018)
12. Obando, D.R., Martinez, J.J., Bérenguer, C.: Deterioration estimation for predicting and controlling RUL of a friction drive system. ISA Trans. **113**, 97–110 (2021). https://doi.org/10.1016/j.isatra.2020.10.013
13. Pour, F.K., Theilliol, D., Puig, V., Cembrano, G.: Health-aware control design based on remaining useful life estimation for autonomous racing vehicle. ISA Trans. **113**, 196–209 (2021)
14. Salazar, J.C., Weber, P., Nejjari, F., Sarrate, R., Theilliol, D.: System reliability aware model predictive control framework. Reliab. Eng. Syst. Saf. **167**, 663–672 (2017)
15. Stengel, R.F.: Optimal control and estimation (1986)

A Predictive Fault-Tolerant Tracking Control for Constrained Dynamic Systems

Norbert Kukurowski(✉), Marcin Mrugalski, and Marcin Witczak

Institute of Control and Computation Engineering, University of Zielona Góra, Szafrana 2, 65-516 Zielona Góra, Poland
{n.kukurowski,m.mrugalski,m.witczak}@issi.uz.zgora.pl

Abstract. The paper proposes a novel fault-tolerant tracking control method based on robust observer for constrained dynamic systems. Accordingly, it is considered that the system may be occupied by the actuator and sensor faults as well as the external disturbances, simultaneously. Moreover, it is assumed that there is already employed nominal predictive controller for the fault-free and disturbance-free reference system. Thus, the additional fault-tolerant tracking controller based on robust observer is proposed, while it is not necessary to change the nominal controller. Additionally, the stability of the described observer and controller is guaranteed by using H_∞ approach. Finally, the laboratory Two-Tank system is used to confirm the correctness and performance of the proposed scheme. Additionally, tanks in the proposed system are physically connected together, while each tank has its own water pump.

Keywords: Robust observer · Predictive fault-tolerant tracking control · Constrained system · Simultaneous faults

1 Introduction

Nowadays, purpose of industry is to increase production speed and quality as well as decrease the final production cost. Thus, objective of industry is to use high performance automation systems. However, the complexity of the every additional system increases the final number of the used actuators and sensors. Regrettably, mentioned situation rises a danger of faults, which can lead into total failures. Additionally, failures may stop the production process as well as can cause dangerous accidents for operators. Thus, there is an essential need for the Fault Diagnosis (FD) [8,15] along with the Fault-Tolerant Control (FTC) [2,7,10].

Furthermore, there are a lot of papers proposing FTC schemes for constrained systems. Accordingly, a lot of them are considering single actuator [1,11] or sensor [4,12] fault, nevertheless, without assuming the case of simultaneous faults along with external disturbances. Li et al. [6] proposed the active FTC based on unknown input observer for a wind turbine system considering only an actuator

The work was supported by the National Science Centre, Poland under Grant: UMO-2017/27/B/ST7/00620.

Z. Kowalczuk (Ed.): DPS 2022, LNNS 545, pp. 410–422, 2023.
https://doi.org/10.1007/978-3-031-16159-9_33

fault. Hamdi et al. [3] presented the FTC with fault estimation based sliding mode observer. Thus, the linear parameter varying (LPV) descriptor system with time delay and without sensor faults is considered. Wang et al. [14] described the linear quadratic predictive FTC for multi-phase batch processes, while the sensor faults are not considered. Tao et al. [13] presented the simplified model predictive FTC for a five-phase permanent-magnet motor. Pham et al. [9] proposed the FTC set-model predictive control for five-phase permanent magnet synchronous generator. In the other hand, Xiao et al. [17] described the robust FTC for dual three-phase permanent-magnet synchronous machine, while only the speed sensor is faulty.

The objective of the paper is to propose a predictive Fault-Tolerant Tracking Control (FTTC) approach based on the robust observer design for the constrained dynamic systems. Accordingly, the proposed system is possibly faulty due to the fact that actuator and sensor faults along with external disturbances may occurred, simultaneously. Thus, the proposed robust observer is capable to estimate real faults and states even of occurred external disturbances. Moreover, the proposed predictive FTTC scheme allows to follow real states even of occurred faults.

The paper is organized as follows: The proposed predictive FTTC approach based on the robust observer design for constrained dynamic systems is described in the Sect. 2. Moreover, it is assumed that the system may be occupied by actuator and sensor faults as well as external disturbances. Section 3 presents the Two-Tank laboratory system, which is used to confirm the performance of the proposed approach. Finally, Sect. 4 concludes the paper.

2 Fault-Tolerant Tracking Controller Design

Let us start with defining a linear system with actuator and sensor fault as well as external disturbances:

$$\boldsymbol{x}_{f,k+1} = \boldsymbol{A}\boldsymbol{x}_{f,k} + \boldsymbol{B}\boldsymbol{u}_{f,k} + \boldsymbol{B}\boldsymbol{f}_{a,k} + \boldsymbol{W}_1\boldsymbol{w}_{1,k}, \tag{1}$$

$$\boldsymbol{y}_{f,k} = \boldsymbol{C}\boldsymbol{x}_{f,k} + \boldsymbol{C}_f\boldsymbol{f}_{s,k} + \boldsymbol{W}_2\boldsymbol{w}_{2,k}, \tag{2}$$

where $\boldsymbol{x}_{f,k} \in \mathbb{X} \subset \mathbb{R}^n$, $\boldsymbol{u}_{f,k} \in \mathbb{R}^r$ and $\boldsymbol{y}_{f,k} \in \mathbb{R}^m$ indicate the state, input and output vectors, respectively. Moreover, the actuator and sensor faults defined with $\boldsymbol{f}_{a,k} \in \mathbb{F}_a \subset \mathbb{R}^{n_a}$ and $\boldsymbol{f}_{s,k} \in \mathbb{F}_s \subset \mathbb{R}^{n_s}$ may occur in the system, simultaneously. Thus, $\boldsymbol{C}_f$ with $\mathrm{rank}(\boldsymbol{C}_f) = n_s$ signifies the distribution matrix of the sensor fault. Additionally, $\boldsymbol{w}_{1,k}$ and $\boldsymbol{w}_{2,k}$ define an exogenous disturbances vectors for the process and measurement uncertainties. Furthermore, $\boldsymbol{W}_1$ and $\boldsymbol{W}_2$ indicate the distribution matrices for the external disturbances.

The primary goal of this work is to develop a FTTC method on the basis of state and faults estimation. Such method is to enable minimization the tracking error according to assumed reference state $\boldsymbol{x}_k$ and the state $\boldsymbol{x}_{f,k}$, which is subject to actuators and sensors faults. Moreover, in the proposed approach the application of the nominal controller is assumed, which will be not changed according to the proposed FTC strategy.

The idea proposed in this paper boils down to designing a predictive controller for the fault-free and disturbance-free reference system, which is expressed by:

$$\boldsymbol{x}_{k+1} = \boldsymbol{A}\boldsymbol{x}_k + \boldsymbol{B}\boldsymbol{u}_k, \tag{3}$$

whilst $\boldsymbol{u}_k \in \mathbb{R}^r$ indicate the control input of the nominal controller. Additionally, $\boldsymbol{x}_k \in \mathbb{R}^n$ is the state vector. Additionally, the system (3) obeys the following constraints:

$$\boldsymbol{F}\boldsymbol{x}_k + \boldsymbol{G}\boldsymbol{u}_k \leq \boldsymbol{1}, \tag{4}$$

where $\boldsymbol{F}$ and $\boldsymbol{G}$ are known matrices of appropriate dimension. Thus, the design of the predictive controller boils down to determining a sequence $\bar{\boldsymbol{u}}_k = \{\boldsymbol{u}_{0,k}, \ldots, \boldsymbol{u}_{N-1,k}\}$ over the prediction horizon N, which minimizes the cost function:

$$J(\bar{\boldsymbol{u}}_k, \bar{\boldsymbol{x}}_k) = \sum_{i=0}^{N-1} \left(\|\bar{\boldsymbol{x}}_{i,k}\|_{\boldsymbol{Q}}^2 + \|\bar{\boldsymbol{u}}_{i,k}\|_{\boldsymbol{R}}^2\right) + \|\bar{\boldsymbol{u}}_{N,k}\|_{\boldsymbol{W}}^2, \tag{5}$$

under constraints (3)–(4), where $\boldsymbol{R} \succ 0$, $\boldsymbol{Q} \succ 0$, $\boldsymbol{W} \succ 0$, $\bar{\boldsymbol{x}}_k = \{\boldsymbol{x}_{0,k}, \ldots, \boldsymbol{x}_{N,k}\}$ and $\boldsymbol{x}_{0,k} = \boldsymbol{x}_k$. Note that $\|\bar{\boldsymbol{u}}_{N,k}\|_{\boldsymbol{W}}^2$ is called the terminal penalty constraint, which is a cost that must be achieved after N predictions. The optimization problem can be efficiently solved by various classical techniques, which can be derived from [5].

Subsequently, the controller for the possibly faulty system (1)–(2) is designed in such a way as to track the reference one (3). This can be achieved using the proposed control strategy, which obeys:

$$\boldsymbol{u}_{f,k} = -\hat{\boldsymbol{f}}_{a,k} + \boldsymbol{K}_c\left(\boldsymbol{x}_k - \hat{\boldsymbol{x}}_{f,k}\right) + \boldsymbol{u}_k, \tag{6}$$

where $\hat{\boldsymbol{f}}_{a,k}$ denotes fault estimate of the actuator and $\boldsymbol{K}_c$ is the gain matrix. It is worth underlining that the observer state $\hat{\boldsymbol{x}}_{f,k}$ can be used when the $\boldsymbol{x}_{f,k}$ cannot be appointed. Now, the following form of the state, sensor and actuator fault estimates can be proposed:

$$\hat{\boldsymbol{x}}_{f,k+1} = \boldsymbol{A}\hat{\boldsymbol{x}}_k + \boldsymbol{B}\boldsymbol{u}_k + \boldsymbol{B}\hat{\boldsymbol{f}}_{a,k} + \boldsymbol{K}_x\left(\boldsymbol{y}_k - \boldsymbol{C}\hat{\boldsymbol{x}}_k - \boldsymbol{C}_f\hat{\boldsymbol{f}}_{s,k}\right), \tag{7}$$

$$\hat{\boldsymbol{f}}_{a,k+1} = \hat{\boldsymbol{f}}_{a,k} + \boldsymbol{K}_a\left(\boldsymbol{y}_k - \boldsymbol{C}\hat{\boldsymbol{x}}_k - \boldsymbol{C}_f\hat{\boldsymbol{f}}_{s,k}\right), \tag{8}$$

$$\hat{\boldsymbol{f}}_{s,k+1} = \hat{\boldsymbol{f}}_{s,k} + \boldsymbol{K}_s\left(\boldsymbol{y}_k - \boldsymbol{C}\hat{\boldsymbol{x}}_k - \boldsymbol{C}_f\hat{\boldsymbol{f}}_{s,k}\right). \tag{9}$$

Substituting (6) into (1) the following relation can be obtained:

$$\boldsymbol{x}_{f,k+1} = \boldsymbol{A}\boldsymbol{x}_{f,k} - \boldsymbol{B}\hat{\boldsymbol{f}}_{a,k} + \boldsymbol{B}\boldsymbol{K}_c\left(\boldsymbol{e}_k - \boldsymbol{e}_{f,k}\right) + \boldsymbol{B}\boldsymbol{u}_k + \boldsymbol{B}\boldsymbol{f}_{a,k} + \boldsymbol{W}_1\boldsymbol{w}_{1,k}, \tag{10}$$

where:

$$e_k = x_k - x_{f,k}, \qquad e_{f,k} = x_{f,k} - \hat{x}_{f,k}, \tag{11}$$

and e_k is the tracking error, $e_{f,k}$ is the estimation error. Now, the expression (11) can be rewritten as follows:

$$x_{f,k+1} = Ax_{f,k} + Be_{a,k} + BK_c(e_k - e_{f,k}) + Bu_k + W_1 w_{1,k}, \tag{12}$$

where, $e_{a,k}$ denotes the actuator fault estimation error. Moreover, the tracking and state estimation error be assumed as follows:

$$e_{k+1} = x_{k+1} - x_{f,k+1} = (A - BK_c)e_k - Be_{a,k} - BK_c e_{f,k} - W_1 w_{1,k}, \tag{13}$$

$$\begin{aligned} e_{f,k+1} = x_{f,k+1} - \hat{x}_{f,k+1} = (A - K_x C)e_{f,k} + Be_{a,k} - K_x C_f e_{s,k} \\ + W_1 w_{1,k} - K_x W_2 w_{2,k}. \end{aligned} \tag{14}$$

Furthermore, the actuator and sensor fault estimation can be defined as follows:

$$\begin{aligned} e_{a,k+1} = f_{a,k+1} - \hat{f}_{a,k+1} = \varepsilon_{a,k} + e_{a,k} - K_a C e_{f,k} - K_a C_f e_{s,k} \\ - K_a W_2 w_{2,k}, \end{aligned} \tag{15}$$

$$\begin{aligned} e_{s,k+1} = f_{s,k+1} - \hat{f}_{s,k+1} = \varepsilon_{s,k} + (I - K_s C_f)e_{s,k} - K_s C e_{f,k} \\ - K_s W_2 w_{2,k}, \end{aligned} \tag{16}$$

where:

$$\varepsilon_{a,k} = f_{a,k+1} - f_{a,k}, \qquad \varepsilon_{s,k} = f_{s,k+1} - f_{s,k}.$$

It is worth noting that the following super-vectors by combining (13)–(16) can be defined:

$$\bar{e}_k = \left[e_k^T \; e_{f,k}^T \; e_{a,k}^T \; e_{s,k}^T\right]^T, \qquad \bar{w}_k = \left[w_{1,k}^T \; w_{2,k}^T \; \varepsilon_{a,k}^T \; \varepsilon_{s,k}^T\right]^T. \tag{17}$$

Finally, on the basis of the above deliberations the following compact form of the state estimation error can be obtained:

$$\begin{aligned} \bar{e}_{k+1} = & \begin{bmatrix} A - BK_c & -BK_c & -B & 0 \\ 0 & A - K_x C & B & -K_x C_f \\ 0 & -K_a C & I & -K_a C_f \\ 0 & -K_s C & 0 & I - K_s C_f \end{bmatrix} \bar{e}_k \\ & + \begin{bmatrix} -W_1 & 0 & 0 & 0 \\ W_1 & -K_x W_2 & 0 & 0 \\ 0 & -K_a W_2 & I & 0 \\ 0 & -K_s W_2 & 0 & I \end{bmatrix} \bar{w}_k, \end{aligned} \tag{18}$$

either in the simpler way

$$\bar{e}_{k+1} = \tilde{A}_1 \bar{e}_k + \tilde{W}_1 \bar{w}_k, \tag{19}$$

while:

$$\tilde{A}_1 = \begin{bmatrix} \bar{A}_1 & \bar{B} \\ 0 & \bar{A}_2 \end{bmatrix}, \quad \tilde{W}_1 = \begin{bmatrix} -W_1 & 0 \\ \bar{W}_1 & \bar{W}_2 \end{bmatrix}, \quad \bar{W}_3 = \begin{bmatrix} 0 & 0 & 0 \\ 0 & I & 0 \\ 0 & 0 & I \end{bmatrix},$$

$$\bar{W}_2 = \bar{W}_3 - \bar{K}\tilde{W}_2, \qquad \tilde{W}_2 = \begin{bmatrix} W_2 \; 0 \; 0 \end{bmatrix}, \quad \bar{W}_1 = \begin{bmatrix} W_1^T \; 0 \; 0 \end{bmatrix}^T,$$

$$\bar{A}_1 = A - BK_c, \quad \bar{B} = \begin{bmatrix} -BK_c \; -B \; 0 \end{bmatrix}, \quad \bar{K} = \begin{bmatrix} K_x^T & K_a^T & K_s^T \end{bmatrix}^T.$$

As it can be seen that observer and fault-tolerant controller may be designed independently because the eigenvalues of $\tilde{A}_1$ depend on the ones of $\bar{A}_1$ and $\bar{A}_2$. Now, let us redefine matrix

$$\bar{A}_2 = \begin{bmatrix} A - K_x C & B & -K_x C_f \\ -K_a C & I & -K_a C_f \\ -K_s C & 0 & I - K_s C_f \end{bmatrix}, \tag{20}$$

as follows:

$$\bar{A}_2 = \bar{A} - \bar{K}\bar{C}, \tag{21}$$

where matrices $\bar{A}$ and $\bar{C}$ has the following forms:

$$\bar{A} = \begin{bmatrix} A & B & 0 \\ 0 & I & 0 \\ 0 & 0 & I \end{bmatrix}, \quad \bar{C} = \begin{bmatrix} C \; 0 \; C_f \end{bmatrix}. \tag{22}$$

In this situation, the observer system can be defined as follows:

$$\tilde{e}_{f,k+1} = \bar{A}_2 \tilde{e}_{f,k} + \bar{W}_2 \bar{w}_k, \tag{23}$$

where $\tilde{e}_{f,k} = \begin{bmatrix} e_{f,k}^T, e_{a,k}^T, e_{s,k}^T \end{bmatrix}^T$. Let us assume the following form of the Lyapunov function:

$$V_{f,k} = \tilde{e}_{f,k}^T P_f \tilde{e}_{f,k}, \tag{24}$$

with $P_f \succ 0$. Now, the stability condition can be written as:

$$\Delta V_{f,k} + \tilde{e}_{f,k}^T \tilde{e}_{f,k} - \mu_f^2 \bar{w}_k^T \bar{w}_k < 0, \tag{25}$$

and finally in below form:

$$\Delta V_{f,k} = V_{f,k+1} - V_{f,k}, \qquad V_{f,k+1} = \tilde{e}_{f,k+1}^T P_f \tilde{e}_{f,k+1}.$$

In order to perform the observer design procedure the following theorem should be invoked:

Theorem 1. *For an assumed μ_f of $\bar{\boldsymbol{w}}_k$, the $\mathcal{H}_\infty$ estimator design problem (1)–(2) can be solved if the following matrices $\boldsymbol{N}_f$ and $\boldsymbol{P}_f \succ 0$ exists for which the following condition is fulfilled:*

$$\begin{bmatrix} -\boldsymbol{P}_f + \boldsymbol{I} & * & * \\ \boldsymbol{0} & -\mu_f^2 \boldsymbol{I} & * \\ \boldsymbol{P}_f \bar{\boldsymbol{A}} - \boldsymbol{N}_f \bar{\boldsymbol{C}} & \boldsymbol{P}_f \bar{\boldsymbol{W}}_3 - \boldsymbol{N}_f \tilde{\boldsymbol{W}}_2 & -\boldsymbol{P}_f \end{bmatrix} \prec 0. \tag{26}$$

Proof. On the basis of (25) the subsequent relation can be assumed:

$$\tilde{\boldsymbol{e}}_{f,k+1}^T \boldsymbol{P}_f \tilde{\boldsymbol{e}}_{f,k+1} - \tilde{\boldsymbol{e}}_{f,k}^T \boldsymbol{P}_f \tilde{\boldsymbol{e}}_{f,k} + \tilde{\boldsymbol{e}}_{f,k}^T \tilde{\boldsymbol{e}}_{f,k} - \mu_f^2 \bar{\boldsymbol{w}}_k^T \bar{\boldsymbol{w}}_k < 0. \tag{27}$$

Substituting (23) into inequality (27) we receive:

$$\begin{aligned} &\tilde{\boldsymbol{e}}_{f,k}^T \left(\bar{\boldsymbol{A}}_2^T \boldsymbol{P}_f \bar{\boldsymbol{A}}_2 - \boldsymbol{P}_f + \boldsymbol{I} \right) \tilde{\boldsymbol{e}}_{f,k} + \tilde{\boldsymbol{e}}_{f,k}^T \left(\bar{\boldsymbol{A}}_2^T \boldsymbol{P}_f \bar{\boldsymbol{W}}_2 \right) \bar{\boldsymbol{w}}_k \\ &+ \bar{\boldsymbol{w}}_k^T \left(\bar{\boldsymbol{W}}_2^T \boldsymbol{P}_f \bar{\boldsymbol{A}}_2 \right) \ddot{\boldsymbol{e}}_{f,k} + \boldsymbol{w}_k^T \left(\bar{\boldsymbol{W}}_2^T \boldsymbol{P}_f \bar{\boldsymbol{W}}_2 - \mu_f^2 \boldsymbol{I} \right) \bar{\boldsymbol{w}}_k \prec 0. \end{aligned} \tag{28}$$

Furthermore, through the definition of vector:

$$\bar{\boldsymbol{v}}_k = \left[\tilde{\boldsymbol{e}}_{f,k}^T \; \bar{\boldsymbol{w}}_k^T \right]^T, \tag{29}$$

it can be shown that (28) can be expressed as:

$$\bar{\boldsymbol{v}}_k^T = \begin{bmatrix} \bar{\boldsymbol{A}}_2^T \boldsymbol{P}_f \bar{\boldsymbol{A}}_2 - \boldsymbol{P}_f + \boldsymbol{I} & \bar{\boldsymbol{A}}_2^T \boldsymbol{P}_f \bar{\boldsymbol{W}}_2 \\ \bar{\boldsymbol{W}}_2^T \boldsymbol{P}_f \bar{\boldsymbol{A}}_2 & \bar{\boldsymbol{W}}_2^T \boldsymbol{P}_f \bar{\boldsymbol{W}}_2 - \mu_f^2 \boldsymbol{I} \end{bmatrix} \bar{\boldsymbol{v}}_k \prec 0, \tag{30}$$

and finally in the following more convenient form:

$$\begin{bmatrix} \bar{\boldsymbol{A}}_2^T \\ \bar{\boldsymbol{W}}_2^T \end{bmatrix} \boldsymbol{P}_f \begin{bmatrix} \bar{\boldsymbol{A}}_2 & \bar{\boldsymbol{W}}_2 \end{bmatrix} + \begin{bmatrix} -\boldsymbol{P}_f + \boldsymbol{I} & \boldsymbol{0} \\ \boldsymbol{0} & -\mu_f^2 \boldsymbol{I} \end{bmatrix} \prec 0. \tag{31}$$

Subsequently, by the application of the Schur complement with multiplication of (31) by diag$(\boldsymbol{I}, \boldsymbol{I}, \boldsymbol{P}_f)$ from left and right, the following inequality can be obtained:

$$\begin{bmatrix} -\boldsymbol{P}_f + \boldsymbol{I} & * & * \\ \boldsymbol{0} & -\mu_f^2 \boldsymbol{I} & * \\ \boldsymbol{P}_f \bar{\boldsymbol{A}}_2 & \boldsymbol{P}_f \bar{\boldsymbol{W}}_2 & -\boldsymbol{P}_f \end{bmatrix} \prec 0. \tag{32}$$

Finally, it follows that:

$$\boldsymbol{P}_f \bar{\boldsymbol{A}}_2 = \boldsymbol{P}_f \bar{\boldsymbol{A}} - \boldsymbol{P}_f \bar{\boldsymbol{K}} \bar{\boldsymbol{C}} = \boldsymbol{P}_f \bar{\boldsymbol{A}} - \boldsymbol{N}_f \bar{\boldsymbol{C}}, \tag{33}$$

$$\boldsymbol{P}_f \bar{\boldsymbol{W}}_2 = \boldsymbol{P}_f \bar{\boldsymbol{W}}_3 - \boldsymbol{P}_f \bar{\boldsymbol{K}} \tilde{\boldsymbol{W}}_2 = \boldsymbol{P}_f \bar{\boldsymbol{W}}_3 - \boldsymbol{N}_f \tilde{\boldsymbol{W}}_2, \tag{34}$$

which completes the proof.

Taking into consideration the above presented theory, the observer design procedure boils down to calculation of the LMIs (26) and then calculation of the gain matrices:

$$\bar{\boldsymbol{K}} = \begin{bmatrix} \boldsymbol{K}_x \\ \boldsymbol{K}_a \\ \boldsymbol{K}_s \end{bmatrix} = \boldsymbol{P}_f^{-1}\boldsymbol{N}_f. \tag{35}$$

Moreover, on the basis of (19), the following form of the controller can be assumed:

$$\tilde{\boldsymbol{e}}_{c,k+1} = \bar{\boldsymbol{A}}_1\tilde{\boldsymbol{e}}_{c,k} - \boldsymbol{W}_1\bar{\boldsymbol{w}}_k. \tag{36}$$

Analogously as it was already shown, let us assume the following Lyapunov candidate function:

$$V_{c,k} = \tilde{\boldsymbol{e}}_{c,k}^T\boldsymbol{P}_c\tilde{\boldsymbol{e}}_{c,k}, \tag{37}$$

where matrix $\boldsymbol{P}_c$ fulfil the condition such that $\boldsymbol{P}_c \succ 0$.

Moreover, the following stability condition for the controller system can be assumed:

$$\Delta V_{c,k} + \tilde{\boldsymbol{e}}_{c,k}^T\tilde{\boldsymbol{e}}_{c,k} - \mu_c^2\bar{\boldsymbol{w}}_k^T\bar{\boldsymbol{w}}_k < 0, \tag{38}$$

where:

$$\Delta V_{c,k} = V_{c,k+1} - V_{c,k}, \qquad V_{c,k+1} = \tilde{\boldsymbol{e}}_{c,k+1}^T\boldsymbol{P}_c\tilde{\boldsymbol{e}}_{c,k+1}.$$

Let us assume corresponding to the observer theorem for the controller design:

Theorem 2. *For an assumed $\mu_c > 0$ of $\tilde{\boldsymbol{e}}_{c,k}$, the $\mathcal{H}_\infty$ controller design problem (6) can be solved if the following matrices $\boldsymbol{N}_c$, $\boldsymbol{U}_c$, $\boldsymbol{P}_c \succ 0$ exists for which the below condition is satisfied:*

$$\begin{bmatrix} -\boldsymbol{P}_c & * & * & * \\ 0 & -\mu_c^2\boldsymbol{I} & * & * \\ \boldsymbol{A}\boldsymbol{U}_c - \boldsymbol{B}\boldsymbol{N}_c & -\boldsymbol{W}_1 & \boldsymbol{P}_c - \boldsymbol{U}_c - \boldsymbol{U}_c^T & * \\ \boldsymbol{U}_c & 0 & 0 & -\boldsymbol{I} \end{bmatrix} \prec 0. \tag{39}$$

Proof. Similar to the proof of Theorem 1 the proof of Theorem 2 can be conducted. However, this proof is comes down to the following inequality:

$$\begin{bmatrix} \bar{\boldsymbol{A}}_1^T \\ -\boldsymbol{W}_1^T \end{bmatrix} \boldsymbol{P}_c \begin{bmatrix} \bar{\boldsymbol{A}}_1 & -\boldsymbol{W}_1 \end{bmatrix} + \begin{bmatrix} -\boldsymbol{P}_c + \boldsymbol{I} & \boldsymbol{0} \\ \boldsymbol{0} & -\mu_c^2\boldsymbol{I} \end{bmatrix} \prec 0, \tag{40}$$

which can be applied to define (39) on the basis of Theorem 1 developed in [16], which proves the theorem.

In summary, the controller design procedure boils down to calculation of the LMIs (39) and:

$$\boldsymbol{K}_c = \boldsymbol{N}_c\boldsymbol{U}_c^{-1}. \tag{41}$$

3 Simulation Results

Let us propose the Two-Tank system to verify the correctness and performance of the proposed fault-tolerant control scheme. Thus, the simplified scheme of the proposed system is presented in the Fig. 1. Moreover, knowing the relationships of the flows of liquid streams and using discretization, the formulas for the value of water height in the tanks were determined:

$$H_{1,k+1} = \frac{T_s}{A_1}\left[F_1 - S_1\left(H_{1,k} - H_{2,k}\right)\right] + H_{1,k}, \tag{42}$$

$$H_{2,k+1} = \frac{T_s}{A_2}\left[F_2 + S_1\left(H_{1,k} - H_{2,k}\right) - S_2\left(H_{2,k}\right)\right] + H_{2,k}, \tag{43}$$

where

- T_s denotes the sampling time,
- A_1, A_2 indicate the cross-section areas of the first and second tank,
- H_1, H_2 signify the water level respectively in the first and second tank,
- S_1 is the coefficient of the stream flowing through the pipe connecting the tanks, while S_2 is the coefficient of the stream flowing out of the second tank,
- F_1, F_2 indicate the 24V DC pumps in the first and second tank.

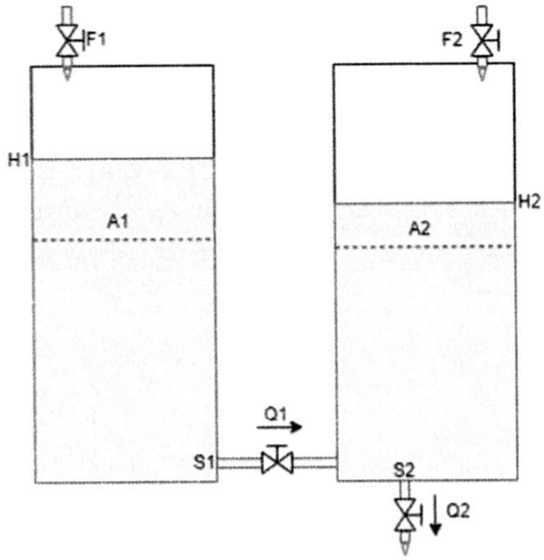

Fig. 1. Scheme of the Two-Tank system

Moreover, in the proposed system, the state $\boldsymbol{x}$ and control $\boldsymbol{u}$ vectors are given as follow:

$$\boldsymbol{x} = \left[H_1\ H_2\right]^T, \qquad \boldsymbol{u} = \left[F_1\ F_2\right]^T, \tag{44}$$

while the reference states are defined as:

$$H_1 = 7\,[\text{cm}], \qquad H_2 = 17\,[\text{cm}]. \tag{45}$$

Thus, in order to verify the performance of the proposed approach, let fault scenario be defined as:

$$\boldsymbol{f}_{a,1,k} = \begin{cases} -0.15 \cdot F_{2,k} & 200 \le k \le 400 \\ -0.65 \cdot F_{2,k} & 500 \le k \le 700 \\ 0 & \text{otherwise}, \end{cases} \tag{46}$$

$$\boldsymbol{f}_{s,1,k} = \begin{cases} 7 + \boldsymbol{y}_{f,k} & 300 \le k \le 600 \\ 0 & \text{otherwise}, \end{cases} \tag{47}$$

along with the sensor fault distribution matrix:

$$\boldsymbol{C}_f = \begin{bmatrix} 1 & 0 \\ 0 & 0 \end{bmatrix}^T, \tag{48}$$

while, the distribution matrices for the process and measurement uncertainties are given as:

$$\boldsymbol{W}_1 = 1 \cdot 10^{-1}\boldsymbol{I}, \qquad \boldsymbol{W}_2 = 1 \cdot 10^{-3}\boldsymbol{I}. \tag{49}$$

Accordingly, it is easily seen that the actuator and sensor faults along with the external disturbances occupied the system, simultaneously. Moreover, the sensor fault occupied the first tank H_1, while the actuator fault occurred in the second tank H_2. The actuator fault is presented in the Fig. 2a, while the sensor fault is illustrated in the Fig. 2b. Thus, the real faults are given with blue dash-dotted line, while their estimates are indicated with red dashed line. Therefore, from this figures it can be concluded that the real faults are properly reconstructed

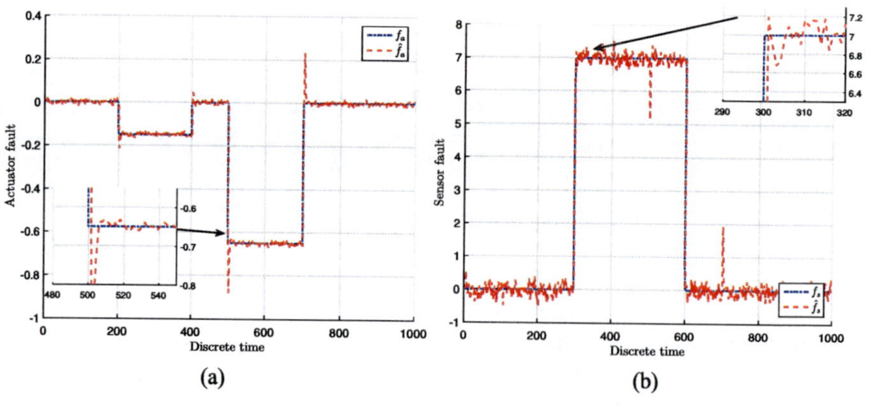

Fig. 2. Actuator $\boldsymbol{f}_{a,1,k}$ (a) and sensor $\boldsymbol{f}_{s,1,k}$ (b) fault

even of existing external disturbances. Moreover, the Figs. 3a–3b illustrate the water level in the first H_1 and second H_2 tank. Accordingly, the real state is given with blue dash-dotted line, while its estimate is indicated with red dashed line. Additionally, the measurement output is illustrated with green solid line. Moreover, the faulty system with the nominal controller is given with magenta dash-dotted line, while the faulty system with the FTTC is presented with black dashed line. Thus, it is easily to observed that the real states are properly estimated even of occurred actuator and sensor faults as well as existing external disturbances. Furthermore, these figures shows that the FTTC is following the real state even of the occurred actuator and sensor faults. Moreover, the actuator fault occurred twice in the same state H_2. Important thing to remind is that the tanks in the system are connected together (see Q_1 in Fig. 1). Thus, in the first case of the actuator fault, the FTTC minimized the tracking error by changing the control law only for the second pump $\boldsymbol{u}_{f,2,k}$. Nevertheless, in the second case, the FTTC changed the control inputs for both water pumps i.e. for the second $\boldsymbol{u}_{f,2,k}$ and first $\boldsymbol{u}_{f,1,k}$ tank, to properly minimize the tracking error in the second tank H_2. In consequence, the FTTC made a steady-state error in the first tank H_1 in order to minimize the tracking error in the second tank H_2. Furthermore, the Figs. 4a–4b indicate the control comparison between nominal and FTTC for the first and second pump. Thus, the nominal controller is given with the blue dash-dotted line, while the FTTC controller is presented with the black dashed line. Thus, from these figures, it can be observed that the FTTC is changing the control law due to the occurred actuator faults, while the nominal controller doesn't react. Moreover, the Fig. 5 illustrate the tracking error for the water level in the first and second tank. Thus, the tracking error for the first tank is given with black dashed line, while for the second tank is presented with magenta dash-dotted line. From these figures, it can be concluded that the tracking error for the first H_1 and second H_2 tank remains longer at a close-zero

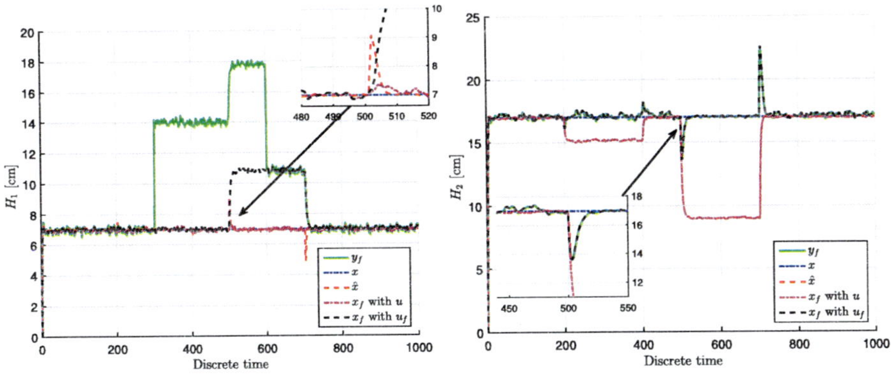

(a) (b)

Fig. 3. Water level in the first H_1 (a) and second H_2 (b) tank

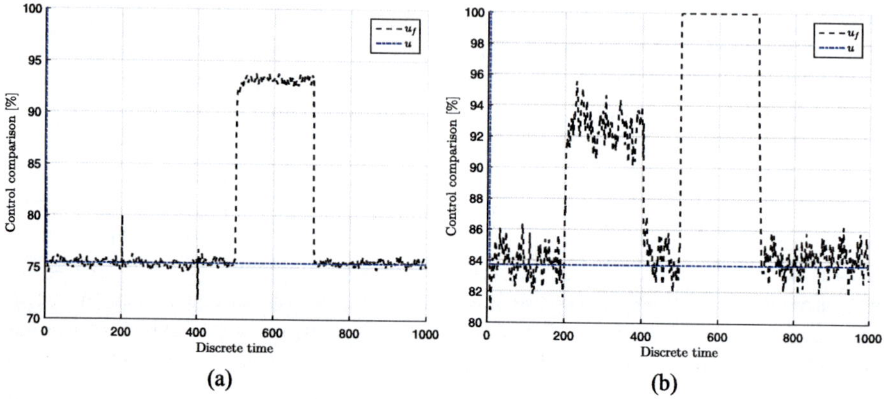

Fig. 4. Control comparison between nominal $\boldsymbol{u}_k$ and fault-tolerant controller $\boldsymbol{u}_{f,k}$ for the first (a) and second (b) control input

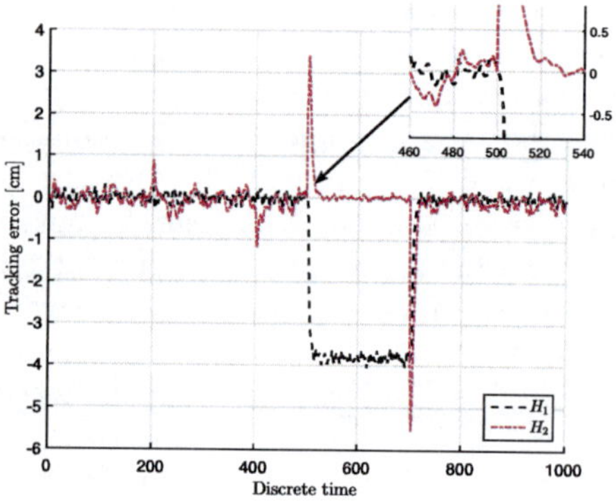

Fig. 5. Tracking error for the water level in the first H_1 and second H_2 tank

level. However, the tracking error of the first tank H_1 is rising, when the second actuator fault occurred in the second water pump.

4 Conclusions

The paper dealt with the novel fault-tolerant tracking control scheme based on robust observer for linear systems. Thus, it was assumed that in the reference system was already employed nominal predictive controller. Moreover, the reference system was fault-free and disturbance-free system. Furthermore, the proposed FTTC scheme was applied to the linear system, which was possibly faulty and may be occupied by actuator and sensor faults as well external disturbances. Accordingly, the stability of the proposed scheme was guaranteed by used H_∞ approach. Moreover, the linear Two-Tank system was used to verify the performance of the proposed FTTC scheme. Concluding, the achieved results clearly confirm the correctness of the proposed approach. Thus, the robust observer accuracy estimate actuator and sensor faults as well as real states. Moreover, the FTTC was following the real states even of occurred simultaneous faults and external disturbances.

References

1. Aouaouda, S., Chadli, M., Righi, I.: Active FTC approach design for TS fuzzy systems under actuator saturation. In: 2019 6th International Conference on Control, Decision and Information Technologies (CoDIT), pp. 483–488. IEEE (2019)
2. Blanke, M., Schörder, J., Kinnaert, M., Lunze, J., Staroswiecki, M.: Diagnosis and Fault-Tolerant Control. Springer, Heidelberg (2006). https://doi.org/10.1007/978-3-540-35653-0
3. Hamdi, H., Rodrigues, M., Rabaoui, B., Benhadj Braiek, N.: A fault estimation and fault-tolerant control based sliding mode observer for LPV descriptor systems with time delay. Int. J. Appl. Math. Comput. Sci. **31**(2) (2021)
4. Kommuri, S.K., Defoort, M., Karimi, H.R., Veluvolu, K.C.: A robust observer-based sensor fault-tolerant control for PMSM in electric vehicles. IEEE Trans. Industr. Electron. **63**(12), 7671–7681 (2016)
5. Kouvaritakis, B., Cannon, M.: Model Predictive Control, vol. 38. Springer, Cham (2016). https://doi.org/10.1007/978-3-319-24853-0
6. Li, S., Wang, H., Aitouche, A., Tian, Y., Christov, N.: Active fault tolerance control of a wind turbine system using an unknown input observer with an actuator fault. Int. J. Appl. Math. Comput. Sci. **28**(1) (2018)
7. Majdzik, P.: Feasible schedule under faults in the assembly system. In: 2020 16th International Conference on Control, Automation, Robotics and Vision (ICARCV), pp. 1049–1054. IEEE (2020)
8. Pazera, M., Witczak, M.: Towards robust simultaneous actuator and sensor fault estimation for a class of nonlinear systems: design and comparison. IEEE Access **7**, 97143–97158 (2019)
9. Pham, H.T., Bourgeot, J.M., Benbouzid, M.: Fault-tolerant finite control set-model predictive control for marine current turbine applications. IET Renew. Power Gener. **12**(4), 415–421 (2018)
10. Puig, V.: Fault diagnosis and fault tolerant control using set-membership approaches: application to real case studies. Int. J. Appl. Math. Comput. Sci. **20**(4), 619–635 (2010)
11. Sun, K., Sui, S., Tong, S.: Optimal adaptive fuzzy FTC design for strict-feedback nonlinear uncertain systems with actuator faults. Fuzzy Sets Syst. **316**, 20–34 (2017)
12. Tabbache, B., Rizoug, N., Benbouzid, M.E.H., Kheloui, A.: A control reconfiguration strategy for post-sensor FTC in induction motor-based EVS. IEEE Trans. Veh. Technol. **62**(3), 965–971 (2012)
13. Tao, T., Zhao, W., Du, Y., Cheng, Y., Zhu, J.: Simplified fault-tolerant model predictive control for a five-phase permanent-magnet motor with reduced computation burden. IEEE Trans. Power Electron. **35**(4), 3850–3858 (2019)
14. Wang, L., Sun, L., Luo, W.: Robust constrained iterative learning predictive fault-tolerant control of uncertain batch processes. Sci. China Inf. Sci. **62**(11), 1–3 (2019). https://doi.org/10.1007/s11432-018-9541-1

15. Witczak, M.: Fault Diagnosis and Fault-Tolerant Control Strategies for Non-Linear Systems. LNEE, vol. 266. Springer, Cham (2014). https://doi.org/10.1007/978-3-319-03014-2
16. Witczak, M., Buciakowski, M., Aubrun, C.: Predictive actuator fault-tolerant control under ellipsoidal bounding. Int. J. Adapt. Control Signal Process. **30**(2), 375–392 (2016)
17. Xiao, L., Zhang, L., Gao, F., Qian, J.: Robust fault-tolerant synergetic control for dual three-phase PMSM drives considering speed sensor fault. IEEE Access **8**, 78912–78922 (2020)

A New Version of the On-Line Adaptive Non-standard Identification Procedure for Continuous-Time MISO Physical Processes

Witold Byrski and Michał Drapała(✉)

Department of Automatic Control and Robotics, AGH University of Science and Technology, Al. Mickiewicza 30, 30-059 Kraków, Poland
{wby,mdrapala}@agh.edu.pl

Abstract. Modern diagnostics and control algorithms rely on physical processes models. Such models have often complicated structure and their synthesis is usually difficult. The approaches based on Partial Differential Equations (PDE) work well for simulation purposes, however their usefulness can be limited in case of industrial applications when a full set of processes data is often inaccessible. The aforementioned problem was the main motivation to propose an adaptive identification method available to work on-line based on processes data. It is based on the Modulating Functions Method (MFM) and utilizes the Exact State Observers. The method was applied for real processes data collected from the industrial glass conditioning installation. The experimental results are presented and discussed in the paper.

Keywords: Modulating functions method · Exact state observers · Adaptive identification · Glass forehearth

1 Introduction

Modern diagnostics and control algorithms are mainly based on models of given physical phenomenas [10,12]. For typical industrial processes consisting of fluid flows and heat transfer, the approach based on Computational Fluid Dynamics (CFD) is often utilized [7,8], however the model synthesis requires extensive theoretical knowledge. For the considered glass conditioning processes, one can also find simplified Partial Differential Equation (PDE) models that work well for some predefined conditions [9]. However, it turns out that the processes dynamics can be often precisely mapped by linear models. There exist many methods of control algorithms synthesis based on them. Hence, the idea of using ordinary lumped-parameters continuous-time linear models is presented in the paper. Of course, the specificity of the analysed problem demands linearisation in a current operating point and forces an adaptive character of the method. The main

Z. Kowalczuk (Ed.): DPS 2022, LNNS 545, pp. 423–436, 2023.
https://doi.org/10.1007/978-3-031-16159-9_34

subject of the paper are adaptive identification procedures devoted to observable and identifiable linear continuous-time Multiple Input Single Output (MISO) systems.

The aforementioned processes of conditioning is the final part of glass melting, before forming containers. The molten glass is gradually cooled down and its temperature is stabilised between adjacent glass streams. It takes place in long ceramic channels called forehearths. An exemplary glass forehearth is presented in Fig. 1. It is divided into several zones, each of which has separate temperature controller. The glass temperature should be adjusted with accuracy up to 1 °C with gas-air mixture burners, cooling valves and dampers for a temperature operating point close to 1180 °C. Efficient tracking of a given temperature profile is very important for production continuity and the knowledge of processes models can be helpful during this control task.

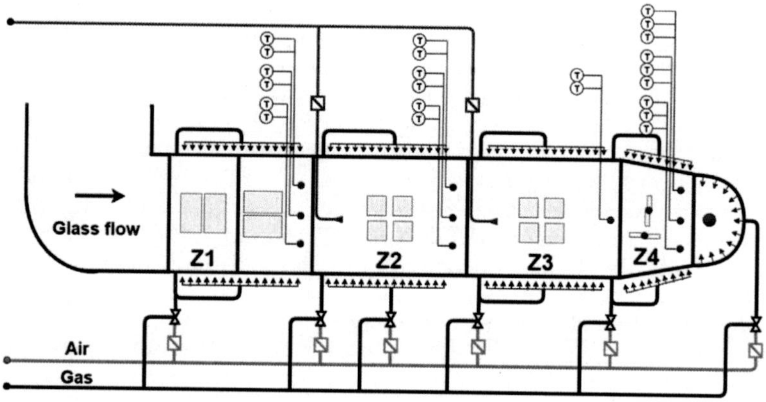

Fig. 1. Typical glass forehearth composed of four zones. In the first three of them molten glass can be heated with gas-air mixture burners or cooled with cooling dampers or cooling air. In the last zone there are only gas burners installed and glass can be only heated.

The paper is organised into several parts. The Sect. 2 presents the theory of the Modulating Functions Method (MFM) applied to the model identification. The idea of obtaining sub-models with different state matrices is also introduced. The use of the Exact State Observers for simulating the predicted system output is explained. The algorithm of adaptive identification in subsequent time intervals is also described. The experimental results, for real historical processes data, are presented in Sect. 3. The last Sect. 4 gives a conclusion.

2 Adaptive Model Identification Method

The developed identification procedure for input/output models consists of several steps. The fundamentals of utilized methods and their modifications are described. Some modifications applied to standard algorithms are presented as well. Finally, the adaptive procedure, applied for the industrial processes is given in details.

2.1 Modulating Functions Method

A differential equation representing a continuous-time linear MISO system of n-th order with K inputs is given as:

$$\sum_{i=0}^{n} a_i y^{(i)}(t) = \sum_{k=1}^{K} \sum_{j=0}^{m_k} b_{kj} u_k^{(j)}(t). \tag{1}$$

It is assumed that only the system input and output signals can be measured. The parameters $\boldsymbol{a}$ and $\boldsymbol{b}$ are usually unknown. A common approach for their identification assumes discretisation of the model, identification of the obtained discrete-time model with one of the known methods and then conversion to the continuous-time form again. This indirect approach can be inefficient especially when measurement disturbances occur.

In contrast to the previously presented methodology, the Modulating Functions Method (MFM) enables to obtain the system parameters at once, without discretisation. It utilizes the concept of convolution transformation of the model differential equation using the rule of integration by parts and was originally presented in [11].

The derivatives of input $u(t)$ and output $y(t)$ functions are convoluted with the known modulating function $\phi(t)$ in the given interval of width h to obtain the set of new functions in the interval $[t_0 + h, T_{ID}]$ for $i = 0, 1, \ldots, n$:

$$y_i = y^{(i)}(t) * \phi(t) = \int_{t-h}^{t} y(\tau)\phi^{(i)}(t-\tau)d\tau,$$
$$u_{kj} = u_k^{(j)}(t) * \phi(t) = \int_{t-h}^{t} u_k(\tau)\phi^{(j)}(t-\tau)d\tau. \tag{2}$$

These functions can already be numerically computed in the given interval $[t_0 + h, T_{ID}]$. In order for Eqs. (2) to be satisfied, for the modulating function $\phi(t)$ the following assumptions have to be made:

- $\phi \in C^n(0, h)$ and satisfies boundary conditions $\phi^{(i)}(0) = \phi^{(i)}(h)$ for $i = 0, 1, \ldots, n-1$,
- $y * \phi = 0 \Rightarrow y = 0$ on the interval $[t_0 + h, T_{ID}]$,
- $h < T_{ID} - t_0$.

Loeb and Cahen functions were utilized in the performed experiments:

$$\phi(t) = t^N(h-t)^M, N < M. \tag{3}$$

The important advantage of the method is that the system initial condition can be unknown. It results from the fact that the modulating function zeroes at the boundaries of the modulation interval. Its width h should be chosen carefully, because it forms a low pass filter (increasing its value causes narrowing the bandwidth).

After the convolution procedure, the differential Eq. (1) can be transformed into the algebraic one:

$$\sum_{i=0}^{n} a_i y_i(t) = \sum_{k=1}^{K}\sum_{j=0}^{m_k} b_{kj} u_{kj}(t) + \epsilon(t), \tag{4}$$

with the same parameters. The difference $\epsilon(t)$ results from modelling inequalities or numerical errors and should be minimised.

Now, two approaches can be used for finding the optimal vector of parameters $\boldsymbol{\theta}$. The most common discrete approach utilizes the Least Squares Method (LSM). Assuming the constraint $a_0 = 1$, the model parameters can be calculated as:

$$(\boldsymbol{M}^T\boldsymbol{M})^{-1}\boldsymbol{M}^T\boldsymbol{y}_0 = \left[\boldsymbol{a}^T \quad \boldsymbol{b}_1^T \ldots \boldsymbol{b}_K^T\right]^T, \tag{5}$$

where the linear regression matrix $\boldsymbol{M}$ consists of column vectors containing sampled modulated signals:

$$\boldsymbol{M} = \left[-\boldsymbol{y}_1 \ldots -\boldsymbol{y}_n \quad \boldsymbol{u}_{10} \ldots \boldsymbol{u}_{1m_1} \ldots \boldsymbol{u}_{K0} \ldots \boldsymbol{u}_{Km_K}\right], \tag{6}$$

and $\boldsymbol{a}, \boldsymbol{b}_1, \ldots, \boldsymbol{b}_K$ are the column vectors of identified parameters. It is worth noting, that despite operating on sampled signals, the obtained parameters refer to the continuous-time model.

In [4] another Equation Error (EE) continuous variant of the MFM was described. In comparison with the previously described procedure, the optimisation is performed for the whole identification interval (not only for the finite number of collected samples). From Eq. (4), the term $\epsilon(t)$ can be written down as:

$$\begin{aligned}\epsilon(t) = \boldsymbol{c}^T(t)\boldsymbol{\theta} = [&y_0(t), \ldots, y_n(t), -u_{10}(t), \ldots, -u_{1m_1}, \ldots, \\ &-u_{K0}, \ldots, -u_{Km_K}(t)]\left[\boldsymbol{a}^T \quad \boldsymbol{b}_1^T \ldots \boldsymbol{b}_K^T\right]^T.\end{aligned} \tag{7}$$

Now, the optimisation problem can be formulated in the space $L^2[t_0+h, T_{ID}]$ as:

$$\min_{\theta} J^2 = \min \|\epsilon(t)\|^2_{L^2[t_0+h,T_{ID}]} = \min \|\boldsymbol{c}(t)^T\boldsymbol{\theta}\|^2_{L^2}. \tag{8}$$

The linear constraint is introduced to avoid the trivial solution:

$$\boldsymbol{\eta}^T\boldsymbol{\theta} = 1, \tag{9}$$

and the minimisation task has the form:

$$J^2 = \langle \boldsymbol{c}^T(t)\boldsymbol{\theta}, \boldsymbol{c}^T(t)\boldsymbol{\theta}\rangle_{L^2} = \boldsymbol{\theta}^T \boldsymbol{G}\boldsymbol{\theta}, \tag{10}$$

where the Gram matrix $\boldsymbol{G}$ is given as:

$$\boldsymbol{G} = \left[\begin{array}{c|c} \boldsymbol{YY} & \boldsymbol{YU} \\ \hline \boldsymbol{UY} & \boldsymbol{UU} \end{array}\right],$$

where:

$$\boldsymbol{YY} = \begin{bmatrix} \langle y_0, y_0\rangle & \cdots & \langle y_0, y_n\rangle \\ \vdots & \ddots & \vdots \\ \langle y_n, y_0\rangle & \cdots & \langle y_n, y_n\rangle \end{bmatrix}, \quad \boldsymbol{UY} = \begin{bmatrix} -\langle u_{10}, y_0\rangle & \cdots & -\langle u_{1m_1}, y_n\rangle \\ \vdots & \ddots & \vdots \\ -\langle u_{1m_1}, y_0\rangle & \cdots & -\langle u_{1m_1}, y_n\rangle \\ \vdots & \vdots & \vdots \\ -\langle u_{K0}, y_0\rangle & \cdots & -\langle u_{K0}, y_n\rangle \\ \vdots & \ddots & \vdots \\ -\langle u_{Km_K}, y_0\rangle & \cdots & -\langle u_{Km_K}, y_n\rangle \end{bmatrix},$$

$$\boldsymbol{YU} = \begin{bmatrix} -\langle y_0, u_{10}\rangle & \cdots & -\langle y_0, u_{1m_1}\rangle & \cdots & -\langle y_0, u_{K0}\rangle & \cdots & -\langle y_0, u_{Km_K}\rangle \\ \vdots & \ddots & \vdots & \cdots & \vdots & \ddots & \vdots \\ -\langle y_n, u_{10}\rangle & \cdots & -\langle y_n, u_{1m_1}\rangle & \cdots & -\langle y_n, u_{K0}\rangle & \cdots & -\langle y_n, u_{Km_K}\rangle \end{bmatrix},$$

$$\boldsymbol{UU} = \begin{bmatrix} \langle u_{10}, u_{10}\rangle & \cdots & \langle u_{10}, u_{1m_1}\rangle & \cdots & \langle u_{10}, u_{K0}\rangle & \cdots & \langle u_{10}, u_{Km_K}\rangle \\ \vdots & \ddots & \vdots & \cdots & \vdots & \ddots & \vdots \\ \langle u_{1m_1}, u_{10}\rangle & \cdots & \langle u_{1m_1}, u_{1m_1}\rangle & \cdots & \langle u_{1m_1}, u_{K0}\rangle & \cdots & \langle u_{1m_1}, u_{Km_K}\rangle \\ \vdots & \vdots & \vdots & \cdots & \vdots & \vdots & \vdots \\ \langle u_{K0}, u_{10}\rangle & \cdots & \langle u_{K0}, u_{1m_1}\rangle & \cdots & \langle u_{K0}, u_{K0}\rangle & \cdots & \langle u_{K0}, u_{Km_K}\rangle \\ \vdots & \ddots & \vdots & \cdots & \vdots & \ddots & \vdots \\ \langle u_{Km_K}, u_{10}\rangle & \cdots & \langle u_{Km_K}, u_{1m_1}\rangle & \cdots & \langle u_{Km_K}, u_{K0}\rangle & \cdots & \langle u_{Km_K}, u_{Km_K}\rangle \end{bmatrix}.$$

The scalar products in the space L^2 are defined as:

$$\langle y_i, u_{kj}\rangle = \int_{t_0+h}^{T_{ID}} y_i(\tau)u_{kj}(\tau)d\tau. \tag{11}$$

The Lagrange multiplier technique is applied to find the optimal vector $\boldsymbol{\theta}$:

$$L = \boldsymbol{\theta}^T \boldsymbol{G}\boldsymbol{\theta} + \lambda(\boldsymbol{\eta}^T\boldsymbol{\theta} - 1). \tag{12}$$

From the necessary optimality condition, the model parameters can be calculated as:

$$\boldsymbol{\theta}^o = \frac{\boldsymbol{G}^{-1}\boldsymbol{\eta}}{\boldsymbol{\eta}^T\boldsymbol{G}^{-1}\boldsymbol{\eta}}. \tag{13}$$

The vector $\boldsymbol{\eta}$ in constraint Eq. (9), represents the weighting coefficients for individual unknown parameters $\boldsymbol{a}, \boldsymbol{b}$ of the model (1). For the experiments presented in Sect. 3, the vector has the value 1 for the parameter a_n, zero values for the other parameters and represents some type of normalisation. In general, different values of weighting coefficients can be assumed.

2.2 Re-identification Procedure for MISO Models

It is easy to notice, that the continuous-time MISO model obtained with the previously described variants of the MFM has the same state space matrices or transfer function denominators for each component SISO model, as presented in Fig. 2. This assumption can be inconvenient especially when time constants of the SISO models vary a lot. The idea of model re-identification in such cases was discussed by many authors [6,13].

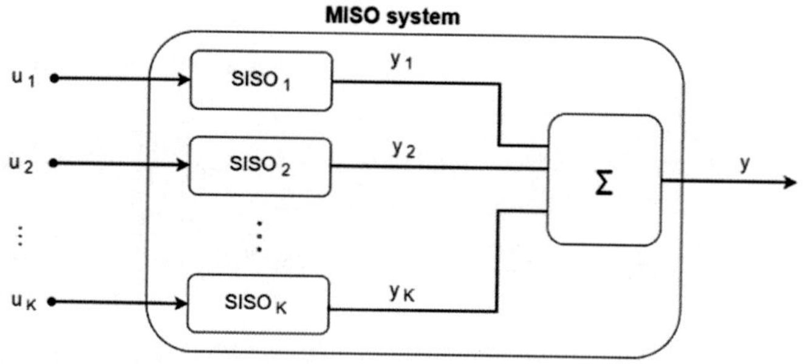

Fig. 2. MISO system composed of several SISO sub-systems with not measured output signals.

It is assumed that the identification procedure is done for each separate k-th SISO sub-model. For this purpose, a virtual output for this model is calculated as a difference between a given physical system output and simulated output signals for other SISO models:

$$y_k(t) = y(t) - \sum_{j=1,\ldots,K:j\neq k} \hat{y}_j(t), \tag{14}$$

where $\hat{y}_j$ denotes the virtual j-th system output obtained by simulation. Having the real system input u_k and the calculated output y_k, the parameters of SISO model can be obtained with one of the MFM versions presented in Subsect. 2.1.

Another important aspect of the algorithm should be mentioned here. It is known that the Least Squares Estimator is unbiased on condition that the disturbance signal has the white noise properties. In real applications, this condition rarely can be met, so the Instrumental Variable Method (IVM) is often implemented instead of the LSM. In the described case, an instrumental variable can easily be generated from the auxiliary model output $\hat{y}_k(t)$. Then, Eq. (5) for the optimal vector of parameters in the discrete MFM version has the following form:

$$(\boldsymbol{Z}^T\boldsymbol{M})^{-1}\boldsymbol{Z}^T\boldsymbol{y}_{k0} = \left[a_{k1}\dots a_{kn} \quad b_{k0}\dots b_{km_k}\right]^T, \tag{15}$$

where the matrices build of sampled signals are given as:

$$\boldsymbol{M} = \left[-\boldsymbol{y}_{k1}\dots - \boldsymbol{y}_{kn} \quad \boldsymbol{u}_{k0}\dots \boldsymbol{u}_{km_k}\right], \boldsymbol{Z} = \left[-\hat{\boldsymbol{y}}_{k1}\dots - \hat{\boldsymbol{y}}_{kn} \quad \boldsymbol{u}_{k0}\dots \boldsymbol{u}_{km_k}\right].$$

In the developed algorithm, the aforementioned procedure can be performed many times, until the overall performance index, defined as a squared difference between the simulated and the real MISO system output, is improved. In the first step, the k-th SISO model with the least value of this parameter is chosen and then, in subsequent iterations, the next sub-models are re-identified for $k = k+1$. If k is greater than K, the first model is selected. The algorithm is stopped if the current performance index value is greater than the previous one. It is worth noting that, while the MFM does not require a zero initial condition for the identified model, this demand is necessary in the re-identification procedure to enable performing simulation of individual SISO sub-models.

2.3 Exact State Observers

In contrast to asymptotic state observers, the exact integral ones guarantee obtaining the real actual value of the observed state $x(T_{OB})$ based on the signals recorded in the observation window $[0, T_{OB}]$. From their theory [1], it is known that they have the form of a sum of two integrals. The general formula for the final state is given as:

$$x(T_{OB}) = \int_0^{T_{OB}} \boldsymbol{G}_1(t)y(t)dt + \int_0^{T_{OB}} \boldsymbol{G}_2(t)u(t)dt, \tag{16}$$

where: $x(T_{OB}) \in R^n$.

Among the set of all admissible observer matrices, the ones that guarantee obtaining the minimal value of the observer norm, should be chosen. In case of disturbances affecting only the system output, the matrices have the following forms:

$$\boldsymbol{G}_1(t) = e^{\boldsymbol{A}T_{OB}}\boldsymbol{M}_0^{-1}e^{\boldsymbol{A}^T t}\boldsymbol{C}^T, \tag{17}$$

$$\boldsymbol{G}_2(t) = e^{\boldsymbol{A}T_{OB}}\boldsymbol{M}_0^{-1}\left[\int_0^t e^{\boldsymbol{A}^T\tau}\boldsymbol{C}^T\boldsymbol{C}e^{\boldsymbol{A}\tau}d\tau\right]e^{-\boldsymbol{A}t}\boldsymbol{B},$$

where:

$$\boldsymbol{M}_0 = \int_0^{T_{OB}} e^{\boldsymbol{A}^T t}\boldsymbol{C}^T\boldsymbol{C}e^{\boldsymbol{A}t}dt.$$

In the developed identification method, the use of the exact state observers enables to obtain the model state for further simulations of the system output with non-zero initial conditions. The state-space matrices for the single k-th SISO subsystem are given as:

$$\boldsymbol{A}_k = \underset{(n\times n)}{\begin{bmatrix} 0 \dots 0 & -\frac{a_{k0}}{a_{kn}} \\ 1 \ddots \vdots & \vdots \\ \vdots \ddots 0 & -\frac{a_{kn-2}}{a_{kn}} \\ 0 \dots 1 & -\frac{a_{kn-1}}{a_{kn}} \end{bmatrix}}, \quad \boldsymbol{B}_k = \underset{(n\times 1)}{\begin{bmatrix} \frac{b_{k0}}{a_{kn}} \\ \vdots \\ \frac{b_{kn-1}}{a_{kn}} \end{bmatrix}}. \tag{18}$$

If the re-identification algorithm was not applied or it was impossible to find the better set of SISO models, the state-space matrices for the overall MISO system has the form:

$$\begin{aligned} \boldsymbol{A} &= \underset{(n\times n)}{\boldsymbol{A}_1, \dots, \boldsymbol{A}_K}, \quad \boldsymbol{B} = \underset{(n\times K)}{\begin{bmatrix} \boldsymbol{B}_1 \dots \boldsymbol{B}_K \end{bmatrix}}, \\ \boldsymbol{C} &= \underset{(1\times n)}{\begin{bmatrix} 0 \dots 1 \end{bmatrix}}, \quad \boldsymbol{D} = \underset{(1\times K)}{\begin{bmatrix} 0 \dots 0 \end{bmatrix}}, \end{aligned} \tag{19}$$

otherwise the other representation is used:

$$\boldsymbol{A} = \underset{(K\cdot n\times K\cdot n)}{\begin{bmatrix} \boldsymbol{A}_1 & \dots & \boldsymbol{0} \\ \vdots & \ddots & \vdots \\ \boldsymbol{0} & \dots & \boldsymbol{A}_K \end{bmatrix}}, \quad \boldsymbol{B} = \underset{(K\cdot n\times K)}{\begin{bmatrix} \boldsymbol{B}_1 & \dots & \boldsymbol{0} \\ \vdots & \ddots & \vdots \\ \boldsymbol{0} & \dots & \boldsymbol{B}_K \end{bmatrix}},$$

$$\boldsymbol{C} = \underset{(1\times K\cdot n)}{\begin{bmatrix}\begin{bmatrix} 0 \dots 1 \end{bmatrix} \dots \begin{bmatrix} 0 \dots 1 \end{bmatrix}\end{bmatrix}}, \quad \boldsymbol{D} = \underset{(1\times K)}{\begin{bmatrix} 0 \dots 0 \end{bmatrix}}. \tag{20}$$

2.4 Adaptive Identification Algorithm

The developed identification algorithm works on the basis of few assumptions. First of all, the system dynamics is described with a continuous-time linear model valid near a defined operating point. Secondly, the procedure should adapt to the change of the current operating point if the recorded signals are almost unchanging in the given interval. Moreover, processes data are registered for intervals of defined width. Batch identification procedure with the use of the MFM is performed for the assumed number of last intervals. For the described processes, long intervals with slight changes of processes variables are common, hence an approach with moving window, often applied for Linear Parameter-Varying (LPV) systems would not succeed here.

The first initial model is identified if the operating point was previously found for the last n_{start} intervals. Additionally, the correlation module between the system input and its output should be greater than tr_{corr} for at least one of the intervals. The model identification procedure itself is performed in two steps: at first the standard MFM algorithm allows to obtain the MISO model with a common state matrix, then a set of separate SISO models can be get with the re-identification procedure utilizing the IVM on condition that the performance index (squared difference between the simulated and the real system output) value is improved. The procedure for the subsequent models is analogous, except the $n_{reident}$ number of intervals used for the identification. During the conditioning, changes in processes parameters are usually more visible with time and the reliable model can be obtained faster.

The identification algorithm performance can be examined by checking if it is possible to predict the system output for the next interval. This kind of experiments is described in the next section. For forward simulation, the knowledge of the system initial state is crucial. The exact state observer (16) is utilized for this purpose. The observation procedure is performed for subsequent time moments at the end of each interval as:

$$t_i = t_{0j} + i \cdot T - (t_{0j} \quad \boldsymbol{modulo} \quad T), \quad i = 1, 2, 3, \ldots, \tag{21}$$

where: t_{0j} is the time moment for the current operating point and T is the single identification interval width.

3 Experimental Results

The simulation experiments were performed for two data sets. The historical data were collected from the real glass forehearth, similar to the one presented in Fig. 1. Each signal was sampled every 1 s because the processes is slowly changing. Then, these historical data were utilized in the procedure simulating an operation of the real processes, in which the model is identified on-line and the future system output is predicted. The aforementioned parameters of the identification method are presented in Table 1. Their values were taken arbitrary on the basis of performed experiments.

Table 1. Parameters of the identification algorithm.

Parameter	Description	Value
T	Width of the single identification interval	250 s
T_{OB}	Width of the observation interval	500 s
n_{start}	Number of intervals for the initial model identification	8
$n_{reident}$	Number of intervals for the model re-identification	4
tr_{corr}	Correlation module threshold for the model identification	$0,5$

The first experiment refers to the data from the third zone of the forehearth. It has two controlled inputs: gas-air mixture pressure and cooling valve position, but only the first of them was changing significantly during the experiment. The average molten glass temperature was the measured system output. The data reflects a standard production change procedure with a temperature set-point increase and simultaneous decrease of a glass pull rate (the amount of glass flowing through the forehearth) from 75 to $68\frac{t}{24h}$.

The system inputs are presented in Fig. 3, while the obtained results in Fig. 4. The measured temperature is compared with the simulated one. Subsequent operating points are denoted with dotted lines. The parameters of identified processes models, as well as the MFM coefficients used for their identification, are given in Table 2. The estimated model parameters have low values, but it results from the fact that they were finally divided by the value of the parameter corresponding with the highest system output derivative in order to make them comparable. The first model refers to the SISO sub-system for which the measured temperature in the previous zone is the input, while the gas-air mixture pressure is the input for the second SISO sub-model.

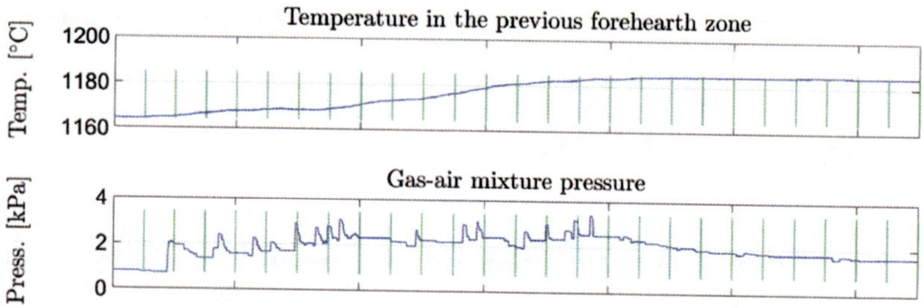

Fig. 3. System inputs during the first simulation experiment.

The input data and results of the second experiment, for the fourth forehearth zone, are given analogously as in the previous case in Figs. 5, 6 and in Table 3.

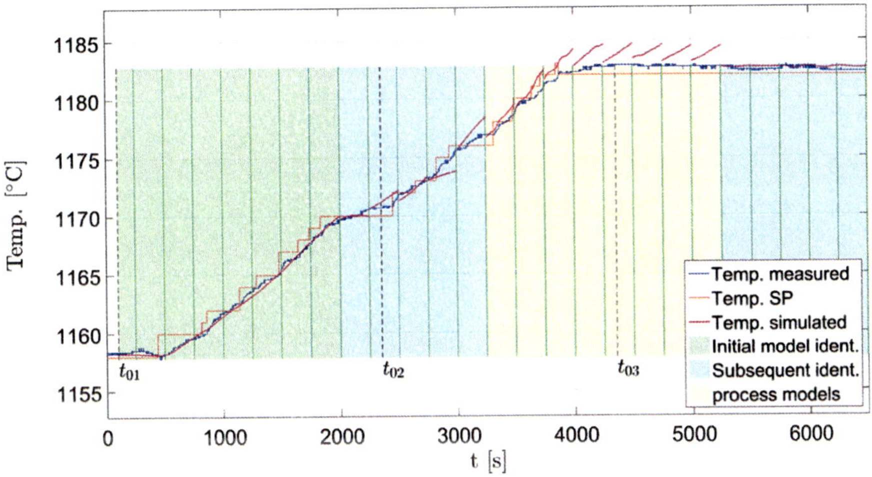

Fig. 4. Results of the first simulation experiment.

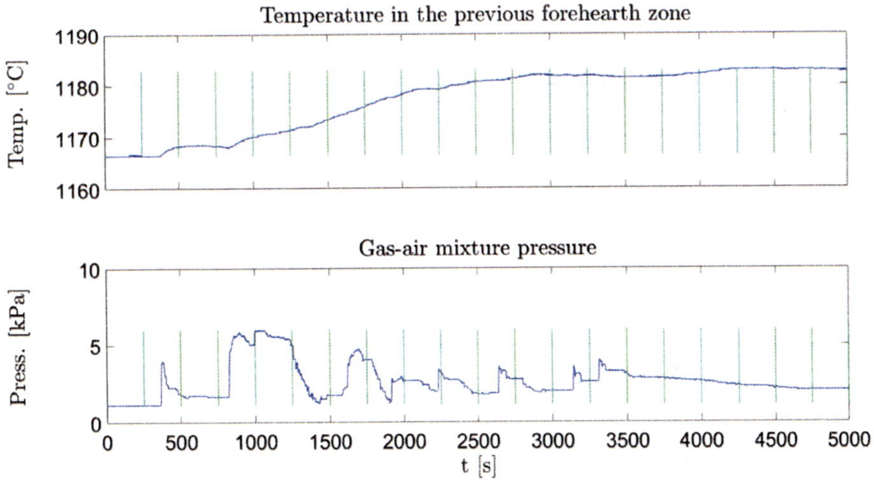

Fig. 5. System inputs during the second simulation experiment.

Table 2. Identified model parameters for the third zone of the forehearth - the first experiment.

Param.	Time [s]		
	102–3250	3250–5250	5250–6500
model nr	1	2	3
h	200	100	200
N, M	5, 6	5, 6	5, 6
a_{10}	2.3105×10^{-5}	3.3232×10^{-6}	4.2725×10^{-4}
a_{11}	3.4900×10^{-3}	2.5591×10^{-3}	6.9158×10^{-3}
a_{12}	1	2.0600×10^{-2}	1
a_{13}	–	1	–
b_{10}	2.6366×10^{-5}	5.2271×10^{-6}	-6.3147×10^{-6}
b_{11}	–	9.8579×10^{-4}	4.2947×10^{-3}
a_{20}	1.1883×10^{-5}	3.6613×10^{-6}	4.2725×10^{-4}
a_{21}	3.6026×10^{-3}	1.8542×10^{-3}	6.9158×10^{-3}
a_{22}	1	5.2237×10^{-2}	1
a_{23}	–	1	–
b_{20}	5.3769×10^{-5}	1.9671×10^{-5}	1.4595×10^{-4}
b_{21}	–	1.6354×10^{-3}	2.1859×10^{-2}

Table 3. Identified model parameters for the fourth zone for the forehearth - the second experiment.

Param.	Time [s]	
	102–4000	4000–5000
model nr	1	2
h	150	100
N, M	5, 6	5, 6
a_{10}, a_{20}	1.6770×10^{-7}	5.8108×10^{-6}
a_{11}, a_{21}	4.6626×10^{-4}	2.2445×10^{-3}
a_{12}, a_{22}	1.0134×10^{-2}	2.3138×10^{-2}
a_{13}, a_{23}	1	1
b_{10}	-4.2159×10^{-8}	1.5537×10^{-5}
b_{11}	1.4940×10^{-4}	8.0438×10^{-4}
b_{20}	1.5175×10^{-6}	2.6142×10^{-5}
b_{21}	1.8843×10^{-4}	5.2284×10^{-4}

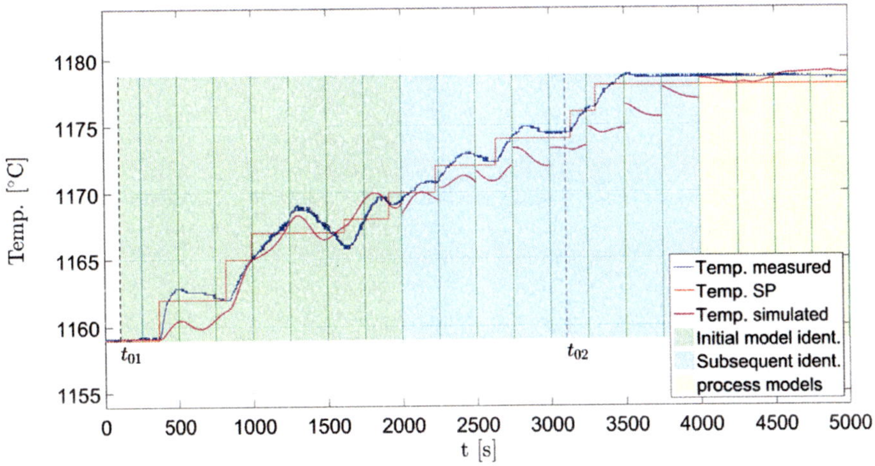

Fig. 6. Results of the second simulation experiment.

4 Summary

Based on the experimental results presented previously, it can be stated that the developed method enables to simulate the real system output, with high accuracy, for the chosen industrial processes of glass conditioning. The first experiment also proves that the re-identification procedure can be beneficial for obtaining a more accurate set of models than in the case of traditional MISO method. Additional results presenting different versions of the adaptive procedure based on the MFM, but without the use of the IVM, for glass conditioning can be found in [2,3]. The problem of application of such linear models in predictive control algorithms was discussed in [5].

Acknowledgements. This work was supported by the scientific research funds from the Polish Ministry of Education and Science and AGH UST Agreement no. 16.16.120.773 and was also conducted within the research of EC Grant H2020-MSCARISE-2018/824046.

References

1. Byrski, W.: Obserwacja i sterowanie w systemach dynamicznych. Uczelniane Wydawnictwo Naukowo-Dydaktyczne AGH im. S. Staszica, Kraków (2007)
2. Byrski, W., Drapałla, M., Byrski, J.: An adaptive identification method based on the modulating functions technique and exact state observers for modeling and simulation of a nonlinear MISO glass melting process. Int. J. Appl. Math. Comput. Sci. **29**(4), 739–757 (2019)
3. Byrski, W., Drapała, M., Byrski, J.: New on-line algorithms for modelling, identification and simulation of dynamic systems using modulating functions and non-asymptotic state estimators: case study for a chosen physical process. In: Paszynski,

M., Kranzlmüller, D., Krzhizhanovskaya, V.V., Dongarra, J.J., Sloot, P.M.A. (eds.) ICCS 2021. LNCS, vol. 12745, pp. 284–297. Springer, Cham (2021). https://doi.org/10.1007/978-3-030-77970-2_22
4. Byrski, W., Fuksa, S.: Optimal identification of continuous systems in L2 space by the use of compact support filter. Int. J. Model. Simul. **15**(4), 125–131 (1995)
5. Drapała, M., Byrski, W.: Continuous-time model predictive control with disturbances compensation for a glass forehearth. In: 2021 25th International Conference on Methods and Models in Automation and Robotics (MMAR), pp. 366–371, 23–26 August 2021
6. Garnier, H., Gilson, M., Young, P., Huselstein, E.: An optimal IV technique for identifying continuous-time transfer function model of multiple input systems. Control. Eng. Pract. **15**(4), 471–486 (2007)
7. Huisman, L.: Control of glass melting processes based on reduced CFD models. Ph.D. thesis, Technische Universiteit Eindhoven (2005)
8. Huisman, L., Weiland, S.: Identification and model predictive control of an industrial glass-feeder. IFAC Proc. Vol. **36**(16), 1645–1649 (2003)
9. Malchow, F., Sawodny, O.: Model based feedforward control of an industrial glass feeder. Control. Eng. Pract. **20**(1), 62–68 (2012)
10. Morari, M., Jay, H.L.: Model predictive control: past, present and future. Comput. Chem. Eng. **23**(4), 667–682 (1999)
11. Shinbrot, M.: On the analysis of linear and nonlinear systems. Trans. Am. Soc. Mech. Eng. J. Basic Eng. **79**, 547–551 (1957)
12. Tatjewski, P.: Advanced Control of Industrial Processes: Structures and Algorithms. Advances in Industrial Control, Springer, London (2007). https://doi.org/10.1007/978-1-84628-635-3
13. Young, P.C.: Optimal IV identification and estimation of continuous-time TF models. IFAC Proc. Vol. **35**(1), 109–114 (2002)

Autonomous Systems Incidentally Controlled by a Remote Operator

Wojciech Moczulski[1,2(✉)]

[1] Silesian University of Technology, 44-100 Gliwice, Poland
wojciech.moczulski@polsl.pl
[2] AuRoVT sp. z o.o., 44-100 Gliwice, Poland
wojciech.moczulski@aurovt.pl
http://www.aurovt.pl

Abstract. Research and development on autonomous systems (AS) focuses the attention of ever more and more countable universities and companies. The autonomy of operation, with respect to the environment and also the complexity of tasks and processes the AS is implemented to operate and solve, requires a knowledge-based approach. However, in many cases, it is very difficult, or even impossible, to equip the system with a complete knowledge base that would allow the flawless operation of AS in any circumstances. Hence, the system can be faced with a situation or a task that it is unable to solve. The approach developed by the author's team expands the idea of Fault-Tolerant systems. We broaden the meaning of a fault including here such situations, that cannot be solved by the AS on its own. If such a fault is detected, the system assesses its severity, basing on its Knowledge Base, and, once it is unable to solve the problem, contacts the remote operator who takes over the control of the system to overcome the problem. To facilitate remote control of the operator, a Virtual Teleportation technology is applied. Then the control is returned to the AS.

This paper deals with the problems of combining autonomy with remote control of the system supported by Virtual Teleportation of the operator. Particular attention is paid to the detection of faults, especially those that are caused by an incomplete Knowledge Base. Additionally, some examples are given.

Keywords: Autonomous system · Virtual Teleportation · Fault-tolerant system · Knowledge Base · Context-based reasoning

1 Introduction

Autonomous systems have started to play an ever more and more important role in the last decades. This is because of the growing necessity to replace humans in

This work is partially carried out in the Grant No. NOR/POLNOR/LEPolUAV/0066/2019-00 in the framework of the Applied Research supported by the Norwegian Financial Mechanism.

Z. Kowalczuk (Ed.): DPS 2022, LNNS 545, pp. 437–448, 2023.
https://doi.org/10.1007/978-3-031-16159-9_35

dangerous, routine and boring tasks. There are several domains and associated examples of introducing AS to everyday work.

The more difficult task and demanding environment, the higher requirements concerning the intelligence of the control system that supports the autonomy. Usually, a knowledge-based approach is required. However, because the Knowledge Base (KB) of the AS cannot contain complete knowledge of all the possible operations to be carried out in each circumstance and condition, there may occur situations in which due to lack of knowledge the system cannot operate safely and reliably. In this situation, a (possibly temporary) support of an external control subject is required.

The paper deals with such situations faced by the AS during its autonomous operation in an environment, where there is a lack of knowledge in the control system that would allow resolving the problem faced with the system. The approach is based on the generalised concept of a Fault-tolerant System (e.g. [2]). The extended FDI component must be able to detect an additional kind of fault – the inability to solve the problem the system is just faced with (or is expected to be faced with in a near future), and to undertake a special kind of a reconfiguration which is taking the control over to the remote operator. To this end, additional functionality of a Virtual Teleportation of the remote operator is being used.

The paper is organised as follows. In Sect. 2 we discuss some issues concerning autonomy. Section 3 introduces the concept of Virtual Teleportation (VT) of the operator. The next Sect. 4 deals with the combination of the autonomy and remote control occurring with the help of the VT. In Sect. 5 some examples of this approach are presented. The paper ends with conclusions and future work.

2 Types of Autonomy and Its Limitations

A historical idea of an autonomous system is shown in Fig. 1 [3]. The autonomous system acquires information about its environment in the form of percepts, created on the basis of signals S received by receptors *Rec*. Percepts are input to a correlator *Kor* that processes them and stores for further usage. A homeostat *Hom* basing on the results of the operation of the correlator allows modification of the system's structure and influences the operation of the correlator. Power P is stored in an accumulator *Ak* and supplies all the parts as well as the effector(s) *Ef* acting back upon the environment.

Depending on the complexity of the environment and tasks to be carried out, there is possible to build a taxonomy presented in Fig. 2. The complexity of the operation of the AS grows from left to right. The dynamic environment of operation of the AS requires the application of more complex sensory systems, and more efficient computing units onboard.

Systems that follow a predetermined program are the simplest kind of AS. A typical example of such a system is AGV working in a factory shop. They must react to the unexpected events when some object arrives at the route of the system playing the role of an obstacle.

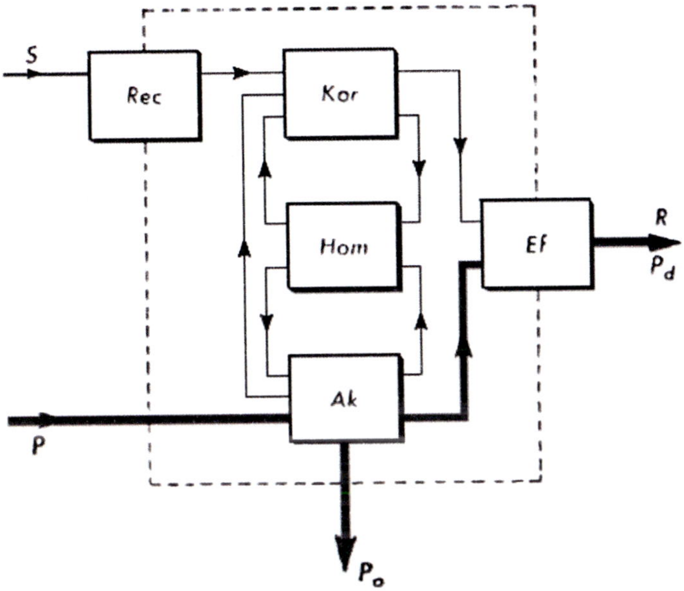

Fig. 1. A model of an autonomous system (descriptions in the text) [3].

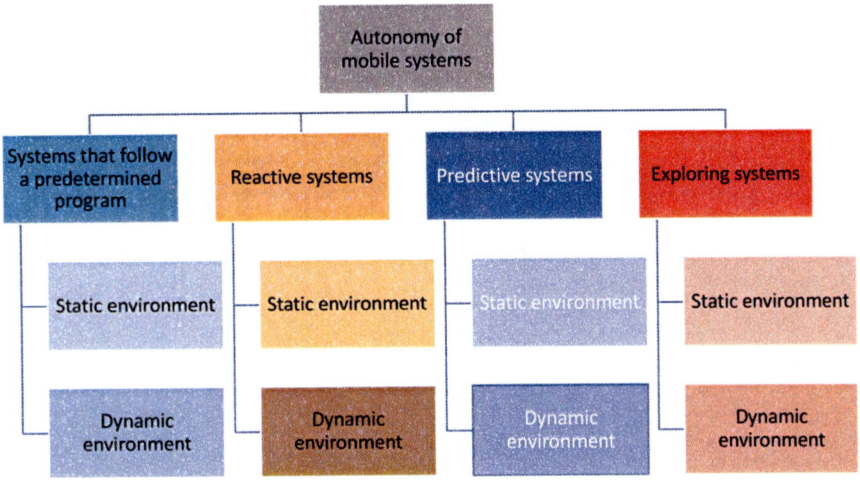

Fig. 2. Taxonomy of mobile autonomous systems.

Reactive systems can accomplish more complex plans and missions. They must be able to detect an unexpected event they are facing during work.

Predictive systems can predict future situations based on the plan of the mission or other external sources of information such as e.g. weather forecasts.

Exploring systems usually do not have any precise plan of the mission and take decisions based upon percepts collected from the environment (and possibly other external sources of information).

The more complex the operation of AS, the more demanding requirements concerning the knowledge base that allows autonomous operation of the system. Furthermore, the more complex situation, the higher probability of occurring unexpected phenomena for which there is no knowledge in the Knowledge Base of necessary reaction and then behaviour by the AS. Hence, it is quite impossible to build a complete knowledge base for AS that would allow autonomous operation in every circumstance, each internal state of the AS, and each variant of the mission.

3 Virtual Teleportation

The concept of Virtual Teleportation in the meaning that is used in this research has been introduced by K. Cyran in 2013. Virtual Teleportation (VT) is a functionality of innovative user interfaces, allowing for full immersion of the user in an environment physically separated from his/her current location, thus creating an impression perceived by the senses of this user, as if he/she were physically present in this environment and could there personally in a passive or active way carry out various activities.

In the applications that develops the author's research group, the real-time VT is used. This approach consists in the immersion of the user in the remote environment that can change during this activity. These changes can be caused either by processes or objects that operate at this remote scene just right now (as an actual traffic in the downtown), or even can be controlled by the remote operator (as an exploring robot that moves through the area of a catastrophic event). This significantly differs our applications from numerous examples allowing virtual walk through remote places (cf. e.g. [11]), which reproduce to the observer the videos or simulations downloaded from the respective database.

3.1 Passive vs Active VT

With respect to the range of activities available to the user, VT may be classified as passive or active (cf. Fig. 3 [4]).

The passive variant of VT corresponds to applications in which the user plays only the role of an observer, i.e. perceives the remote scene by his/her own senses. The goal of the passive VT is to fully immerse the user into the remote environment, to finally achieve the feeling as if he/she were personally present in this environment and could perceive it using his/her senses. This approach allows personal participation in the reality reproduced by the interface the given person uses.

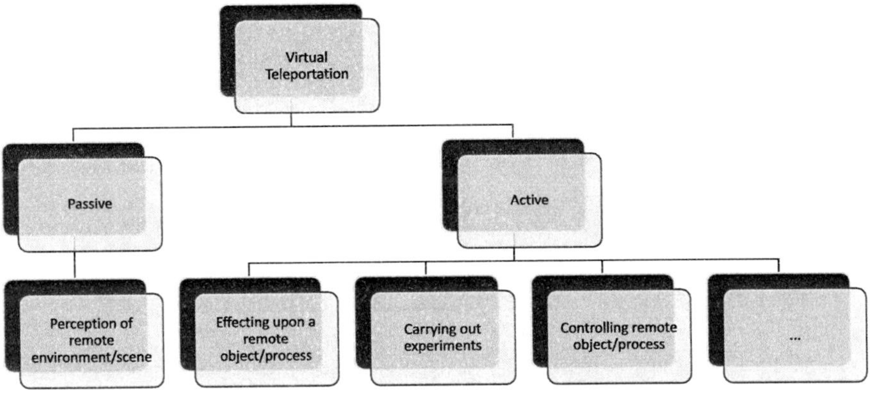

Fig. 3. Taxonomy of virtual teleportation [4].

Apart from perceiving the remote environment and activities that take place therein, the active VT allows the user to interfere with the observation at the remote scene, or even affect processes and/or objects present in the remote scene. The user's reactions are acquired to at least allow him/her controlling sensors located in the remote scene, but in the most advanced applications allow the user to operate actuators located at this scene to affect objects or processes in real-time. The simplest application can consist in tracking the eye movement of the user by the interface, and then directing a remote camera to follow the eye movements of the operator. The more advanced application can take advantage of the gestures and voice commands of the operator to control the remote system at the scene.

The first implementation of an active VT system was demonstrated in the TeleRescuer project [1]. The remote operator – a rescuer taking part in the rescuing action in the coal mine roadway – controls the robot of high mobility, which serves as a carrier of cameras, IR-cameras, and different sensors, allowing remote inspection of the area affected by the catastrophic event. An advanced HMI allows the rescuer to control the movement of the robot of 6 DOFs. The User Interface takes advantage of a stereovision to allow immersion of the operator in the remote scene.

3.2 Subtasks to Implement VT

To achieve a very realistic VT several steps are necessary, depending on whether we deal with a passive VT, or an active one (cf. Fig. 4 [4]). The upper row in this figure corresponds to the passive VT. Percepts are collected at the remote scene using sensors and cameras located there. Then percepts are transmitted to the location where the operator works. Since we are dealing with real-time VT, the latency caused by transmission links can play an extremely important role, depending on the applications. Then the percepts are reproduced to the operator using respective HMI.

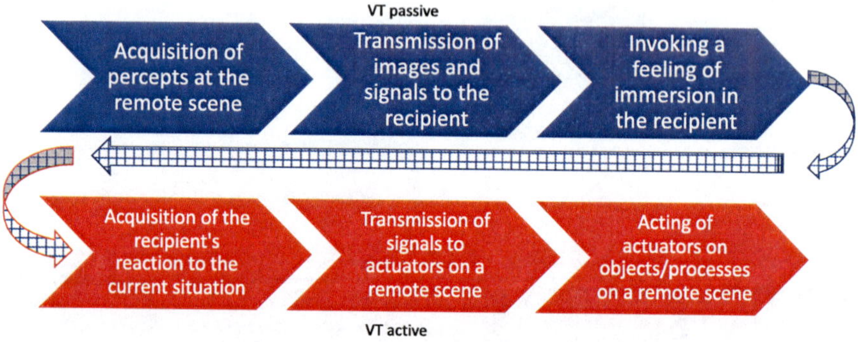

Fig. 4. Subtasks of operation of an VT system [4].

Recent applications of VT usually rely upon sight and, rarely, upon hearing. However, to achieve more advanced immersion of the operator into the remote scene, additional senses can be engaged. Table 1 [4] presents more possibilities concerning methods of acquiring percepts.

Table 1. Acquisition of percepts at the remote scene [4]

Kind of percepts	Data form	Acquisition method
Image of the scene and objects	Stereo image	Stereo camera/ omnicamera
Sound	Stereo sound	Lattice of microphones
Fragrances	Vector of components of fragrances	Fragrances acquisition system
Touch, mechanical resistance	Pressures, forces	Artificial leather, strain gauges

The second line in Fig. 4 is connected with the active VT. First, the reactions of the operator to the perceived information are acquired by the user's interface. Depending on the application, a number of specialised interface systems can be applied, starting from a computer mouse, joystick, CAD manipulator, gamepad, driving wheel associated with pedals, to flight controllers. Furthermore, special interfaces allowing tracing of eye movement, cameras acquiring gestures or body language can be applied.

In the second step, signals carrying information on the operator's reactions and control actions must be transmitted to the controlled system located at the remote scene, whose task is to evoke actions that correspond to the operator's behaviour. Then the factual operation of the remote device corresponding to the operator's request is performed. A variety of actuators and sensors may be applied. The simplest variant is a camera attached to the turnable platform, that can follow the movement of an object located at a remote scene, while the turn of the platform is controlled by the eye movements and head turn of the operator. Other possibilities are the control surfaces of a UAV, which are remotely controlled by the operator.

It is worth stressing that there is a kind of feedback in the active variant of VT: controls performed by the remote operator are sensed back by the sensory system that can influence the operator's behaviour.

4 Autonomy Combined with Virtual Teleportation

As it has been mentioned in Sect. 2, autonomous operation of any system requires a complete Knowledge Base. This completeness means that the KB contains all chunks of knowledge that are sufficient in respect to the control situation that can appear in any circumstances, internal state of AS, external state of its environment, and possible plans to be completed by AS. However, it is easy to prove that building a Knowledge Base that would allow flawless operation in any circumstances is quite impossible. Therefore, there is an important problem what the autonomous system has to do in case of lacking knowledge in the Knowledge Base. In our research, we use a combination of autonomy with remote control via Virtual Teleportation for this purpose.

The general idea of this approach is the following. The autonomous system must detect if in the next step it is able to still operate autonomously. If not, it must preserve a safe waiting mode (e.g. park in a safe place) and contact the remote operator to allow him/her to take over the control of this system. This algorithm is similar to the generalized algorithm of operation of Fault-Tolerant Systems. The situation or event whose solution is too difficult for the AS due to a lack of knowledge in the Knowledge Base (KB) can be understood as a special kind of fault. The detection of such a fault and then contact with the remote operator can be interpreted as a special kind of the system's reconfiguration. Hence, the methodology for the development of Fault-Tolerant systems [2,7] can be applied for this special autonomy/VT combination.

The horizon of detecting the inability to operate autonomously is worth discussing. In reactive autonomous systems, the fault detection module works on data collected recently. In the predictive autonomous systems the detection module, basing not only on recent percepts and inner sensors readings but also on additional data such as operation plans, maps, weather forecasts, external information etc. predicts future situations in which the help of the external operator will be necessary and can warn this operator of the necessity to taking over the control by him/her in near future.

4.1 Knowledge Base for an Autonomous System

There are different methods for representing knowledge in a Knowledge Base. In the following, we assume that the Knowledge Base of an autonomous system is represented by *rules of operation*. The general syntax of the rule is as follows (Eq. 1):

$$\textbf{if} < premise > \textbf{then} < operation > . \tag{1}$$

The premise can have the following format:

$$cond_1 \wedge cond_2 \wedge \ldots \wedge cond_n, \tag{2}$$

where conditions can be formulated for nominal, ordinal, interval and ratio variables of the parameters. Furthermore, the premises can be formulated as fuzzy relations, similarities and so on. A special kind of condition may concern similarities for realizations of signals in the time domain, or for trajectories in the state space. The parameters involved in the conditions may concern the internal state of the autonomous system, conditions of its operation, the state of the environment, and other external factors and objects that could affect the operation of the autonomous system. It is reasonable to group these conditions into contexts, that could facilitate the usage of the Knowledge Base during the operation of the autonomous system.

The $< operation >$ part of the rule contains the pointer or call of a procedure to be applied if the conditions are met. To this end, different forms of representation can be applied, such as a computer program in a binary form, or a ladder diagram for a PID controller.

The rules governing the operation of the autonomous system can be organized in the form of a decision table that facilitates the work of a knowledge engineer responsible for building this knowledge base. Additionally, the rule matching can be speed up if the rules are grouped in the framework of predefined contexts, corresponding to the values of context variables [10].

To speed up the reasoning of the required operation in the next step, it is recommended to add rules of operation of the following form (that would be analysed at the beginning of the inference process):

$$\textbf{if} < AS\, is\, unable\, to\, solve\, the\, problem > \textbf{then} < contact\, the\, operator > . \tag{3}$$

Contacting the operator results in that the remote operator will take control to solve the problem too difficult to solve by AS itself.

4.2 Detection of Inability to Operate Autonomously

The purpose of this stage of the autonomous system's operation is to detect the need to transfer control of the system to a remote operator. In the following part, only one variant will be considered, in which the autonomous system requires the remote operator to take control in order to solve a situation that is too difficult for the system to solve on its own in the autonomous mode, or for which the autonomous operation of the system is (so far) too risky.

There are two possible solutions:

1. KB does not contain rules type (3);

2. KB contains such rules.

In the first version of KB, the control system in each step checks if there is a rule in the KB whose premises are matched with the actual values of the parameters. If no such rule is found, the system contacts the operator. If the KB contains many rules, this approach can take a significant amount of time to conclude that the remote operator must help.

Therefore, the second possibility is promising, in that there is possible to find faster that the system cannot solve the problem by itself and the remote operator must interfere to solve the problem.

The other possibility for the autonomous system would be to ask the remote operator for help in the form of suggestions to the control system on how to solve the problem on its own. However, this possibility will not be discussed in this paper.

4.3 Learning from the Remote Operator

The goal of constructors of autonomous systems is to reduce the number of situations when autonomy fails, to the very rare cases. Therefore, the KB shall be constantly improved. To this end, the autonomous system can learn from examples that are acquired by the system itself by the observation of the remote operator during his/her intervention in the operation of the autonomous system. The consecutive remote control steps are characterised by means of several parameters of control operations and then represented in a procedural form. Then, additional rules of operation can be formulated by the system and added to the knowledge base. However, to reduce the risk of inducing improper rule(s), it is expedient to flag new rules and not use them until the external expert/-knowledge engineer approves them.

5 Example of the Application

The method described in Sect. 4 is applied for controlling a high-altitude long-endurance UAV for profiling atmospheric pollutions [5,6,9].

The concept of UAV *TwinStratos* is presented in Fig. 5. This UAV is devoted to long-range flights at high altitudes in the autonomous mode that includes also landing at previously selected airfields. However, in case of unexpected development of a mission, an emergency landing at an unknown terrain must be taken into consideration.

In the following let us assume that the KB of the control system uses knowledge representation by means of rules of operation assigned to contexts represented by context identifiers. There can be distinguished such contexts, as: *TakeOff*, *FlightToWaypoint*, *SideWind*, *SuddenDescending*, *ScheduledLanding*, *EmergencyLanding* and other, defined by the knowledge engineer.

Let us consider the situation in which the UAV is forced to land in an unknown terrain (context *EmergencyLanding*). Additionally, let us assume that

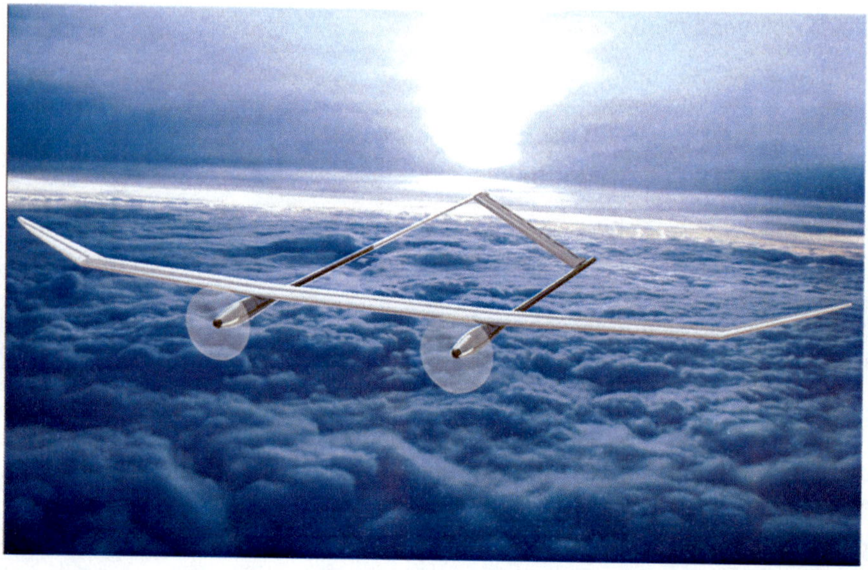

Fig. 5. A HALE UAV as an object for implementing the methodology (courtessy of W. Skarka)

a wideband GSM connection with the drone is available assuring little latency (20–50 ms). The landing speed of the drone of 12–15 m/s allows in this case remote controlling of the drone.

Before landing, the autonomous control system of the TwinStratos must assess the terrain where the drone has to land. It can be done during circular flight above the terrain and recognising the area by means of video cameras and a respective pattern recognition system [8]. The features of the terrain and values of the parameters of the internal state of the drone are compared with premisses of operating rules contained in the Knowledge Base. If no matching is found, the autonomous system contacts the remote operator, who basing upon video transmitted in real-time to the ground station of the operator, takes over the remote control of the drone during landing operation.

In the discussed case, some landings of the drone will necessarily require the assistance and control of the remote operator. One can point out here:

- Emergency landing at the coronas of the trees. Switching to the RC mode may be triggered by exemplary rule:

  if EmergencyLanding **and** AreaCoveredByTrees
  then CallRemoteOperator

- landing by strong sidewind:

  if EmergencyLanding **and** SideWind > 5 ms
  then CallRemoteOperator

To allow checking completeness of the Knowledge Base, and to make the decision of the FDI subsystem clear, these situations can be also represented in the Knowledge Base together with the operation part: "Necessarily contact the remote operator with the request to take over the control".

6 Conclusions

A new approach to autonomy incidentally supported by the remote operator by means of the Virtual Teleportation technology has been discussed. By expanding the definition of a fault to the situations in which the autonomous system lacks the sufficient knowledge to operate on its own, the way of proceeding becomes similar to the operation of Fault-Tolerant Systems. The reconfiguration of the system faced with too difficult situation can be appended by the taking the control of the system to the remote operator, who, thanks to a very innovative interface that allows his/her immersion into the scene of operation of an autonomous system, is able to solve the problem and then return the control to the autonomous system.

The self-learning component of an autonomous system promises to constantly broaden the extent of situations in which the AS is able to operate without any intervention of a remote operator. However, the Knowledge Base of the AS that is modified by the self-learning component must be periodically verified and validated. To this end, a methodology and tools must be developed. One of the solutions would be to carry out this verification in a virtual environment, where a tester that would be a specialised software tool could generate different situations, which would be then solved by a virtual model of a real AS. The results of verification could then be reviewed by the knowledge engineer and domain expert for final acceptance.

Our further work will focus on the implementation of the described approach in the control system of the TwinStratos drone capable of flying autonomously, as well as on other autonomous systems developed in our research group at the Silesian University of Technology. Additionally, the self-learning component of the system will be investigated.

Acknowledgements. The Virtual Teleportation technology is a valuable Know-How of AuRoVT sp. z o.o. (former SkyTech Research sp. z o.o.), a spin-out of the Silesian University of Technology. Basic research on VT has been carried out in the framework of the project "System for virtual TELEportation of RESCUER for inspecting coal mine areas affected by catastrophic events" partially financed by the Research Fund for Coal and Steel and the Ministry for Science and Higher Education of Poland. The methodology of combining autonomy with remote control of the operator supported by VT is developed in the framework of the project "Long-endurance UAV for collecting air quality data with high spatial and temporal resolutions", carried out under Grant No. NOR/POLNOR/LEPolUAV/0066/2019-00 in the framework of the Applied Research supported by the Norwegian Financial Mechanism.

References

1. Cyran, K., Moczulski, W., Myszor, D., Paszkuta, M., et al.: Immersive human-machine interface for controlling the operation of the telerescuer robot. In: Proceedings of International Conference on Computing, Communication and Information Technology - CCIT, Zurich, Switzerland, 02–03 September, 2017, pp. 99–103. Institute of Research Engineers and Doctors, New York (2017)
2. Koren, I., Krishna, C.M.: Fault-Tolerant Systems, 2nd edn. Morgan Kaufmann, Burlington (2020)
3. Mazur, M.: Cybernetyczna teoria układów samodzielnych (in Polish). PWN, Warszawa (1966)
4. Moczulski, W.: Autonomous systems control aided by virtual teleportation of remote operator. In: 11th IFAC Symposium on Intelligent Autonomous Vehicles IAV 2022, Prague, 6–8 July 2022, (2022, submitted)
5. Moczulski, W., et al.: Long-endurance UAV for collecting air quality data with high spatial and temporal resolutions (2019). Grant No. NOR/POLNOR/LEPolUAV/0066/2019-00
6. Moczulski, W., Skarka, W., Myszor, D., Adamczyk, M.: Zastosowania bezzałogowych statków powietrznych do diagnozowania zanieczyszczeń atmosfery (in Polish). In: Diagnostyka Maszyn : XLVII Ogólnopolskie Sympozjum, Wisła, 1.03-5.03. 2020. Streszczenia, p. 53. Silesian University of Technology (2020)
7. Noura, H., et al.: Fault-Tolerant Control Systems. Springer, London (2009). https://doi.org/10.1007/978-1-84882-653-3
8. Pęszor, D., Wojciechowski, K., Szender, M., Wojciechowska, M., Paszkuta, M., Nowacki, J.P.: Ground plane estimation for obstacle avoidance during fixed-wing UAV landing. In: Nguyen, N.T., Chittayasothorn, S., Niyato, D., Trawiński, B. (eds.) ACIIDS 2021. LNCS (LNAI), vol. 12672, pp. 454–466. Springer, Cham (2021). https://doi.org/10.1007/978-3-030-73280-6_36
9. Skarka, W., Jałowiecki, A.: Automation of a thin-layer load-bearing structure design on the example of high altitude long endurance unmanned aerial vehicle (HALE UAV). Appl. Sci. **11**(6), 1–22 (2021)
10. Timofiejczuk, A.: Identification of diagnostic rules with the application of an evolutionary algorithm. Eksploat. i Niezawodn. **37**(1), 11–16 (2008)
11. https://www.youvisit.com/tour/machupicchu . Accessed 10 Apr 2022

Author Index

Z. Kowalczuk (Ed.): DPS 2022, LNNS 545, pp. 449–450, 2023.
https://doi.org/10.1007/978-3-031-16159-9